Biomass Energy Development

Biomass Energy Development

Edited by
Wayne H. Smith
Institute of Food and Agricultural Sciences
University of Florida
Gainesville, Florida

Springer Science+Business Media, LLC

Library of Congress Cataloging in Publication Data

Southern Biomass Energy Research Conference (3rd: 1985: Gainesville, Fla.)
   Biomass energy development.

   "Proceedings of the Third Southern Biomass Energy Research Conference, held March
12–14, 1985, in Gainesville, Florida"—T.p. verso.
   Bibliography: p.
   Includes index.
   1. Biomass energy—Congresses. I. Smith, Wayne H. II. Title.
TP360.S67   1985   662′.8   85-31172
ISBN 978-1-4757-0592-8   ISBN 978-1-4757-0590-4 (eBook)
DOI 10.1007/978-1-4757-0590-4

Proceedings of the Third Southern Biomass Energy Research Conference,
held March 12–14, 1985, in Gainesville, Florida

PREFACE

The energy short-falls of the 1970's made clear both individual and national vulnerability when reliance is on a single energy source such as petroleum. Further, the situation was a sharp reminder that fossil fuels are finite and that their supplies will become limiting. Whether short supplies are for political, economical, or geological reasons, they signal the need to plan for a future with diverse options so that needs can be met without disruptive consequences when supplies of one source become limiting. Conservation plays a major role in any strategy; however, it should be recognized that such steps can only "buy time" by extending the supply -- not create new energy supplies. Proven petroleum reserves in the United States currently are stated to be equivalent to only $8\frac{1}{2}$ years consumption at current rates assuming no imports or additional petroleum discoveries.

Another important strategy is to develop biomass, solar and other alternative energy supplies and integrate their use into contemporary society. A fundamental difference between fossil fuels and biomass involves the concept of stock versus flow resources. For example, the amount of crude oil is basically a fixed or stock level which is being rapidly mined. On the other hand, biomass is growing or renewing annually, so it represents a continually flowing resource. Fossil fuels are, of course, a result of biomass produced millions of years ago and converted by nature over eons into oil, gas, and coal. With today's technologies biomass is converted from lower forms of energy into more usable or premium fuels by processes which take minutes or hours rather than millions of years. With diverse options available, the mix of energy sources for the future will allow flexibility when constraints occur.

Rapid escalation of petroleum energy prices in the 1970's without ready alternatives was a major force triggering many catastrophic events, especially run-away inflation and eventually world-wide recession. During this period, crash programs were implemented to increase energy production and to effect conservation. Thus, a massive energy supply "glut" resulted when the recession came and energy consumption plummeted. During the haste to add production, many alternative energy projects were implemented that failed for a variety of reasons. A major reason for many of the failures was because the technology available was either inadequate or was developed for other purposes, and was not suitable to an energy economy. Over-zealous advocates promoted alternate energy installations with such enthusiasm and publicity that their failures have created a "credibility gap!" Both the "energy glut" and consequent leveling or lowering of energy costs as well as the alternate energy "credibility gap" have caused serious problems in sustaining research and development programs to provide the scientific and engineering knowledge that will be essential for meeting energy needs when the world economy is again at full throttle and the surpluses become deficits -- events that experts project to occur before year 2000.

This book is about biomass energy development, just one of the energy sources that will comprise the diverse array of energy supplies to fuel the 21st century. Biomass is the oldest energy form used by civilization with wood once used as singly as were coal and then petroleum. While biomass is not the major energy source used today (14% of the world total), it is the major energy source for the most people in the world (100% dependence in some rural areas of developing countries). The standing crop of biomass in the world is vast (approximately equal to the proven reserves of fossil energy--coal, oil, and natural gas) and annual production by photosynthesis is great (about eight times the world's annual energy consumption). But because of its dispersion, low energy density and the extensive and intensive energy consumption characteristics in the developed world, biomass cannot be expected to provide all or even the majority of the energy needed in these countries. Nevertheless, on a local or regional basis biomass can contribute significant quantities of energy to meet specific needs. In spite of these constraints on biomass, its current contribution to the world's energy mix is in excess of six times the energy provided by nuclear and hydro sources combined. In the United States, energy derived from biomass is approximately equal to energy derived from either hydroelectric or nuclear sources.

The frustration to the biomass community of scientists, engineers, and practitioners is the difficulty in maintaining research and development momentum while support is restrained. Our American society tends to be short run oriented, and so while energy prices are not escalating rapidly, consumers tend to rapidly forget the lines at gasoline stations in the 1970's and the cost impacts on family budgets and the national economy. Our political system tends to mirror public moods and hence funds needed to conduct effective research for the longer term are scarce. It seems much more appropriate to systemmatically develop alternative energy technologies now than to face the next crisis without alternative energy sources available. Crash programs tend to be expensive and lead to a high failure rate. Admittedly, biomass energy is not glamourous, it has no high-tech, highly capitalized constituency who can organize well-funded, effective lobbies to assure a reasonable share of the research and development funds.

Biomass requires additional education as well as research and development to approach its potential energy contributions in the United States. Over the past several centuries, fossil fuels have been adequate to meet energy needs; but as evidenced recently, supplies are diminishing and political boundaries are impacting energy resources. Biomass resources development for the past several centuries has focused on producing food, feed, and fiber. Crops with much greater net energy potentials should be possible by selecting and improving varieties for their energy value using similar scientific methods employed to develop crops for food, feed and fiber values. Likewise, conversion technologies developed for spirits industry or waste management are inadequate as energy producing technologies. These factors and the potential contributions of biomass must be articulated to the public. At the same time, we must demonstrate that long-term research and development programs properly planned, funded, and managed can be successful just as the well-funded nuclear, photovoltaics and other alternative energy programs promise to be.

The southern USA is a region especially well-suited to biomass energy development. It lies in the humid portion of the Sun Belt where biomass production can be high. Demographic patterns favor biomass development and the soils in this region are not highly-prized for food crops and not over-committed to other uses. American farmers, including those in the south, easily over-produce many of the domestic crops now grown. Thus, farmers should welcome new crops that are in demand and can be grown at a profit. Similarly, new energy sources for the future and new industries should be of value to all citizens. This book describes strategies to develop biomass energy -- capture and use waste when possible; select and improve plant species as energy crops adaptable to both terrestrial and aquatic environments; advance both biological and thermochemical conversion technologies to produce needed fuel forms (solids, liquids, or gases); and adapt these to compatible utilization options. Each major section in this book is introduced with a paper outlining the state-of-the-science followed by papers on recent research advances and concluded with papers from practitioners on the state-of-the-art for biomass technologies. Abstracts for all other presentations made at the conference follow the papers. While the final answers are not here, at least a process for biomass energy development is illustrated. With fiscal and administrative support to a body of scientists and engineers with a vision toward contributing to solutions of problems in the 21st century, the effort can be successful. Integrated systems approaches based on biomass feedstocks to produce multiple products offer much in terms of wise use of natural resources, environmental improvement, and economic development.

Compiling this book for these purposes resulted from efforts of a number of people. I wish to acknowledge, first among these, the Planning Committee for the Third Southern Biomass Energy Research Conference--the source of papers for this volume. They helped plan the program, procure participants and, most importantly, served as Associate Editors to assure high quality manuscripts. Members of the Planning Committee, in addition to myself, were: William Bulpitt, Georgia Institute of Technology; J. L. Butler, U. S. Department of Agriculture; Milton J. Constantin, Phyton Technologies Inc.; Nathan Dean, University of Georgia; Mike J. Giamalva, Louisiana State University; Read Holland, The University of Alabama; I. A. Jefcoat, The University of Alabama; Thomas J. Laughlin, Southern Research Institute; Thomas A. McCaskey, Auburn University; Joe Roetheli, Tennessee Valley Authority; and Robert I. Van Hook, Oak Ridge National Laboratory. Arrangements support was provided by William Bowden, Office of Conferences and Institutes, University of Florida, IFAS. Because of their attentiveness, the conference went forward smoothly. Also, I wish to acknowledge the authors who chose to prepare manuscripts in accordance with reviewer requirements and to share their technical knowledge and experiences. They were ex- tremely patient and tolerant of our requirements for developing final copy.

Finally, I wish to thank all of my staff in the IFAS Center for Biomass Energy Systems who assisted in many ways as the book developed. Special appreciation goes to Cathy L. Ritchie who played a major role in every step of the process from preparing program announcements and call for papers to the final typed, reviewed, revised and edited manuscripts ready for the publisher. She worked effectively and tirelessly with each of the over 50 authors, with staff of Plenum Publishing Corporation, and with her colleagues in the Center. To all, I say "Thanks," and I hope we all can say, "It was worth it."

Wayne H. Smith, Editor

CONTENTS

AGRICULTURAL FEEDSTOCKS DEVELOPMENT

MARINE AND AQUATIC FEEDSTOCKS DEVELOPMENT

CONVERSION PROCESSES -- THERMOCHEMICAL

Abstracts

RESEARCH AND DEVELOPMENT OPPORTUNITIES

RESOURCES ASSESSMENT

WOODY FEEDSTOCKS DEVELOPMENT

MARINE AND AQUATIC FEEDSTOCKS DEVELOPMENT

CONVERSION PROCESSES -- BIOLOGICAL GASIFICATION

BIOMASS, BIOENERGY AND BIOTECHNOLOGY:

A FUTURISTIC PERSPECTIVE

Ralph P. Overend

Division of Energy
National Research Council of Canada
Ottawa, Canada K1A OR6

ABSTRACT

The decade of biomass energy or bioenergy research since the first of the two oil shocks has created a new perspective on the availability of agricultural and forest crops. Whereas previously there was knowledge only of the grain products and the merchantable part of the tree, now for many regions of the world we finally have a reasonable estimate of the availability of total plant biomass. Despite research into new energy crops and novel plants, it is postulated that the actual biomass to be used will either be through diversion of predominantly starch from grains or from the efficient utilization of the large stock of lignocellulosics which exist today either as wood or as plant residues.

In the industrialized nations, agriculture is now so productive that the major problem is, in fact, that of dealing with surpluses produced. With the advent of the new biotechnologies it is likely that the OECD (Organization for Economic Development) countries will move to increasing production of industrial rather than food products from the crops presently grown under subsidy conditions. The bioenergy opportunity is as the residual in the cascade of values attributable to the different plant fractions. Elsewhere, as a consequence of "green revolutions" and shifting market conditions, the concept of whole plant utilization will also become feasible.

Keywords. Biomass, bioenergy, lignocellulosics, starch, liquid fuels, chemicals, energy policy, agricultural policy, biotechnology.

INTRODUCTION

The role of biomass in the energy sector has been the subject of considerable study since the first oil shock of the 1970's. In real terms, the
consumption of bioenergy has shown considerable growth both in the industrialized and the developing countries, not only in the liquid fuel production
programs such as Proalcool in Brazil or Gasohol in the USA but also in the
conventional sectors such as forestry and agricultural product production
where combustion of biomass serves to substitute for fossil fuels.

Today's situation is, however, very different from that of the early
seventies and has 3 salient features:

(1) The energy supply situation has stabilized. This is due to a
number of factors among which are: the reduced growth in energy demand due
to both conservation and economic recession; the effect of substitution of
oil by natural gas and by electricity from heavy investments in nuclear
capacity that were undertaken in the late 1960's in anticipation of high
growth rates in consumption; and finally but not least, the emergence of new
oil production fields stimulated by the high OPEC prices.

(2) In the predominantly OECD (Organization for Economic Development)
member countries the agricultural production system is in considerable
surplus due in part to applications of technology as well as economic stim-
ulation through policies such as the EEC common agricultural policy. Similar technological improvements in developing countries, the so-called green
revolution, have improved production in a number of countries. Nevertheless, for a majority of the world's population, there is evidence of considerable malnutrition and it is well known that famine is widespread in the
Sahel where a decade long drought has enabled the Sahara to advance further
southward.

(3) The creation of the new biotechnologies, based on advances in
biology and molecular understanding, promises to lead to new products and
revolutionary industrial processes based mainly on glucose as a substrate.

Taken together, the first two factors lead to an almost paradoxical
situation. On the one hand, we can see demonstrable, global, renewable

2

resource-depletion in the form of deforestation and loss of agricultural production in much of the world, while on the other, it is evident that the presence of an oil "glut" suggests non-exhaustion of a non renewable resource.

The Primary Production of Agricultural and Forestry Products

The terrestrial land surface of 133 million km^2 is not all suitable for · the growth of crops, and the utilization of the land is approximately as shown in Table 1.

The totals given are likely to be underestimates since they are based on an incomplete statistical base supplied to the FAO and, particularly with respect to firewood and plant residues, it is evident that large quantities are taken from pasture and open woodlands. This factor will be examined in a little more detail in the discussion of bioenergy. Nevertheless, examination of the FAO crop and forestry statistics, along with factors that relate the crop residue to the crop weights, enables the construction of the following Table 2 which illustrates the total anhydrous production of both crops and their accessible residues.

Examination of the composition of the primary products shows the preponderance of starch as the primary agricultural product, while a similar consideration of the residues shows the main component in that fraction to lignocellulosics. The primary product is, of course, grains and these are mainly grown for their protein content which, in general, is about 10-12% by weight.

The research into the use of biomass for energy has, for many countries, created a large scale knowledge of crop and residue availability and, in effect, created not only an opportunity for bioenergy but also the possibility of a new industrial product and food or fibre based agriculture. The futuristic perspective of this paper takes, as its starting point, the third promise, that advances in biotechnology will increase the demand for biomass feedstocks and will lead to more integrated utilization of the natural resource base than hitherto in a context of widespread availability of other material and energy resources. In offering this perspective, I will structure the discussion on the three aspects of my title: the biomass resource base, the true role of bioenergy today, and the impact of biotechnologies.

Table 1. Global land utilization and approximate annual crop production.[*]

Land Use Pattern	Area (10^6 km^{-2})	Product yield (Pg^{-1})
Agriculture:		
Cultivation (Crops)	15	2.0
(Residues)	–	2.2
Pasture/Grazing	40	–
Forestry:		
Closed Forests	29	1.15
Open Woodland	11	?
Totals	95	5.35+

[*]Agricultural data from Buringh[1] and FAO[2] with the residue data as developed in Table 2 below. Forestry data from FAO[3] with correlations as discussed by Overend and Silversides.[4]

Table 2. Crop and Residue Production by Class.[*]

Crop	Crop (Tg^{-1})	%	Residue Tg^{-1}	%
Agriculture:				
Sugar	255.4	8.7	87	3.1
Root Starch	118.9	4.1	26	0.9
Grains	1454	49.6	2025	72.3
Oil Seeds	161	5.5	191	6.8
Sub-Total	1989.3		2329	
Wood:				
Industrial	433	14.8	471	16.8
Fuel	507	17.3	–	–
Sub-Total	940			
Totals	2929.3	100	2800	100

*Production Data from FAO.[2,3] Residue factors derived from Table 4.2 of Smil.[5]

BIOMASS

The last decade of biomass/bioenergy research has covered many differ-ent options ranging from field residue recovery of today's agricultural and forest plants through to proposed energy plantations composed of novel terrestrial species or alternatively, both marine and freshwater cultivation of various algae on a large scale. At this point one thing is clear, that while the productivity of some of the proposed cultivars is quite exciting and indeed promising, the time taken and the investment required to bring new crop species into large scale production will not result in the large scale implementation of such schemes in any period much less than two to three decades. It thus falls to today's terrestrial systems to provide the resources for bioenergy either from residues or the diversion of existing crops into the fuel stream. Examples of the latter are of course the maize based ethanol production in the USA and the sugarcane based production in Brazil; while the former option is illustrated by the increased utilization of forest residues and thinnings for energy as in Sweden.

To simplify discussion, this paper will take a restrictive approach to the definition of biomass. Contemporary use of the word includes, along with agricultural and forest crops, animal production and their wastes, wastes in food processing and often includes municipal solid wastes (MSW). Within the context of the discussion here, the emphasis is on the primary production of biomass and not on what could be referred to as conservation technologies, particularly in the case of MSW utilization. This in no way diminishes their present day contribution or future potential and, where the comments are appropriate, can be seen to be included in the treatment of primary biomass production. As a further simplification the total crop and residue production is considered in order to establish the broad composition of the different streams available for food, fibre, fodder, fuel and chemicals production. Clearly, selection of crops and the degree to which the residues can be ecologically, socially and economically utilized are choices based on country and site specific criteria, and it is these factors that will finally determine the contribution that biomass will make.

The State of the Biomass Production System

A discussion of the state of the world's forests and of its agriculture could be the subject of an entire conference; indeed, the issue of the relationship of agriculture and energy alone justified a major United Nations University conference[6] on the topic during 1982. The relationship between agriculture and bioenergy has also been examined in detail by the OECD.[7]

On a global basis, agriculture and the food supply system are not able to meet the food requirements of the world's population. The annual food requirement per person is in the neighbourhood of 145 kg whilst the present average production is 100 kg. Food shortages and famines extend to more countries than the Sahel. According to the World Food Program of the FAO and the U.N.[8] there were 38 countries (26 in Africa) listed as critical with respect to food supply in 1983.

Worse still, if the low level of agriculture practised today, i.e. traditional seeds, no fertilizer or chemicals, no soil conservation and the continuance of present day cropping patterns, is maintained, then almost 60 countries will be unable to maintain their projected year 2000 populations.

In a major IIASA study, Shah and Fischer[9] have shown that reasonable assumptions of technology improvements similar to those of the green revolution could lead to there being only a handful of countries in today's situation by the year 2000. Africa, again, is the most critical of the continuents considered. There are, however, more than technological barriers to ameliorating the world foodstuff supply. To quote the Economist editorial of April 14th, 1984:

"Since 1945 by fixing food prices in rich countries artificially high and prices in poor lands artificially low, governments have contrived to keep food output down where it is most needed and keep it up where farmers could be doing other things more profitably."

Thus, in the mainly OECD countries, the agricultural productivity has risen to extremely high levels, such that less than 10 percent (probably only 4% in the USA and Canada) of the population work on the land while providing one-quarter of the world's food and the major portion of the exportable surplus. In the remainder of the world, perhaps 50 percent of the people are still on the land growing three-quarters of the world's food output and in many countries there is starvation and malnutrition. The existence of the CAP in Europe has generated butter mountains and wine lakes, while full granaries in the U.S.A. produced last year's PIK program. In fact, it is part of the US agricultural lands policy to have an extensive acreage "set aside" from grains production. This productivity suggests that it would be feasible to consider non-food products from the agricultural sector through the new biotechnologies as we follow in the tracks of such bioenergy programs as Gasohol. In fact, the question behind the OECD[7] study of economic and policy issues concerning biomass for energy was phrased as follows:

"After the two oil shocks, energy production from biomass was considered an element of energy policy. Today, with the agriculture sector of OECD countries facing serious overproduction and income problems, can it become an instrument of agricultural adjustment policy?"

From the forestry perspective, the situation as with agriculture depends on the world region under consideration. In many OECD countries the forests have in fact stabilized under varying degrees of intensive management. In Scandinavia, the USA, and Canada, for example, the forest land area is undergoing relatively little change and current removals are less than the annual allowable cut. For these countries, the phase of rapid population growth and expansion of agricultural lands is more or less completed. Elsewhere in areas often described as developing countries, the forest land base continues to diminish. The present decrease in forest area is estimated to be 10^5 km^2 per year, and area approximately the size of Ontario. The main reason for this is land clearance for agriculture, especially shifting cultivation in tropical areas. It is not clear what, if any, effect this will have on the production of industrial wood since many countries are moving towards the use of intensively managed plantation forestry, especially in Brazil. With respect to fuel wood, the resource depletion is often from open forests on savannah margins such as the Sahel, and it is here that the effects on the ecosystem are most deleterious.

BIOENERGY

The commercial energy consumption of the world is close to 300 EJ in SI units, (for Imperial measure 1EJ $\simeq$ 0.95 Quad). The total mass of agricultural crop production, of industrial wood and their residues can be considered to be a proxy for the sustainable annual production of biomass. This figure of 5 Pg (Table 1) is of course much less than the estimate of standing biomass which is between a factor of 100 and 1000 times larger. At a combustion equivalence of 16kJg^{-1} the total energy content is only of the order of 80 EJ which serves to remind us that the bioenergy potential is such that it never could replace fossil fuels on a global basis. In fact "Quadromania" is not the means by which rational utilization of biomass for energy will be realized. The potential has to be assessed on a regional basis and due attention paid to the ecological impacts of such use.

In actual fact, the world energy comsumption is probably underestimated by around 25% because it is based, in the main, on the energy vectors that are commercially traded, whilst much of bioenergy use in both the developing and industrialized countries is not accounted for in this way. As with the discussion of the biomass resource base, a distinction has to be made between the role of bioenergy in the industrialized world, and that of the developing world, though in both instances figures are hard to come by for the reasons previously discussed. Some insight into the discrepancy for the non-industrialized countries can be shown by examining the fuelwood figures of the FAO in comparison with the results of studies and surveys conducted on behalf of OLADE[10] for South and Central America (Table 3). For comparison purposes, data for those countries having less than 1 PJ annual fuelwood consumption have not been included.

Even this relatively small sample indicates an underestimate of the contribution of bioenergy by a factor of almost 2 on the basis of all sources of biomass and approximately 30% on firewood and charcoal. Fortunately the error for the largest contributor - Brazil - is relatively small.

Table 3. Bioenergy in Latin and Central America.

| COUNTRY | FUELWOOD PJ^{-1} | | TOTAL BIOENERGY PJ^{-1} |
	OLADE	FAO	OLADE
Mexico	502	31	502
Central America			
Costa Rica	18	22	24
El Saldavor	57	17	64
Guatemala	79	62	86
Honduras	45	23	48
Nicaragua	24	13	30
Panama	12	9	17
Caribbean			
Haiti	47	27	59
Jamaica	0	0	10
Dominican Republic	19	1	63
Andean Region			
Bolivia	29	24	33
Columbia	124	222	135
Chile	56	20	56
Ecuador	33	26	41
Peru	111	16	134
Venuzuela	0	47	1
South East Region			
Argentina	8	38	64
Uruguay	22	10	23
Brazil	868	964	1115
TOTAL	2055	1570	2504

Within the IEA member countries bioenergy contributes about 4% overall to the primary energy supply. This average conceals contributions of 8-10% in Scandinavia (16% in Finland), 6% and 4% in Canada and the USA, respectively, and of course, negligible contributions for the Netherlands and Belgium. Almost all of this contribution is in the form of residue utilization in the pulp and paper sectors of these countries. A number of countries have extensive use of bioenergy in district heating and in the residential and commercial sectors.

Thus, it is generally true that in the IEA and OECD nations the use of bioenergy is in the form of residue utilization within processes to produce fibre and food products. In the developing world, there is a much greater reliance on firewood in the rural residential and commercial sectors, with charcoal serving in peri-urban and urban settings.

Biofuels Other Than Direct Combustion

The dominant public impression of bioenergy is of the liquid fuel contributions of the Proalcool and Gasohol programs. These are, however, a fraction of the thermal contribution in both Brazil and the USA at 20% and 3%, respectively. These are important developments because they bring bioenergy into the commodity market and export the energy out of the agricultural and forestry regions. There is little doubt that captive use of residues will become increasingly important both with respect to energy use and environmental protection. One technology that will gain large scale use will be anaerobic digestion in the captive circumstances of manure treatments and for wet residue treatment in the food processing industries. Extensive research and demonstration in the last decade has resulted in a reliable technology which is likely to be valued as much for its environmental benefit as for its energy production.

While alcohol from fermentation is already a commercial scale operation from both sugars and starches, the attempts to utilize lignocellulosics have not yet reached the same stage. There has of course been considerable research into the production of liquid fuels from lignocellulosic residues. There are essentially 3 strategies available:

    1. Indirect Liquefaction via syn-gas production and methanol or FT gasoline systhesis.

    2. Direct Liquefaction

3. Hydrolysis/Fractionation to glucose and ethanol/alcohols
production by fermentation.

The syn-gas production route can be said to have left the laboratory
with a range of demonstration units under construction around the world.
One of these for example is in Canada - the BIOSYN Project - and is expected
to process 10 Mg h^{-1} (maf) at a total oxygen pressure of 2 MPa. This unit
is presently in the early stages of operation.

The direct liquefaction route offers a superior efficiency to that of
the indirect route; this is offset by the chemical composition of the products obtained to date. In general, the product has residual oxygen and
aromatic character which qualifies it as neither a gasoline nor a diesel.
Further up-grading work is required to make this option viable in the fuel
market even though it may be a source of heavy organic chemicals in the long
run, especially in phenolic and BTX production.

The hydrolysis fractionation route has a long history with both dilute
and strong acid processes existing at around the time of the 2nd world war.
Current research encompasses organosolv pulping technologies and steam/water
reactions which will be discussed further under the biotechnology options.

For bioenergy outside of the captive markets such as those at lumber
and pulp mills, the economic barriers are quite severe. The process technology is normally subject to scale limitations due to the bulky nature of
biomass both with respect to its storage and transportation. The final
energy products are not normally direct substitutes for existing petrochemical fuels and neccessitate expensive conversions for equipment to take the
fuel. In the final analysis, competition with crude oil derived fuels is
unlikely to be effective until there is an increase in the cost of oil by at
least a factor of two. The near term possibilities for ethanol are not so
much as a direct fuel but more as a co-solvent for other fuel substitutes
such as methanol from fossil resources or as an octane enhancer as lead is
phased down. In both of these cases, the value is considerably higher than
that calculated only on the energy content of the ethanol.

BIOTECHNOLOGY

Historically the processing of many biomass substrates has been through
fermentations such as in brewing and wine production. Such processing is

known today as biotechnology and already is making inroads into food processing and bioenergy. The present day commercial activity is almost entirely in the area of starch processing from corn (maize). As could be seen from the biomass production figures of Table 1, the availability of starch from grain is many times that from conventional sugar production either from cane or sugar beet. The production of high fructose corn syrup (HFCS) from corn has gone from almost zero in the mid 1970's, to almost 4 Tga^{-1} today. This is a major application of the new biotechnology where an enzyme is used to convert alpha D glucose from hydrolysed corn starch, quantitatively to fructose (iso-glucose) for use as a replacement for cane sugar in sweetener applications. The price of starch is more reliable than that of sugar and the HFCS is able to replace sucrose at concentrations considerably less for the same degree of sweetness.

Fermentation Alcohol

I introduced this section with the history of HFCS because it is part of the evolution of the US ethanol for fuel industry. Prior to the 1970's, the market share of fermentation alcohol had been declining both relatively and absolutely in the face of low cost production from the hydration of ethylene, a co-product of natural gas and Naphtha processing. The reversal of the energy prices and the evolution of the wet milling of corn for HFCS and corn oil recovery has resulted in the ability to swing production between HFCS and ethanol as the market dictates. Thus, the cost of producing ethanol from ethylene has moved towards 30¢ L^{-1}, when it is produced from approximately 0.5 kg of ethylene as a cost of $.60 kg^{-1}. Maize at $3.50

Table 4. Yields of products from the wet milling of corn

INPUT: 1000 kg Corn

OUTPUTS: kg

Gluten Feed	290
Gluten Meal	70
Corn Oil	27
Starch	610 (for HFCS or Ethanol Production)

EITHER HFCS production
   HFCS 900 kg in Fructose concentrations ranging from 80 to 90%

OR Ethanol production
   Ethanol 385 litre and 300 kg of Carbon Dixoide

11

Bu^{-1} has a cost of around \$140 Mg^{-1} and will yield about 430 Lt^{-1} for a feedstock price component of around 32.5¢ L^{-2}. The advent of large scale maize processing for corn oil and HFCS as well as ethanol has resulted in both economies of scale and the creation of markets for the appreciable quantities of by-products that are produced. The product balance is shown in Table 4.

At a large scale the production of alcohol or processed products from corn will result in changes as to how agriculture is practised. For one thing, the protein by-products imply that at the margin 1 ha of corn will result in the withdrawal of possibly 0.5 ha of soybean from production. Already the subsidy structure of fuel ethanol in the USA has resulted in European allegations of dumping of Gluten products in the EEC. Nevertheless, the initial steps towards a new sucro-chemistry based on starch can be seen to be emerging.[11]

The markets for fermentation alcohol are by no means limited to the fuel substitution markets that have been so useful in creating the initial infrastructure. Already in Brazil, the catalytic dehydration of ethanol to provide ethylene has been re-introduced to provide a source of chemical feedstock for the Brazilian petrochemicals industry. It is also likely that traditional solvent applications of synthetic ethanol will be economically substituted by biomass derived material.

Lignocellulosics

From many perspectives, the real prize for biotechnological processes is the extremely large scale lignocellulosic resource. The obstacles are of course many. In the first place, the lignocellulosic complex is an extremely tight package requiring significant pretreatment in order for the lignin, hemicellulose and cellulose to be made accessible to further processing. Much of the energy sponsored research has been to improve the availability of cellulose to enzymatic hydrolysis to produce glucose and ethanol. Once the three major fractions are made available, the choice of products and co-products to make a viable commercial process is not an easy one. Initial attempts at low cost pretreatment were those targeted during the 1960's towards the generation of fodder substitutes from hardwoods and bagasse. These substitutes were to be energy sources for ruminant animals which are able to utilize cellulose directly through the activity of cellulase producing microflora in the rumen. With the shift to an energy emphasis, a major portion of the research effort has been to find ways and means of generating

12

low cost glucose from the cellulose polymer. This, however, addresses only a fraction of the opportunity in that the hemicellulose and lignin fractions together are probably equal in weight to the cellulose in a given resource and successful markets are required for these also.

The availability of fractionation procedures either organosolv based or aqueous/steam based techniques will lead to the challenge of successfully marketing the 3 major fractions of the material: Cellulose; Hemicellulose; and Lignin. These can enter the market place in several ways:

* As substrates for further conversion to petrochemical replacements
* As direct replacements for traditional organic polymers and chemicals
* As functional replacements for traditional products
* Incorporated into completely new materials of unusual properties

The use of cellulose as a source of glucose for fermentation to other products such as ethanol, is an example of the application as substrate. The use of celluloses obtained from the new fractionation procedures as pulp for paper production or as a feedstock for the cellulose acetate industry is an example of the direct replacement potential as is the pyrolysis and production of furfural from the hemicellulose stream. The extension of adhesives by the addition of process derived lignin is an example of the functional replacement strategy where the final product has almost the same final properties even though the substitute has totally different composition and characteristics to those of the petrochemical polymers being replaced. The completely new materials market is exemplified by the use of glucose as a substrate in the production of polyhydroxybutyrate (PHB) using <u>Alcaligenes eutrophus</u>. The product PHB is produced in the cells in concentrations of up to 80% and has thermoplastic properties superior to propylene homopolymer. The high cost of starch derived glucose is presently a barrier, though it is obviously the ambition of the lignocellulosics fractionation technique to reduce this cost significantly.

CONCLUSIONS

The oil shock stimulated interest in bioenergy has resulted in the increased use of biomass for energy with both positive and negative aspects when viewed from the differing perspectives of the OECD and developing countries. The negative aspects of deforestation and land erosion have always been present for traditional agriculture and forestry and as experience has shown in Europe and elsewhere, these can be ameliorated by investment in land improvement and the adoption of improved agricultural and silvicultural methods. Indeed with large scale investment it has almost been too successful and therefore today's problems are those of handling the excess production.

The decade of work on bioenergy has shown that it is a constrained resource that can only totally replace fossil fuels in exceptional circumstances of low population density and large productive land areas. The identification of the limits of the resource has opened the door to a new perspective on the biomass resource base. Before the energy interest, nearly all of the agricultural and forestry statistics were in the form of the merchantable component of the plant; be it the corn yield or the merchantable timber harvest. Of necessity the need to know the residue production has resulted in the creation of biomass statistics for many crops in several world regions, the knowledge of the whole plant and its component yields is what will enable the emergence of whole plant utilization industries. Just as the biomass potential is now better quantified in many regions, so it is that the fossil fuel availability is now also more assured than was expected during the mid 1970's. What has changed dramatically, has been the price to be paid for the fossil fuels; this price can be viewed as approaching the true replacement cost and thus it is likely that much more rational utilization of biomass for energy will evolve where it is generated as residues during fibre and food processing.

The interaction of a total plant perspective and the new science of biotechnology together open up the possibility of increased industrialization of the agriculture and forest products of today. Using the illustration of corn and the production of HFCS or alcohol as an example, the basis of a future total use of the whole plant can be seen. There is a natural economic hierarchy for the plant components which can be viewed as the sequence: Energy, Animal Feeds, Fibres, Human Foods, and Pharmaceuticals. Each has its market size limitations but there is no reason why whole plant utilization technologies should not evolve.

Biotechnology offers the ability to transform the biomass products into direct replacements for traditional fossil resource derived products or alternatively to provide a functional replacement as well as to generate materials with novel properties. HFCS is an example of a functional replacement of sugar from sugarcane.

14

Despite the study of novel energy crops, the real biomass resource for the next two or three decades will be the biomass embodied in today's crops. While the biomass of choice is dictated by regional and local cultural and climate considerations, the choice of substrate is likely to be a starch if the crop is chosen; or a lignocellulosic if the residue is utilized. While starch utilization is commercial today, that of lignocellulosics awaits the successful evolution of a low cost fractionation process.

REFERENCES

1. P. Buringh. Limits to the productivity capacity of the biosphere in future sources of organic raw materials - CHEMRAWN I. Eds. L. E. St-Pierre and G. R. Brown. Pergamon Press. Oxford. 1979.
2. FAO. FAO Production Yearbook. Food and Agriculture Organization of the United Nations. Rome. 1983.
3. FAO. 1979 Yearbook of Forest Products 1968-1979. FAO-Rome. 1981.
4. R. P. Overend and C. R. Silversides. Energy from forest biomass - a Canadian perspective. NRCC 18064. Ottawa, Canada. 1981.
5. V. Smil. Biomass energies: resources, links, constraints. Table 4.2, p.166. Plenum Press New York and London. 1983.
6. M. Levy and J. L. Robinson. Energy and agriculture: their interacting futures - policy implications of global models. 371 p. United Nations University and Harwood Academic Publishers. 1984.
7. OECD. Biomass for energy: economic and policy issues. OECD Paris. 1984.
8. C. Brisset. Africa heads table of victims. Manchester Guardian Weekly. p. 12. December 9, 1984.
9. Shah and Fischer. People, land, and food production: potentials in the developing world. Options p. 1, 5. 1984/2. IIASA Vienna. 1984.
10. G. Sanchez-Sierra and A. Umana-Quesada. Quantitative analysis of the role of biomass within energy consumption in Latin America. p. 21-42. Biomass: An International Journal. Vol. 4., No. 1. Elsevier Applied Science Publishers. 1984.
11. P. J. Sicard. A new suchrochemistry from starch. In La Biomasse Source d'intermediares industriels. Eds. H. Heslot and R. Villet. p. 229-256. Adeprina, Paris. 1983.

REGIONAL EFFORTS TO PROMOTE BIOMASS ENERGY TECHNOLOGIES

Phillip C. Badger

Southeastern Regional Biomass Energy Program
Tennessee Valley Authority
Muscle Shoals, Alabama 35660

ABSTRACT

The Tennessee Valley Authority's Biomass Fuels Program includes resource assessments, conversion of lignocellulosics to ethanol, end-product
evaluations, and other laboratory research. However, this paper focuses on
the Southeastern Regional Biomass Energy Program (SERBEP) activities, one of
the Department of Energy's (DOE) Biomass Energy Technology Division regional
biomass energy programs established to promote the development of effective
uses of biomass for energy. SERBEP includes the states of Alabama, Arkansas, Florida, Georgia, Kentucky, Louisiana, Mississippi, Missouri, North
Carolina, South Carolina, Tennessee, Virginia, and West Virginia.

The regional energy program goal is to encourage the production of
biomass feedstocks and their conversion to fuels by the private sector
through support of regionally specific biomass energy projects. Specific
objectives include: (1) establishing the availability of biomass resources
within the defined regions through resource assessment studies, (2) enabling
industry to match local resources with conversion technologies that will
permit private sector investments in biomass energy technologies, (3) transferring results of research and development to the private sector, and (4)
establishing a partnership with industry through cost-shared projects that
will build private sector confidence in adopting biomass energy technologies.

Major program thrusts are in three categories: (1) resource assessments, (2) furthering production and conversion technologies, and (3) trans-

ferring information and technologies to potential users. During fiscal 1984, a regional woody resource assessment was completed; plans were formulated for a regional information exchange system; 20 cooperative application/demonstration projects were funded; and cooperative projects were set up with most states in the region. Initial emphasis has been on near-term technologies, such as direct combustion and gasification, and on wood resources due to the abundance of wood in the region.

Keywords. Regional Biomass Program, biomass, biomass energy, biomass energy demonstration, biomass information exchange, wood resource assessment, biomass assessment, Tennessee Valley Authority.

INTRODUCTION

In late 1980, the Tennessee Valley Authority (TVA) organized a Biomass Fuels Program which blended multidisciplinary renewable fuels activities into a systems-oriented program designed to coordinate TVA biomass fuel activities. The program has grown significantly and current activities range from resource assessments to end-product evaluations, from laboratory research on ethanol from cellulosic materials, to technical monitoring of commercial ethanol plants receiving U.S. Department of Energy (DOE) loan guarantees, from grain crops to trees as feedstocks, and from local technical assistance to international bioenergy applications. The program is structured under a matrix management organization. Funding for the program is derived from congressional appropriations to TVA and from work conducted for DOE and the Agency for International Development (AID) on a reimbursable contract basis by TVA; no electricity ratepayers' funds are used in the Biomass Fuels Program. Current projects include:

1. Wood supply information in terms of inventory of hardwoods by geographic location in the 201-county TVA area, methods to reduce harvesting and transportation costs, and management of forests for energy.

2. Methods to effectively harvest small diameter wood (short-rotation, intensive-culture tree crops; rights of way clearings; etc.) for energy (contract with Oak Ridge National Laboratory).

3. Effective processes to convert underutilized, low-grade hardwoods to ethanol via acid hydrolysis and subsequent fermentation of both five- and six-carbon sugars to ethanol.

4. Conversion of nonwoody cellulosic material to ethanol and other products in an experimental plant designed to process about 4 tons per day of corn stover, wheat straw, or other nonwoody cellulose (contract with DOE).

5. Technical monitoring of the DOE loan guarantee program. Ten commercial operations were given conditional approval to develop detailed plans to build commercial ethanol plants. Two firms have converted their approvals from conditional to actual loan guarantees: New Energy of Indiana and Tennol (contract with DOE).

6. Economic and marketing assessments for the research and development activities to assist in identifying major areas for process improvement and to determine market potential and channels for products and by-products.

7. Staff and technical assistance to AID's Bioenergy Project to assist in implementing bioenergy projects in developing countries. Currently, a gasification installation is being pursued in Costa Rica and numerous coordination activities are underway (contract with AID).

8. The Southeastern Regional Biomass Energy Program (SERBEP), managed in conjunction with DOE (contract with DOE).

The main purpose of this paper is to present information on the regional energy program. The program's goal is to promote the development of effective uses of biomass for energy. The authority to establish the program was provided by Congress in the report of the Senate Committee on Appropriations· to the Energy and Water Development Act of 1983 (P.L. 97-88). Congress provided additional guidance in the report of the Senate Committee on Appropriations to the Energy and Water Development Act of 1984 (P.L. 98-50). The committee directed that four regional programs be established with TVA to manage the program in the Southeastern United States and that these programs "carry out activities related to technology transfer, industry support, resource assessment, and matching local resources to conversion technologies." These regional programs provide a method of concentrating on biomass-for-energy technologies best suited to regional biomass resources and regional energy needs. They also provide direction and focus for research and development projects, facilitate communications among the numerous groups involved in biomass research and development, and provide a means of transferring results of research and development to potential users.

Regional programs have been established for the Northeast, Great Lakes area, Pacific Northwest and Alaska, and Southeast. The Southeastern region includes the states of Alabama, Arkansas, Florida, Georgia, Kentucky, Louisiana, Mississippi, Missouri, North Carolina, South Carolina, Tennessee, Virginia, and West Virginia.

PROGRAM MANAGEMENT

The regional programs are administered through the Biomass Energy Technology Division of DOE. The Southeastern Regional Biomass Energy Program (SERBEP) is managed within TVA's Biomass Fuels Program in the Office of Agricultural and Chemical Development. A regional planning committee with representatives from federal and state agencies, universities, and private industry provides advice and assistance in establishing major program thrusts and areas of activity. Regional constituency groups have also been formed to provide further technical advice and assistance in reviewing and selecting projects for program funding.

REGIONAL PROGRAM GOAL, OBJECTIVES, AND OVERVIEW

The goal of the regional energy programs is to encourage the production of biomass feedstocks and their conversion to fuels by the private sector through support of regionally specific biomass energy projects.

Specific objectives in the SERBEP include: (1) establishing the availability of biomass resources within the defined regions through resource assessment studies, (2) enabling industry to match local resources with conversion technologies that will permit private sector investments in biomass energy technologies, (3) transferring results of research and development to the private sector, and (4) establishing a partnership with industry through cost-shared projects that will build private sector confidence in adopting biomass energy technologies.

Major program thrusts during fiscal 1984 have been in three categories: (1) resource assessments, (2) furthering recovery and conversion technologies, and (3) transferring information and technologies to potential users.

20

Three avenues for implementation have been targeted: establishment of an
information exchange system, solicitation of proposals for selected activities, and cooperative projects with state agencies. Cost sharing is sought
in most projects.

PROGRAM ACTIVITIES

General Activities

An assessment of the woody biomass resources in the 13-state region was
completed in 1984. The assessment describes the woody biomass resources by
ownership, species, and type, and includes annual projections of the amount
of woody biomass potentially available. A summary and a complete report
entitled "Woody Biomass Analysis For 13 Southeastern States" was issued in
early 1985 by TVA's Division of Land and Economic Resources, Office of
Natural Resources and Economic Development, Knoxville, Tennessee (Technical
Note B52, TVA/ONRED/LER-84/6, December 1984).

Georgia Tech Research Corporation (GTRC) will prepare technical briefs
for two SERBEP projects involving direct application of wood combustion
technology. The first brief covers the SERBEP project with GTRC to perform
demonstration testing on a self-cleaning, rotary-recycling, off-gas separator for removing effluents from wood pyrolysis and gasification processes.
The second brief will cover the SERBEP project with Ala-Tenn Industries,
Inc., Sheffield, Alabama, to install a wood combustor in a system that steam
conditions lumber prior to chemical treatment. Upon project completion,
SERBEP will examine the efficiency of using such briefs as a tool for the
transfer of technology to the private sector.

Aerospace Research Corporation (ARC) has been working for several years
on a project to produce electricity using a wood fuel. In their process,
wood fuel is pulverized, gasified in a controlled-combustion furnace, and
the hot gases used to drive a turbine coupled to a generator. After initial
testing, the entire gas turbine system was moved from Roanoke, Virginia, to
Red Boiling Springs, Tennessee, in 1984. Completion of system shakedown,
connection to the TVA electrical grid, and full-scale operation for the 3-MW
wood-to-electricity facility is scheduled for early 1985.

Regional Biomass Information Exchange System

As part of activities to make available and transfer biomass information, a computerized information network is being established. Several bulletin board software packages and hardware configurations have been studied for use in the online segment of the biomass information exchange system. The initial system will use IBM, or IBM compatible, personal computers to run a teleconferencing/bulletin board package. This package can be extensively customized and includes a data base management system.

Data entry is in progress using a temporary system. Arrangements have been made to establish the data base for harvesting equipment manufacturers and vendors. Contacts are being made with state energy offices, Oak Ridge National Laboratory, and others to provide input to the exchange system.

An information systems specialist is working full time to develop and implement the system which is scheduled to begin full operation by the spring of 1985. TVA is coordinating the development of the system with Biomass Energy Research Association and other organizations that collect and disseminate biomass information.

Other Technology Transfer Activities

A Biomass Information Consolidation and Sharing Workshop brought together academic, private sector, and government representatives in March, 1985, to compile existing information on biomass research and development projects and existing industrial applications. The workshop also identified ways to better utilize existing information and determine information needs in the information exchange system and technology areas.

Articles on SERBEP activities and biomass energy projects have been submitted quarterly to Biologue. Biologue is a monthly newsletter sent nationwide to potential biomass producers and users and includes biomass energy news and information. It is published by the National Wood Energy Association in cooperation with the four regional biomass energy programs and private industry. Four special editions each year feature the regional biomass programs. Approximately 2,500 people in the 13-state region were receiving Biologue at the end of 1984.

A typical example of program outreach approach is the Southeast Industrial Biomass Exposition sponsored by SERBEP and others held in late November, 1984, in Atlanta, Georgia, in conjunction with the Seventh World Energy Engineering Congress. Approximately 150 people nationwide attended the technical sessions designed to provide updated biomass technology. Approximately 40 vendors of biomass-energy-related products participated in the trade show.

1984 Request for Proposal (RFP) Projects

Twenty projects were funded out of about 60 proposals submitted by universities, governmental agencies, private industry, and others. SERBEP funding for individual projects ranged from $9,700 to $50,000. Total for all 20 projects was about $735,000 with an additional $1,000,000 provided by the contractors or other sources. The projects cover a wide range of biomass fuel applications as follows:

Auburn, Alabama. Donald L. Sirois, USDA/Forest Service, Auburn University. A roll-splitting process in harvesting to speed up field drying and reduce transportation costs. Process appears well suited for harvesting small-diameter stems growing on utility rights of way and other areas containing small wood.

Sheffield, Alabama. Chester McKinney, Jr., Ala-Tenn Industries, Inc. Installation of a combination natural gas/wood burner to generate steam for wood preservative treatment processes.

Gainesville, Florida. R. A. Nordstedt, University of Florida. Demonstration using swine waste and biomass as feedstocks for a methane generation unit.

Atlanta, Georgia. Z. Redkevitch, Georgia Tech Research Corporation. A self-cleaning, rotary-recycling, off-gas separator for the effluent products resulting from wood-burning and gasification processes.

Macon, Georgia. J. Fred Allen, Georgia Forestry Commission. Formulas to predict how much logging residues (trash, brush, treetops, etc.) will remain after stand of trees is harvested for traditional forest products. Study will be used to project economics of harvest.

Pineville, Louisiana. James P. Barnett, USDA/Forest Service. Green and dry timber from loblolly pines to develop equations to predict expected yields per acre.

Jackson, Mississippi. Tal Bankston, Mississippi Department of Energy and Transportation. Wood-heating system at correctional center housing unit near Parchman, Mississippi. Wood-heating system will replace resistance electric heaters now being used.

Mississippi State University, Mississippi. Philip H. Mitchell, Mississippi State University. Superheated steam method of drying wood particles from four selected hardwood species. Study will test various drying rates and temperatures as well as different particle sizes.

Rolla, Missouri. O. C. Sitton, University of Missouri. Use of higher sugar concentration levels in the production of ethanol.

Durham, North Carolina. P. Aarne Vesilind, Duke University. Existing combustion research facilities to determine what happens when wood and solid waste are burned at the same time. Study will address effect wood has on unburned sludge.

Durham, North Carolina. P. Aarne Vesilind, Duke University. Formula to predict size distributions for biomass material (wastepaper, cornstalks, wastewood) that has been shredded.

Raleigh, North Carolina. Douglas J. Frederick, North Carolina State University. With U. S. Forest Service, measure trees of different ages and sizes and develop formulas to predict total biomass in harvest as well as nutrients removed via whole-tree harvest.

Raleigh, North Carolina. Larry G. Jahn, North Carolina State University. Costs and operating characteristics of different methods of harvesting, transporting, and combusting wood. Results of the study will provide the basis for an information guide for those interested in using wood as a fuel.

Clemson, South Carolina. Charles H. Gooding, Clemson University. More efficient means of dehydrating ethanol using three American-made membranes and a novel evaporation process for separation.

Clemson, South Carolina. Fred A. Payne, Clemson University. Gasification of wood chips as an economical alternative to natural gas and liquid propane for small-scale agricultural and industrial applications.

Columbia, South Carolina. Erwin G. Lambrecht, South Carolina Forestry Commission. Wood-fired system to heat forest tree-seedling greenhouse at Creech Seed Orchard near Wedgefield, South Carolina. Study will compare wood-burner system with present propane gas heating system.

Knoxville, Tennessee. David Ostermeier, The University of Tennessee. An intensive study of market constraints for the energy utilization of woody biomass fuels.

Nashville, Tennessee. Robert W. House, Vanderbilt University. Needs, methods, and costs of transferring existing biomass technology to industry.

Blacksburg, Virginia. Wolfgang G. Glasser, Virginia Polytechnic Institute and State University. Lignin, a natural by-product of ethanol production, as a replacement for phenol, a chemical compound used to make formaldehyde-type resin.

Blacksburg, Virginia. Dennis R. Jaasma, Virginia Polytechnic Institute and State University. New, computer-controlled combustor that will burn any kind of biomass with less pollution than any combustor now available. Technology could have immediate application in tobacco curing and brooder house heating.

1985 RFP Projects

A second RFP was mailed in January, 1985, with requests for proposals in the following areas: conversion and utilization applications, economic and financial factors, environmental factors, and information transfer. Total funding for the solicitation was approximately $400,000 with funding of 15 to 20 proposals anticipated. Proposals will be selected in early spring, 1985, and contracts awarded by early summer, 1985.

1984 State Cooperative Projects

Most of the 13 states in the Southeastern region have submitted proposals and have started work on cooperative projects with SERBEP. Summaries of ongoing or proposed projects as of January 1, 1985, follow.

Alabama. Ralph Stanford, Department of Economic and Community Affairs. A wood-fired heating system will be installed at a state prison facility to displace the use of natural gas. The wood-fired system will also demonstrate the economic and technical feasibility of wood-energy systems and encourage in-state industry acceptance of biomass energy fuels.

Arkansas. Ed Davis, Arkansas Energy Office. Major long-term markets for wood energy will be identified and their potential for development assessed. Sources of wood-energy technical assistance and service will be matched with sources of financial assistance and services. Wood-energy technologies most applicable to state industry needs will be identified along with research work needed to ensure an adequate fuelwood supply is maintained for future generations.

Florida. Harold Draper, Governor's Energy Office. County-wide inventories of wastes and residues will be developed along with projections of the potential for supplemental energy crops. Resources will be classified according to their adaptability to conversion technologies and mapped throughout the state.

Georgia. J. Fred Allen, Georgia Forestry Commission. Operating problems of an industrial-scale biomass gasification system will be resolved so that the technology can be transferred throughout the region. Performance of the Rome, Georgia, gasifier will be evaluated to determine operating characteristics and modifications necessary to achieve design performance.

Kentucky. John Stapleton, Kentucky Energy Cabinet. Information packets, technical briefs, factsheets, and case studies will be developed to promote industry awareness of biomass energy conversion opportunities. A survey of existing and potential wood-energy users will be conducted and the desirability of information transfer workshops evaluated. Also, an air gasifier fueled with corn-cob residues will be demonstrated for small-scale farm applications.

Louisiana. John Rogers, Department of Natural Resources. A multiphase program will be conducted to promote the use of wood as a fuel in new construction and in renovation, to stimulate technology transfer to industry, and to provide reliable biomass energy information to users. The program will also improve coordination and capabilities among the various state agencies involved with biomass energy products and applications.

Missouri. Norman Lenhardt, Department of Natural Resources. The economic and technical feasibility of a portable corn-cob gasifier will be demonstrated at a farm site. An in-state grain drying equipment company will provide major support. The demonstration will be coupled with work-shops to encourage small state and local industries to utilize available biomass energy resources.

North Carolina. Ellen Pulaski, Commerce Department. Information on existing wood-fuel users in the state will be systematically collected and analyzed. A directory will be published to enable potential wood-energy users to contact existing industrial users. An information guidebook will also be prepared for residential wood-fuel users and will include infor-mation on wood supply equipment design and operation, and safety.

South Carolina. Leon Thompson, South Carolina Energy Research and Development Center. Pee Dee Biomass, Inc., will work with Agripen of Kings-tree, South Carolina, to provide sole source wood fuel for a 5.3 MW (18,000 lb steam h^{-1}) wood-burning boiler to be installed at an Agripen corn wet-milling facility. The boiler will utilize wastewood within a 50 km (30mile) radius of the plant in the form of wood chips and sawdust.

West Virginia. Tom Reishman, Coal Development Authority. Case studies of sawmills with cogeneration/dry kiln facilities fueled by mill residues will be developed and will include cost and return data for each element in the residue-fired system. A computer model will also be developed for evaluating the economic feasibility of investment in mill-residue-fueled cogeneration/dry kiln facilities and will be used to develop workshops to stimulate industry interest.

SUMMARY

SERBEP, although less than two years old, has already made a signifi-cant impact on the use of biomass energy in the region. State governments in Southeastern states are either implementing biomass programs or strength-ening existing ones. Commercial and industrial companies, especially among the nonwood industry, are increasing their use of biomass resources as they become more aware of resource potential and utilization technologies. Future program growth will depend on a variety of factors including conven-tional fuel costs. However, it is anticipated that the large amount of unused biomass resources will play a significant role in meeting the energy requirements of the Southeastern region in the near future.

CHARACTERISTICS OF INDUSTRIAL WOOD ENERGY USERS

David M. Ostermeier, Timothy M. Young and David F. Walsh

Department of Forestry, Wildlife and Fisheries
The University of Tennessee
Knoxville, Tennesee 37901

ABSTRACT

Characteristics of wood energy users from a national survey are analyzed and discussed. Of particular emphasis are: incentives and constraints of using wood as an energy fuel; and user characteristics that will
help define the most competitive segment of the energy market for wood fuel.
The survey is part of an overall study whose objectives are to identify
constraints of using wood energy and evaluate alternatives to remove or
decrease these constraints.

Survey results indicate that wood energy users produce process steam
via direct combustion in relatively small boilers. Low price of wood fuel
is the principle incentive these producers have in using wood fuel. Most
respondents indicated numerous constraints that had to be overcome before
making the decision to use wood fuel including: wood supply, handling and
storage; proper operation and maintenance of boiler systems; capital costs
of installation; pollution control; and management's skepticism regarding
wood fuel.

INTRODUCTION

In this paper, certain characteristics of wood energy users are presented. These characteristics result from the first part of a study which
evaluates the seemingly low rate of adoption of wood energy by commercial
and industrial users outside the forest products industry. The overall

study hypothesis is that market and institutional constraints exist which prevent non-forest products industries, and commercial establishments and institutions, from using wood energy. The three study objectives are: first, to identify the competitive segments of the industrial wood energy market; second, to identify market and institutional constraints to wood enery use; third, to identify policy for modifying or removing constraints to wood energy production.

The forest products industries have significantly increased their use of wood energy in recent years. In the last ten years, the lumber and wood products industry had doubled its use of wood energy.[1,2] In 1982, the industry derived 78 percent of its energy needs from wood.[2] In 1983, the pulp and paper industry derived 50 percent of its energy needs from wood based fuels.[3] Given the widespread use of wood energy by the forest products sector, future increases in wood energy production by this sector are assumed to be small relative to present use levels.

In contrast, wood energy use by non-forest products industries has been negligible. Wood fuel consumption among these industries (e.g. textile industry, chemical and allied products industry, etc.) is estimated to be between two and three percent (Personal Communication: Dr. Collin High, Dartmough College. 1984. Conversation regarding results from D.E. Hewett's Ph.D. dissertation. Mitre Corporation. 1980. TVA's wood for energy program, integrated environmental assessment. McLean, Virginia. 63 p.). This low rate of market adoption of wood energy indicates that the greatest potential for increased wood energy production exists among non-forest products industries and commercial establishments.

This paper presents the characteristics of wood energy users that will be used to help identify the competitive market segments sought in the first objective.

METHODS

The approach used follows contemporary marketing theory by applying the concept of market segmentation to focus on a particular part of a market.[4,5,6,7] First, institutions and non-forest products industries are identified that use wood fuel in producing at least some of their energy requirements. Second, characteristics (e.g. boiler size, location, type of industry, etc.) of these organizations are identified and analyzed. These characteristics will eventually be used to help define the energy market segments that appear to be most competitive for using wood.

Data collection was obtained via mail survey of non-forest products industries and institutions currently using wood energy. The survey was conducted from November 1984, to March 1985, and the study area was the continental United States.

Industries and institutions that were surveyed were selected from the National Emissions Data System (NEDS). [The U.S. Environmental Protection Agency National Emissions Data System is a computerized data handling system that accepts, stores, and reports on the latest available information relating to sources of air pollutant emissions of particulates, SO_x, NO_x, CO and hydrocarbons.[8]] An extensive review of other information sources was made before selecting NEDS as the population of energy users. The review process consisted of reviewing the content and coverage of data bases maintained by state agencies, the U.S. Department of Energy, The Energy Information Administration, The American Boiler Manufacturers Association, and Dunn and Bradstreet, Inc. These data bases were either not as complete as NEDS, or were not obtainable because of excessive cost or proprietary considerations. NEDS was selected as the best data base given the requirements of the study for a data base consisting of both wood and conventional (e.g. oil, gas, and coal) energy users. NEDS was not combined with state information sources to prevent biasing the national analysis towards those states with good information of wood energy users. Information on NEDS which is useful for the study are: type of industry (4-digit Standard Industrial Classification code), location (street address), contact person, boiler size (MJ hr^{-1}),[8,9] and annual fuelwood consumption (tons).

Eighty-five respondents have completed the questionnaire, representing a relatively high response to a mail survey of 71 percent. [NEDS contains information on 54,405 facilities in the United States. There are 1,225 forest products industries (SIC codes 24, 25, and 26), and 119 non-forest products industries using wood energy as listed in NEDS. The 119 non-forest products industries and installations were selected for the mail survey, representing a complete sample of the subpopulation of wood energy users from the NEDS population of energy users.] Fifty-nine percent of the respondents are currently using wood fuel. Eight percent of the respondents have used wood fuel, but have converted to an alternative fuel. The remaining 33 percent of the respondents have never used wood as an energy source, or were no longer in business. The non-wood energy users in the mail survey are misclassified as wood energy users in the National Emissions Data System.

Survey Procedures

The mail survey incorporates the concepts of the Total Design Method (TDM). [The Total Design Method is a concept in mail and telephone surveys which identified each aspect of the survey process that may affect response quantity or quality and shapes them in a way that will encourage good response. The TDM rests on both a theory of response behavior and an administrative plan to direct its implementation.[10]] The purpose of the survey was to collect information in three areas: (1) factors influencing the decision to use wood energy; (2) source of wood fuel; and (3) type of wood energy system.

Two types of data were sought concerning the factors influencing the decision to install a wood-fueled energy system. First, data were collected to help identify incentives perceived by current wood energy users of using wood energy. It was assumed that most wood energy users utilize wood because it is a cheaper fuel. Consequently, other incentives were addressed. Second, data were collected to help determine the major constraints of using wood energy.

Information was sought in the second section of the survey on sources of wood fuel. Respondents were asked the sources of wood fuel, form of wood fuel, average and longest hauling distances, contract length and term regarding wood fuel supply.

In the third section of the survey, the desired information pertained to the engineering specifics of the wood energy systems. Three types of data were collected: (1) type of combustion process and energy produced from the system; (2) the number of wood-fueled and multi-fueled energy producing units; and (3) the percentage of total annual fuel consumption supplied by wood energy.

RESULTS

Incentives and Constraints to Using Wood Energy

The respondents did not consider any of the incentives listed in the questionnaire to be important in the decision to use wood energy (Table 1). The only incentive which had a consistently high ranking was an open-ended "other" category. In this category, respondents listed readily available wood fuel supply and low cost of wood fuel as important incentives for using wood energy. These results indicate that for the wood energy users in this survey, cheap fuel is the principle incentive. In addition, if other incentives exist and if the users are aware of these, they aren't important in their decision to use wood.

The respondents were asked to rate the importance of constraints that had to be overcome before they adopted wood energy. These were classified in the questionnaire in four groups: (1) institutional; (2) technological; (3) financial; and (4) environmental (Table 2).

Among the institutional constraints, both "finding a wood fuel supply source" and "availability of long-term supply" appear to be inconclusive. Both had high response rates in the categories of "not important," and "extremely important." Some of the respondents used wood parts in their production process. (These were companies that produced products like wood toys, bowling pins, burial caskets, musical instruments, etc., and are not normally included as part of the forest products industries.) These companies generally used their own wood residues to produce wood energy. When these companies were deleted from the sample, a much larger portion of the remaining respondents indicated that wood supply was a constraint. More specifically, approximately 3/4 of the remaining respondents indicated that these two factors were moderately to extremely important constraints (compared to approximately 50 percent when the wood users were included). Additionally, only 10 percent of this smaller sample indicated that wood supply was "not important," whereas 30 to 40 percent indicated "not important" when wood users were included. In essence, the availability of wood supply has been a constraint, sometimes a very important one, to the wood energy users surveyed in this study that did not use their own residues.

The other factor in the institutional group of constraints was "management's acceptance of wood fuel system." This was considered to be a moderately to very important constraint to overcome. Again, if the wood users were deleted from this sample, the respondents indicated a higher level of importance. This would seem logical that when wood supply is less certain, management will be more cautious about installing a wood dependent system.

Table 1. Incentives to using wood energy and the level of importance as perceived by current wood energy users.

Incentive	Importance Level					Mean** Rank
	Not Important	Slightly Important	Moderately Important	Very Important	Extremely Important	
	(percent)					
(a): Sale of Excess Electricity to Public Utilities	91	0	3	3	3	1.14^a
(b): Technical Assistance from State and Federal Programs	81	11	3	5	0	1.31^{ab}
(c): Public Financing Incentives	82	3	9	6	0	1.38^{abc}
(d): Private Financing Incentives	76	9	9	0	6	1.60^{bcd}
(e): State and Federal Tax Incentives	65	16	11	8	0	1.62^{bcd}
(f): Other*	0	0	13	20	67	4.38

* The majority of responses were the "competitive cost of wood fuel."

** Mean values based on a 5-point response format, where 1 = not important and 5 = extremely important. Means with the same superscript are not significantly different at the 0.05 level. For example, the incentives (a), (b) and (c) are not significantly different at the 0.05 level.

Table 2. Constraints that had to be overcome among current wood energy users and the level of importance.

Constraints	Importance Level				
	Not Important	Slightly Important	Moderately Important	Very Important	Extremely Important
Institutional:					
Finding a wood fuel supply source	40	10	14	10	26
Availability of long-term supply	30	14	9	14	33
Management's acceptance of system	19	7	32	27	15
Technological:					
Proper operation of system	2	5	16	42	35
Maintenance of system	4	11	23	48	14
Wood fuel handling	7	17	26	31	19
Wood fuel storage	10	16	37	23	14
Efficiency of system	24	21	24	19	12
Market:					
Capital cost of installation	12	9	16	28	35
Environmental:					
Cost to comply with air pollution regulations	11	14	23	25	27
Cost and problems of ash disposal	25	40	16	14	5

Most of the "technological" factors that were addressed were perceived to be important constraints among respondents. "Proper operation" and "maintenance of the wood energy system" were considered to be very or extremely important among 77 and 62 percent of the respondents, respectively. (These percentages are calculated from data in Table 2. These represent cumulative additions across importance levels. For example, for "proper operation of system" 35 percent indicated "extremely important" and 42 percent said "very important". Hence, this represents a cumulative total of 77 percent that indicated "very" or "extremely important". In some cases, cumulative totals are across three importance levels.) "Wood fuel handling" and "storage" were considered to be moderately to extremely important among 76 and 74 percent of the respondents, respectively. Hence, these four factors were constraining factors that needed to be overcome in order to use wood energy.

The remaining technological factor was evidently not as constraining. "Efficiency of the wood fuel system" was considered to be moderately to extremely important among 55 percent of the respondents. The remaining 45 percent indicated that "efficiency" is either only slightly important (21%) or not important at all (24%).

The only financial constraint listed in the questionnaire was "capital cost of installation relative to traditional energy systems." Sixty-three percent of the respondents considered this to be very to extremely important; seventy-nine percent considered it moderately to extremely important.

The environmental constraint category, "costs to comply with air pollution regulations" were considered to be moderately to extremely important by 75 percent of the respondents. This may vary by age of the wood energy system, location and other factors. Generally, most respondents felt this to be at least a moderate constraint. The "cost and problems of ash disposal" was considered to be only "slightly important" or "not important" by 65 percent of the respondents. Alternatively, most of the remaining 35 percent felt it was either moderate or very important.

Wood Fuel Supply

Data collected on wood fuel supply were: form of wood fuel; percent of total wood fuel by supply source; transportation distance; and length and source of wood fuel contract.

The majority of the respondents used hogged fuel. (Hogged fuel is defined as wood which has been processed to a specific size, i.e. wood chips.[11] A hog is a machine used for reducing the size of wood slabs, edgings, bark, and other materials.[12]) Of those using hogged fuel, a large share used hogged fuel with sawdust and mill residues. Of those respondents that process on-site, 84 percent were hogging wood fuel. The majority of the respondents' total wood fuel supply came from wastewood from another company's production of wood products.

Respondents were asked their "average" and "longest" distance that their wood fuel was hauled. The mean distance for the "average" haul was 53 km (33 miles), while the mean distance for the "longest" haul was 142 km (88 miles).

Fifty-six percent of the respondents had a contract to purchase their wood fuel. Of those that had a contract, fifty-seven percent had contracted directly with a forest products company. Another twenty-seven percent of those having contracts, did so with a wood fuel broker. The remaining sixteen percent had agreements with numerous types of organizations.

Of those that had contracts, the majority (68%) had contracts of less than three years. The remaining 32% had longer term contracts of three years or more.

Type of Wood Energy System

Ninety-two percent of the respondents use wood energy for direct combustion. The majority of respondents (71%) produce either process steam or some combination of process steam, space heat and process heat.

Sixty-two percent use one or more single-fuel wood-fired boilers as the primary energy system. Twenty-eight percent use one or more multi-fueled boilers as the primary energy source. The remaining six percent use single fuel oil, gas, or coal boilers as the primary energy system. Respondents with multi-fueled boilers use natural gas or oil as alternative fuels. The average number of boilers per respondent were three units.

Respondents with multi-fueled boilers indicate that, on average, seventy-five percent of annual fuel consumption is supplied by wood. Fifty percent of the respondents burn green wood fuel exclusively, and 42 percent burn dry wood fuel exclusively. The remaining eight percent burn a combination of green and dry wood fuel.

Thirty-four percent of the respondents have used wood fuel for less than five years. Twenty-six percent have used wood fuel for more than 20 years (Figure 1). The most recent adopters of wood energy (less than 5 years) were the Electric, Gas and Sanitary Services (SIC 49) and institutions (SIC 80 and 82).

Boiler capacity information from NEDS, for the respondents to the survey, indicates that the distribution of boiler capacities among respondents are significantly skewed toward smaller capacities. Seventy-eight percent of the respondents have boiler capacities of less than (52,800 MJ hr^{-1}) (Figure 2). The data from NEDS, for respondents to the survey, seems to indicate that a market segment where wood fuel is competitive is among small boiler sizes.

Industry Type

As given in NEDS, two-digit Standard Industrial Classification (SIC) codes are used to classify the respondents. The greatest concentration

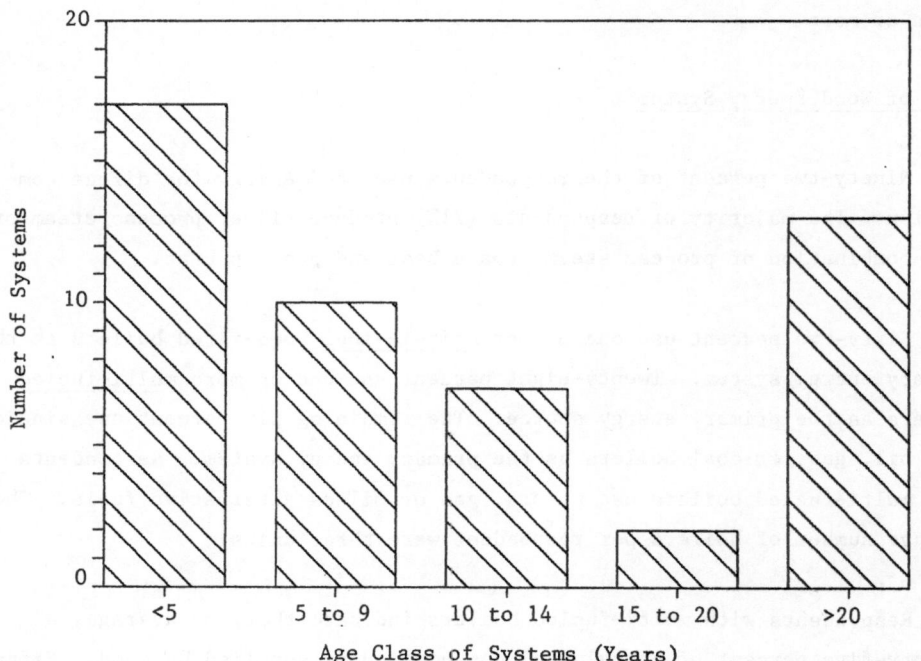

Figure 1. Number of wood energy systems by age class for respondents to survey 1.

among respondents is in the Miscellaneous Manufacturing Industries (SIC 39) (Table 3). The Miscellaneous Manufacturing Industries are further disaggregated in the following 4-digit SIC code categories: musical instruments, games and toys, sporting and athletic equipment, pencils and art goods, brooms and brushes, and burial caskets. The second highest concentration among respondents by 2-digit SIC code is the Electric, Gas, and Sanitary Services (SIC 49). The third highest concentration among respondents by 2-digit SIC code is tied among three categories: Health Services (SIC 80); Educational Services (SIC 82); and Chemicals and Allied Products (SIC 28). It is important to note that respondents in SIC code 28 are concentrated in the Gum and Wood Chemicals Industry (SIC 2861).

Table 3. Distribution of Respondents by 2-digit Standard Industrial Classification (SIC) Code

SIC Code	SIC Code Description	Frequency
01	Agricultural Prod. (Horticultural Specialties)	2
20	Food and Kindred Products	4
22	Textile Mill Products	2
28	Chemicals and Allied Products	5
33	Primary Metal Industries	1
34	Fabricated Metal Products	2
35	Machinery, Except Electrical	2
37	Transportation Equipment	1
38	Instrument & Related Products	1
39	Miscellaneous Manufacturing Industries[*]	13
49	Electric, Gas, and Sanitary Services	6
80	Health Services	5
82	Educational Services	5
91	Executive, Legislative and General	1
	Total:	50

[*]Respondents among the miscellaneous manufacturing industries category are subdivided by: musical instruments, games and toys, sporting and athletic goods, pencils and art goods, brooms and brushes, and burial caskets.

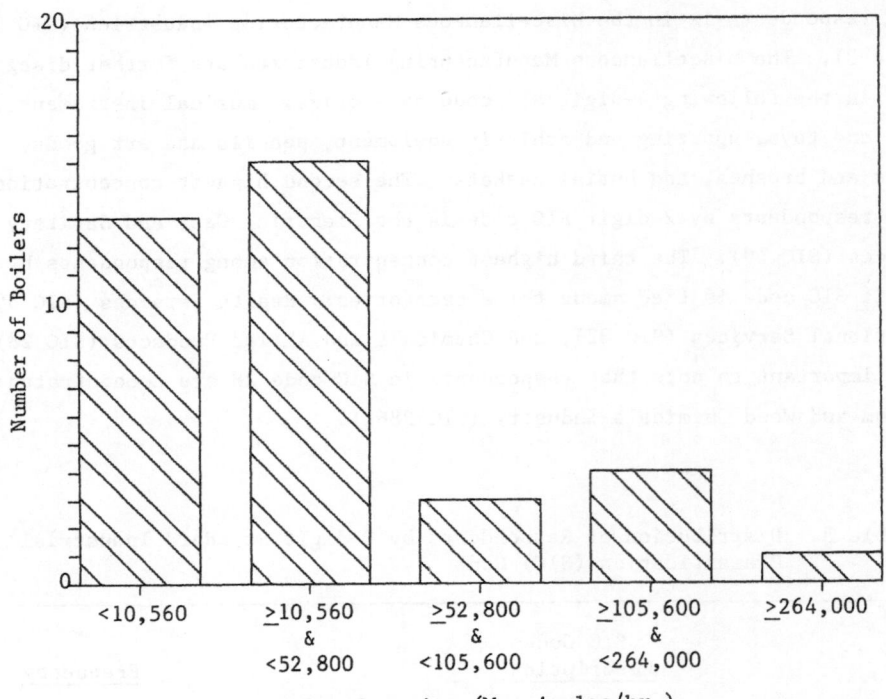

Figure 2. Number of wood-fired boilers by boiler size class for respondents to survey 1 (capacity data taken from the NEDS data base).

CONCLUSIONS

The cost of wood fuel under some conditions can be relatively low when compared to traditional fuels. These conditions can relate to both wood supply, such as proximity to wood residues, and prices of alternative fuels. The energy users consulted in this survey indicated that the low cost of wood fuel was the principle reason that they relied on wood energy. Additionally, they also indicated that other incentives were not important, when compared to low fuel cost.

Conventional wisdom indicates that the use of wood for producing energy is a risk-taking investment. Traditional sources such as natural gas, oil and coal are generally more established and convenient, do not have as significant on-site handling problems, and have good histories of reliable long term supply. Generally, capital cost of a wood fuel system is higher than traditional systems.

The analysis of this survey seems to generally support this conventional wisdom. The respondents reported a number of factors that tended to act as constraints to the adoption of wood energy. Wood supply was of concern to those that do not produce their own wood wastes. This appears to be true both in terms of identifying a reliable wood supply and in the continued availability of wood fuel. In addition, wood fuel handling and storage were constraints to many wood energy users.

There are also technological problems of using a non-uniform fuel like wood. The respondents indicated that the proper operation and maintenance of wood-fueled systems can be critical and are important constraints.

Initial capital costs were also identified as constraints. Both the cost of installation and the cost to comply with air pollution regulations were identified as important constraints. In addition, it appears that due to a number of uncertainties, management is generally skeptical of committing to a wood-fueled system.

Current wood users surveyed in this report are producing direct combustion in relative small boilers (less than 52,800 MJ hr^{-1} or 50 MMBTU hr^{-1}). Approximately 2/3 of these boilers use only wood fuel, with most of the rest using a combination of wood and natural gas or oil. These multi-fueled boilers meet 3/4 of their annual fuel needs with wood.

The respondents represented numerous industries with the most noted concentration being in the "miscellaneous manufacturing industries". This seems to contradict conventional wisdom in particular states or regions that there is a relatively high degree of concentration of wood energy users. For example, in some states brick manufacturers are major new entrants whereas in other regions, higher concentrations are in utilities and institutions. Given the national scope of this survey and the data problems encountered, these local concentrations generally did not surface.

In summation, the price of alternative fuels appears to be the main incentive or constraint to wood energy use. However, where delivered wood fuel prices give wood energy a good competitive advantage, the decision to use wood is complicated by having to overcome a number of constraints. In the words of one west coast respondent "...the wood energy industry is still in an early stage of evolution. We have a number of problems that need to be worked out. Yet it appears to be a significant energy cost savings."

REFERENCES

1. D. Salo, L. Gsellman, D. Medville and G. Price. Near-term potential of wood as a fuel. Metrek Division of MITRE Corporation Report HCP/T4101-02, UC-61. McLean, Virginia. 65p. 1978.
2. A. Goetzl. Energy consumption and efficiency improvement. Memorandum to participating companies, NFPA energy reporting program, NFPA committee on energy. National Forest Products Association. Washington, D.C. 11 p. 1983.
3. American Paper Institute, Inc. Paper industry energy efficiency: second quarter and first six months 1984. Unpublished manuscript to energy coordinators - pulp, materials and technology group. New York, New York. 4 p. 1984.
4. H. Assael and M. Roscoe, Jr. Approaches to segmentation analysis. Jounal of Marketing. 13(1): 67-76. 1976.
5. M. L. Bell. Marketing concepts and strategy. Third edition. Houghton Mifflin Company. Boston, Massachusetts. 595 p. 1979.
6. D. W. Cravens, G. E. Hills and R. B. Woodruff. Marketing decision making - concepts and strategy. Richard D. Irwin, Inc. Homewood, Illinois. 573 p. 1980.
7. P. Kotler. Marketing management - analysis, planning and control. Fifth edition. Prentice-Hall, Inc. Englewood Cliffs, New Jersey. 794 p. 1984.
8. U. S. Environmental Protection Agency. NEDS - national emissions data system information. Office of Air Quality Planning and Stardards. EPA-45014-80- 013. Research Triangle Park, North Carolina. 99 p. 1980.
9. E. S. Miyata, H. M. Steinhilb and L. A. Coyer. Metric conversions for foresters. USDA Forest Service. North Central For. Exp. Stn. St. Paul, Minnesota. (loose-leaf). 1981.
10. D. A. Dillman. Mail and telephone surveys - the total design method. John Wiley and Sons. New York, New York. 325 p. 1978.
11. Georgia Institute of Technology. The industrial wood energy handbook. Developed by The Technology Applications Laboratory. Van Nostrand Reinhold and Company, Inc. New York, New York. 240 p. 1984.
12. North Carolina Department of Commerce - Energy Division. Wood energy information guide. Raleigh, North Carolina. 76 p. 1982.

DOMESTIC FUELWOOD USE IN LOUISIANA

Victor A. Rudis

Southern Forest Experiment Station
U.S. Forest Service
Starkville, Mississippi 39759

ABSTRACT

A telephone survey of Louisiana households and commercial vendors of
domestic fuelwood was conducted in 1984 to assess domestic fuelwood use and
sources of production. Twenty-two percent of the households surveyed used
fuelwood during the 1983-84 heating season. Domestic fuelwood production
amounted to 981,000 m^3, an amount comparable to 10% of Louisiana's 1982
pulpwood production. The survey also pointed out differences in total
production by type of producer, condition and species of trees used, land
use class, ownership class, and geographic regions. Of the total fuelwood
produced, 71% was cut or collected directly by households, while the remaining 29% was cut or collected by tree service companies, other companies, and
other individual vendors. The majority of fuelwood came from live trees,
primarily oaks, and was cut from private woodlands. Production was greatest
in northern parishes near major cities.
Keywords. Consumption, producers, woody biomass, firewood.

INTRODUCTION

In the South, as in other parts of the United States, interest in wood
as a household fuel has been increasing. A marked rise in the drain on wood
products by domestic fuelwood producers is of obvious concern to industrial
wood energy producers as well as timber producers. Fuelwood demand affects
tree species composition, land ownership, and timberland management, and it
creates an economic market for marginally commercial wood resources.

Four times more fuelwood is burned in American homes than burned 10 years ago.[1] In the South, 29% of the households use fuelwood. Consumption averages 1.16 m^3 (0.55 cords) per household, or 3.99 m^3 (1.9 cords) per fuelwood-using household.[1] Recent studies have examined domestic fuelwood use with regard to state-level forest resources in Colorado[2], Wisconsin[3], and South Carolina.[4] However, the current impact of domestic fuelwood use on state-level timber resources in the South and detailed data for Gulf Coast states are largely unknown.

This study takes a look at the amount of fuelwood used by Louisiana households for the 1983-84 heating season. Sources of wood cut or collected are presented by (a) type of producer, (b) land use class, (c) condition and species of trees used, (d) ownership class, (e) region of the state, and (f) parish (i.e. county) location. The impact of domestic fuelwood production on timber resources is examined by determining the amount harvested from standing live trees in woodland areas.

Excluded from this investigation is domestic fuelwood derived from wood residues and wood consumed for industrial fuel. Terms and conversion factors used in the analysis are listed in the Appendix. Domestic fuelwood production varies annually, and further research is needed to establish trend information.

METHODS

Households were contacted by telephone, and the interviewer collected information using a standardized questionnaire. If households cut or collected their own wood, questions were asked about the amount and proportion of wood obtained from various sources. If wood was purchased, households were questioned about the amount purchased and burned within the past 12 months.

The volume cut, collected or burned by each household was asked as an open-ended question. Conversion from truck loads, face cords, and pounds of green wood to cubic meters was made as necessary. Most responses were in cords of fuelwood. A copy of the questionnaire used (OMB No. 0596-0009) and further details are available from the author.

The state was divided into 8 regions based on state planning boundaries and telephone directory service areas (Figure 1). Each region is designated

by its largest city. Names and telephone numbers were selected at random
from the residential listings of telephone books of each region.

Calls were made until a quota of interviews was completed from each
phone book. The quota was based on a random sample representing 0.064% of
the households in the state, and amount sufficient to obtain statewide
fuelwood use estimates with a 10% standard error. The quota was subdivided
by region and the number of residential listings in each phone book. If
there was no response, if no person in the.household knew about fuelwood
used, or if the person reached asked to be called back at another time, that
listing was recontacted. Listings not reached were called on subsequent
days (up to 5 different days and times of day) until reached or until the
quota was satisfied for that phone book. Another listing was selected at
random if the household refused to be interviewed.

Except for households recontacted, calls were made on different days
and different times of the day between February 16 and March 7, 1984. Most
calls were made between 5 and 9 pm during weekdays and 1 to 9 pm on week-
ends. Calls were made in April, 1984, to households that (a) asked to be

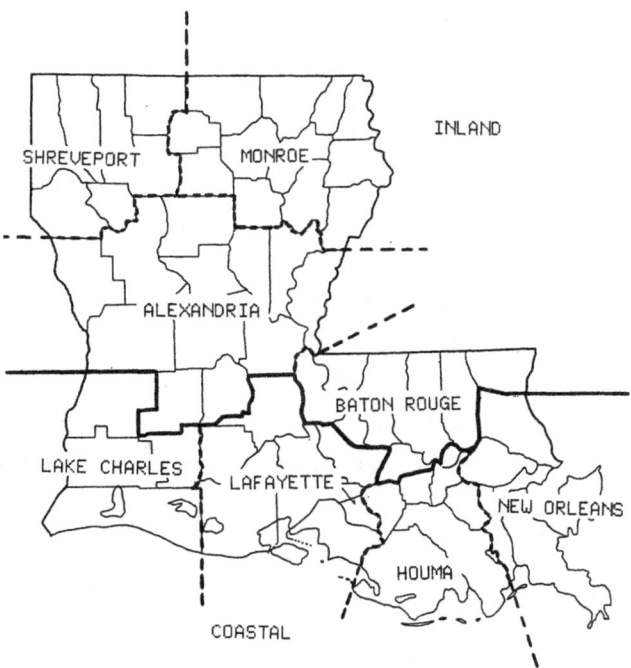

Figure 1. Louisiana telephone survey regions

called back but not reached by March 7, (b) burned wood but did not collect it themselves, and (c) had incomplete or questionable responses.

Of the 1,474 listings selected, 937 were reached. Of these, 911 interviews were completed and 26 were refused. The response rate was 62%. The 537 nonresponses represent households where the residents are seldom home. These households might be expected to burn less wood than households contacted. There was also the chance that households without phones and of lower socioeconomic class might be expected to burn more wood than households contacted. Results were biased if fuelwood was significantly different in quantity of sources from households interviewed. On balance, however, the data from the 911 completed interviews were believed to be representative of Louisiana households.

A survey of all known commercial vendors of domestic fuelwood was used to estimate purchased wood by sources. Lists of potential domestic fuelwood vendors were obtained from commercial telephone directories of fuelwood operators, loggers, wood products dealers, and from classified ads in newspapers for firewood or fuelwood sales. Of 234 contacted, 67 qualified as commercial vendors of domestic fuelwood. Of these, 63 completed interviews. Commercial vendors were grouped into three types based on the name of the enterprise (e.g. John Smith Tree Service = "tree service," Smith Corp. = "other company," John Smith = "other individuals" (not households).

These 63 commercial vendors of domestic fuelwood accounted for 2.35% of the volume burned but not cut or collected by households. Tree service companies accounted for 19 interviews and 0.78% of the volume; other companies, 21 interviews and 0.75% of the volume; and other individuals (not households), 23 interviews and 0.82% of the volume. Results may not be representative of all commercial vendors. The 97.65% of domestic fuelwood sold but not accounted for was believed to originate from part-time operations associated with tree service companies, logging companies, and other individuals. Results from interviews with known vendors were assumed to be representative of the sources of wood produced by all commercial vendors.

Fuelwood production exported from Louisiana was not determined. Imports reported by Louisiana households and commercial vendors were assumed to be equal in magnitude to exports. For the purposes of accounting at the parish level, the assumed "exports" were divided equally and allocated to each of the 23 parishes bordering Texas, Arkansas, and Mississippi.

RESULTS

Twenty-two percent of Louisiana households burned, cut, or collected fuelwood for domestic use during the 1983-84 heating season. Consumption averaged 0.697 m^3 (0.332 cords) per household (Table 1). The standard error of the estimate is 10%. Fuelwood-using households average 3.19 m^3 (1.52 cords) per household. Of these, 68% cut or collect their own wood and use 3.30 m^3 (1.57 cords) per household. The remaining 32% that burn wood but obtain it through vendors use 2.76 m^3 (1.31 cords) per household. The averages are comparable to that reported for the South Central states from a 1980-81 national survey by Skog and Watterson.[1]

Regional differences within Louisiana are apparent (Table 1). The geographic pattern appears to be associated with climate and household density. The average per household is greater for inland regions than coastal regions and lower for the more densely settled regions, e.g. Baton Rouge and New Orleans.

Table 1. Number of households, density of households, number of samples, and average fuelwood burned, cut, or collected per household, Louisiana 1983-84 heating season, by region.

Region	Number of households	Density of Households	Number of samples	Average (standard error)
	thousands	km^{-2}		cubic meters
INLAND				
Alexandria	144.0	5.2	95	1.508 (0.353)
Monroe	120.0	6.5	85	1.496 (0.357)
Shreveport	157.5	11.7	88	1.149 (0.291)
Baton Rouge	239.7	16.7	152	0.647 (0.119)
All Inland	661.2	8.9	420	1.119 (0.132)
COASTAL				
Lake Charles	79.3	7.1	54	0.734 (0.246)
Lafayette	163.6	12.6	107	0.398 (0.098)
Houma	81.5	8.7	107	0.263 (0.088)
New Orleans	426.2	55.9	223	0.246 (0.066)
All Coastal	750.6	18.2	491	0.336 (0.050)
STATEWIDE	1,411.8	12.2	911	0.697 (0.068)

[a]1980 U.S. Census.

The amount of domestic fuelwood produced for the 1983-84 heating season is 981,000 m^3. Considering that the state's pulpwood production for 1982 totals 9,780,000 m^3,[5], domestic fuelwood production amounts to about 10% of the state's pulpwood production.

Seventy-one percent of domestic fuelwood production, 699,000 m^3, is cut or collected by households; of this amount, 27,000 m^3 comes from adjacent states. The remaining 29% (282,000 m^3) is produced by commercial vendors. Of the volume from commercial vendors, 13,200 m^3 comes from adjacent states.

Woodlands comprise the major source of domestic fuelwood supply. More than 604,000 m^3 are derived from woodlands, but only 443,000 m^3 come from standing live trees on woodlands. Households obtain the majority from woodlands, but only 38% comes from standing live trees on woodlands (Figure 2a). Commercial vendors obtain 64% of their wood from standing live trees (Figure 2b). A closer examination of commercial vendors suggests important differences in wood sources. Tree service companies obtain less than 10% of their fuelwood from woodlands, while other companies and other individuals obtain 90% of their fuelwood from standing live trees on woodlands (Figure 3).

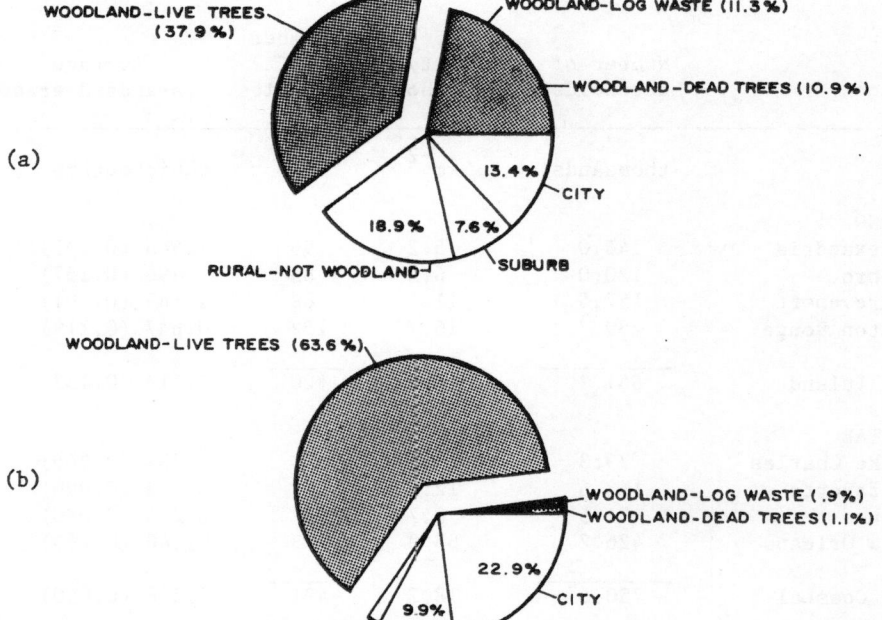

Figure 2. Land use class of domestic fuelwood by (a) households and (b) commercial vendors

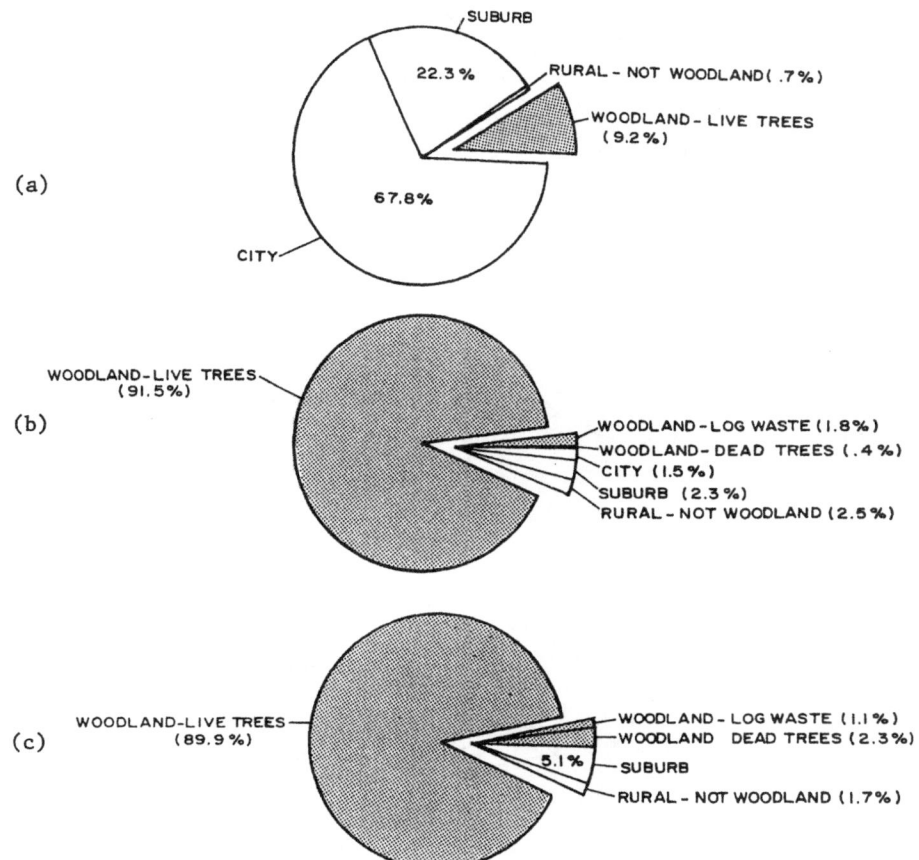

(a)

SUBURB

22.3%

RURAL – NOT WOODLAND(.7%)

WOODLAND – LIVE TREES
(9.2%)

67.8%

CITY

(b)

WOODLAND–LIVE TREES
(91.5%)

WOODLAND–LOG WASTE (1.8%)
WOODLAND–DEAD TREES (.4%)
CITY (1.5%)
SUBURB (2.3%)
RURAL – NOT WOODLAND (2.5%)

(c)

WOODLAND–LIVE TREES
(89.9%)

WOODLAND – LOG WASTE (1.1%)
WOODLAND DEAD TREES (2.3%)
5.1%
SUBURB
RURAL – NOT WOODLAND (1.7%)

Figure 3.   Land use class of domestic fuelwood by commercial vendors: (a)
tree service companies, (b) other companies and (c) other
individuals (not households)

Domestic fuelwood is virtually all hardwoods. Oaks account for 693,000
m^3 or 71% of the fuelwood produced. Approximately 338,000 m^3 are derived
from standing live trees on woodlands. Oaks comprise between 63% and 69% of
the fuelwood harvested by households, tree service companies, and other
individuals. Oaks comprise better than 90% of the fuelwood harvested by
other companies (Figure 4).

In terms of ownership, most of the wood comes from private land.
Forest industry land accounts for 95,000 m^3; other private land, 844,000 m^3;
and public land, 31,000 m^3. Examination by type of producer suggests that
forest industry land is important mainly to other companies (Figure 5).

Domestic fuelwood production by parish is presented in Figure 6.
Production is greatest in northern parishes near major cities. The inland
regions of the state produce more than twice as much wood as the coastal

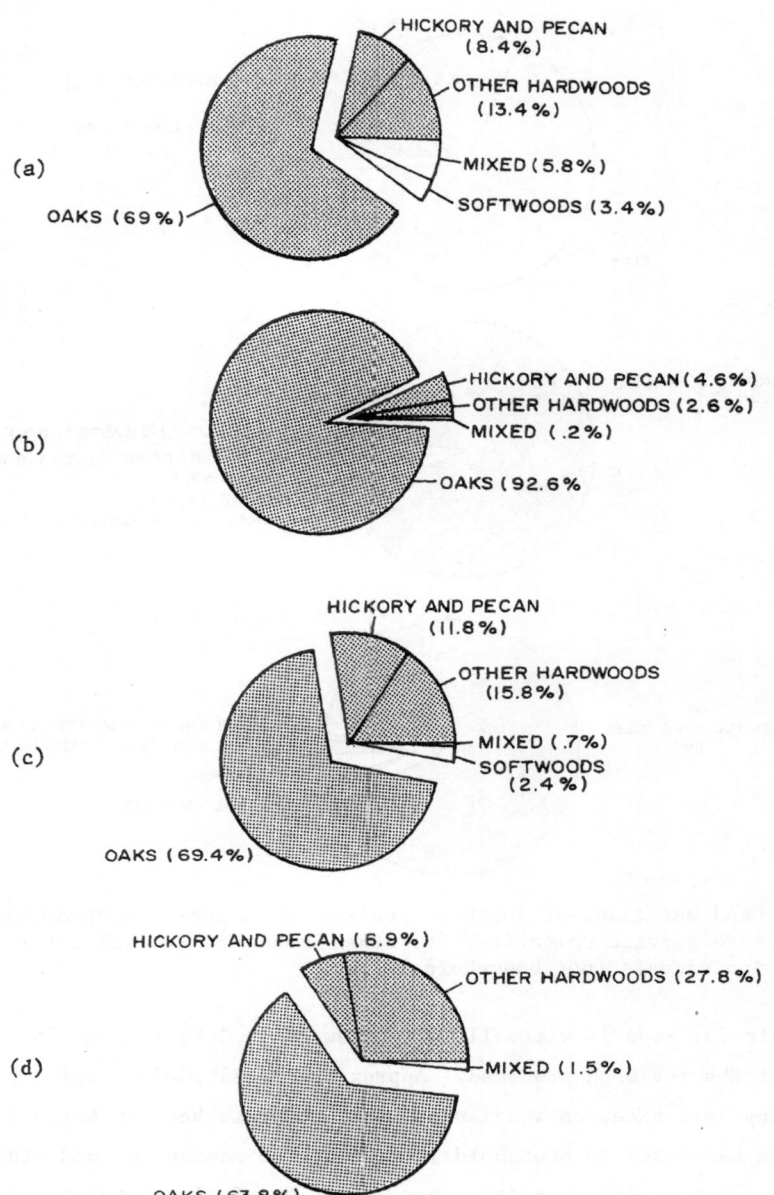

Figure 4. Species of domestic fuelwood produced by type of producer: (a)
all households (b) other companies (c) tree service companies and
(d) other individuals (not households)

regions. Many of the differences are attributable to the greater use of
wood by northern households. The inland regions are generally colder, more
rural, and contain the majority of woodland in the state (6). Woodland is a
source for greater proportion of fuelwood production in the inland regions
of the state for both households and commercial vendors (Figure 7).

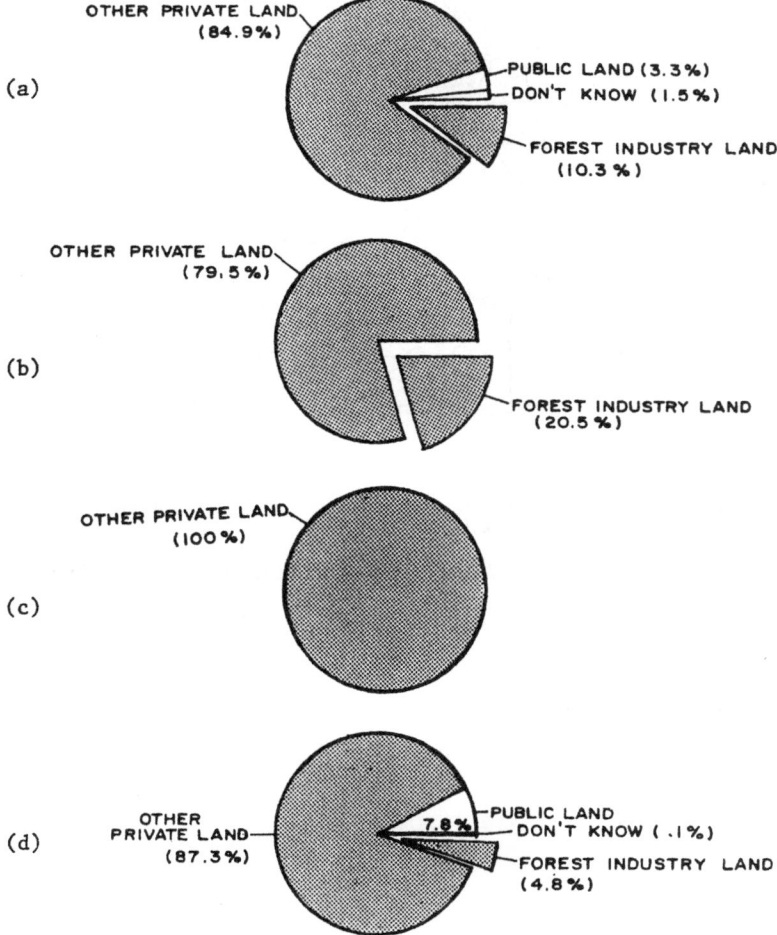

Figure 5. Ownership of land upon which domestic fuelwood was harvested by type of producer: (a) all households (b) other companies (c) tree service companies and (d) other individuals (not households)

Annual domestic fuelwood production, principally hardwoods, amounts to 443,000 m^3 of the standing live tree volume on woodlands. Hardwood growing-stock volume associated with biomass production has been estimated at 69% of standing live hardwood trees in Louisiana.[7] If an assumption is made that 2/3 of the standing live tree volume cut for fuelwood is hardwood growing stock, then the drain on timber resources amounts to 295,000 m^3 annually.

The growing-stock portion of domestic fuelwood production is equal in magnitude to 2% of the state's 18,300,000 m^3 roundwood harvest, or 8% of the 3,800,000 m^3 hardwood roundwood harvest for 1983.[8] The impact is somewhat more pronounced if one compares domestic fuelwood production with pulpwood production. The 295,000 m^3 is equal to 3% of the state's 9,780,000 m^3

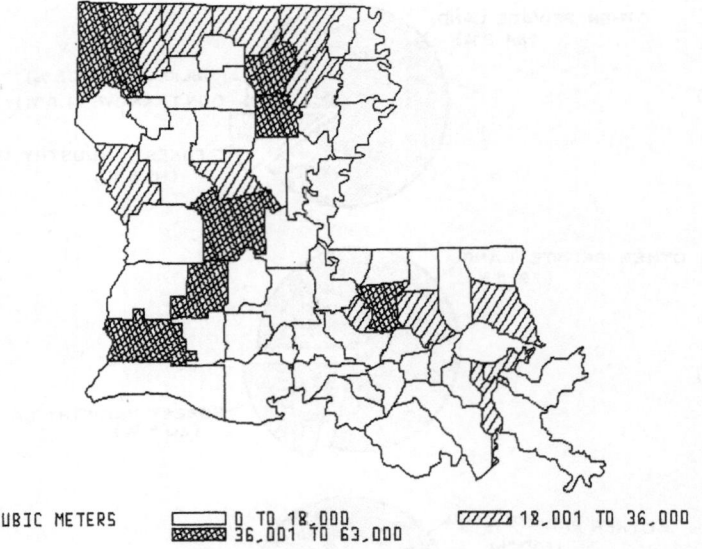

CUBIC METERS   ☐ 0 TO 18,000   ▨ 18,001 TO 36,000
                ▩ 36,001 TO 63,000

Figure 6. Louisiana domestic fuelwood production by Parish, 1983-84 heating
season

pulpwood production (roundwood and residues), or 19% of the 1,530,000 m^3
hardwood roundwood harvested for pulp in 1982.[5]

CONCLUSIONS

Louisiana's timber industry derives most of its raw wood material from
pulpwood (largely softwoods) and softwood lumber. Hardwoods represent less
than 1/4 of the state's timber harvest.[8] The domestic fuelwood production
impact is limited, but may increase, particularly if households obtain more
wood from commercial vendors, or if a greater fraction of fuelwood is cut
from growing-stock trees. Survey results suggest competition, chiefly for
oaks, is most likely near metropolitan areas in the northern part of the
state.

APPENDIX

City- Inside city or village limits.

Forest industry land - Land owned by companies or individuals operating
primary wood-producing plants.

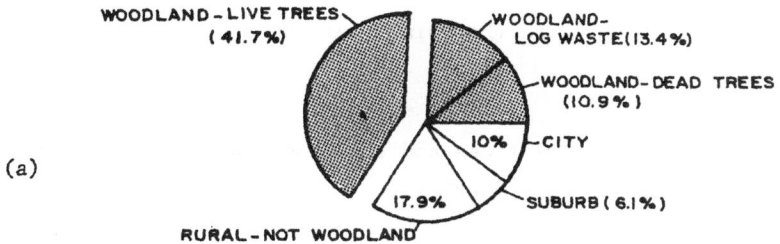

(a)

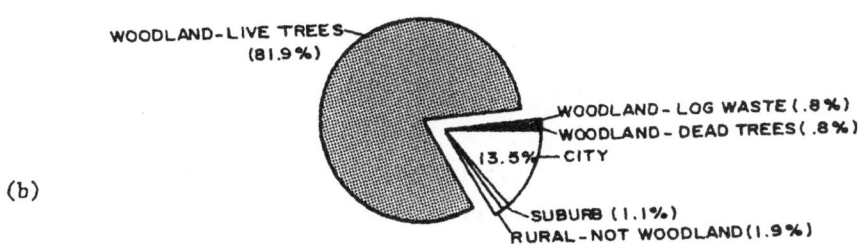

(b)

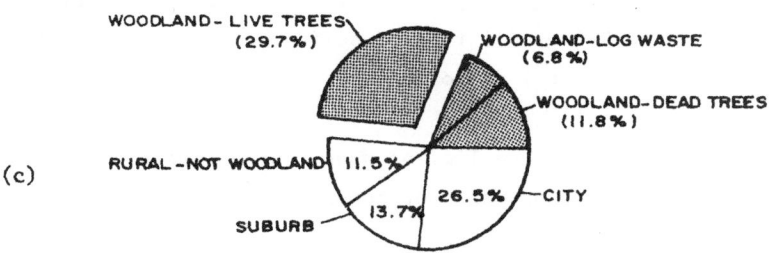

(c)

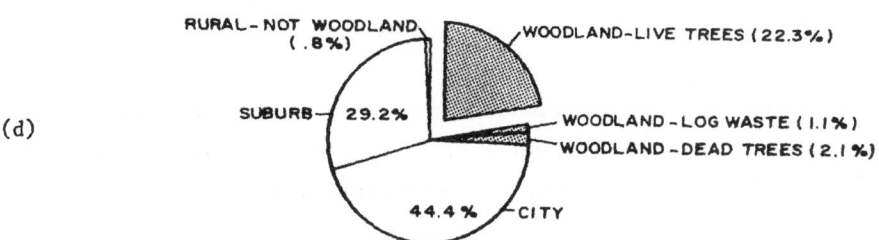

(d)

Figure 7. Land use class of domestic fuelwood production for inland and
coastal regions of Louisiana by households and commercial
vendors: (a) all households – inland regions (b) commercial
producers – inland regions (c) all households – coastal regions
(d) commercial producers – coastal regions

Growing-stock volume - Volume of sound wood in the bole of commercial trees 127 mm (5.0 inches) diameter at breast height from a 0.305 m (1 foot) stump to a minimum 101.6 mm (4.0 inch) top diameter outside bark of the central stem or to the point where the central stem breaks into limbs.

Logging waste - Noncommercial trees and the unused portions of trees cut or killed by logging on timberland.

Residues - Wood wastes derived from primary wood-producing plants.

Roundwood - Logs or other round sections cut from trees required to produce lumber, plywood, wood pulp, paper, or other similar products.

Rural--not woodland - Outside city or village limits; wood sources include trees on pasture or cropland.

Standard cord - A pile of logs 4 by 4 by 8 feet (128 ft^3). A cord of fuelwood contains 2.1 m^3 (75 ft^3) of solid wood. A cord of pulpwood contains 2.24 m^3 (80 ft^3) of solid wood.

Suburb - Outside city or village limits; wood sources include trees in fencerows, windbreaks, or yards of homes.

Timberland - Forest land that is producing, or is capable of producing, crops of industrial wood and not withdrawn from timber utilization by administrative statute or regulation.

Woodland - Woodland areas outside city or village limits, not including fencerows, windbreaks, yards of homes, pasture, or cropland. In this study, "woodland" is assumed to be timberland.

Commercial vendor of domestic fuelwood - An individual or business operation that sells logs or other round sections of trees to households for use as fuelwood.

REFERENCES

1. K. E. Skog and I. A. Watterson. Residential fuelwood use in the United States, 1980-81. 55 p. + Appendices. USDA-Forest Service, Forest Products Laboratory, Madison, WI. 1983.

2. D. R. Betters, W. R. Wilcox and P. P. Ryan. Domestic fuelwood utilization and resource management in Colorado. J. For. 82(8):484-487. 1984.

3. J. E. Blyth, M. E. Bailey and W. B. Smith. Fuelwood production and sources in Wisconsin, 1981. Resource Bulletin NC-75. 54 p. USDA-Forest Service. North Central Forest Experiment Station, St. Paul, MN. 1984.

4. A. Marsinko and T. E. Wooten. Firewood consumption in South Carolina. 14 p. + Appendix. Dept. Forestry, For. Res. Series No. 36, Clemson University, SC. 1983.

5. J. F. Rosson, Jr. Southern pulpwood production, 1982. Resource Bulletin SO-99. 23 p. USDA-Forest Service, Southern Forest Experiment Station, New Orleans, LA. 1984.

6. P. A. Murphy. Louisiana forests: status and outlook. Resource Bulletin SO-53. 31 p. USDA-Forest Service, Southern Forest Experiment Station, New Orleans, LA. 1975.

7. USDA Forest Service. Tree biomass-a state of the art compilation. Gen. Tech. Rep. WO-33. 34 p. Washington, D.C. (Table 3, p. 11). 1981.

8. Louisiana Department of Natural Resources, Office of Forestry. Unpublished. Timber severed and tax receipts for 1983. Baton Rouge, LA.

CLUSTER SAMPLING AND CONSTRUCTION OF BIOMASS REGRESSIONS:

RESULTS OF A SIMULATION STUDY

T. Cunia and A. J. Gillespie

SUNY College of Environmental Science and Forestry
State University of New York
Syracuse, New York 13210

ABSTRACT

Tree biomass regressions are ordinarily constructed from samples of
trees selected by cluster rather than random sampling. The least squares
techniques are then used to estimate the regression functions, techniques
which ordinarily ignore the cluster effect. The results of a simulation
study are reported whereby (i) samples of trees were selected by cluster
sampling, (ii) the biomass regression functions were estimated by the ordinary least squares regression techniques and by techniques that were modified to take the cluster effect into account and (iii) the estimates of the
average biomass per acre was compared to the known true value of the population. It has been found that (i) the ordinary and the modified techniques
generate average biomass estimates that are essentially the same with about
the same precision, (ii) the bias, if any of the average estimates is very
small (less than .5 percent), is about the same for all procedures and most
of the time is not significantly different than zero, and (iii) the precision as estimated from the sample data is grossly overstated if the ordinary
regression techniques are used, but approximately correct if the modified
regression techniques are used instead.

Keywords. Biomass, forest inventory, regression functions, cluster
sampling.

INTRODUCTION

The standard sampling design for the biomass inventory of a given
forest area is that of a double sampling. A large sample of trees is first

selected and the trees are measured for their diameter and possibly height. These trees are not measured for biomass. A second, much smaller sample of trees is also selected and its trees are measured for biomass in addition to diameter and height. The regression function of tree biomass on diameter (or diameter and height) as estimated from the second sample is then applied to the trees of the first sample to estimate the biomass of each tree and the average biomass per hectare.

The biomass regression functions are generally estimated by the ordinary weighted least squares method which assumes, among other things, that the sample trees are selected independently of each other. As the sample trees are generally selected by methods other than simple random sampling (with replacement) this assumption is seldom satisfied. As shown by Cunia,[1,2] Kotimaki and Cunia[3] and Briggs and Cunia[4] this assumption may become critical. Then, one must modify the least squares procedure to take the sampling method into account, if valid inferences about the regression are desired.

The sampling procedure considered here is cluster sampling whereby (i) the trees of the population are divided into clusters (plots), (ii) several clusters are selected at random and (iii) the trees from the sample clusters (or a subsample) are measured for biomass, diameter and height. A major advantage is the large decrease in the average sampling costs per tree; the movement of men and machines is greatly reduced. The disadvantage is that the average amount of information per tree is also reduced; trees growing close to each other have the tendency to be more similar than trees growing farther apart.

Cunia[2] suggested three approaches to modify the least squares method for cluster sampling. The first, using ratio estimators models, has been applied by Kotimaki and Cunia[3] to two cluster samples of trees. The second approach uses linear regression models and has been applied to the same two samples by Briggs and Cunia.[4] In both cases it was found that taking the cluster effect into account has no great effect on the biomass estimates; the error of the estimates, however, was greatly underestimated. The third approach as well as an additional approach using generalized least squares techniques have never been applied to our knowledge.

The objectives of the present study is to verify by simulated sampling the conclusions reached by Kotimaki and Cunia[3] and Briggs and Cunia.[4] In their study, the parameters of the population were unknown and the various estimators derived under different model assumptions, could not be compared to the true values of the estimators; they had to be compared to the estimator for which the model was thought to be the best. On the other hand, by simulated sampling one could repeatedly (i) select trees from a known population by a variety of sampling procedures, (ii) estimate, for each sample, biomass regression functions for a variety of statistical models, (iii) apply the resulting regression to the population trees to derive biomass estimates and (iv) compare the estimates, for bias and precision, to the true known parameter values.

The population of trees we have used consists of 22,753 trees contained in 667 non-empty and 260 empty permanent one-fifth acre (one acre = .405 ha) sample plots, defined here as the population clusters, selected from the state of New York. The trees were measured in the field for their species, diameter and merchantable height. The total height and total biomass (green weight above ground) were generated for each tree by Monte Carlo techniques described in detail by Cunia and Michelakackis[5,6,7] and Cunia, Michelakackis and Lee.[8] In generating total height and biomass, the techniques took into account (i) the effect of species, diameter, merchantable height, site, geographical region, cluster and (ii) the probability distribution of total height and biomass about their regression functions.

Our sampling procedure is the following. Fifty clusters were selected at random (without replacement) from the 667 non-empty clusters of the population, and thirty percent of the trees from each of these clusters were selected at random (without replacement) and measured for diameter, total height and biomass. When thirty percent represented a fraction, not an integer, a Monte Carlo procedure was used to decide whether a tree associated with the fractional part of the number of trees to sample should or should not be selected. As 100 such simulated samples were already available from the work of each of two doctoral students (John Michelakackis and Sueh-Fang Hsu, graduate students – SUNY College of Environmental Science and Forestry, Syracuse, N.Y.), we did not have to actually simulate the sampling process; we simply used these samples.

ESTIMATION PROCEDURES AND RESULTS

The population parameter we shall estimate is μ = average biomass per hectare (not per tree) defined as the total biomass of all trees divided by the total area, A = (927 plots) x .081 ha (1/5 acres per plot) = 75.09 ha

(185.4 acres), of the population 927 plots. The true value μ is known to be 129514 kg ha^{-1}. To estimate μ we shall use two sets of models. The first set consists of fourteen ratio estimators models, the second, of twenty linear regression models. These models are summarized below. For more details the reader is referred to Kotimaki and Cunia[9] for the first and Briggs and Cunia[1] for the second set.

Set 1: Ratio Estimators Models. These models assume the existence of the relationship y = Rx, where y is the tree biomass, x is some variable highly correlated with y and R is the ratio τ_y / τ_x of the totals of y and x in the population of trees. In our study x = d^2 or x = d^{2}h, where d = tree diameter and h = total tree height. Using first x = d^2 and letting μ_1 = τ_x/A = average "sum of d^2" per hectare, we define the following seven ratio estimators models, where z, r and V denote respectively the estimators of μ, R and variance of z as calculated under each model assumptions. Unless otherwise stated, Σ denotes summation over all sample trees.

Model 1 - It is assumed that the true regression functions of y on x is y = Rx, the conditional variance of y given x is homogeneous, and the sample trees are selected by simple random sampling. Then,

$$r_1 = \Sigma\, xy/ \Sigma x^2, \quad z_1 = r_1\, \mu_1 \text{ and}$$

$$V_1 = (\, \Sigma y^2 - (\, \Sigma xy)^2/ \Sigma x^2)\ \mu_1{}^2 \big/ (n-1)\, \Sigma x^2$$

Model 2 - Same as Model 1 but it is assumed that the conditional variance of y given x is proportional to x. Then,

$$r_2 = \Sigma y/ \Sigma x, \quad z_2 = r_2\, \mu_1 \text{ and}$$

$$V_2 = (\, \Sigma(y^2/x) - (\Sigma y)^2/ \Sigma x)\ \mu_1{}^2 \big/ (n-1)\, \Sigma x.$$

Model 3 - Same as Model 1 but it is assumed that the conditional variance of y given x is proportional to x^2. Then,

$$r_3 = \Sigma(y/x)/n, \quad z_3 = r_3\, \mu_1 \text{ and}$$

$$V_3 = (\, \Sigma(y/x)^2 - (\, \Sigma(y/x))^2/n)\ \mu_1{}^2 \big/ n(n-1).$$

Model 4 - Ratio-of-means model where it is assumed that the sample trees are selected by simple random sampling.
Then, $\quad r_4 = r_2 = \Sigma y/ \Sigma x, \quad z_4 = r_4\, \mu_1 = z_2$ and

$$V_4 = (S_{yy} - 2r_4 S_{xy} + r_4{}^2 S_{xx})\ \mu_1{}^2 \big/ n\ \bar{x}^2,$$

where S_{yy}, S_{xy}, S_{xx} and $\bar{x}$ are the usual sample variances, convariance and mean respectively of y and x.

Model 5 – Mean-of-ratios model – Same assumptions as model 4 but
$$r_5 = r_3 = \Sigma(y/x)/n, \quad z_5 = r_5 \; \mu_1 = z_3 \text{ and } V_5 = V_3.$$

Model 6 – Ratio-of-means model where it is assumed that the trees are selected by random cluster sampling. Then,

$$r_6 = r_2, \; z_6 = z_2 \text{ and}$$
$$V_6 = (S_{uu} - 2r_6 S_{uv} + r_6^2 S_{vv}) \; \mu_1^2 / m \; \bar{v}^2,$$

where $u = \Sigma y$ and $v = \Sigma x$ are cluster variables (sums within given clusters), Σ means summation over the trees of a given cluster, m is the number of sample clusters and S_{uu}, S_{uv}, S_{vv} and $\bar{v}$ are the usual sample variances, covariance and mean respectively of u and v.

Model 7 – Mean-of-ratio model where it is assumed that the trees are selected by random cluster sampling. Then,

$$r_7 = r_3, \; z_7 = z_3 \text{ but}$$

$$V_7 = m \; (S_{ww} - 2r_7 S_{wt} + r_7^2 \; S_{tt}) \; \mu_1^2 / n^2,$$

where $w = \Sigma(y/x)$, t = number of trees in a given cluster, and S_{ww}, S_{wt} and S_{tt} are the usual sample variances and covariance of w and t.

Note that these models define three rather than seven estimators of μ. As their model assumptions differ, however, we have six rather than three estimators of their variances (the fifth and the third estimators being identical in all but assumptions).

Models 8 to 14 are identically defined except that $x = d^2h$. These formulas are not repeated.

Set 2: Linear Regression Models. All twenty models assume that the true regression function of the tree biomass y on diameter d and total height h is of the form
$$y = \beta_1 x_1 + \beta_2 x_2 + \text{---} + \beta_p x_p = [\beta]' \; [x]$$
where [ ] are [ ]' denote matrices and transposed matrices and variables x_j are functions of d and h.

We have used five regression equations that differed by the order and definition of [x]. More specifically, the vector [x] of regression 1, 2, .., 5 is defined respectively as

$$\begin{bmatrix} 1 & d & d^2 \end{bmatrix}', \quad \begin{bmatrix} 1 & d^2 \end{bmatrix}', \quad \begin{bmatrix} 1 & d^2h \end{bmatrix}', \quad \begin{bmatrix} d^2h \end{bmatrix}' \text{ and}$$

$$\begin{bmatrix} 1 & d & d^2 & d^2h & dh & h \end{bmatrix}',$$

where in the last equation only those variables x are kept for which the estimates of the corresponding coefficients are significantly different from zero. To estimate $[\beta]$ we have used four approaches. Approach 1 is the ordinary least squares (OLS) applied to individual tree data. Approach 2 is the ordinary weighted least squares (OWLS) applied to individual tree data with the conditional variance of y assumed proportional to d^4 (if only d is used in regression) or d^4h^2 (if both d and h are used in regression). Approach 3 is the modified least squares (MLS) applied to individual cluster variables, with the conditional variance of cluster biomass assumed proportional to cluster size (number of trees in the cluster). Approach 4 is the modified weighted least squares (MWLS) applied to individual cluster variables, with the conditional variance of cluster biomass assumed proportional to Σd^4 or Σd^4h^2 as the case may be, where Σ denotes summation over the trees of a given cluster.

For a detailed description of Approach 3 and 4, the reader is referred to Cunia[2,9] and Briggs and Cunia.[4] For a summary description, we shall assume that (i) sample trees are measured for y and x, (ii) true regression function is $\hat{y} = [\beta]'$ [x], (iii) conditional variance of y given [x] is equal to $a_i \sigma_{yy}$ where a_i but not σ_{yy} is known for every tree i in the sample, and (iv) all other assumptions of the least squares linear regression are satisfied, except that, due to cluster sampling, the various values y within a given cluster are not statistically independent. Note that in tree biomass regressions, the known value of a_i is the value d^4_i or $d^4_i h^2_i$ of the i-th tree.

Summing up within each cluster the tree variables y, x_1, x_2,..., x_p we define the <u>cluster</u> variables u = Σy, $v_1 = \Sigma x_1$, $v_2 = \Sigma x_2$,..., $v_p = \Sigma x_p$, where Σ is taken over the trees of a given cluster. Making a few assumptions about the probabilistic behavior of y <u>within</u> the clusters, we can make the new, reasonable assumption that the conditional variance of u is proportional to $(\Sigma a_i) \sigma_{yy}$, where again Σ is taken over the trees of the cluster. Because now μ and [v] of the various clusters are statistically independent, the OWLS method can be applied to the regression of u on [v]. When $a_i = 1$ for all trees, we have Approach 3, while for $a_i = d_i^4$ (for equations 1 and 2 or $a_i = d_i^4 h_i^2$ (for the remaining equations) we have Approach 4.

In this way, twenty new models are defined by combinations of five regression equations with four least squares approaches. More specifically, we have Model 15: approach 1 applied to equation 1, Model 16: approach 1 applied to equation 2,..., Model 19: approach 1 applied to equation 5, Model 20: approach 2 applied to equation 1,..., Model 34: approach 4 applied to equation 5.

For any given linear regression model i, i = 15, 16,..., 34, the estimators z_i and V_i are given by the formulae

$$ z_i = \left[b_i \right]' \left[\mu_i \right] \quad \text{and} \quad V_i = \left[\mu_i \right]' \left[S_{b_i b_i} \right] \left[\mu_i \right] $$

where $[b_i]$ is the estimator of $[\beta_i]$, $[S_{b_i b_i}]$ is the estimator of the covariance matrix of $[b_i]$ and $[\mu_i]$ is the vector of the averages per acre μ_{ij} of variables x_j model i, calculated by the formula

$$ \mu_{ij} = (\text{total of population values } x_j)/\text{total forest area (acres)} $$

Note that Model 18 is identical to Model 8 and Model 23 is identical to Model 10. With Models 3 and 10 identical respectively to Models 5 and 12, we have only 30 rather than 34 estimators.

For each estimation procedure (model) i and each simulated sample j of trees, i = 1, 2, -, 34 and j = 1, 2, --, 100 we have calculated the estimators z_{ij} and V_{ij}. In addition we have noted whether μ fell below, within, or above the 95 and 99 percent confidence intervals $(z_{ij} \pm t\sqrt{V_{ij}})$. For each estimation procedure i we have also calculated the following summary statistics, where Σ stands for summation over the 100 samples.

$$ \bar{z}_i = \Sigma z_{ij}/100 = \text{estimator of } \mu $$

$$ S_{z_i z_i} = \left(\Sigma z_{ij}^2 - (\Sigma z_{ij})^2/100 \right) \Big/ 99 = \text{estimator of the variance of } z_i $$
as calculated from the differences between the 100 samples values z_{ij}; the assumptions of model i are not used.

$$ \bar{V}_i = \Sigma V_{ij}/100 = \text{estimator of the variance of } z_i \text{ as calculated from the} $$
100 individual estimates V_{ij}; each estimate V_{ij} is calculated from the

Table 1. Basic summary statistics for the first set of 100 simulated samples and the number of times μ was below (b), within (w), or above (a) the confidence intervals: $\bar{z}$ is expressed in thousands kg ha^{-1} and S_{zz} and $\bar{V}$ in millions of kg^2 ha^{-1}.

Model	$\bar{z}$	$\bar{z}-\mu$	S_{zz}	$\bar{V}$	t	b	w	a	b	w	a
Ratio Estimators: $x = d^2$											
1	136.42	6.90	36.469	2.677	11.44	79	16	5	72	24	4
2	129.29	−0.22	13.424	2.823	−.60	16	63	21	11	78	11
3	120.82	−8.70	10.533	2.709	−26.80	0	6	94	0	9	91
4	129.29	−0.22	13.424	5.555	−.60	8	81	11	3	90	7
5	120.82	−8.70	10.533	2.709	−26.80	0	6	94	0	9	91
6	129.29	−0.22	13.424	16.005	−.60	2	95	3	0	100	0
7	120.82	−8.70	10.533	12.895	−26.80	0	6	94	0	9	91
Ratio Estimators: $x = d^2 h$											
8	126.52	−2.99	15.594	1.183	−7.59	11	31	58	6	43	51
9	129.67	0.16	5.726	1.475	.64	20	66	14	12	78	10
10	134.93	5.41	7.751	2.053	19.45	81	19	0	73	27	0
11	129.67	0.16	5.726	2.928	.65	11	79	10	4	92	4
12	134.93	5.41	7.751	2.053	19.45	81	19	0	73	27	0
13	129.67	0.16	5.726	8.251	.65	0	100	0	0	100	0
14	134.93	5.41	7.751	9.795	19.45	40	60	0	12	88	0
OLS Linear Regression Estimators											
15	129.51	0.00	12.111	4.965	0	9	80	11	3	91	6
16	129.45	−0.07	12.548	5.096	−.20	9	80	11	2	91	7
17	129.72	0.20	5.908	2.800	.85	12	77	11	4	91	5
18	126.52	−2.99	15.594	1.183	−7.59	11	31	58	6	43	51
19	129.79	0.28	6.113	2.628	1.15	14	74	12	7	87	6
OWLS Linear Regression Estimators											
20	129.67	0.16	12.344	4.622	.45	13	74	13	5	93	2
21	128.15	−1.37	12.656	3.645	−3.83	7	64	29	3	80	17
22	128.94	−0.57	6.501	2.794	−2.26	7	76	17	3	87	10
23	134.93	5.41	7.751	2.053	19.45	81	19	0	73	27	0
24	129.76	0.25	6.506	3.054	.96	12	80	8	4	94	2
MLS Linear Regression Estimators											
25	129.61	0.10	12.104	14.481	.28	2	96	2	0	99	1
26	129.50	−0.01	12.371	14.646	−.04	1	97	2	0	100	0
27	129.74	0.22	6.039	7.974	.90	2	97	1	0	100	0
28	128.67	−0.84	10.156	5.981	−2.64	3	85	12	0	97	3
29	129.84	0.33	7.435	7.491	1.19	2	95	3	0	99	1
MWLS Linear Regression Estimators											
30	129.57	0.07	11.822	17.344	.18	1	97	2	0	100	0
31	129.42	−0.09	12.608	14.790	−.026	1	97	2	0	100	0
32	129.85	0.34	6.807	9.368	1.28	2	97	1	0	100	0
33	130.89	1.38	7.001	7.478	5.21	5	95	0	3	97	0
34	129.79	0.28	6.873	9.366	1.05	1	98	1	0	100	0

Table 2. Basic summary statistics for the second set of 100 simulated samples and the number of times μ was below (b), within (w), or above (a) the confidence intervals: $\bar{z}$ is expressed in thousands kg ha^{-1} and S_{zz} and $\bar{V}$ in millions of kg^2 ha^{-1}.

Model	$\bar{z}$	$\bar{z} - \mu$	S_{zz}	$\bar{V}$	t	b	w	a	b	w	a
Ratio Estimators: $x = d^2$											
1	137.45	7.94	34.417	2.673	13.53	76	23	1	73	26	1
2	130.07	0.56	13.799	2.824	1.50	20	65	15	13	80	7
3	121.73	-7.78	10.651	2.696	-23.85	0	12	88	0	14	86
4	130.07	0.56	13.799	5.654	1.50	13	80	7	8	89	3
5	121.73	-7.78	10.651	2.696	-23.85	0	12	88	0	14	86
6	130.07	0.56	13.799	16.029	1.50	2	95	3	0	100	0
7	121.73	-7.78	10.651	12.626	-23.85	0	39	61	0	68	32
Ratio Estimators: $x = d^2 h$											
8	126.66	-2.85	18.780	1.132	-6.58	12	30	58	10	36	54
9	129.96	0.44	7.816	1.475	1.58	27	59	14	14	75	11
10	135.68	6.17	8.593	2.119	21.05	88	12	0	82	18	0
11	129.96	0.44	7.816	2.853	1.58	12	82	6	7	91	2
12	135.68	6.17	8.593	2.119	21.05	88	12	0	82	18	0
13	129.96	0.44	7.816	8.216	1.58	6	92	2	2	98	0
14	135.68	6.17	8.593	10.354	21.05	45	55	0	21	79	0
OLS Linear Regression Estimators											
15	130.10	0.58	13.342	5.203	1.60	15	78	7	9	88	3
16	129.98	0.47	13.312	5.297	1.28	13	79	8	8	89	3
17	130.02	0.50	7.626	2.785	1.83	14	80	6	9	89	2
18	126.66	-2.85	18.780	1.132	-6.58	12	30	58	10	36	54
19	130.10	0.59	7.561	2.633	2.14	16	79	5	9	89	2
OWLS Linear Regression Estimators											
20	130.23	0.71	13.926	4.589	1.91	16	76	8	11	84	5
21	128.84	-0.68	12.695	3.623	-1.90	11	72	17	8	81	11
22	129.09	-0.42	7.836	2.822	-1.51	7	77	16	3	89	8
23	135.68	6.17	8.593	2.119	21.05	88	12	0	82	18	0
24	130.10	0.59	7.709	3.209	2.13	15	79	6	5	94	1
MLS Linear Regression Estimators											
25	130.15	0.64	13.878	15.043	1.72	3	94	3	1	99	0
26	130.01	0.49	13.929	15.312	1.32	3	94	3	1	99	0
27	130.08	0.57	7.976	8.312	2.02	5	94	1	0	100	0
28	128.45	-1.06	18.542	6.035	-2.47	7	80	13	3	87	10
29	130.10	0.59	7.856	7.730	2.09	5	93	2	1	99	0
MWLS Linear Regression Estimators											
30	130.11	0.60	13.031	17.084	1.66	2	96	2	0	100	0
31	129.93	0.42	14.121	14.821	1.11	4	93	3	0	99	1
32	130.10	0.59	8.587	9.563	2.00	4	94	2	0	100	0
33	131.42	1.91	9.017	7.750	6.35	10	90	0	2	98	0
34	130.13	0.62	8.736	9.500	2.09	3	95	2	0	100	0

data of the sample j, under the assumptions of the model i, without reference to the known values μ or $\bar{z}_i$.

$(\bar{z}_i - \mu)$ = estimator of bias (if any) of z_i

$t_i = (\bar{z}_i - \mu)/\sqrt{S_{zizi}/100}$ = value t to test the null hypothesis that the bias of z_i is equal to zero. Finally, we have counted the number of times μ fell below, within or above the 95 and 99 percent confidence intervals for model i, as calculated under its assumptions.

These basic statistics are listed, by model number, in Table 1 for the first set and Table 2 for the second set of 100 simulated samples. Additional summary statistics have also been used in the analysis of the results but are not listed.

ANALYSIS OF RESULTS

Set 1: Ratio Estimators

In their study, Kotimaki and Cunia[3] concluded that (i) r_1 and r_3 estimators are generally poor but r_2 seems to be consistently good, (ii) the error of r_2 is grossly underestimated when the cluster effect is ignored as in models 4 and 11 and (iii) a reliable estimator of the precision of r_2 is obtained from models 6 and 13. Examination of Set 1 models confirms these findings.

(1) The ratio-of-means estimators $(r_2, r_4, \ldots)$ are unbiased for both $x = d^2$ and $x = d^2h$. The estimated bias, if any, is very small (between .2 and .4 percent of μ) and, in general, not significantly different from zero. On the other hand, the estimators r_1, $r_3 = r_5 = r_7$ and r_8, $r_{10} = r_{12} = r_{14}$ are all biased, with the estimated biases varying from 2 to over 7 percent.

(2) The estimate S_{zz} of the variance of z (as calculated from the differences between the 100 individual values z) is (i) very high for r_1 and r_8, (ii) relatively average for ratio-of-means and mean-of-ratios estimators and (iii) because of the large bias, the accuracy of the mean-of-ratios models is very poor. This implies that the ratio-of-means estimator is by far the most precise and accurate.

66

(3) The estimate $\bar{V}$ of the variance of z (the average of 100 estimates V calculated within the sample under the given model assumptions) is generally much lower than the corresponding estimate S_{zz}, whenever the cluster effect is being ignored. When the cluster effect is taken into account, (models 6, 7, 13, 14) $\bar{V}$ is slightly higher. Because of the large bias of the mean-of-ratio estimators (models 7 and 14), the only acceptable estimate V of the precision of z is that provided by models 6 and 13.

In conclusion, the only acceptable estimator is the ratio-of-means with the cluster effect accounted for (models 6 and 13); the bias is small, if any, the error is low, and its estimate V seems correct, except maybe for a slight tendency of overestimation. If the ratio $\sqrt{\bar{V}/S_{zz}}$ of the standard errors is taken, we find the values 1.08 and 1.20 for the first, and 1.08 and 1.03 for the second set of 100 samples respectively. This means that, on the average, the confidence interval is probably overestimated, for models 6 and 13 by about 10 percent. Note also, that the only set of confidence intervals that seems correct is that of models 6 and 13; for all other sets, the confidence levels are quite different than the one desired.

Set 2: Regression Estimators

In their study of linear regression models, Briggs and Cunia[4] concluded that (i) the ordinary and modified weighted least squares methods produced biomass tables that are essentially the same and (ii) the error of the biomass tables as estimated by the OWLS was grossly underestimated. A close look at lines 15 to 34 of Tables 1 and 2 gives the following similar conclusions.

(1) The linear models 18, 23, 28 and 33 (with the regression function $\hat{y} = \beta\, d^2 h$) are generally poor and should be eliminated from further analysis. They are of the ratio estimators type and two of them, models 18 and 23, are identical to models 8 and 10 already analysed and discarded as poor. All four models give erratic results with large biases and small accuracies; they are no longer considered here.

(2) The other 16 models consisting of the combinations of 4 estimation approaches by 4 regression equations are generally good. The bias, if any, is generally small, even when, in a few cases is significantly different from zero. The estimate S_{zz} of the precision of z is about the same for the models based on the regression equations 1 and 2 (biomass on diameter d

alone) on one hand and similarly for the models based on the regression equations 3 and 5 (biomass on diameter d and height h) on the other.

(3) The statistic V underestimates the variance of z, whenever the OLS and OWLS models ignoring the cluster effect are used. MLS, on the other hand, provides better estimates of the error of z. Because it was original-ly thought that the MWLS methods yield best estimates of the variance of z, let us compare its statistics $\bar{V}$ and S_{zz}.

As the reader can verify, $\bar{V}$ seems to overestimate the variance. If we use the ratios $\sqrt{\bar{V}/S_{zz}}$ of the standard errors we find the values 1.21, 1.08, 1.17 and 1.17 for models 30, 31, 32 and 34 of the first set, and 1.14, 1.02, 1.06 and 1.04 for the same models of the second set of 100 samples. If we use an average value of about 1.10, we may conclude that the MWLS models taking the cluster effect into account may slightly overestimate the error of z. Note that one reason for the overestimation of the variance of z is the fact that our formulae ignore the effect of the finite population cor-rection factor. Because 50 out of 667 clusters are selected without re-placement, the approximate correction for $\sqrt{\bar{V}/S_{zz}}$ should be about 4 percent.

Analysis of the number of times μ is found to fall within the 95 and 99 percent confidence intervals confirms these conclusions. The OLS and OWLS models yield 95 and 99 percent confidence intervals that are about 75 and 85 percent respectively. The corresponding percentages are close to 95 and 99, as they should be, for the modified least squares techniques.

SUMMARY COMMENTS

The main objectives of the present study were to verify the conclusions reached by Kotimaki and Cunia[3] and Briggs and Cunia[4] that, when the trees are selected by cluster sampling, (i) the estimates of the biomass tables are about equally good when the cluster effect is taken or not taken into account and (ii) if valid estimates of the error of biomass tables are desired, one must take into account the cluster effect; otherwise, the error is grossly underestimated.

We have found that these conclusions are approximately correct. The only difference may be in the fact that the modified weighted least squares method seems to overestimate the true error of z. This may be due to the fact that (i) the effect of the finite population correction factor is ignored, (ii) we may have simulated samples that on the average, yield estimates of the errors that is higher and (iii) the assumptions used by the MWLS are slightly off relative to the population used in the simulated sampling.

We have used a sampling method which can be defined as random cluster sampling with subsampling and without replacement. More research is needed to investigate what would be the conclusions if (i) sample clusters are selected by stratified or two-stage sampling, (ii) the trees within the sample clusters are selected with probability proportional to some measure of tree size (say basal area), (iii) the number of sample clusters is relatively small, say 3 or 4 and (iv) the number of trees selected from a sample cluster is fixed or proportional to cluster size.

REFERENCES

1. T. Cunia. On tree biomass tables and regression: some statistical comments. In: Forest Resource Inventory Workshop Proceedings. Vol. 2, W. E. Frayer (Ed.), Colorado State University. Fort Collins, CO. 1979.
2. T. Cunia. On sampling trees for biomass tables construction: some statistical comments. In: Forest Resource Inventory Workshop Proceedings. Vol. 2. W. E. Frayer (Ed.). Colorado State University. Fort Collins, CO. 1979.
3. T. Kotimaki and T. Cunia. Effect of cluster sampling in biomass tables construction: ratio estimators models. Canadian Journal of Forest Research 11:475-486. 1981.
4. E. F. Briggs and T. Cunia. Effect of cluster sampling in biomass tables construction: linear regression models. Canadian Journal of Forest Research 12:255-263. 1982.
5. T. Cunia and J. Michelakackis. A method to construct a forest biomass population model. In: Renewable Resource Inventories for Monitoring Changes and Trends. J. F. Bell and T. Atterbury (Eds.), Oregon State University, Corvallis, OR. 1983.
6. T. Cunia and J. Michelakackis. A Monte Carlo technique for generating total height of forest trees. School of Forestry Miscellaneous Publication No. 4, SUNY College of Environmental Science and Forestry. Syracuse, NY. 1984.
7. T. Cunia and J. Michelakackis. Constructing forest biomass populations for simulated sampling. School of Forestry Miscellaneous Publication No. 5. SUNY College of Environmental Science and Forestry. Syracuse, NY. 1984.
8. T. Cunia, J. Michelakackis and S. Lee. Generating total tree heights by a Monte Carlo technique. In: Proceedings of the 1983 Southern Forest Biomass Workshop. R. F. Daniels and P. H. Dunham (Eds.) U.S. Department of Agriculture Forest Service. Southeastern Forest Experiment Station, Asheville, NC. 1984.
9. T. Cunia. Cluster sampling and tree biomass tables construction. In: Interdivisional Proceedings. 17th IUFRO World Congress. Kyoto, Japan. 1981.

RESEARCH ON SHORT-ROTATION WOODY CROPS IN THE SOUTH

J. W. Ranney, J. L. Trimble, L. L. Wright and R. D. Perlack

Oak Ridge National Laboratory
P. O. Box X
Oak Ridge, Tennessee 37831

ABSTRACT

National priorities for research on short-rotation woody crops as an energy feedstock are set to achieve economic competitiveness with alternative energy and forest product feedstocks. Present estimates of production costs are about \$3.50 per GJ before conversion and up to \$8.00 per GJ after conversion. Based on research results, short-rotation-derived feedstocks may soon be competitive in some locations with most other fuels, except wood wastes and coal. However, further cost reductions under operating conditions are still necessary for wide concept utilization. These reductions can best be accomplished through genetic improvement of species, development of more efficient harvesting methods and equipment, and utilization of optimum cultural techniques. These topics are discussed with some specific references to the South to more specifically portray the state of the art of short-rotation woody crops for energy. Ultimate costs of production could be \$2.00 to \$3.00 GJ^{-1} before conversion or about \$15/green Mg^{-1} (\$30/dry Mg^{-1}), while productivity rates could reach 9 to 15 dry Mg ha^{-1} yr^{-1} on many sites.

Keywords. Biomass, energy crops, trees, economics, woody crops, intensive culture.

INTRODUCTION

Short-rotation intensive culture (SRIC), as defined in the U.S. Department of Energy's Short Rotation Woody Crop Program, is the application of agricultural principles to woody plants to attain the highest productivity per unit land in the least amount of time and at a competitive cost. It is emerging as the single most important technology that can avert direct competition between the existing forest products industries and a growing wood energy industry. With the application of SRIC, our forest resources will be capable of meeting future combined demands placed on them, including energy.

This paper broadly summarizes economic and technical research progress on SRIC of woody plants sponsored by the U.S. Department of Energy's Biomass Energy Technology Division and cooperative research with industry. The technology may already be economically competitive under unusual local conditions. However, many risks remain, and considerable improvements are still needed to make the technology viable.

ECONOMICS OF SHORT-ROTATION INTENSIVE CULTURE

Examination of major cost components of SRIC (Figure 1) as identified in BIOCUT, a microcomputer cost accounting model,[1] shows which components are important in cost determinations. Harvesting accounts for 33% of system costs, the most expensive component of the SRIC system. Planting (including plantlet costs), site preparation, and cultural management together comprise another 27%. A necessary profit margin of 15%, a net tax cost of 10%, and the cost of land, items over which little control can be exercised, account for 31% of the total cost of producing SRIC biomass. These calculations do not include the cost of wood drying and are based on field research sites in

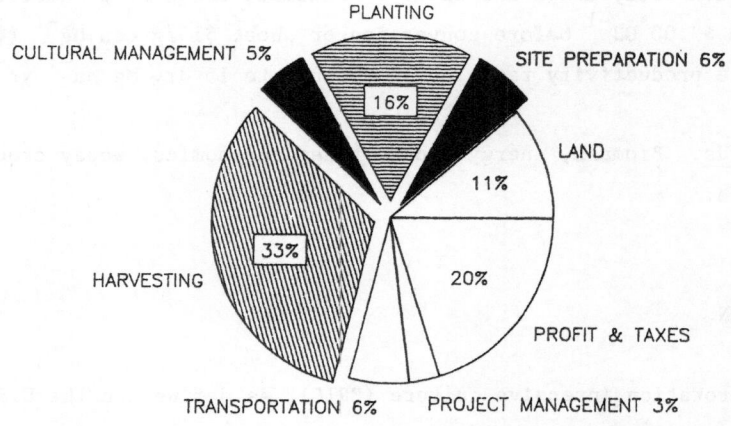

Figure 1. Major cost components of short-rotation intensive culture

central Pennsylvania,[2] northern Wisconsin,[3] southern and northern Florida,[4] Hawaii[5], and eastern and central Kansas.[6]

These cost categories, because of their significance, have been the focus of research and cost-reduction studies. A sensitivity analysis of these costs was conducted by Perlack et al.[7] Each category was evaluated for its change in overall production costs.

Figure 2 shows only the categories having the highest sensitivities. The analysis suggests that the same categories as mentioned earlier are of most concern in reducing costs.

These economic evaluations do not assess the relative importance of increased growth or genetic improvement on costs per GJ. Figure 3 represents the effect of changes in productivity alone on costs per GJ. This is an average from the same seven projects in the United States using the same economic assumptions. Percent changes in productivity prove to be even more significant than harvest costs. SRIC research is focused on the most important links between cost components and productivity and their effect on costs of production.

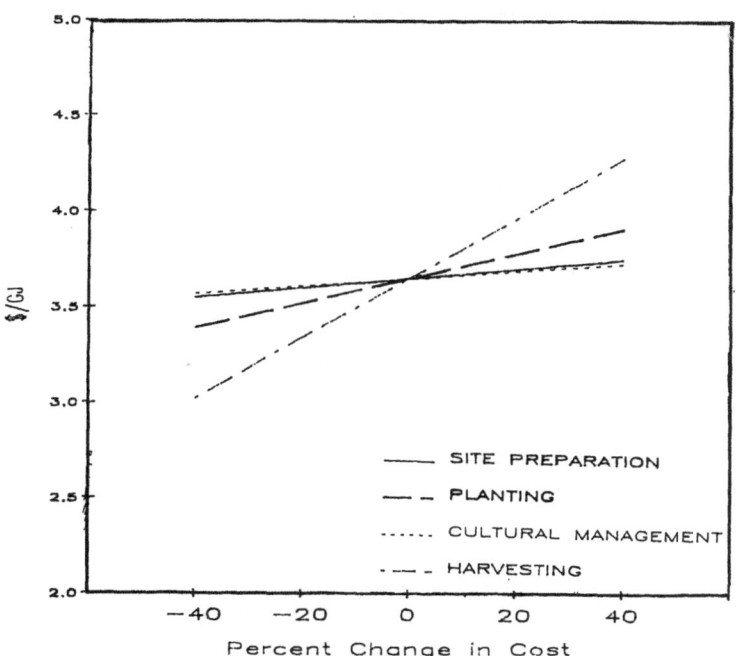

Figure 2. Sensitivity analysis of SRIC costs from best species and sites

Allowing for profits, taxes, and conversion inefficiencies of wet wood
but not allowing for declines in productivity rates as one moves from research plots to operational plantations, the present cost of SRIC production
is about $3.50 per GJ. When compared to alternative fuels on cost per GJ
after conversion, SRIC is $5.00 to $8.00 per GJ (see Figure 4) and appears
economically attractive. These cost evaluations do not include the benefits
of coppicing which should reduce costs another $0.40 GJ^{-1}. Of course, there
are many other pros and cons to the use of the technology which, for the
most part, must be determined on a site-specific basis. For example, tree
genotypes should be developed for local climates and soils, fertilizer
prescriptions must be customized for local soils, and, to a lesser extent,
planting and harvest systems should be adapted to local terrain. However,
the major point is that the technology shows very real potential today and
that major improvements in the technology (to be discussed later) are still
possible for significantly reducing SRIC costs. Future competitiveness is
unknown, but SRIC production costs are expected to decline as advances in
research are realized and as costs of other nonrenewable fuels climb, as
expected (see Figure 5).

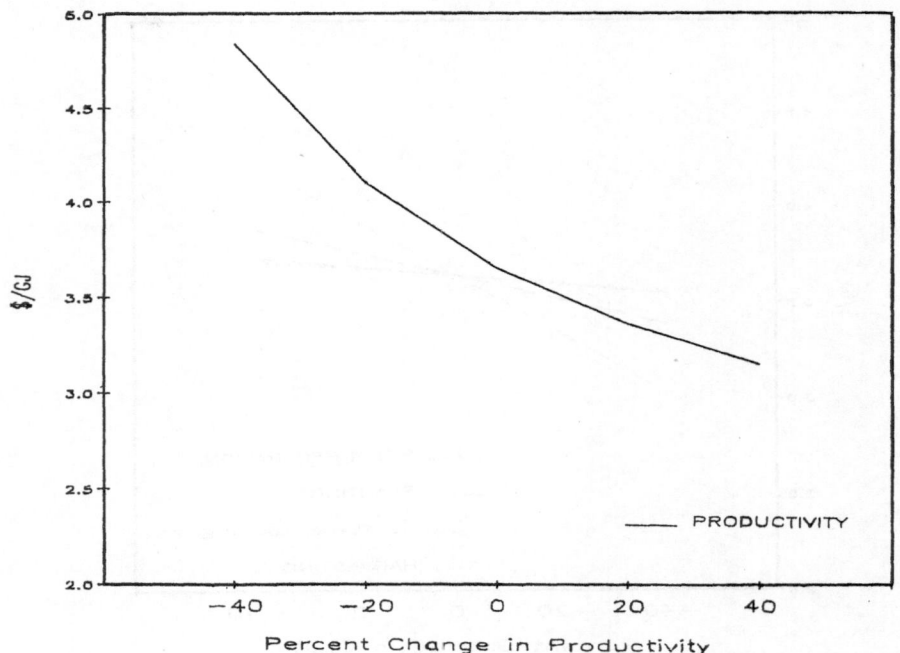

Figure 3. Sensitivity anaylses of SRIC productivity from best species and
sites

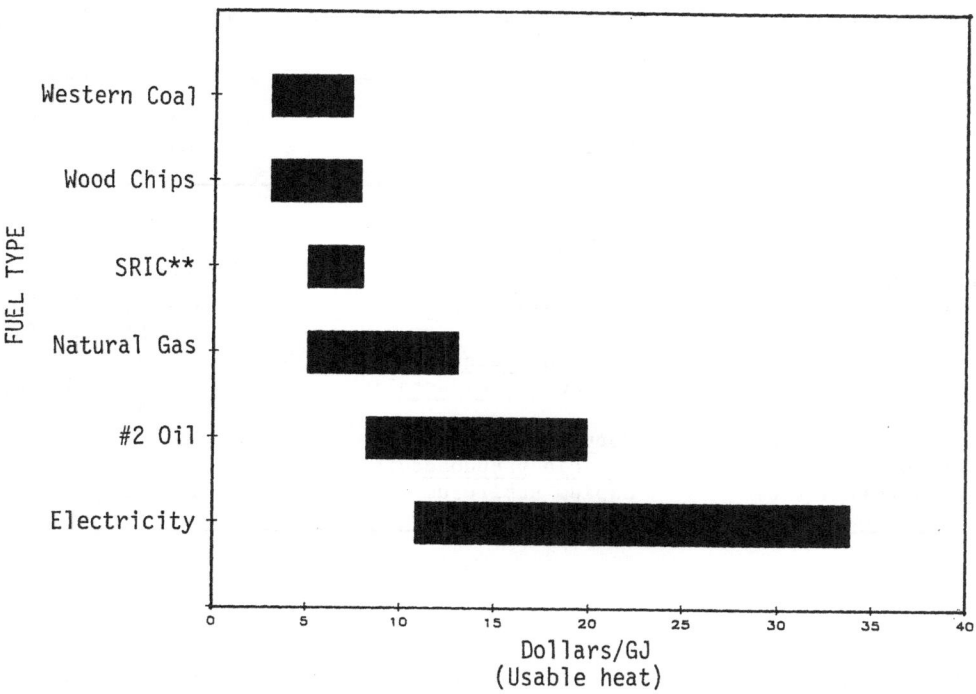

Figure 4. Range in energy costs for 1984.[8]

RESEARCH ACTIVITIES IN SRIC

Species Screening and Breeding

Nationally, 22 hardwood species have been recognized for their potentially superior performance in closely spaced short-rotation monocultures.
Many more species are under preliminary evaluation, and since 1978 over 125
species have been evaluated. DOE-supported research is concentrating on
hardwoods due to the serious lack of prior knowledge on hardwoods. There
are more species being tested in the South than in any other region, due
mainly to the rich diversity of species naturally existing in the region
(Table 1) and the variety of sites available for SRIC plantations. Nationally and regionally, species evaluations are generally at the stage of
range-wide evaluations of genetic variation in preparation for selection and
breeding for improved productivity in monocultures, for disease resistance,
and for tolerance-to-site stresses such as drought, poor drainage, or frost.
In the case of American sycamore (unpublished master's thesis at the University of Tennessee) and hybrid poplars,[9] very specific parameters related
to growth dynamics and leaf display are under investigation to understand
the relationships among tree geometry, competition, and growth in dense

Table 1. Nine species under energy evaluation in the mid-South and Southeast for SRIC

Common name	Scientific name	States where testing has occurred
European alder	Alnus glutinosa	GA, NC
Eucalyptus	Eucalyptus spp.	FL, GA, AL, NC
Green ash	Fraxinus pennsylvanica	NC, VA
Sweetgum	Liquidambar styraciflua	GA, NC, AL, SC, LA, MS, VA, TN
Leucaena	Leucaena leucocephala	FL, TX
American sycamore	Platanus occidentalis	GA, NC, MS, AL, VA, TN
Eastern cottonwood	Populus deltoides	MS, AL, NC, TN
Black locust	Robinia pseudoacacia	GA, KY, TN
Chinese tallow tree	Sapium sebiferum	FL, TX

hardwood monocultures. It is becoming apparent in the Pacific Northwest that different rotation lengths require different clones of hybrid cottonwoods to achieve the highest productivity rates. Similar trends can soon be expected for the South.

It is difficult to define the potential improvements in productivity that can be achieved, but some estimates have been made. Figures compiled from DOE-supported research indicate that possible genetic gains (parameters vary from site to site) appear to be on the order of 20-40% (Table 2). Estimates of potential productivity improvement based on the best single family or clone performance over mean group performance is somewhat higher at roughly 20 to 60% or more (Table 2). Such gains could be achieved only by cloning the best families. Achievement of genetic gains either through breeding or cloning will require sustained tree improvement programs. Improvement of trees for SRIC could progress faster than most tree improvement programs since rotation ages are short and propagation/gene manipulation technologies are particulary applicable to SRIC research. Attainment of higher productivity levels will have extremely important positive impacts on the economic viability of SRIC for energy production or other purposes.

Cultural Management

Effects of cultural management practices and site quality on productivity must be understood before realistic economic evaluations can be per-

Table 2. Information currently available on predicted genetic gains for promising hardwood species for southern SRIC plantations

Species	Parameter	Predicted biomass gain[a] (%)	Potential for improvement[b] (%)	Reference
Eucalyptus grandis	Coppice Volume	100	1000[c]	(4)
Eucalyptus grandis	Volume	–	19-63	(5)
Eucalyptus saligna	Volume	–	26-61	(5)
Fraxinus pennsylvanica	kg tree^{-1} yr^{-1}	29[d]	–	(10)
Platanus occidentalis	Mg ha^{-1}	17-22	34-65	(11)
Platanus occidentalis	Height, DBH	29-37	–	(12)
Prosopis spp.	kg tree^{-1}	–	480[d]	(13)
Prosopis alba and hybrids	Mg ha^{-1}	–	72	(13)

[a]Predicted biomass gain values reported by U.S. Department of Energy subcontractors. Values represent gains achievable in one generation except for Eucalyptus grandis which shows additive effects of selection over four generations.

[b]These values were calculated by determining the percent increase of the best single provenance, family, or clone over the mean of the group tested. They roughly represent maximum, site-specific improvements that could be gained by cloning the best candidates of the genetic material currently being evaluated.

[c]This high figure is based on differences in coppice growth between 529 progenies harvested during August at one site.

[d]Comparison of selected clonal Prosopis alba with native Texas mesquite tested at one site in Texas.

formed. Yield responses to management investments are being evaluated to determine optimum management strategies. Responses are specific to species and site conditions. Results from several sites were compiled to assess likely general improvements in productivity expected from various management options (Table 3). Expected improvement from, say, fertilizer addition should not be construed as cumulative with other site investments such as improved spacing/rotation age, although simultaneous treatments should produce more than either treatment alone.

Average productivity from experimental plots throughout the United States is 6 to 12 dry Mg ha^{-1} yr^{-1}. Improvements over this average from utilization of improved cultural treatments together should be about 10 to

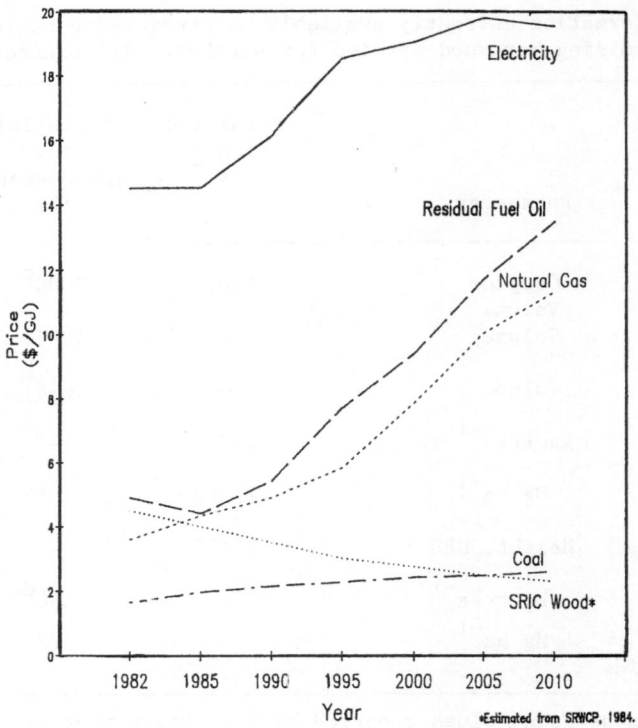

Figure 5. 1982 NEPP price forecasts for the industrial market by fuel type

20%. Considering coppice productivity and genetic improvements, combined
productivity responses should increase 40 to 60%, while costs remain level
or decline slightly from increased efficiency in use of fertilizers and
other resources. Plantlet costs, however, are a major expense (at $0.15 to
$0.35/propagule) that will require significant technological breakthroughs
to reduce. Using these figures, SRIC for energy has potential for being
economically competitive with most other fuels and lead the way as perhaps
the best renewable fuel feedstock technology available for some time to
come.

     Some concerns with SRIC technology not yet accurately considered are
(1) site limitations for plantation, (2) reduced productivity resulting from
larger-scale nonresearch operations, and (3) risks of crop failure or productivity declines from unanticipated problems such as those associated with
monocultures and environmental effects. First, hardwoods with the capacity
to coppice and grow rapidly at an early age demand rather high quality
sites, or amendments to sites, to improve their quality. Sites no worse in
quality than those currently considered marginal for agriculture are appropriate for SRIC technology. Field equipment necessary for SRIC management

imposes the same limitations. Tests conducted by Douglas Frederick[12] and Klaus Steinbeck[14] on southern piedmont and coastal plain sites, and tests and surveys conducted by Donald Rockwood in Florida,[4] are helping to delineate the land resource potentially suitable for SRIC technology. As a result of these investigations, agriculturally depleted piedmont clay soils are no longer considered a resource for SRIC plantations except for black locust. Productivity figures from these sites have not exceeded 4 to 5 dry Mg ha^{-1} yr^{-1}, even with fertilization and irrigation. Agricultural-quality clay soils, however, appear to be very suitable for SRIC. Also, significant areas of sandy coastal plain soils and peat soils appear appropriate for SRIC plantations.

Table 3. Differences between lowest and highest mean annual increments indicate likely yield improvements as a result of using best cultural treatments or best sites

Species	Age (years)	Mean Annual Increment (dry Mg ha^{-1} yr^{-1})	Variable	Reference
CULTURAL MANAGEMENT EFFECTS				
Alnus rubra	9	5.9-7.4	Spacing	(15)
Alnus rubra	5	5.1-7.2	Fertilizer	(16)
Eleagnus umbellata	2-4	2.6-9.2	Coppice	(17)
Eucalyptus grandis	2	5.5-15.7	Spacing	(4)
Platanus occidentalis	2	4.1-6.3	Spacing	(12)
Populus hybrid NE 388	4	5.4-9.6	Fertilizer	(2)
Populus hybrid NC 5331	4	5.7-7.0	Spacing	(3)
SITE EFFECTS				
Acer saccharinum	6	5.8-8.5	Rainfall and soil	(6)
Eucalyptus saligna	1.3-3.4	9.3-24.6	Rainfall and soil	(5)
Populus deltoides	6	5.6-13.9	Rainfall and soil	(6)
Robinia pseudoacacia	6	7.2-10.3	Rainfall and soil	(6)

Second, 20 to 40% declines in productivity are expected in operational conditions. It is not known if efficiencies associated with scaled up operations (i.e., reduced costs ha^{-1} or Mg^{-1}) will fully compensate for such expected productivity declines. Few experiments are addressing this concern, although private industry and the U.S. Department of Energy are beginning to evaluate larger plantings.

Third, the effect of SRIC monocultures on site quality is not known and is now under study. The effects in large part are believed to depend on initial site quality and on management techniques. Observations indicate that SRIC should not be as nutrient demanding or as erosion prone as agricultural practices because crops are not removed annually (soil is not disturbed annually), and a significant root and litter layer would be in place 70 to 90% of the time. Other risks with pests, diseases, and infrequent climatic events are not well defined because past agricultural and forestry techniques are not sufficiently similar to predict SRIC responses. It is imperative that larger-scale SRIC plantations be established to assess these risks and to further define likely productivity rates under operational situations.

Although there are well over 100 experimental plantings of SRIC for energy, only one of these exceeds 40 ha, a 280-ha eucalyptus plantation in Hawaii. This plantation is currently at harvest age, so coppicing information is just becoming available. Productivity rates in the range of 20 to 30 dry $Mg\ ha^{-1}\ yr^{-1}$ have been measured. Similar productivity rates have been recorded for eucalyptus on organic soils in southern Florida. Plantations supported by industry are becoming much more numerous, especially in the South. Most are affiliated with the North Carolina State University Hardwood Research Cooperative, the University of Georgia, the University of Florida, or the USDA Forest Service station in Stoneville, Mississippi. One of the most active and interesting of these industrial trials has been on good clay piedmont land in Alabama owned by the Scott Paper Company. Productivity of sycamore on this site reached over 12 dry $Mg\ ha^{-1}$ in two years on an old-field site. Experimental 2-ha harvests using innovative equipment from Canada are now under way to better assess total SRIC energy production costs.

Although there is an important need for larger scale testing of SRIC by industry to perform the most effective evaluations of the concept, there are important distinctions to realize between the forest products industry and the nonforest products industry. Mainly, the forest products industry has a much more sophisticated appreciation for genetics research and long-range improvements, whereas the nonforest products industry perceives SRIC with little understanding of such things as site limitations, genetics, and biological constraints on the flow of boiler feedstocks from the field. They are relying on popular literature and the limited number of experienced researchers that are knowledgeable on the concept. It is important that larger scale testing of the concept be conducted, but it will require the involvement of experienced people. The forest products industry is presently the best alternative for such testing.

Harvest Technology

Although harvesting accounts for the greatest single cost component in SRIC production, little research has yet been conducted on reducing these costs for plantations of densely spaced, small-diameter trees. With few exceptions, research of this type is associated with government support, since no significant market appears to exist for appropriate harvesting equipment. However, this situation is beginning to change.

The Tennessee Valley Authority (TVA) has joined DOE in an investigation of small-diameter harvesting equipment suitable for SRIC plantations and powerline rights-of-way—a significant market. Acting only to stimulate interest from manufacturers, this team is seeking strong industrial involvement, facilitating information exchange, and examining the possibilities for some significant cost-sharing projects.

An assessment of equipment concepts by the U.S. Department of Energy is to be completed in June 1985 and will include recommendations for the most attractive harvest concepts and some general design standards useful to equipment manufacturers. Trials are now being conducted in SRIC stands in the South and Pacific Northwest with forest products industries, using prototypes, modified existing logging equipment, modified agricultural harvesting equipment, and various combinations of unmodified logging equipment. Most activity in the South is with Scott Paper Company, Westvaco, and Georgia Pacific. Concepts proving most acceptable use circular saw devices, or modifications of the chainsaw principle, that harvest trees while the unit moves continuously at a constant speed. However, it appears that equipment must move at about 5 km h^{-1} rather than the present 2 km h^{-1} to significantly reduce harvest costs per megagram. Present costs are in excess of $20 dry Mg^{-1}. It is necessary to reduce this cost if SRIC material is to be competitive with coal as an energy feedstock.

Most of the cultivation experiments for SRIC energy plantations have been completed, resulting in identification of required procedures for achieving early high yields. Research is showing that productivity can be increased about 30% over present levels through genetic improvement. Rotation ages of 2 to 8 years have proven most economical while maintaining high productivity. Harvesting costs can possibly be reduced by 25 to 40% with the development of small-tree plantation harvesters, but the appropriate equipment does not yet exist. Plant material and cultural treatment improvements combined may be able to cut production costs about 20 to 30%. These possibilities are achievable with continued research and greater interaction with industry.

It would be inaccurate to suggest a total energy potential for SRIC-derived energy, because its potential is based simultaneously on its economic outlook and on the resources (land quality and distribution) that must support it. The cost of SRIC production now appears to be between $3.50 and $8.00 GJ^{-1} (expressed in terms of usable heat from direct combustion), which is competitive with natural gas, oil, and electricity. If research continues to improve the technology, genetic research combined with cultural improvements can reduce production costs to about $3.00 to $5.00 GJ^{-1}. Further cost reductions are possible through harvest improvements to the extent that production costs could be lowered to $2.50 to $4.00 GJ^{-1} of usable heat. Except under ideal site conditions, larger-scale plantations and a concomitant reduction in productivity (below research plot levels) will likely raise the production costs to $2.75 to $5.00 GJ^{-1} of usable heat. These figures should be recognized as estimates based on information summarized in this paper. The figures presented reflect estimated costs of production only, not product value. In terms of costs of production, our goal is to attain about $30 dry Mg^{-1}. Estimates indicate that this can be attained through continued research and utilization of abandoned and marginal agricultural land, where rainfall is not usually limiting and productivity rates of 9 to 15 dry Mg ha^{-1} yr^{-1} can be attained.

Conversion Factors

GJ = gigajoule, which equals 0.95 million Btu.

Mg = megagram or metric ton, which equals 1.10 tons.

ha = hectare, which equals 2.47 acres.

Mg ha^{-1} = 0.446 tons $acre^{-1}$.

ACKNOWLEDGMENTS

Research supported by the Biomass Energy Technology Division, U.S. Department of Energy, under Contract No. DE-AC05-84OR21400 with Martin Marietta Energy Systems, Inc. Publication Number 2512, Environmental Sciences Division, Oak Ridge National Laboratory.

REFERENCES

1. S. Das, R. D. Perlack, W. F. Barron and P. Kroll. BIOCUT: A microcomputer economic evaluation model for wood energy plantations, model description and users' guide, ORNL/TM-9576. Oak Ridge National Laboratory, Oak Ridge, TN (in press). 1984.
2. P. R. Blankenhorn, T. W. Bowersox, C. H. Stauss, S. C. Grado, C. Hornicsar and M. L. DiCola. Net energy and economic analyses for producing Populus hybrid under four management strategies. Unpublished annual report submitted to the U.S. Department of Energy's Short Rotation Woody Crops Program. Pennsylvania State University, University Park. 162 p. 1983.
3. E. A. Hansen and H. Niestaedt. Establishment of Populus energy plantations. Unpublished report submitted to the U.S. Department of Energy's Short Rotation Woody Crops Program. USDA Forest Service, Forestry Sciences Laboratory, Rhinelander, WI. 22 p. 1984.
4. D. L. Rockwood, C. W. Comer, S. V. Kossuth and G. F. Meskimen. Eucalyptus for biomass production in Florida. Unpublished report submitted to the U.S. Department of Energy's Short Rotation Woody Crops Program. University of Florida, Gainesville. 29 p. 1984.
5. T. B. Crabb, C. D. Whitesell and T. H. Schubert. Eucalyptus plantations for energy production in Hawaii. Unpublished report submitted to the U.S. Department of Energy's Short Rotation Woody Crops Program. BioEnergy Development Corporation, Hilo, Hawaii. 121 p. 1984.
6. W. A. Geyer. Great Plains energy forest. Unpublished report submitted to the U.S. Department of Energy's Short Rotation Woody Crops Program. Kansas State University, Manhattan. 101 p. 1984.
7. R. D. Perlack, J. W. Ranney and L. L. Wright. An economic evaluation of the competitiveness of short-rotation intensive culture for energy. IN Energy from Biomass and Wastes, Conference Proceedings, D. L. Klass (Ed.). Institute of Gas Technology, Chicago, IL (in press).
8. Minnesota Department of Natural Resources. Status of fiber fuel use in Minnesota with emphasis on automated systems. Minnesota Department of Natural Resources, St. Paul. 40 p. 1984.
9. USDA Forest Service. Energy and wood from intensively cultured plantations: Research and development program. USDA Forest Service General Technical Report NC-58. USDA Forest Service, North Central Forest Experiment Station, St. Paul, MN. 28 p. 1980.
10. J. W. Hanover. Tree species and management strategies for biomass production in the Lake States. Unpublished annual report submitted to the U.S. Department of Energy's Short Rotation Woody Crops Program. Michigan State University, East Lansing. 167 p. 1984.
11. S. B. Land. Genetic selection of American sycamore for biomass production in the mid-South. Final Report to DOE. ORNL/Sub/81-9051/1. Oak Ridge National Laboratory, Oak Ridge, TN. 1982.
12. D. J. Frederick, R. C. Kellison and R. Lea. Species selection and silviculture systems for producing fuels from woody biomass in the southeastern United States. Unpublished annual report submitted to

the U.S. Department of Energy's Short Rotation Woody Crops Program. North Carolina State University, Raleigh. 41 p. 1984.

13. P. R. Felker, S. K. Beck, I. Reyes and D. Smith. Production of woody biofuels from mesquite (*Prosopis* spp.). Unpublished report submitted to the U.S. Department of Energy's Short Rotation Woody Crops Program. Texas A&I University, Kingsville. 80 p. 1983.

14. K. Steinbeck, C. L. Brown, H. E. Sommer and B. Bongarten. Increasing the biomass production of short-rotation forests. Unpublished annual report submitted to the U.S. Department of Energy's Short Rotation Woody Crops Program. University of Georgia, Athens. 72 p. 1983.

15. D. S. DeBell, C. S. Harrington, M. A. Radqan, D. L. Reukema, R. F. Stettler and R. O. Curtiss. Increasing the biomass production of alder plantations in the Pacific Northwest. Unpublished report submitted to the U.S. Department of Energy's Short Rotation Woody Crops Program. USDA Forest Service, Forestry Sciences Laboratory, Olympia, WA. 79 p. 1984.

16. L. S. Dolan and P. Schroeder. The cultural treatment of selected species for woody biomass production in the Pacific Northwest. Unpublished final report submitted to the U.S. Department of Energy's Short Rotation Woody Crops Program. Seattle City Light Department, Seattle, WA. 206 p. 1984.

17. G. L. Rolfe and K. A. Majerus. Woody biomass production through solar energy conversion. Unpublished report submitted to the U.S. Department of Energy's Short Rotation Woody Crops Program. University of Illinois, Urbana. 123 p. 1984.

DEVELOPMENT OF WOODY BIOMASS CULTURAL SYSTEMS FOR FLORIDA

D. L. Rockwood

Department of Forestry
University of Florida/IFAS
Gainesville, Florida 32611

ABSTRACT

Eucalyptus grandis, slash pine, and sand pine have considerable promise for silvicultural biomass farming. Eucalyptus amplifolia, Liquidambar styraciflua, Platanus occidentalis, and Sapium sebiferum are additional species having potential for northern Florida. Currently selected E. grandis clones vary considerably in height and cold tolerance; only five are recommended for further testing. Winter harvests are preferred for E. grandis as summer harvests result in greatly decreased coppice productivity. Coppicing of other species harvested during the winter is good. Fifth-year responses of sand and slash pines to site amendments are similar to those at three years. Research on further development of cultural systems is continuing.

Keywords. Eucalyptus grandis, Eucalyptus amplifolia, Liquidambar styraciflua, Platanus occidentalis, slash pine, sand pine, Sapium sebiferum, coppice, genetic variation, clones, fertilization.

INTRODUCTION

Initial research on the development of silvicultural biomass farms for the various climatic and edaphic regions of Florida emphasized eucalypts, sand pine, and slash pine for short-rotation, intensive culture. Experimental-scale and pilot-scale studies, such as those in Table 1, assessed productivity in response to a range of cultural options, led to preliminary management guidelines, and provided estimates of net economics and energetics.

Table 1. Ongoing woody biomass production field studies at the University of Florida

Estab. Date	Location	Species	Management	Genetics	Spacing	Spacing x Genetics	Biomass	Lysimeter
Experimental-Scale Studies:								
7/77	Glades Co.	E. grandis		X				
3/79	Glades Co.	3 eucalypts	X	X				
1/80	Alachua Co.	Slash pine	X	X	X	X	X	X
1/80	Gilchrist Co.	Slash pine	X	X	X	X	X	
1/80	Calhoun Co.	Sand pine	X	X	X	X	X	X
1/80	Marion Co.	Sand pine		X				
5/80	Palm Beach Co.	7 species	X	X	X			
8/81	Alachua Co.	2 species					X	X
8/81	Bradford Co.	2 species					X	X
8/82	Glades Co.	E. grandis		X				
6/85	Glades Co.	4 eucalypts	X	X	X			
6/85	Alachua Co.	2 species	X	X	X			
6/85	Polk Co.	4 species	X	X	X			
7/85	Palm Beach Co.	4 eucalypts	X	X	X			
Pilot-Scale Studies:								
7/82	Glades Co.	E. grandis	X	X	X		X	
1/83	Bradford Co.	Slash pine	X	X	X		X	
2/83	Taylor Co.	Sand pine	X	X	X	X	X	
2/83	Taylor Co.	Slash pine	X	X	X	X	X	
1/84	Liberty Co.	Sand pine	X				X	
Species Screening Trials:								
4/81	Alachua Co.	11 species	X	X				
8/82	Palm Beach Co.	12 species		X			X	
6/83	Highlands Co.	16 species		X			X	
6/83	Polk Co.	22 species		X			X	
6/83	Polk Co.	13 species		X				
6/83	Polk Co.	8 species		X				
3/84	Polk Co.	4 species	X	X	X		X	
7/82	Glades Co.	5 eucalypts		X				
9/84	Alachua Co.	4 eucalypts		X				
Growth and Yield Studies:								
2/82	Gulfport, MS	Slash pine	X	X	X	X	X	
6/82	Statewide	Sand pine	X		X		X	
1/83	Alachua Co.	Slash pine	X	X				X
1/85	Dublin, GA	Slash pine	X	X			X	
1/85	Jackson Co.	Slash pine	X	X	X	X		
1/85	Alachua Co.	Slash pine	X	X	X		X	X
2/85	Poplarville, MS	Slash pine	X	X	X	X		

Eucalyptus _grandis_ is without peer for productivity in southern Florida.[1,2] For frost-frequent areas, _E_. _grandis_ offers higher productivity than other species but at the risk of frost damage and resulting growth loss. Cultural practices for maximizing biomass production on "palmetto prairie" soils include 2 m spacing between planting beds, ground rock phosphate applied before bedding, 1 m spacing between trees on beds, and summer planting.[1] On organic muck soils, no bedding is required on drained sites, planting density should be as great as permitted by the planting equipment, with a 1 x 1 m spacing as a desirable goal, and a wider range of planting times may be possible. At higher planting densities on muck soils, rotations as short as 18 months may produce biomass at a cost of between \$11 and \$15 per dry Mg, depending on interest rate.[3] Associated energy ratios are approximately 24 to 1. Considerable gains have been achieved from an intensive genetic improvement program, and advanced generations retain appreciable variability that can be further exploited to increase productivity.[4,5]

Slash pine and sand pine have potential for short rotation culture in the northern part of the state.[2,6,7,8] On "flatwoods" sites, slash pine at 10,000 trees ha^{-1} can achieve a productivity of 10 Mg $ha^{-1}yr^{-1}$ in six years. Sand pine, the best choice for short-rotation biomass production on sandhills, has produced 8 Mg $ha^{-1}yr^{-1}$ of stemwood. Considerable genetic variation exists within both species.[9] Each species is tolerant of high planting density and may be harvestable within 10 years following establishment at 1 x 1 m.

Ongoing studies, including screening trials to identify additional species and growth and yield studies with sand and slash pines for broad, long-term assessments of productivity, periodically provide results to modify and expand preliminary conclusions. Recent results depicting continuing development of appropriate cultural systems are reported here on identification of suitable species/genotypes, estimation of coppice productivity and harvest time, and evaluation of site amendments.

MATERIALS AND METHODS

Species Screening

Eleven species (_Alnus_ _glutinosa_, _Cinnamomum_ _camphora_, _Populus_ sp., and _Quercus_ _shumardii_, plus the seven listed in Table 2) were established in a demonstration planting in Alachua County from April to December 1981. Each species was represented by a 100-tree square plot at a 1 x 1 m spacing. Tree height, diameter breast height (DBH), and survival were measured at the end of the 1984 growing season; earlier measurements were available on seedling and/or coppice performance of three species. Biomass yields were estimated, when possible, with previously obtained tree dry weight predictive equations or from equations derived from trees harvested in 1984.

Genetic Improvement of Eucalyptus

Each of 55 fast-growing, reputedly frost-hardy, and well-formed E. grandis selected in research studies and commercial plantations was propagated as rooted cuttings and entered into single-tree plots at a Glades County site in August 1982. A completely randomized design was utilized with a 4 x 1.8 m spacing; the number of ramets per clone ranged from one to fourty-two. In April 1984, height, survival, and cold score on a -1 to 10 scale (-1 = killed to ground, 10 = undamaged) were evaluated to assess clonal performance following a hard freeze.

Seed sources of four eucalypts considered promising for northern Florida were outplanted in Alachua County in September 1984. Sources, each generally a bulked seedlot from several trees in a portion of a species' range, were obtained from CSIRO and other seed suppliers in Australia. Sources for E. amplifolia(12), E. dunnii(12), E. nitens(2), and E. viminalis(51), respectively, were represented in 10-tree row plots with three replications in a randomized complete block design. Spacing was 1 x 1 m. Survival was determined in April 1985 after an exceptional freeze in January.

Coppice Productivity and Management

Commercial plantations of E. grandis and E. robusta established in 1973 and putative E. grandis x E. robusta hybrids planted in 1974 in Glades County were harvested monthly from March 1979 to March 1980 to study the effect of harvest date. For each species and the hybrids, the 13 harvest plots were installed sequentially within the plantation and were 100-tree blocks. Spacing approximated 3 x 2 m. Coppice regrowth was measured in December 1983 for height and DBH of the three largest stems on each stool and the number of stems per stool greater than or equal to one-half the DBH of the largest stem. Data were adjusted to a common age of 48 months after

Table 2. Performance of seven best species in a woody biomass demonstration planting in northern Florida

Species	Age	Ht. (m)	DBH (cm)	Surv. (%)	Biomass Yield (Mg ha^{-1}yr^{-1})
Eucalyptus amplifolia	8-mo. Seedlings	1.6	–	81	–
	23-mo. 1st. Coppice	7.9	8.2	94	22.8
	10-mo. 2nd. Coppice	3.7	2.0	100	9.7
Liquidambar styraciflua	4-yr-old Seedlings	2.9	2.2	99	–
Pinus clausa var. immuginata	4-yr-old Seedlings	2.8	3.1	83	2.8
Pinus elliottii var. elliottii	4-yr-old Seedlings	4.5	5.4	92	8.4
Pinus taeda	3-yr-old Seedlings	2.9	–	88	–
Platanus occidentalis	33-mo. Seedlings	5.6	3.4	97	–
	10-mo. 1st. Coppice	3.1	1.5	100	–
Sapium sebiferum	33-mo. Seedlings	5.7	4.3	100	–
	10-mo. 1st. Coppice	2.5	1.4	100	8.3

harvest. Coppice stem dry weights were estimated by appropriate equations,[10] summed over the inner 60 trees in a plot, and converted to a per ha basis.

Site Amendments for Pines

In January 1980, slash pines were planted in Alachua and Gilchrist counties and sand pines in Calhoun County. At each location three, 100-tree square plots at 1 x 1 m spacing were installed for each site amendment. The slash pine studies used a completely randomized design; the sand pine study employed a randomized complete block design. A control was maintained in each study; three N/P fertilizer combinations (50/50, 150/50, and 200/100 kg ha^{-1}) were included in the slash pine studies, while the first two N/P treatments only were applied in the sand pine study. Two sewage sludge treatments were additionally evaluated in the Alachua and Calhoun counties studies. Fifth-year observations included height, DBH, and survival.

RESULTS AND DISCUSSION

Species Screening

The demonstration planting at Gainesville typifies exploratory evaluations of alternative species for northern Florida sites. Seven of 11
species (Table 2) have performed well. Of these seven, E. amplifolia, L.
styraciflua, P. occidentalis, and S. sebiferum are previously untested
species that appear to be suitable for short-rotation intensive culture and
warrant continued research. The four poor-performing species in the planting, in combination with similar performance in other northern Florida
screening trials, have been eliminated from further consideration.

Genetic Improvement of Eucalyptus

Within the current E. grandis base population, outstanding progenies
and exceptional individuals can be selected.[5] These exceptional individuals, candidates for vegetative propagation to capture their full genetic
potential, must be evaluated in clonal tests, however, before being used
operationally. For example, 55 superior Eucalyptus clones previously
selected varied significantly in cold resistance and early height growth but
were similar for survival (Table 3).

Overall, cold resistance was weakly correlated with height. The high
variability for height is surprising in view of the strong size differential each ortet had, but the range in cold resistance reflects the fact
that the clones had not been subjected to cold as severe as that of the
December 1983 freeze. While 10 clones displayed acceptable cold tolerance, no
clones were completely undamaged; six clones had some ramets without frost
damage. Five of the 10 most cold tolerant clones also were excellent in
growth and are recommended for expanded propagation and testing.

New genetic materials are being evaluated for several eucalypts so that
genetic base populations may be expanded. For northern Florida, E. amplifolia consistently outsurvived three other previously promising species
following the rigorous January 1985 freeze (Table 4). None of the E.
dunnii, E. nitens, or E. viminalis sources had acceptable survival. Several
of the E. amplifolia exceeded 50% survival. Subsequent performance of these
sources will identify valuable germplasm for E. amplifolia.

Table 3. Performance of 55 Eucalyptus clones through 1.4 years

Trait	All Clones		10 Frost Resilient Clones	
	Mean	Range	Mean	Range
Height (m)	2.4	1.5-3.8	2.8	1.8-3.8
Survival (%)	95.5	80-100	99.5	95-100
Cold Score	2.1	0.0-8.0	5.8	4.0-8.0

Table 4. Seven-month survival of seed sources of four Eucalyptus species at a northern Florida test site

Species	No. of Sources	Mean (%)	Range (%)
E. amplifolia	12	47.7	30.0-70.0
E. dunnii	12	1.9	0.0-16.7
E. nitens	2	0.0	0.0-0.0
E. viminalis	51	0.8	0.0-26.7

Table 5. Coppice productivity of three Eucalyptus 48 months after monthly harvests

Date of Harvest	E. grandis	E. robusta	E. grandis x E. robusta
		$(Mg\ ha^{-1}yr^{-1})$	
03/79	2.42	1.36	1.28
04/79	2.72	1.22	0.83
05/79	1.67	1.52	0.93
06/79	1.41	1.97	1.00
07/79	0.89	1.50	0.91
08/79	0.22	0.95	0.26
09/79	0.73	1.86	0.74
10/79	0.96	1.84	1.04
11/79	1.26	2.05	1.12
12/79	1.84	2.11	0.93
01/80	1.54	1.56	1.05
02/80	1.33	1.76	1.81

Coppice Productivity and Management

Coppice productivity, essential to multi-rotation management, is influenced by time of harvest (Table 5). For E. grandis, winter harvests were best whereas summer harvests produced poor coppice regrowth. Eucalyptus robusta had more consistent coppicing throughout the year. The E. grandis x E. robusta hybrid did not have consistency or good growth. Due to the wide spacing in this study and some frost damage, coppice yields here were less than what has been observed under intensive culture.[11]

Coppice growth of three northern Florida species appears good (Table 2). Very adequate seedling growth of E. amplifolia, P. occidentalis, and S. sebiferum was, in fact, exceeded by coppice growth following dormant season harvests. Two-year coppice productivity of E. amplifolia greatly surpassed one-year coppice growth and was essentially equal to that of some E. grandis coppice in southern Florida.[2]

Site Amendment for Pines

Fifth-year responses to fertilization (Table 6) confirm earlier indications that slash pine responds dramatically to amendments on poor to average sites and that sand pine growth is minimally influenced by normal fertilizer levels.[2] Comparison to treatment differentials against the control for third-year height shows all treatment responses for fifth-year height have decreased in relative size. Therefore, application of traditional rates of fertilizers for sand pine appears unnecessary, but high rates of sewage sludge may be beneficial. Fertilization of slash pine, including use of sewage sludge, should be matched to nutrient deficiencies of the site.

FUTURE RESEARCH

Continuing research to refine guidelines for silvicultural biomass farming will utilize all studies listed in Table 1. Research with eucalypts in central and southern Florida will address yield of rotation length coppice stands and comparison with seedling stands; timing and method of harvest; management of coppice stands; genetic improvement, use of hybrids, and clonal propagation; biomass properties of coppice stands; and economics/energetics of multi-coppice systems. Eucalyptus amplifolia and Sapium sebiferum will be further evaluated for productivity and biomass quality in

Table 6. Mean height, DBH and survival for 5-year-old slash and sand
pine in three fertilizer tests.

Species	Treatment[1] (N/P)	Height Mean	Height Vs. Control	DBH Mean	DBH Vs. Control	Survival Mean	Survival Vs. Control
	(kg ha^{-1})	(m)	(%)	(cm)	(%)	(%)	(%)
Slash Pine	0	2.8		3.8		96	
	50/50	2.6	−7	2.8	−26	89	−7
	150/50	3.9	+39	4.2	+11	96	0
	200/100	4.2	+50	4.3	+13	95	−1
	S(470/165)	3.7	+32	4.7	+24	94	−2
	S(945/335)	5.1	+82	6.0	+59	91	−5
Slash Pine	0	5.0		6.0		95	
	50/50	5.1	+2	6.4	+7	96	+1
	150/50	5.5	+10	6.8	+13	88	−7
	200/100	5.0	0	6.2	+3	92	−3
Sand Pine	0	3.5		3.0		100	
	50/50	3.6	+3	3.1	+3	100	0
	150/50	3.6	+3	3.1	+3	99	−1
	S(175/135)	3.5	0	2.9	−3	100	0
	S(340/265)	3.9	+11	3.6	+20	99	−1

[1] N=Nitrogen; P=Phosphorus; S=Dry sewage sludge.

northern Florida. Yields of slash and sand pines will be modeled as a
function of planting density, genotype, genotypic competition, cultural
intensity, age, and site.

ACKNOWLEDGEMENTS

Research reported here is supported by Oak Ridge National Laboratory
under subcontract No. 19X-09050C, by a cooperative program between the
Institute of Food and Agricultural Sciences of the University of Florida and
the Gas Research Institute entitled "Methane from Biomass and Waste," and by
the U. S. Forest Service under cooperative agreements A8fs-9,961, suppl.51
with the Southeastern Forest Experiment Station and No. 19-83-049 with the
Southern Forest Experiment Station. Journal Series paper
No. 6539 of the Florida Agricultural Experiment Station.

REFERENCES

1. D. L. Rockwood, D. R. Dippon and C. W. Comer. Potential of Eucalyptus
    grandis for biomass production in Florida. Proc. BioEnergy 84.
    1984.

2. D. L. Rockwood, C. W. Comer, L. F. Conde, D. R. Dippon, J. B. Huffman, H. Riekerk and S. Wang. Final Report of Energy and Chemicals from Woody Species in Florida. ORNL/Sub/81-9050/1. 1983.

3. D. R. Dippon, D. L. Rockwood and C. W. Comer. Cost sensitivity of Eucalyptus woody biomass systems. Proc. 3rd. S. Biomass Energy Res. Conf. 1985.

4. D. L. Rockwood, R. C. Kellison, E. C. Franklin and G. F. Meskimen. Operational advanced generation improvement programs for minor species in the South. Proc. S-23 Workshop. 1984.

5. K. V. Reddy, D. L. Rockwood, C. W. Comer and G. F. Meskimen. Genetic improvement of Eucalyptus grandis for biomass production in Florida. Proc. 3rd. S. Biomass Energy Res. Conf. 1985.

6. M. S. F. Campbell, C. W. Comer, D. L. Rockwood and C. Henry. Biomass productivity of slash pine in young, heavily-stocked stands. Proc. 1983 S. For. Biomass Wrkshp.:77-82. 1983.

7. D. R. Dippon, D. L. Rockwood and C. W. Comer. Economic and energetic analyses of short-rotation culture of slash and sand pine. Proc. 7th S. For. Biomass Wkshp. 1985.

8. D. L. Rockwood, L. F. Conde and R. H. Brendmuehl. Biomass production of closely spaced Choctawhatchee sand pine. USDA For. Serv. Res. Paper SE-293. 1980.

9. L. J. Frampton, Jr. and D. L. Rockwood. Genetic variation in biomass traits of sand and slash pines. Silvae Genetica 31(2-3): 18-23. 1983.

10. D. L. Rockwood, T. F. Geary and P. S. Bourgeron. Planting design and genetic influences on seedling growth and coppicing of eucalypts in southern Florida. Proc. 4th S. For. Biomass Wkshop. 1982.

11. C. W. Comer and D. L. Rockwood. Screening of Eucalyptus species for coppice productivity. Proc. 6th S. For. Biomass Wkshp.:95-97. 1984.

BIOMASS PRODUCTION AND NUTRIENT REMOVAL

BY LEUCAENA IN COLDER SUBTROPICS

A. B. Othman and G. M. Prine

Department of Agronomy
University of Florida/IFAS
Gainesville, Florida 32611

ABSTRACT

Leucaena (Leucaena spp.) is a tropical tree legume with potential for use as biomass and forage in subtropical and mild temperate climates. From 373 accessions planted at Gainesville, Florida in mid-1979, 62 and 53 accessions were evaluated for biomass after tops were killed by freezes in 1982 and 1983, respectively. Each accession was planted in a five plant plot arranged in a completely randomized design with two replications. The above-ground dry matter yields ranged from 5.6 to 39.8 Mg ha^{-1} in 1983, the fifth growing season. The 12 top-yielding accessions, selected for chemical analyses on stem and branch material, did not differ significantly within each season in the biomass yields with means of 29.3 and 24.7 Mg ha^{-1} for 1982 and 1983 seasons, respectively. Nitrogen concentrations in these 12 accessions (1982 season) varied from 0.55 to 0.86 percent, while phosphorus concentrations did not differ significantly with a mean of 0.09 percent. Potassium, calcium, and magnesium concentrations varied between accessions ranging from 0.49 to 0.72, 0.22 to 0.41, and 0.05 to 0.11 percent, respectively. The energy contents also ranged from 19,360 to 20,000 J g^{-1} for 1982 and 19,570 to 20,110 J g^{-1} for 1983. The energy content of K8 (PI 263695) leucaena, the highest yielding accession, was equivalent to 17.9 and 11.8 Mg oil ha^{-1} in 1982 and 1983, respectively. The higher biomass yielding leucaena accessions showed great promise for annual harvesting of biomass on well-drained soils of colder subtropical and warm temperate regions of southeastern USA.

Keywords. Leucaena spp., Energy content, forage, coppice, legume, accession.

INTRODUCTION

The consequences of the energy supply shortage are well realized throughout the world. The monetary inflation that has caused various difficulties has mainly been attributed to increasing price of oil or energy. Fossil fuel is not readily regenerated in natural situations and requires millions of years to develop. Meanwhile, the demand for energy continues to increase as the world population increases. On realizing these factors, efforts have been initiated in many countries to try to make possible uses of alternative sources of energy.

One alternative source of energy is biomass, which is the solar energy trapped through the process of photosynthesis. However, biomass has some disadvantages. Being bulky, biomass is difficult to transport and store, contains a lower amount of energy per unit basis in comparison to the fossil fuel, and biomass production varies depending on climate and soil fertility. On the other hand, among many other advantages, biomass poses no problem of acid rain, water pollution, or harmful radioactivity. Biomass could be converted into less bulky forms, with higher energy densities that are more convenient for transport, storage, and use.[1,2,3]

In many developing countries, biomass in the form of wood is the source of fuel. In the U.S., the use of wood as an energy source is also on the increase.[4] Florida has started a program of biomass production for energy purposes in 1979[4], focusing on woody plants. Slash pine (Pinus elliottii), sand pine (P. clausa), and Eucalyptus viminalis were included in the program for northern Florida, while for southern Florida, Eucalyptus grandis, E. robusta, Melaleuca quinquenervia, Casuarina spp., and Taxodium distichum were emphasized. Several other species including Leucaena [(Leucaena leucocephala (Lam.) de Wit.)], were investigated less intensively. Leucaena, a tropical leguminous tree, probably was not emphasized because of lack of evidence as to the potential biomass production in this subtropical state and because the shoots are sometimes killed in freezes. Leucaena must be harvested annually for biomass in upper subtropical and mild temperate climates. This report describes the biomass productivity and energy content of leucaena, and the major nutrients removed through harvesting at Gainesville, a north central Florida location.

MATERIALS AND METHODS

On a 0.5-ha site, 373 accessions of leucaena, originating in many countries, were planted in June and July, 1979, on the University of Florida campus. Five plants were established per accession in a row in 6 x 1 m plots, replicated in two blocks. The soil was characterized as a Lakeland fine sand, a member of the thermic, coated Typic Quartzipsamments with a pH of 5.6 to 6.0. In preparation for this study, 0-10-20 fertilizer with microelements was broadcast on soil surface at a rate of 500 kg ha^{-1} in April, 1982 and 1983.

For the 1982 growth season, 62 accessions which visually showed high biomass production were selected and cut at four to five cm above the ground in February, 1983. By February, the leaves had been killed and most had fallen off in response to earlier freezing winter temperatures. The total above-ground biomass per plot, excluding the fallen leaves, was weighed at harvest and a random sample of three to five stems, including branches, were chipped. A subsample of the resulting chips was oven dried at 60°C to determine dry matter. In 1983 season, only 53 of the 62 accessions were evaluated since nine accessions were used for another study.

The energy contents of the 12 top yielding accessions, based on their dry matter yield in the 1982 season, were determined by oxygen bomb calorimetry. Nitrogen (N) and phosphorus (P) concentrations in the biomass were determined with an autoanalyzer, while potassium (K), magnesium (Mg), and calcium (Ca) were determined by the atomic absorption spectrophotometry, after being digested by the micro-Kjeldahl and dry-ashing methods, respectively.

RESULTS AND DISCUSSION

Under the humid, subtropical climate of Gainesville, leucaena plant tops die every year because of winter freezing temperatures that usually occur in December and January. By December in both 1982 and 1983, the plants have reached a mean height of about 5 m with a mean diameter at breast height (DBH) of about three cm, though some reached four cm. After the winter months, usually in March, each plant sprouted into six to 10 stems of which four or five produced the main bulk of the above-ground biomass, similar to that observed in Hawaii.[5]

Biomass (Energy) Productivity

The existing numbers of plants per plot were not the same for all plots. A covariance analysis indicated that the covariate (number of plants per plot) was highly significant (P = 0.0026 for 1982 and P = 0.0036 for 1983). The paired difference analysis for the 62 and 53 accessions as provided by Statistical Analysis System (SAS) was conducted for the biomass yields.[6] Comparison between yields in 1982 and 1983 seasons was not made because of severe damage from the insect jumping plant louse (Heteropsylla sp. (Psyllidae:Homoptera) in the 1983 season, which probably caused a lower yield in 1983. Similarly, no comparison of energy contents of the biomass was made between the two years of study.

The dry matter yields of the 62 accessions ranged from 5.6 to 39.8 Mg ha^{-1} in 1982 and of 53 accessions in 1983 from 6.7 to 30.7 Mg ha^{-1}. The 12 highest yielding accessions were selected for further energy and nutrient study.

A covariance analysis of the top 12 yielding accessions indicated that the numbers of plants per plot did not affect the yield significantly (P > 0.05), and the biomass yields of these 12 accessions did not differ significantly for either the 1982 or 1983 seasons (Table 1) with means of 29.3 and 24.7 Mg ha^{-1}, respectively. The world-wide-known accession, K8 or PI 263695 proved to be a top yielder also under Gainesville conditions with production of 39.8 and 26.5 Mg ha^{-1} of above-ground biomass for the 1982 and 1983 seasons, respectively.

Comer and Rockwood[7] produced 25.2 Mg ha^{-1}yr^{-1} of dry biomass from a two-year coppice of Eucalyptus amplifolia at 1 x 1 m spacing grown at Gainesville, Florida. The next year the 1-year old coppice produced 9.7 Mg ha^{-1} yr^{-1}. The better accessions of leucaena promise to have a higher biomass production rate on the moderately- to extremely-drained soils where leucaena is adapted.

The energy contents varied between accessions, ranging from 19,360 J g^{-1} for accession PI 443541 to 20,030 J g^{-1} for accession PI 443696 in the 1982 season crop. For the 1983 season, the range of energy contents was from 19,570 J g^{-1} for accession PI 443541 to 20,100 J g^{-1} for accession PI 443482. Virtucio[9] believed that energy contents of leucaena may vary according to the percentage of heartwood. However, in conditions like that of Gainesville, plant tops die every year from the freezing temperatures in

Table 1. Above-ground biomass yields and energy contents of the 12 highest yielding accessions from the biomass study of Leucaena spp. in 1982 and 1983 growing seasons at Gainesville, Florida.

Accession PI number	Growth season			
	1982		1983	
	Biomass	Energy	Biomass	Energy
	(Mg ha^{-1})	(J g^{-1})	(Mg ha^{-1})	(J g^{-1})
263695	39.8a*	19,820abc	26.5a	19,730bc
443696	35.2a	20,030a	30.7a	19,820abc
443541	31.8a	19,360d	20.6a	19,570c
281607	31.7a	19,610abcd	24.1a	19,860bc
286295	28.5a	19,690abcd	30.0a	19,780bc
281608	28.5a	19,820abc	28.6a	19,650bc
288005	28.1a	19,690abcd	27.9a	19,940ab
443674	27.6a	19,780abcd	19.9a	19,860abc
443482	26.6a	19,570bcd	21.0a	20,110a
443610	25.8a	19,820ab	21.6a	19,780bc
370749	24.4a	19,400cd	25.4a	19,780bc
443483	24.3a	19,650abc	19.9a	19,730bc
Avg. of 12 accessions	29.3	19,690	24.7	19,820

*Values within columns followed by the same letters are not significantly different at the 5% level according to DMRT.

winter and such variation of energy contents over the years are not expected. Since oil contains 44,000 J g^{-1},[8] it could be calculated, for example, that on an energy basis, the 1982 and 1983 yields of accession K8 were equivalent to 17.9 and 11.8 Mg oil ha^{-1}, respectively.

Major Nutrients Removal

The N concentration varied from 0.55 percent for accession PI 263695 (K8) to 0.86 percent for accession PI 370749 (Table 2). However, the P concentrations did not vary significantly among accessions with a mean of 0.09 percent. The concentrations of K varied from 0.49 percent for accession PI 263695 to 0.72 percent for accession PI 370749, even though a K containing fertilizer was applied to all plots. The Ca concentrations varied from 0.22 percent for accessions PI 263695 and 263696 to 0.41 percent for accession PI 370749. These variations in nutrient concentrations may have been caused by the variation in soil minerals available to the root system, in addition to the possibility of genetic differences in growth rates and different capacities of accessions to absorb nutrients. The P from the applied fertilizer and in the experimental soil was probably

Table 2. Nutrient concentrations in the dry matter of the 12 highest
yielding accession from the biomass study of <u>Leucaena</u> spp. in
1982 growing season at Gainesville, Florida.

Accession PI numbers	Nutrient				
	N	P	K	Ca	Mg
	------------------------------%------------------------------				
370749	0.86a*	0.11a	0.72a	0.41a	0.10ab
286295	0.85a	0.10a	0.64ab	0.30ab	0.08abc
443610	0.84ab	0.11a	0.68ab	0.26b	0.08cde
281608	0.83ab	0.08a	0.58ab	0.32ab	0.11a
288005	0.76abc	0.09a	0.61ab	0.24b	0.08abc
443674	0.75abc	0.08a	0.66ab	0.30ab	0.09abc
443482	0.72abc	0.09a	0.68ab	0.31ab	0.10abc
443483	0.70abc	0.08a	0.66ab	0.32ab	0.07cde
443541	0.66abc	0.10a	0.58ab	0.32ab	0.07cde
281607	0.62abc	0.08a	0.58ab	0.24b	0.08bcd
443696	0.57bc	0.09a	0.50b	0.22b	0.05e
263695	<u>0.55c</u>	<u>0.08a</u>	<u>0.49b</u>	<u>0.22b</u>	<u>0.06de</u>
Avg. of 12 accessions	0.73	0.09	0.61	0.29	0.08

*Values within columns followed by the same letters are not significantly
different at the 5% level according to DMRT.

adequate to result in no significant differences being obtained between its
concentrations in the biomass of different accessions. The correlations
coefficients (R) between these nutrient concentrations in the biomass of the
1982 season were all low (ranging from 0.09 to 0.77).

Though there was variation of nutrient concentrations between accessions, the amount of each nutrient removed through harvesting the above-
ground biomass did not differ significantly (P>0.05) between accession
(Table 3). The mean amounts of nutrients removed were 210 kg N ha^{-1}, 30 kg
P ha^{-1}, 180 kg K ha^{-1}, 80 kg Ca ha^{-1}, and 20 kg Mg ha^{-1}. Leucaena being a
legume, the nitrogen would be supplied through nitrogen fixation from
<u>Rhizobium</u> bacteria. On many well-drained soils in Southeastern USA, liming
and fertilizations of leucaena may be needed to furnish the nutrients
necessary for high annual biomass yields. This fertilization and/or lime
would be applied after biomass harvest in winter and preferably before
regrowth begins in late winter or spring.

Table 3. Nutrient removal by harvesting the above-ground biomass of Leucaena from the 1982 season after leaves had fallen, at Gainesville, Florida.

Accession PI number	Nutrient				
	N	P	K	Ca	Mg
	------------------------------%------------------------------				
286295	240a*	30a	190a	80a	20a
281608	240a	20a	170a	90a	30a
263695	220a	30a	190a	90a	20a
288005	220a	20a	170a	70a	20a
443610	210a	30a	180a	60a	20a
443674	210a	20a	180a	80a	20a
443541	210a	30a	190a	100a	20a
370749	200a	30a	180a	100a	30a
443696	200a	30a	180a	70a	20a
281607	200a	30a	180a	70a	20a
443482	200a	20a	180a	80a	20a
443483	170a	20a	160a	80a	20a
Avg. of 12 accessions	210	30	180	80	20

*Values within columns followed by the same letters are not significantly different at the 5% level according to DMRT.

ACKNOWLEDGEMENTS

Research was in partial fulfillment of the M.S. degree by senior author and was funded through an agreement between the University of Florida, Institute of Food and Agricultural Sciences, and the Gas Research Institute.

REFERENCES

1. R. T. Bailey and P. R. Blankenhorn. Caloric and porosity development in carbonized wood. Wood Sci. 15:19-28. 1982.
2. P. F. Bente. The dawning of sunfuels. In International Bio-energy Directory. Bio-energy Council, Washington, DC. p. xi-xvii. 1981.
3. E. S. Lipinsky. Fuels from biomass--integration with food and materials systems. Science 199:644-651. 1978.
4. W. H. Smith and M. L. Dowd. Biomass production in Florida. J. Forestry. 79:508-515. 1981.
5. M. Takahashi and J. C. Ripperton. Koa hoale (Leucaena glauca): its establishment, culture and utilization as a forage crop. Hawaii Agric. Exp. Stn. Bull. 100. 56 p. 1949.
6. A. B. Othman. Evaluating leucaena introductions for biomass and forage production. M. S. Thesis, Univ. of Fla., Gainesville, FL. 1984.
7. C. W. Comer and D. L. Rockwood. Screening of Eucalyptus species for coppice productivity. Proc. 1984 S. Forest Biomass Workshop. Athens, Ga. June 507, 1984. p. 95-97. 1984.

8. L. P. White and L. G. Plaskett. Biomass as Fuel. Academic Press, London. 211 p. 1981.
9. F. D. Virtucio. Management of ipil-ipil wood production. In International Consultation on Ipil-Ipil Research. Nat. Acad. Sci. Washington, DC. p. 69-75. 1977.

GENETIC IMPROVEMENT OF <u>EUCALYPTUS GRANDIS</u>

FOR BIOMASS PRODUCTION IN FLORIDA

K. V. Reddy, D. L. Rockwood, C. W. Comer and G. F. Meskimen

Department of Forestry
University of Florida/IFAS
Gainesville, Florida 32611

ABSTRACT

<u>Eucalyptus grandis</u> is a promising species for biomass production in southern Florida. A genetic base population (GPOP77) composed of 529 progenies, representing four generations of selection, planted in July, 1977, at a density of 1,916 trees ha^{-1} was partially harvested in August, 1978. Regrowth through December, 1983, was evaluated to assess genetic improvement potential for coppice productivity. Four generations of selection have produced impressive genetic gains. At 64 months after harvest, 1st, 2nd, 3rd, and 4th generation coppice averaged 3.9, 13.34, 16.18, and 23.16 dry Mg ha^{-1}, respectively. Fourth generation families also had the best frost resilience and coppice quality. The family and individual tree heritabilities for 64-month coppice height, DBH, and volume were 0.65, 0.65, and 0.59, and 0.31, 0.32, and 0.27, respectively. Genetic correlation between seedling height at 7-months and coppice height at 3-months was 0.27; and between 64-month height and DBH, was 0.98. Genetic gains that could be achieved through different selection strategies increased by more than 100% for combined selection of three trees per family of the top 100 families. Similar gains were predicted through clonal propagation of selected families. High genetic variation exists in <u>Eucalyptus grandis</u> and can be utilized for increasing woody biomass production in Florida.

<u>Keywords</u>. <u>Eucalyptus grandis</u>, frost resilience, heritability, genetic gain, genetic correlations, coppice, energy crop.

INTRODUCTION

Eucalyptus grandis Hill ex. Maid offers considerable promise as a fast growing tree crop in Florida. In southern Florida eucalypts may be planted on wet organic soils, histosols.[1] The use of the species, originally for pulpwood, now includes biomass production for energy use.[2] Considerable variation in growth is observed in the species, and breeding has been successful in increasing the growth rate.[3] Several of the progenies of E. grandis tested in Florida appear to be well suited for short rotation intensive culture, biomass production.[4]

Heritability and genetic correlations are important parameters in genetic studies and are used as indices for genetic variation. Heritability expresses the portion of the total variance that is attributable to the effects of genes, but the most important role of heritability is its predictive role, expressing the reliability of phenotypic value as a guide to its genetic value. It is used to predict gains that can be achieved through selection. A high heritability suggests the effectiveness of mass selection and, hence, no need for progeny testing.[5] A high genetic correlation between two traits suggests pleiotropy.[6] Hence, selection for one trait will cause simultaneous selection to be carried out for the other trait.

Vegetative propagation of clones selected for biomass production permits the utilization of selected lines in the first generation. Several methods, including propagation via rooted cuttings[7] and propagation by tissue culture,[8] have been studied for vegetative propagation of eucalypts and some of them were effective. This paper reports the gains that could be achieved through different selection strategies, including clonal propagation of selected individual.

MATERIALS AND METHODS

The population (GPOP77) selected for this study was planted in July, 1977, LaBelle, Florida. GPOP77 consists of 31,725 trees representing 529 open-pollinated families on 17.3 hectares. Spacing was 1.8 m between trees on paired beds spaced 2.3 m within pairs and 3.5 m between pairs, for a density of 1,916 trees ha^{-1}. Nine percent of the families are fourth generation, twenty-four percent are third generation, forty percent are second generation and twenty-seven percent are first generation families. Geographically, thirty-seven percent of the families trace their maternal

origin back to New South Wales, Australia, thirty-two percent originated
from Queensland, Australia, twenty-one percent came from South Africa, four
percent from other nations, and six percent of the families cannot be traced
beyond Florida.

The experiment was layed out in 141 rows X 225 columns as completely
randomized single tree plots. The southern half of the plantation was
harvested in August, 1978, to obtain the coppice information. The tallest
coppice shoot on each stool was selected to obtain the coppice data. Total
stem height and diameter at breast height (DBH) were measured at seven
months after planting and at three-months, 30 months, and 64 months after
harvest. Additionally, at 64 months after cutting, the trees were scored
for coppice quality and frost resilience. Frost resilience and coppice
assessment were ranked from 0 to 3, with 0 being the most desirable and 3,
the least. The factors considered for coding frost resilience were the
extent of frost damage and percent recovery of original height from frost.
The coppice assessment was based on coppice defects, including forking,
crooks, etc. Appropriate samples were taken and dried at 65°C for expressing biomass yields as dry weights. The genetic gains obtained were derived
from heritabilities and selection itensities; the latter were obtained from
the tables provided by Namkoong and Snyder.[9]

RESULTS AND DISCUSSION

Mean height, DBH, and volume show gains with each successive generation
of improvement (Tables 1 and 2). At 64 months after harvest, the average
first generation tree contained 7.04 dm^3 of stemwood, second generation
trees were 206% larger at 21.54 dm^3, the third generation had 20% more wood
over the second with 25.91 dm^3 and the fourth generation trees had yet
another 55% gain up to 40.16 dm^3. The fourth generation trees also had the
best frost resilience and coppice quality when compared to the other three
generations. The fourth-generation families contained the best frost resilience, suggesting the selection of these for future biomass plantings in
Florida where frost occurs frequently in winter.

Geographic origin of E. grandis may be important in developing good
coppicing trees for south Florida (Table 3). For early seedling growth,
sources were generally similar in each generation. Progeny introductions
from South Africa (crosses among highly selected parents in that region)
were slightly taller when grown first in Florida, but their selected off-

Table 1. Eucalyptus grandis generation means for individual tree per-
         formance in GPOP77 through 64 months after harvest.

| Trait/Age | Overall | Generation* | | | | 200 Best |
		1	2	3	4	Trees
Seedlings						
7 months						
Height (m)	1.75	1.60^a	1.79^b	1.81^b	1.83^c	2.93
Coppice (tallest stem)						
3 months						
Height (m)	0.63	0.56^a	0.63^b	0.67^c	0.71^d	1.69
30 months						
DBH (cm)	3.32	2.46^a	3.37^b	3.74^c	4.19^d	8.13
Height (m)	4.30	3.43^a	4.34^b	4.71^c	5.28^d	7.09
Dry Weight (kg)	0.87	0.29^a	1.02^b	1.24^b	2.00^c	4.16
Volume (dm^3)	2.58	1.56^a	2.85^b	3.24^b	4.58^c	8.39
64 months						
Height (m)	7.36	5.00^a	7.75^b	8.42^c	9.39^d	16.79
DBH (cm)	6.66	4.03^a	7.09^b	7.75^c	9.18^d	17.49
Volume (dm^3)	20.16	7.04^a	21.54^b	25.91^c	40.16^d	130.25
Frost Resilience	1.47	1.75^a	1.39^b	1.34^c	1.25^d	0.00
Coppice Assessment	0.54	0.77^a	0.48^b	0.41^c	0.40^c	0.00

*Generation means not sharing the same letter are significantly dif-
ferent at the 5% level.

spring were virtually of the same height as other sources in the second
generation. For coppice height, sources from Queensland, Australia, improved considerably with each generation of selection beyond the initial
introductions and surpassed third and fourth generation means for other
sources by more than one meter. Should this trend continue with new introductions from Queensland, the prospects for developing suitable E. grandis
are bright.

    Per hectare biomass productivity increased after each generation of
selection (Table 4). Yields per hectare per year increased considerably
from 30 to 64 months after cut. The first generation trees produced 0.7 Mg
ha^{-1} yr^{1-} of stem and branches; the second generation trees, 2.5 Mg ha^{-1}
yr^{-1}; the third generation trees, 3.0 Mg ha^{-1} yr^{-1}; and the fourth, Mg ha^{-1}
yr^{-1}. The 200 best trees in the population produced an average of 25.70 mg
of wood. These selected trees with excellent coppicing ability and frost
resilience have the potential of being clonally propagated to obtain excellent biomass yields.

Table 2. Percent gains for Eucalyptus grandis generations compared to the preceding generation for different traits.

Trait	Age (mos)	Generation 2	Generation 3	Generation 4	200 Best Trees
Seedling Ht.	7	12	1	1	60
Coppice Ht.	3	13	6	6	138
Coppice Ht.	30	27	9	12	34
Coppice DBH	30	37	11	12	94
Tree Volume	30	83	13	42	83
Coppice Ht.	64	55	9	12	79
Coppice DBH	64	76	9	19	91
Coppice Volume	64	206	20	55	224

Table 3. Comparison of origin of Eucalyptus grandis progenies in GPOP77S.

Trait	Gen.[**]	NSW n[***]	NSW $\bar{X}$	QLD n	QLD $\bar{X}$	SA n	SA $\bar{X}$	Other n	Other $\bar{X}$	Unknown n	Unknown $\bar{X}$
7 months	1	4	1.60bB	91	1.57aB	25	1.71aA	24	1.61aB	–	–
Seedling	2	66	1.81aA	35	1.82bA	85	1.77aA	24	1.83bA	1	1.71B
Height	3	117	1.81aA	6	1.73bA	–	–	3	1.79bA	–	–
(m)	4	11	1.84aA	37	1.83bA	–	–	–	–	–	–
64 months	1	4	5.23aA	89	4.85aA	25	4.92aA	24	5.44aA	–	–
Coppice	2	66	7.82bA	35	8.24bA	85	7.52bA	24	7.67bA	1	7.64A
Height	3	117	8.36cB	6	9.76cA	–	–	3	8.35bB	–	–
(m)	4	11	7.40dB	37	9.86cA	–	–	–	–	–	–

[*] NSW = New South Wales, QLD = Queensland, SA = South Africa, Other = mainly south Florida.

[**] Origins within generations not sharing same uppercase letter are significantly different at the 5% level; generations within origin not sharing same lowercase letter are significantly different at the 5% level.

[***] Number of progenies.

Estimates of individual tree and family heritabilities of five traits, namely, 7 month seedling height, 3 and 64 months coppice growth, are not significantly different. The individual tree heritabilities range from 0.27 to 0.36 and the family heritabilities from 0.59 to 0.75 (Table 5). High family heritabilities suggest that good gains can be achieved through family selection.

The genetic gains that can be achieved by different selection strategies are given in Table 6. The greatest gain in volume (107%) over the

107

Table 4. Summary of per hectare coppice productivity of <u>Eucalyptus</u> <u>grandis</u> progenies in GPOP77 at 30 and 64 months after harvest.

Trait	Overall	Generation*				200 Best Trees
		1	2	3	4	
30 months						
Vol. (m^3)	2.97	1.79C	.28B	3.72B	5.27A	9.65
weight Mg	1.00	0.33C	1.18B	1.43B	2.30B	4.92
64 months						
Vol. (m^3)	23.18	8.09D	24.76C	29.79B	46.17A	243.88
weight Mg	12.51	3.90D	13.34C	16.18B	23.16A	137.24

*Generations not sharing same letter are significantly different at the 5% level.

whole population can be achieved when combined selection is performed, selecting the top 100 families and three trees per family. Similar gains can be achieved by selecting one tree from the top 300 families. A gain of 94% can be realized by individual selection of the top 200 trees.

The genetic correlations among the four traits, 7-month seedling height, 3- and 64-month coppice height, and 64-month DBH, range from 0.27 to 0.98 (Table 7). The high correlation between height and DBH at 64-months after cutting suggests pleiotropy, where the same gene is involved in controlling both these characters. Consequently, if this gene segregates, it causes simultaneous variation in both height and DBH. The positive correlations indicate that selecting for one trait has no negative effect on the other trait.

CONCLUSIONS

Impressive gains have been achieved through four generations of selection in E. grandis in south Florida. While improvement over generations was observed, the variability among progenies within each generation should result in meaningful increase in biomass productivity. When the best trees were rated against fourth generation means, nearly two-fold differences in traits such as DBH and height were observed.

In volume production, the best trees were more than three times larger at 64-months after harvest. This suggests an obvious potential for

Table 5. Heritabilities of Eucalyptus grandis traits by and over generations.

Trait	Individual Tree Heritability					Family Heritability				
	Generation					Generation				
	1	2	3	4	Overall	1	2	3	4	Overall
Seedling Height (7 months)					.36					.75
Coppice Height (3 months)					.32					.67
Coppice Height (64 months)	.62	.30	.28	.31	.31	.75	.59	.58	.64	.65
Coppice DBH (64 months)	.56	.33	.28	.28	.32	.73	.61	.58	.61	.65
Coppice Volume (64 months)	.37	.27	.23	.22	.27	.63	.56	.54	.54	.59

Table 6. Predicted genetic gains in 64-month coppice volume for alternative improvement strategies with Eucalyptus grandis.

Selection Strategy	Genetic Gain (%)
200 best trees	+94
300 best trees	+82
10 top families, 30 trees per family	+48
30 top families, 10 trees per family	+72
100 top families, 3 trees per family	+107
300 top families, 1 tree per family	+102

Table 7. Genetic correlations for individual tree heights and diameter at breast height for Eucalyptus grandis through 64 months.

Trait/Age	3 month Coppice Height	64 month	
		Height	DBH
7 month Seedling Height	0.27	0.29	0.42
3 month Coppice Height	—	0.53	0.49
64 month Height	—	—	0.98

improvement of biomass productivity through clonal propagation of the best families to capture full genetic potential. The genetic correlations obtained indicate the presence of pleiotropy and, hence, selecting for and improving height will simultaneously improve DBH and, hence, biomass production. The fourth generation families scored well for frost resilience and coppice quality, suggesting the use of these families for energy farms in south Florida.

With continued introductions to expand the genetic base and selection, the future looks bright for E. grandis as one of the prime woody species in Florida for biomass productivity.

ACKNOWLEDGEMENTS

Research reported here is supported by Oak Ridge National Laboratory under Subcontract no. 19x-0905C and a cooperative program between the Institute of Food and Agricultural Sciences of the University of Florida and the Gas Research Institute, titled "Methane from Biomass and Waste".

REFERENCES

1. FAO. Eucalypts for planting. FAO Forestry series No. 11, FAO of the U.N., Rome, Italy. 676 p. 1979.
2. T. F. Geary, G. F. Meskimen and E. C. Franklin. Growing eucalyptus in Florida for industrial wood production. USDA For. Serv. Res. Paper, SE-23, 43 p. 1983.
3. D. L. Rockwood and T. F. Geary. Genetic variation in biomass productivity and coppicing of intensively grown Eucalyptus grandis in southern Florida. Proc. 7th N. Amer. For. Biol. Wkshp: 400-405. 1982.
4. K. G. Eldridge. Genetic improvement of eucalypts. Silv. Gen. 27: 205-208. 1978.
5. S. Kedharnath and R. K. Vakshasya. Estimate of components of variance, heritability and correlation among some growth parameters in Eucalyptus tereticornis. Proc. 3rd World Conf. For. Tree Breed: 667-676. 1977.
6. D. S. Falconer. Introduction to quantitative genetics. Ronald press, New York, 340 p. 1982.
7. E. Camphinos, Jr. and Y. K. Ikemori. Mass propagation of Eucalyptus spp. by rooted cuttings. Silvicultura, IUFRO/SBS Symposium, Vol. 4(32): 770-775. 1983.
8. V. J. Hartney and P. K. Baker. Vegetative propagation of the Eucalyptus by tissue culture. Silvicultura, IUFRO/SBS Symposium, Vol. 4(32): 791-793. 1983.
9. G. Namkoong and E. B. Snyder. Accurate values for selection intensities. Silv. Gen. 18: 172-173. 1969.

MICROPROPAGATION OF EUCALYPTUS CLONING

CANDIDATES IN FLORIDA

Gary P. Howland

Clonal Products, Inc.
1056 Captiva Point
Lakeland, Florida 33801

George Meskimen

Forest Research Consultant
4459 Riverside Drive SE
Ft. Myers, Florida 33905

Milton J. Constantin

Phyton Technologies, Inc.
7327 Oak Ridge Highway
Knoxville, TN 37931

ABSTRACT

Severe freezes in recent years have provided the selective pressure
which enabled researchers to select 451 cloning candidates representing 14
eucalyptus species and hybrids from experimental stands in central and south
Florida. At least 1000 more cloning candidates could be selected from
commercial plantations of Eucalyptus grandis in southwestern Florida.
Tissue culture techniques and production schedules have been developed for
the micropropagation of cloning candidates. Successfully cultured species
included E. camaldulensis, E. grandis, E. robusta, E. rudis, and various
hybrid eucalypts. Cultures have been started from nodal explants of juvenile, basal sprouts either induced by girdling the ortet's bole, or arising
after freeze damage. Cloned ortets ranged in age from 8-23 years. A tissue
culture clone bank is proposed to preserve ortet genotypes and to provide
limited planting stock for clone testing, start-up cultures for mass propagation, and foundation stock for breeding future cloning candidates.

Keywords. Eucalyptus, tissue culture, micropropagation, cloning,
freeze damage, juvenility.

111

INTRODUCTION

Why Clone Eucalypts?

Eucalyptus species have been identified as the most promising woody
species for short rotation biomass energy crops in Florida and in California.[1,2,3,4] The potential for eucalypt plantations on a variety of
Florida sites has been reported.[1] There are at least 200,000 hectares
(50,000 acres) of prime eucalyptus sites in southwestern Florida alone. An
operational eucalyptus tehnology exists in south Florida, developed by U. S.
Forest Service researchers working cooperatively with state, industrial and
private interests. Between 1972 and 1981, landowners planted 8.8-million
seedlings of Eucalyptus grandis Hill ex Maiden on 6,475 hectares in southwest Florida. Every outplanted seedling was grown from genetically improved
seed collected from local seed orchards.

Efficient vegetative propagation can magnify the realized gains from
commercial plantations of genetically superior eucalypts.[5] In south
Florida, even the most genetically advanced seedling plantations of E.
grandis still exhibit excessive heterogeneity in tree size, coppicing,
freeze tolerance, and susceptibility to basal cankers. Breeding to improve
multiple traits may require several generations to stabilize gains, whereas
vegetative propagation captures large gains in the first generation.

Eucalyptus seedling populations include a great diversity of genotypes
with individuals potentially adapted to particular environmental niches
(e.g., frost pockets, mine spoils, saline or alkaline soils). However, it
is difficult to replicate desirable genotypes through seed production.
Interspecific crosses often show hybrid vigor, but F_1 hybrid seed has not
been produced in commercial quantities. Conversely, trees suffer inbreeding
depression when seed results from self-fertilization.[6]

Successful commercial establishment of clonal eucalyptus plantations
involves the following critical steps:

a) Selection of elite phenotypes growing under commercial conditions;
b) Vegetative capture of the cloning candidates;
c) Small-scale vegetative production to test the cloning candidates on
   representative sites; and
d) Mass propagation of proven clones recommended for planting.

The advantages of vegetative propagation to produce eucalyptus planting stock are now generally acknowledged.[1,3,7,8,9,10,11,12,13] Classic approaches have included grafting, layering, and rooting of stem cuttings,[14] but these methods are limited in their applicability here in the U. S. by low productivity and consequent high cost. As an alternative, workers around the world have studied the potential of eucalyptus tissue culture. Tissue culture techniques are being used more frequently in commercial propagation of both herbaceous and woody plants.

The key to successful vegetative propagation of eucalyptus has been the recognition of the requirement for rejuvenation of the shoot.[14,15,16,17] Secondly, selections must be done with great care, since the vegetatively propagated plants are only as good as the original genetic stock placed in culture. It is possible to evaluate most of the selection criteria only in adult trees. Capturing the genotypes of adult trees, which have demonstrated superiority, requires an induced reversion from adult to juvenile physiology.[8,14,17]

Juvenile shoots (coppice) develop on the stump following felling of an adult tree, or in the region directly below a bark girdle. Alternatively, when an adult scion is grafted to a juvenile rootstock, juvenile sprouts emerge, and this effect can be amplified by regrafting these shoots back to juvenile rootstock. Reversion to juvenile physiology can be encouraged by spraying the developing coppice sprouts with a cytokinin solution.[16] Also, epicormic or water sprouts appear on adult trees following certain types of trauma such as fire or freeze damage.[14]

Several different groups have reported successful organ culture of eucalypt nodes using juvenile sprouts of mature trees or by rejuvenation of adult scion serially grafted to seedling rootstocks.[8,9,14,16,17,18,19,20,21] The basic method for micropropagation of adult eucalypt selections involves the following steps:

-induce juvenile sprouts;
-disinfest and culture nodes (Stage I);
-induce rapid proliferation of shoots from cultured nodes (Stage II)
-index for presence of known microbial contaminants;
-multiply shoots to required level of production;
-elongate shoots (Stage IIB);

-root shoots (Stage III);

-transfer rooted shoots to greenhouse (Stage IV); and

-finish containerized nursery stock to outplanting size.

This approach has already been used to produce thousands of eucalypts now growing in the field.[8,17,20,21,22]

MATERIALS AND METHODS

Cloning Candidates

In winter 1982, U. S. Forest Service researchers selected 451 cloning candidates scattered over 40 sites in 17 counties in central and southwest Florida. This selection effort followed back-to-back freezes which caused serious damage in central Florida during January 1981 and in southwest Florida during January, 1982. Compared to their plantation populations, selections combined exceptional vigor and form with either conspicuously less freeze damage or extraordinary resilience after freeze damage. Cloning candidates were girdled 1 m above the ground to induce juvenile, basal sprouts which were sectioned into cuttings and rooted in mist beds by the Florida Division of Forestry. Of 451 girdled selections, 244 produced sufficient rooted ramets to outplant on one or more of five test sites planted in central and southwest Florida in summer 1982.

The historic freeze of Christmas 1983 damaged ortets on the selection sites as well as ramets on the clone test sites. Resultant juvenile, epicormic freeze sprouts provided explants for the establishment of the tissue cultures reported here. Many of the sprouts had matured past the optimal juvenile physiology by the time we collected them on May 10th, 138 days after the freeze. Still, two-thirds of the capture attempts resulted in successful cultures of E. camaldulensis and E. rudis.

Establishment of Stage I Cultures

Since collecting explants from field-grown plants often results in high rates of Stage I contamination losses, sprouts were collected in clean plastic bags and held in a cooler until brought back to the laboratory. Shoots were then trimmed to approximately 3-6 cm above and below each node, and the leaves removed, leaving the petioles attached. Twenty to 30 nodal sections were disinfested in approximately 500 ml of full-strength chlorine bleach (5.25 percent sodium hypochlorite) plus several drops of Tween-20 detergent as a wetting agent. Exposure times of three to seven minutes, depending upon the diameter of the cuttings, have yielded successfully disinfested cuttings without extensive tissue damage. The exposed, naked axillary buds and tender shoot tips do not survive, but deeper-lying pre-ventitious buds are far less sensitive to the chlorine.

Following several rinses in sterilized water, the cuttings were asepti-cally trimmed to a final size of about 10 mm above and below the node, and the petiole was cut to one-half its original length. The nodes were then planted in culture tubes containing Stage I medium (Table 1).

The charcoal in the medium is especially important to absorb phenolic compounds which are produced by the cuttings and released into the medium.[8,17] Alternatively, a medium without charcoal can be employed, provided that the initial cultures are incubated in the dark.[15,16,17] Shoots which began to develop after 1 week were excised and transferred to fresh Stage II medium (Table 1). Species and individuals can vary greatly in the intensity of production of these growth-inhibiting phenolics.

Nodal explants usually developed preventitious sprouts within a week following introduction into culture. Transfer of these in vitro sprouts to an appropriate medium resulted in rapid shoot multiplication. Successfully cloned ortets ranged in age from 8 to 23 years.

Stage II Multiplication of Shoots

The Stage II multiplication medium described in Table 1 has yielded high rates of shoot multiplication in vitro. To stimulate axillary bud proliferation, the concentration of cytokinin, BAP (benzylamino purine), must be adjusted between 0.1 and 1 ppm to suit the individual clone. Shoot proliferation increases with increasing BAP concentration, but care must be taken to avoid toxic levels of cytokinin in the medium. Rates of increase on the order of ten-fold or more were obtained with a 3 to 4 week culture cycle.

Stage IIB Elongation of Shoots

The shoots which proliferated rapidly under the influence of Stage II medium tended to have thin stems with cotyledon-like leaves, often with very short internodes. The Stage IIB medium (Table 1) allowed development of

seedling-type leaves and a sturdier, elongated stem that could better survive the hardening off process. The inclusion of activated charcoal in the
medium is thought to reduce the concentration of cytokinin (i.e. BAP) available to the shoots, and the gibberellic acid (GA-3) stimulates elongation
growth.

Rooting of Shoots in Stage III

The routine, reliable development of roots on cultured explants is
often the most difficult phase of any vegetative propagation program with
woody shoots. The Stage III medium used for rooting cultured eucalypt
shoots is shown in Table 1. Shoots cultured for 1-2 weeks on this medium or
on a modification of this medium initiated several roots per stem. Plants
held for a longer period developed long roots which had to be cut back
before transfer.

Refinement of the Greenhouse Program

Well-rooted plants from Stage III culture were gently washed free of
agar medium, and transplanted into nursery flats containing a commercial

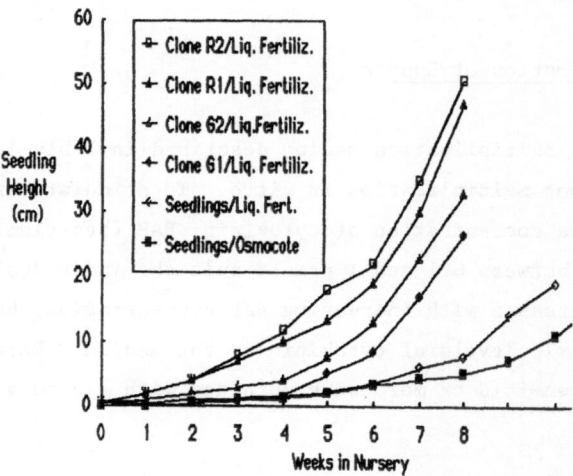

Figure 1. Comparative growth rates of micropropagated E. robusta and E.
grandis clones vs. E. grandis seedlings to reach shipping size
(20 cm).

peat mixture. Plants were maintained under intermittent mist and shade for -the first 1-2 weeks, followed by transfer to an area of higher light intensity for an additional 6-7 weeks. Fertilization with 20-20-20 (N-P-K) was begun at this time as a constant application with every watering (~50 ppm N) or as a weekly application (~150 ppp). Experience indicates that greenhouse/nursery time can be reduced from the 12 weeks needed to finish seedlings to less than 8 weeks (Figure 1) for micropropagated eucalypts.[17]

CONCLUSION

It is clear to us that only cloning can bring commercial eucalyptus forestry to central Florida within the next decade. The requisite combinations of cold tolerance, growth rate, and site adaptation are too rare and elusive in seedling populations to support operational planting. In south Florida, seedling silviculture with E. grandis is an operational reality, but clonal planting stock promises massive gains in wood production, freeze tolerance, and coppice regeneration.

In the present work, only 10 clones have been cultured, but hundreds of worthy cloning candidates have been identified while at least a thousand more are there for the finding in commercial plantations. Ramets of captured clones will be available for further field evaluations and pilot projects.

We believe that the gains inherent in these cloning candidates can best be realized by capturing them in a tissue culture clone bank, providing small number of ramets for field testing, and subsequent mass propagation of proven clones.

REFERENCES

1. T. F. Geary, G. F. Meskimen and E. C. Franklin. Growing eucalypts in Florida for industrial wood production. Gen. Tech. Rep. SE-23. Asheville, NC: U. S. Department of Agriculture. Forest Service, Southeast Forest Experiment Station. 43 pp. 1983.
2. R. Hunt and B. Zobel. Frost-hearty Eucalyptus grow well in the Southeast. South. J. Appl. For. 1:6-10. 1978.
3. C. B. Low, G. H. Matson and R. M. Sachs. Propagating fast-growing eucalypts for energy crops. Calif. Agric. 34:18-20. 1983.

4. R. B. Standiford and F. T. Ledig. Proceedings of a workshop on Eucalyptus in California. June 14-16, 1983. Sacramento, California. Gen. Tech. Prep. PSW-69. Berkely, CA: Pacific Southwest Forest and Range Experiment Station. Forest Service, U. S. Department of Agriculture. p. 128. 1983.

5. J. M. Bonga. Vegetative propagation in relation to juvenility, maturity, and rejuvenation. In: Tissue Culture in Forestry. J. M. Bonga and D. J. Durzan, Ed. The Hague/Boston/London: Martinus Nijhoff/Dr. W. Junk Publ. p. 387-412. 1982.

6. K. G. Eldridge. Genetic improvement of eucalypts. Silva Genetica 27:5. 1978.

7. E. Campinhos, Jr. and Y. K. Ikemori. Mass production of eucalyptus spp. by rooting cuttings. In: Proc. IUFRO Conf. on Fast Growing Trees. Aguas de Sao Pedro. S.P. Brazil. 17 pp. 1980.

8. R. Durand-Cresswell, M. Boulay and A. Franclet. Vegetative propagation of eucalyptus. In: Tissue Culture in Forestry. J. M. Bonga and D. J. Durzan, Eds. The Hague/Boston/London: Martinus Nijhoff/Dr. W. Junk Publ. p. 387-412. 1982.

9. T. F. Geary and W. G. Harding. The effects of leaf quantity on rooting success with Eucalyptus camaldulensis Dehn. cuttings. Commonw. For. Res. 63:225-230. 1984.

10. B. Martin and G. Quillet. Bouturage des arbres forestiers au Congo. Rev. Bois et Forets des Tropicques. 154:41-57. 1974.

11. B. Martin and G. Quillet. Bouturage des arbres forestiers au Congo. Rev. Bois et Forets des Tropicques. 155:15-33. 1974.

12. B. Martin and G. Quillet. Bouturage des arbres forestiers au Congo. Rev. Bois et Forets des Tropicques. 156:39-61. 1974.

13. B. Martin and G. Quillet. Bouturage des arbres forestiers au Congo. Rev. Bois et Forets des Tropicques. 157:21-40. 1974.

14. V. J. Hartney. Vegetative propagation of the eucalypts. Aust. For. Res. 10:191-211. 1980.

15. A. Franclet and M. Boulay. Micropropagation of frost resistant eucalypt clones. XIII Intl. Bot. Congr., Sidney. 1981.

16. A. Franclet and M. Boulay. Micropropagation of frost resistant eucalypt clones. Aust. For. Res. 13:83-89. 1982.

17. G. P. Howland. Vegetative propagation of eucalypts. Working Document No. 2. In: Methanol Production from Eucalyptus Wood Chips. H. H. Fishkind, Ed. U. S. Department of Energy (DOE/RA/50316, Attachment II). 56 pp. 1982.

18. P. K. Gupta, A. F. Mascarenhas and V. Jagannathan. Tissue culture of forest trees-Clonal propagation of mature trees of Eucalyptus citrodora Hook, by tissue culture. Plant Sci. Letters 20:195-201. 1981.

19. V. J. Hartney. Tissue culture of Eucalyptus. Proc. Intl. Plant Prop. Soc. 32:98-109. 1982.

20. G. P. Howland. Eucalyptus biomass and tissue culture studies. Session in Depth--Energy from Biomass. 32nd Annual Meeting of the Tissue Culture Association. Washington, D.C. 1981.

21. G. P. Howland. Large scale methanol production from clonal eucalypt plantations. In: In Vitro Cultivation of Forest Tree Species. AFOCEL. 1981.

22. G. P. Howland. Unpublished Observations.

23. T. Murashige and F. Skoog. A revised medium for rapid growth and bioassays with tobacco tissue cultures. Physiol. Plant 15:473-497. 1962.

HARVESTING OPTIONS FOR SHORT ROTATION CROPS

William B. Stuart

Department of Forestry
Virginia Polytechnic Institute and State University
Blacksburg, Virginia 24061

ABSTRACT

The selection of a harvesting option for short rotation crops is constrained by the methods available for severance, aggregation, processing,
and transport of biomass material. The capabilities and limitations of
sickle bars, flails, chain saws, circular saws, augers, and hydraulic shears
for severance are compared for various sizes and types of biomass. Agricultural devices for aggregation such as curtain conveyors, opposing belts, and
reels or gates appear to be limited to relatively small and uniform material, whereas grasping devices similar to those used in conventional forest
harvesting operations lend themselves to larger stems. Processing options
range from severance only to the production of a product meeting the specifications of the conversion facility. Bundling, baling, cutting material
into billets, crushing, and chipping are all processing options which may
take place in the field, at a remote facility, or at the conversion plant.
Transport of uncompacted biomass material remains a major influence on
harvesting costs. Transport of material directly from the field to an
adjacent processing plant is an attractive alternative to compaction and
highway transport. At present there appears to be no major effort to develop special equipment for the harvesting of biomass. Until a market or
the capital required to support such an effort is forthcoming, biomass plantation harvesting will be the major limiting constraint on the acceptance of
the biomass concept.

Keywords. Biomass, harvesting, severance, aggregation, processing,
transport, plantation, coppice, baling, and billeting.

INTRODUCTION

When a tree is planted, the options and opportunities of final product marketing strategies are infinite. Unfortunately, however, these myriad options are often greatly constrained by the reality of harvesting technology and marketing opportunities. Short rotation forest crops are no different from conventional forestry or agricultural crops in having to face these realities.

A key element in discussing these problems lies in identifying or defining what is meant by short. Extremely short rotations which have very close spacings--tens of thousands of stems per acre with annual or biennial harvests--are closely aligned with agriculture cropping. Rotations based on 7500 stems ha^{-1} (3,000 stem $acre^{-1}$) or less and rotations longer than 8 to 10 years are entering the realm of conventional forestry. The realism in between these extremes is the true terra incognita from a harvesting standpoint.

Harvesting is further constrained by stand management requirements of minimizing stool damage and disturbance of the root system, harvesting during the dormant season to assure maximization of growth potential and coppice, while efficiently accommodating the winter weather and road conditions. The ability to meet these goals is, of course, influenced by the tract size to be serviced per machine visit.

The moisture content of the material on the stump is high--in the range of 100 to 120 percent dry weight basis for most coppicing species in the juvenile stages. Any contribution this water makes to the fuel supply system is negative. It increases product weight, allows living cells to continue to respire after severance with the attendant losses in dry matter, and makes the material vulnerable to infestation by molds and decay fungi. The method and location of drying will have a major influence on the transport and conversion costs of the product.

Harvesting and transport will be the largest cost items associated with short rotation forest management. Management factors which can influence these costs must be identified and accommodated throughout the planning and management process.

SEVERANCE

The first step in any harvesting process is cutting the tree free from
its rootstock. (There has been work on removing the major share of the
rootstock with the tree, but it has tended to retard coppicing.) The alternatives available for severance are sickle bars, rotatory flails, chain
saws, circular saws, augers, and shears.

Sickle bars are commonly used in agriculture. A common form consists
of a reciprocating bar carrying a number of sharpened triangular teeth or
"sections" working with a comb of "guard teeth" which separate the crop and
serve as the anvil for shearing. The relationship between the "sections"
and "guard teeth" is such that the crop being cut is sheared as the bar
moves back and forth within a machined channel. Sickle bars have the advantage of a long history of development and are readily available. Design
swath widths are limited only by the carrier structure's ability to carry
the bar and sufficient power to operate it. Row organization requirements
are minimal. Disadvantages associated with sickle bars include vulnerability to damage from rocks and other obstacles, sensitivity to the size and
structural density of the material, and controlling the vibration associated
with the reciprocating action. The most important of these is the structural density of the material. Sickle bars work best on grasses and other
monocots with a fibrous external sheath surrounding a soft pith with a
diameter only a small fraction of its cross-section. The forces required to
harvest larger organized stems are difficult to transmit through a simple
push bar arrangement. Consequently, the sickle bar may only find application in very short rotation (one growing season or less) crops. Some crushing of the stem at the point of severance will be experienced and could lead
to limited mortality, but the survival rate in these short rotation high
density crops should be adequate.

Rotary flails or bush hogs depend upon a shaped, but not necessarily a
sharpened, knife or chain attached to the end of a rotating disk or bar.
Severance is accomplished by "beating" or abrading the stem away. The
approach is simple, robust, and relatively power efficient. The technology
has evolved for agriculture and right-of-way maintenance application, and a
variety of machine types are available. The major disadvantages of this
approach are the damage to the stool during severance and the difficulty of
capturing the material for further processing. The fraying and subsequent
survival of individual stools will likely be less critical for those species
with a strong tendency for suckering or root sprouting, but the ability to

capture the material before it is pulled under the flail and lost to mulching remains a problem. The power requirements and shock forces transmitted through the machine, when it encounters larger material, restricts the application to stems less than two inches in diameter.

Chain saws occupy a dominant position worldwide as a tool for severance in commercial forest crops. Modern chain technology has resulted in a durable and easily maintained tool available in sizes to suit nearly every need. Unfortunately, chain sawing seems to be best adapted to motor-manual systems. Efficient operation depends, to a great extent, on operator feel or the ability to closely control cutting rate to match tool requirements. Efforts toward adapting the saw to mechanical felling have been only marginally successful because of the difficulty of maintaining control circuits which can reproduce the element of feel. Consequently, it is likely that the chain saw will find application in those situations where the trees are of a size comparable to the conventional pulpwood. The cut produced is clean, with little impact on the coppicing ability. The capital costs are low, and operating costs are acceptable and especially suitable to small plantings and steep or broken terrain.

Circular saws actually predate the chain saw as an economical severance tool. The bicycle saw was one of the first practical mechanical felling devices. Technology took an alternative path, but circular saws have been undergoing an extensive review in recent years as an alternative to shearing in frozen timber. The mechanical principles involved are simple, and the present technology is well practiced in many applications. The use of this technology on standing trees is less well-refined. The saws have moderate power requirements which vary as a function of tree size and frequency of use. A variety of tooth forms are commercially available to meet specific needs. The quality of cut can vary from very rough to very smooth as a function of tooth design, saw rim speed, gullet capacity, and feed rate, as shown by Marley.[1] Major disadvantages include the rapid dulling or damage from encounters with mineral soil or rocks, the fact that the cutting width is only half the saw diameter, and the difficulties in shielding the trailing edge of the saw. The material being severed is well-presented for capture and subsequent processing, but should not be left to ride on the saw surface. Saws are very sensitive to vines and dense, flexible weeds in the understory which can foul rotating parts.

Augers or milling heads remove wood through repetitive cuts parallel to, rather than across, the grain, as in sawing. The cutting device is

either a rotating bar with spiral flutes, as in the U.S.F.S. short rotation
harvester[2], or a rotating tool bar equipped with detachable straight cutters.[3] The bar is mounted parallel to the ground, across the front of the
swathing device, and mills or machines a wide kerf out of the standing
biomass as the machine moves forward. The cut produced is of a good
quality, and power requirements are moderate. The milling approach results
in a compact tool and drive assembly. These tools, like saws, are very
susceptible to dulling from contact with rocks or mineral soil. Sharpening
tends to be a slow process, and back up units are expensive at current
levels of production. Bending is probably a more crucial problem. Any
misalignment or imbalance in the tool can result in serious damage to the
tool, drive units, and the carrier machine.

Shears, which pinch the tree off between two hydraulically driven
moveable blades or between one blade and an anvil, are commonly accepted in
the harvesting of medium-size softwood timber. They are robust, easy to
maintain, adaptable to a wide variety of carriers, and simple to operate.
The major disadvantages for short rotation crops include relatively low
cycle times (10 seconds to one minute per tree), weight, damage to the
stool, and difficulties with adequately capturing wide-spreading multiple
stems. Shears offer potential in those stands with individual stems with
7.6 to 10 cm (3 to 4 in) groundline diameters arranged in recognizable rows.

A probable classification of these devices according to their most
likely application is shown in Table 1. This classification assumes a
direct transfer of existing technology with only minor adaptations.

AGGREGATION

Capturing the stems after severance and accumulating or directing them
from subsequent processing or transporting is the usual second stage in the
process. Again the processes considered are a mix of agricultural and
conventional harvesting practices depending on stem size, stand organization
and the site. The alternatives include curtain conveyors, opposing belts or
chains, reels or gates, and grasping devices.

Curtain conveyors are as old as McCormick's drop reaper. The crop is
forced to lie flat on a wide rubber or canvas belt which transports the
material away from a swathing severance device. These conveyors have the
advantages of low capital cost, low power requirements, and easy maintaina-

Table 1. Severance Devices and Possible Adaptations

Device	Application
Sickle Bars	Very high density, very short rotation, uniform stands on well-maintained sites.
Rotary Flails	One or two year rotations of natural or only partially controlled regeneration of mixed species on irregular, rocky sites.
Chain Saws	Single tree felling heads on feller bunchers or in motor-manual systems on difficult sites.
Circular Saws	Multiple saw, swath felling, heads for rotations resulting in multiple stems per stool of 10.1 cm (4 in.) or less on prepared sites or single tree felling heads for felling larger timber.
Augers	Similar to circular saws, moveable mounts may find application in stands, containing a wider range of diameters.
Shears	Feller-buncher heads for harvesting trees 10.1 cm (4 in) or larger at ground line on suitable but unprepared terrain.

bility. This method is best suited for short, homogeneous crowns that tend
to lie flat on the conveyor, which are uncommon plant characteristics in
short rotation crops. Their effectiveness can be increased through the use
of rolls to separate and direct the crop and top conveyors, augers, and
other devices. These infeeds seem best when the material is to be fed top
first directly into a processing unit of approximately the same width as the
swather.

An alternate approach involves mounting the processing device close to
the felling head and transferring the material directly by the use of reels,
spring-loaded fingers, augers, and other devices. These usually require
that the material be fed butt first with the top pushed ahead of the machine
into the standing crop. The advantages offered are compact design, low
centers of gravity, and minimization of machine components. The disadvantages include tendency of crowns to tangle, high front end weight concentrations, and limited surge capacity between machine components and poor visi-
bility of the severance device (unless the processor throughput rate is such
that the linear feet processed per foot of forward movement is greater than
the expected maximum tree height).

The opposing chain or belt technology is borrowed from corn and sugar cane harvesting equipment. The stem or group of stems is grasped between one or more opposing sets of belts or chains immediately before or at the time of severance. The stems are held vertically and transported to the collector or processing device in that position. The control of the individual piece must be positive, with the upper belt or chain located at or above the center of gravity of the stem. This approach appears to have promise for trees in the 2.5 to 7.5 cm (1-3 in.) size range and less than 9.1 m (30 ft) tall. The structural and power requirements to control and move larger stems appear to be incompatible with field equipment. Additional problems may arise from high variation in tree sizes and vines in the crop.

Reels or gates, passing material into a cage or support frame which accumulates it in an upright position, have been incorporated in a variety of machine types, including an early version of the Propst harvester, the Rome tree combine, and the Canadian poplar harvester. The stem is swept immediately from the severance device into the capture device by the reel or gate. The system works well on crops with a single well-defined stem per plant and good self-pruning. Leaning, crooked, or forked stems and heavy lower limbs can result in missed captures or blockages.

Grasping devices usually resemble a pulpwood grapple rotated vertically. The stem is grasped tightly immediately before cutting with shears, or immediately after cutting in the case of saws near the butt. The hold on the tree is sufficient to hold it in an upright attitude. Multiple stems may be collected by adding an accumulating finger to a grapple. A variety of feller-buncher heads are available for a wide range of base machines. Productivity rates vary from one tree per minute through 8 to 10 trees per minute depending upon head type, carrier, tree size, and stand organization.

These devices tend to work best in stands with a well defined tree form with stems weighing over 45 kg (100 lbs) each. Definite row definition can be useful for the tree-to-tree carriers. Currently, machines with felling heads which move to each tree, offer opportunities for faster production rates than those with a felling head mounted on the end of a swinging boom. The major disadvantages are the amount of machine movement in the stand required to accumulate an efficient package for transport and the sensitivity to tree size. The adaptation of feller-forwarder technology to smaller trees may offer opportunities for reducing the movement required.

Table 2. Aggregation Devices and Possible Applications

Device	Application
Curtain Conveyor	Behind sickle bar or other suitable swather handling small, very light 0.45-4.5 kg (1-10 lb) material on fairly level, even terrain.
Opposing Belts	Behind single or multiple saw severance devices handling material between 1.2 and 9.1 m (4 to 30 ft) tall.
Reels or Gates	As a means of passing variable-sized material from a swather type felling head into a closely coupled processing device, or as a means of passing a tree-sized stem from a felling head to a vertical accumulating cage.
Grasping Devices	Incorporated into conventional feller-bunchers or feller-forwarders for individual trees or well-defined clumps weighing 45 kg (100 lb) or more.

PROCESSING

Processing in harvesting is the reduction of the severed material into a more manageable form for subsequent operations: onsite transport, storage, drying, or market transport. The options range from no reduction to the production of a product meeting the specifications of the conversion facility. Locations and combinations for processing include severance and processing on a single machine, or a second machine working from bunches or windrows at the tract boundary in conjunction with recovery, or from prepared storage areas in conjunction with market transport. Processing options include severance only, bundling, billeting, crushing, chunking, and chipping.

No processing other than severance is, of course, the simplest option. Breakdown becomes the responsibility of the converting firm. The difficulty lies in the low bulk density and handling difficulties associated with the crop. Very few, if any, of the species-rotation-age combinations put forward result in a product sufficiently dense to create an economic load size for loading and transport equipment. In instances where the conversion facility is located on the plantation and all transport can be accomplished using specially designed off-highway equipment, this poses no problem. Leaving the material in full tree form offers advantages in speeding the drying of moisture in the wood and that from subsequent precipitation.

Bundling involves compressing a group of stems together to form a sheaf, module, or bale. The technology has been borrowed again from agriculture. A sheaf is a bundle of aligned stems bound by one or more ties. This form of materials handling was very common in agriculture through the mid-20th century. It works best with uniform material up to 3 to 3.6m (10 to 12 ft) long with a stem form which yields to lateral compression. A module is an aggregate of material formed by compression in a mobile die. No binding or strapping is used. This technology has been used successfully in crops as diverse as cotton and corn. Module builders are most effective in materials with low resiliency--a limited tendency to spring back after being compressed. This would tend to limit the application to succulent materials or those which had been previously broken down. A bale is a round or rectangular solid subjected to greater compaction forces than used in module building and restrained by banding materials to return that compression. Baling forces may range from approximately equivalent to those used in module building through pressures sufficient to squeeze free water from the cells. Each approach has the advantage of arranging stems or pieces of stems in a roughly uniform package for subsequent handling while increasing the bulk density of the material.

Billeting involves cross cutting the material into uniform lengths shorter than one stem length. Lengths proposed have ranged from 6 inches to 2 feet. The working prototype incorporating this method used rotating shears to do the cross cutting. Billeting has the advantage of making the length of the material uniform while retaining the bark sheath and low bulk densities to avoid rewetting and speed air drying during field storage.

Crushing involves passing the stem between two counter-rotating rollers or drums with a predetermined clearance. Crushing breaks the bark sheath, expediting drying, and destroys the natural resiliency of woody materials. Roll clearances may be set closely to squeeze some of the free water from the stem during the crushing process. A variety of roll designs have been experimented with, resulting in splitting, crimping, and spreading actions. The process has again to be adapted from agriculture and can leave the resultant product in a form compatible with conventional forest agricultural equipment.

Chipping is the main means of product reduction in conventional forest harvesting and widely considered for short rotation crops as well. Chipping has the advantage of relying on an existing processing and transport technology and being less sensitive to variations in piece size than some of the more sophisticated processing options. Power requirements tend to be high, and the effort required to feed large numbers of low-volume stems into the chipper can prove prohibitive. Agricultural versions of chippers in the form of forage harvesters have been used successfully on very short rotation materials. The technology of chipping is almost too familiar to demand the needed refinements of the method. Relatively little attention has been paid to optimizing chip size relative to power requirements and conversion equipment.

Chunking systems being developed by the Forest Service Lab in Houghton, Michigan,[4] use an innovative processing device and produce large-size chips. The larger chips require less power and permit better air circulation for drying and combustion while resulting in bulk densities comparable with conventional chips for transport. The technology is new, undergoing refinements at present, and may emerge as a most promising alternative for small tree-sized biomass.

Selecting the appropriate processing method is only half the battle—where the processing is to occur is equally critical. Four basic options exist: on the machine at the time of severance, on a subsequent machine recovering material piled in the stand, at roadside or on-site concentration area, or at a central conversion facility. Each has advantages which may best serve individual situations.

Processing at the time of severance can result in a compact one-pass machine paralleling agricultural harvesters. The material would be cut, processed, transported to the conversion facility, and stored as a single coordinated activity. While this approach has a philosophic appeal, the advantages must be considered in light of machine versatility and size and power requirements, required support equipment, and the coordination effort, and the storage needs at the central facility. The one machine, one pass, operation could be quite efficient for large plantings located near the conversion site, but smaller scattered plantings would likely require segmenting the operation involving piling the material down in the stand or at roadside to reduce the transportation needs and plant site storage areas. This approach also provides little opportunity for passive drying of the material.[4,5]

Processing with a recovery unit presupposes a severance machine which piles or organizes the material. Again, parallels abound in agriculture—especially in haying. This approach results in a segmentation of the operation, allowing the use of potentially simple single-function machines. Because woody biomass deteriorates slowly, the severance and processing activities could be separated by periods of weeks or even months during the dormant season. This approach would permit field drying and allow the use of the site for temporary inventorying of goods in process.[6]

Processing at roadside may be done in conjunction with severance or from piles formed at the time of harvest. Processing with loading equipment as the material is cut offers the advantage of having the onsite operations completed quickly and of simplifying inventory control. The interaction between harvesting functions and harvesting and transport can be reduced by careful scheduling. Processing from storage piles at roadside allows removal from the growing space at the time of severance to satisfy biological requirements, delays the added costs of processing until near the time of conversion, permits some field drying of the material in the round, and permits further segmentation of the operations. This latter attribute may be particularly important in those climates where good field working conditions may not be synonymous with good transport conditions. This approach would also permit the use of several less productive severance machines piling material for a single highly productive set of processing and transport equipment.

Processing at the conversion facility has the advantage of transferring one of the most expensive components of harvesting to the conversion facility, where costs would perhaps be reduced by working a limited set of equipment on a multiple-shift, day-in-day-out basis. This arrangement assumes a transport system which can move whole-tree material economically over public roads or a private road system from plant to plantation. In addition, the conversion facility would need sufficient inventory space to support several weeks' or months' demand of unprocessed materials. The difficulty of transporting whole-tree material is a major factor in retarding the full development of this option. Economic factors, particularly those related to the additional capital costs of the processing equipment and woodyard construction coupled with the inventory carrying costs, would slow development, even if the transport problem could be solved.

TRANSPORT

Transport remains a major influence on harvesting and harvesting costs. At the current stage in the evolution of biomass plantations, it appears

that most applications will require public road transport of the material
between the plantation and the conversion facility. The economics of the
system will require that the fuel value per load be optimized within the
weight and dimension restrictions imposed on the haul truck. This will
require that the material have a bulk density of at least 240 kg m^{-3} (15 lb
ft^{-3}), roughly three to four times that of loose, uncompacted, small trees
at a field moisture content of 100% dry weight basis. Any pre-transport
drying of the material will increase the fuel value but decrease the bulk
density of the material. Alternate forms of transport--rail, barge, and
pipeline--may offer opportunities in selected locations. Attention to
minimizing costs will yield benefits regardless of transportation method.

SUMMARY

Harvesting and transportation of woody biomass will continue to constitute a major share of the delivered cost of fuels from short rotation biomass. Development of appropriate harvesting equipment and systems can
reduce these costs, but it is unlikely that any exceptional cost advantages
will arise from the plantation characteristics of the material.

In the short run, the plantation characteristics may, in fact, remain a
significant disadvantage. Harvesting systems have evolved to harvest agricultural crops or conventional forest crops. Direct application of these
systems in biomass plantation may be possible in some instances by forcing
the machines to work beyond their normal design limits with the attendant
loss in productivity and increase in operating costs.

Developing special systems for these crops is a slow and expensive
process. Agricultural systems have evolved over a period of 100 years. The
evolution has been aided by a cooperative effort between plant scientists
and engineers to develop plants best suited to mechanical harvesting and
efficient equipment for the harvesting and processing. Forestry equipment
has had a shorter evolutionary period and is still in the latter stages of
sorting out feasible silvicultural systems and economic harvesting systems.
As the variation in silvicultural systems is reduced, the possibility of
designing efficient harvesting machines and systems increases.

The variability in species, spacings, and rotation ages and the uncertainty about markets and market prices have held the development of commercial harvesting equipment in a state of abeyance. The development and

testing of concepts have been on an uncertain basis during most of the period of interest in short rotation crops. The magnitude of investment required to mount a credible effort frightens most proponents of biomass production.

There are no miracles, no Henry Ford working in an unheated garage in the Midwest assembling the ultimate harvester that will revolutionize plantation management. Today's equipment arises from the assignment of design and manufacturing teams who are provided with a set of design criteria and have the freedom to select the most promising alternative concepts from those above and others. The manufacturer then decides if sufficient market exists to warrant the refinement of the concept into prototypes and a marketable machine. The process is slow and costly. Until a market or the capital required to support such an effort is forthcoming, biomass plantation harvesting will be the major limiting constraint on the acceptance of the biomass concept.

REFERENCES

1. D. S. Marley. An evaluation of existing and conceptual short rotation energy plantation harvesting machines and systems. Unpublished M.S. Thesis, VPI & SU, Blacksburg, VA - 213p. (1982)
2. J. A. Mattson. Forest service short rotation harvester. Personal Communication. 1982.
3. P. Koch and T. E. Savage. Development of swath-felling mobile chipper. Journal of Forestry 78(1):17-21. 1980.
4. R. A. Arola, S. A. Winsauer, R. C. Radcliffe and M. R. Smith. Chunk-wood Production: A New Concept. Forest Products Journal 33(7/8): 43-51. 1983.
5. J. A. Berard. Energy biomass harvesting equipment being developed in Canada. Paper presented at 1981 Winter Meeting of the American Society of Agricultural Engineers. Chicago, IL Dec. Paper No. 81-1598. 1981.
6. R. W. Bryan. Mobile shear-chipper delivers clean pulp chips to forwarders. Forest Industries 107(2):28-30. 1980.

COSTS OF HARVESTING FOREST BIOMASS ON STEEP SLOPES WITH A SMALL

CABLE YARDER: RESULTS FROM FIELD TRIALS AND SIMULATIONS

John E. Baumgras and Chris B. LeDoux

USDA Forest Service
Northeastern Forest Experiment Station
Morgantown, West Virginia 26505

ABSTRACT

Cable yarding can reduce the environmental impact of timber harvesting on steep slopes by increasing road spacing and reducing soil disturbance. To determine the cost of harvesting forest biomass with a small cable yarder, a 13.4 kW (18 hp) skyline yarder was tested on two southern Appalachian sites. At both sites, fuelwood was harvested from the boles of hardwood trees 10 to 36 cm (4 to 14 inches) in dbh. The volume of pieces yarded ranged from 0.01 to 0.63 m^3 (0.2 to 22.4 ft^3). With a crew of four on a small clearcut block and piece volumes averaging 0.14 m^3 (5.1 ft^3), yarding costs were \$12.03 per m^3 (\$33.70 per cunit). With a crew of two on a site previously harvested for sawlogs, it cost \$6.78 per m^3 (\$19.00 per cunit) to yard pieces averaging 0.21 m^3 (7.5 ft^3). Because productivity was generally constrained by the yarder's 429 kg (1,150 lb) mainline pull capacity, the two-person crew proved the most efficient. Production and cost analyses integrating field studies with computer simulation showed that the total cost of yarding biomass with a two-person crew could range from \$5.50 to \$11.00 per m^3 (\$15.00 to \$31.00 per cunit), depending upon average piece volume. This analysis also revealed a tradeoff between biomass utilization and total yarding cost: costs can be reduced by limiting the minimum piece volume yarded.

Keywords. Timber harvesting, cable yarding, biomass, logging cost, simulation.

INTRODUCTION

The large cable yarders that were once used on Appalachia's steep slopes disappeared along with the old-growth timber, but interest in cable yarding has been renewed by the concern over the environmental impact of harvesting timber on steep slopes with conventional ground-based systems. A study conducted in north-central West Virginia found that 1 km of skidroad was required to harvest 5 ha of steep ground with rubber-tired skidders (20 acres mile^{-1} of road), whereas 20 ha per km of road could be harvested by a skyline yarder (80 acres mile^{-1} of road).[1] Increased road spacing resulting from the use of cable yarder systems improves the aesthetic quality of harvest areas and significantly reduces the potential for soil erosion. Cable yarding also reduces ground disturbance on nonroad portions of harvest areas.[2]

Because of the high cost, expensive rigging, and long setup time of medium- and large-capacity cable yarders, their use in the eastern mountains has been limited to removals of high-value sawlogs or high-volume biomass.[3] In this region, there are also numerous opportunities for low-volume biomass harvests on steep slopes. These include regeneration cuttings on poor sites, thinnings on better sites, and recovery of harvesting residue.

Economic yarding of low-volume removals on steep slopes may require low-cost, highly mobile yarders such as the Bitterroot Miniyarder (Figure 1). Developed by the USDA Forest Service's Missoula Equipment Development Center, this 13.4 kW (18 hp) yarder can be mounted on a small truck or trailer and operated by forest crews.[4] Reports by Brown and Bergvall,[5] and Cubbage,[6] indicate that this yarder has potential for harvesting small trees on steep slopes. This paper presents results of field and simulation studies of harvesting forest biomass with the Bitterroot Miniyarder.

YARDING SYSTEM

The yarder studied was rigged as a live skyline, as shown in Figure 2. The skyline could be slackened or tightened as required during yarding. Yarding was done uphill, and the carriage returned by gravity when the mainline was slackened. When the carriage engaged the moveable stop, it clamped onto the skyline and released the mainline that fed through the carriage. The logs were attached to the mainline with wire-rope chokers. When the mainline drum was engaged and the end of the mainline reached the

134

Figure 1. The bitterroot miniyarder

carriage, the skyline clamp released. The carriage and logs then returned
to the landing. The lift provided by the skyline was generally sufficient
to keep the front of the logs off the ground during yarding.

YARDING STUDIES

    The Bitterroot Miniyarder was studied on two steep sites in the south-
ern Appalachians to determine its yarding cost and evaluate the effects of
crew size and site conditions on yarding cost. Time and motion studies were
conducted to measure and record its performance. All volume measurements
and weight estimates included both wood and bark.

    The field sites evaluated had slopes ranging from 20 to 45 percent.
On both sites, bolewood was harvested from trees 10 to 36 cm (4 to 14
inches) dbh and utilized for fuelwood. Much of this material could also
have been utilized for roundwood pulpwood or small diameter logs for low-
quality sawn products. Before yarding, all trees were limbed and topped
with chainsaws. Although harvesting limbwood and topwood would have significantly increased biomass yields, the small yarder could not deliver enough
wood to justify a costly whole-tree chipper.

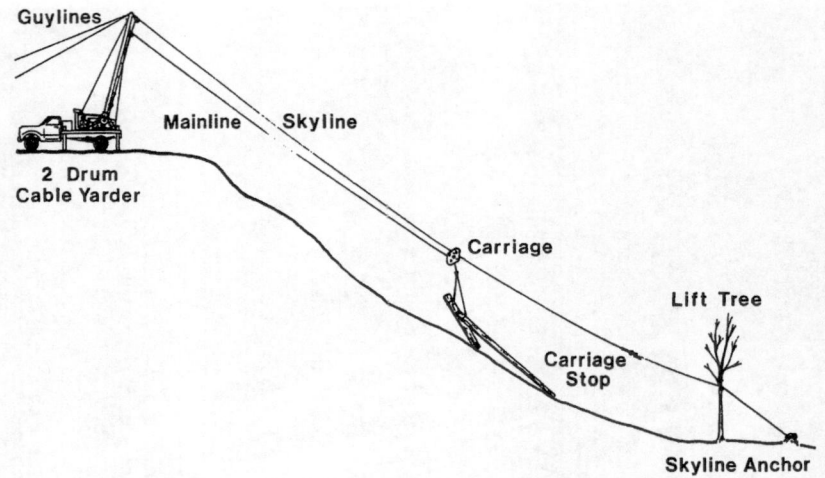

Figure 2. Cable yarder rigged as live skyline

The 429 kg (1,150 lb) mainline pull capacity of the small yarder[7] limited turn volumes on both sites to 0.68 m^3 (24 ft^3). Therefore, trees larger than 25 cm (10 inches) in dbh were bucked to keep bole sections within the yarder's capacity.

Site 1 was a 0.81 ha (2.0 acre) fuelwood clearcut that yielded 94 m^3 per ha (1,350 ft^3 acre^{-1}). This fan-shaped unit was yarded to a single landing, using four skyline corridors. Yarding was done with a crew of four: a yarder operator, one person unhooking logs at the landing, and two chokersetters. A small crawler tractor was used to clear the landing and deck the fuelwood. Trees harvested from this hardwood site were predominantly oaks and hickories, averaging 19.6 cm (7.7 inches) in dbh.

On Site 2, two 0.61 ha (1.5 acre) units yielded 77 m^3 per ha (1,100 ft^3 acre^{-1}). On each of these rectangular units, one corridor was used with a crew of two, one operating the yarder and unhooking, the other hooking chokers. All wood was decked under the skyline, so no tractor was needed to swing logs away from the yarder. Sawlogs had previously been harvested from this site, and all residual trees had been felled to promote the regeneration of desirable tree species. Residual trees harvested averaged 19.1 cm (7.5 inches) in dbh. The predominant tree species were oak, hickory, and yellow-poplar. Bolewood residue from sawtimber trees was harvested together with the bolewood from residual trees felled during site preparation.

The yarding conditions and yarding cycle characteristics from both sites are summarized below:

Yarding Conditions	Site 1		Site 2	
	Mean	Range	Mean	Range
Piece weight – kg	136	5–653	187	15–537
– lb	300	12–1439	412	32–1183
Piece volume – m^3	0.14	0.01–0.63	0.21	0.01–0.62
– ft^3	5.1	0.2–22.4	7.5	0.5–22.0
Slope yarding distance – m	63	8–107	87	14–165
– ft	208	25–350	285	45–540

Yarder Cycle Characteristics

	Site 1		Site 2	
Productive time – min	4.25	1.7–7.6	3.78	1.6–9.5
Delay time – min	0.95	0–8.3	1.09	0–11.8
Pieces yarded	2.3	1–5	1.5	1–5
Volume yarded – m^3	0.3	0.08–0.67	0.3	0.03–0.67
– ft^3	11.6	2.7–24.0	11.1	1.0–24.0
Weight yarded – kg	308	82–635	280	23–518
– lb	680	180–1400	618	50–1143

COST AND SIMULATION ANALYSIS

Based upon 1983 equipment costs, the estimated daily cost of the yard-ing operation was $270.00 per 8-hour day with a four-person crew, or $166.00 for a crew of two.[7] The cost estimates assume wage rates of $5.00 per hour plus 30 percent payroll costs; and include fixed and operating costs for the yarder, two chainsaws, and a radio for communication between the yarder operator and the chokesetters. The total stump-to-landing yarding cost at site 1 was $12.03 per m^3 ($33.70 cunit^{-1}); at site 2 it was $6.78 per m^3 ($19.00 cunit^{-1}). These costs estimates are based on the production rates sampled at each site, and include the cost of moving the yarder and changing yarding corridors. The unit cost on site 1 does not include the tractor and operator used to move yarded wood from the yarder deck.

The difference in yarding costs between sites 1 and 2 can be attributed to both crew size and yarding conditions. To compare crews under equal yarding conditions, THIN--a cable yarder simulation model[8,9] --was used with the yarder cycle-time regression equations developed from the time-study data.[10] The simulation results showed that under similar operating conditions, productivity is constrained by the capacity of the small yarder. Consequently, the cost of additional crew members could not be offset by increased production, and for most applications, the two-person crew would be most efficient.

Computer simulation was also used to explore the effects of site factors such as average piece volume and slope yarding distance on production and cost. These results showed average piece volume to be the most important variable affecting yarding cost. The analysis of piece volume effects assumed a crew of two, a yarding corridor 168 m (550 ft) long by 37 m (120 ft) wide, and a volume hrvested of 84 m per ha (1,200 ft^3 $acre^{-1}$). When average piece volume was increased from 0.05 to 0.40 m^3 (2 to 14 ft^3), yarding costs declined from $11.00 per m^3 ($30.80 $cunit^{-1}$), to $5.50 per m^3 ($15.40 $cunit^{-1}$) (Figure 3). Figure 3 was developed to show the general nature of the relationship between average piece size and total yarding cost when volume per acre is held constant. These results agree with practical experience, which shows that bigger pieces are cheaper to yard.

For a specific harvest unit, the effect of piece size can be exploited by designating a minimum piece size. Although this will reduce total yarding cost, it will also reduce the amount of biomass harvested. To demonstrate this, the yarding of unit 2 was simulated using successively larger minimum piece volume limits. The results in table 1 show that as the minimum piece volume was increased from 0.01 to 0.28 m^3 (0.5 to 10.0 ft^3), the portion of available bolewood that was harvested declined to 46.8 percent. Yarding costs also declined from $6.82 per m^3 ($19.09 $cunit^{-1}$), to $6.11 per m^3 ($17.10 $cunit^{-1}$). The cost was minimal with a piece-volume limit of 0.14 m^3 (5.0 ft^3), which utilized 85 percent of the available bolewood. Although the landowner's objectives may preclude the use of piece-size limits, it is important to recognize its value in harvesting small pieces of biomass.

An alternative approach to minimizing residue yarding cost proposed by LeDoux,[11] is to establish piece-size limits by yarding distance zones. This approach permits increased utilization close to the yarder, but imposes

Table 1. Simulation results illustrating the effect of minimum piece volume
limits on total yarding cost for the Bitterrtoot Miniyarder and
percent of bolewood biomass utilized on unit 2

Minimum Piece volume[a/]		Bolewood volume harvested	Total yarding cost	
m^3	cubic feet	percent	dollars/m^3	dollars/cunit
0.01	0.5	100.0	6.82	19.09
0.05	2.0	97.3	6.43	18.01
0.08	3.0	94.8	6.22	17.43
0.11	4.0	90.3	6.42	17.97
0.14	5.0	85.6	6.08	17.02
0.17	6.0	79.2	6.27	17.56
0.20	7.0	72.6	6.12	17.13
0.22	8.0	64.6	6.18	17.29
0.25	9.0	57.0	6.14	17.19
0.28	10.0	46.8	6.11	17.10

[a/]Volume of smallest piece yarded, not average piece volume.

successively larger piece-size limits as yarding distance increases. The
tradeoff of piece volume and slope distance helps reduce the effects of
slope yarding distance on yarding cost.

The effect of slope yarding distance was determined using the piece
volume distribution and corridor width from unit 2. The length of the
corridor was incremented from 61 to 244 m (200 to 800 ft). Because average
yarder cycle times were longer, total yarding costs increased approximately
$1.50 per m^3 ($4.20 cunit^{-1}) (Figure 4). The impact of longer yarding
distances was partially offset by the increased area and volume yarded per
corridor, which reduced the unit cost of moving the yarder between
corridors.

ECONOMIC FEASIBILITY

The economic feasibility of low-volume forest biomass removals depends
on the total stump-to-market costs and the market values. The relationships
shown between yarding cost and piece volume also hold true for felling,
limbing, and loading: it costs more to process small pieces than large
pieces. This is especially true on terrain too steep for the mechanized
felling and bunching machines often used to harvest small trees. With
favorable yarding conditions, the estimated total stump-to-mill cost of

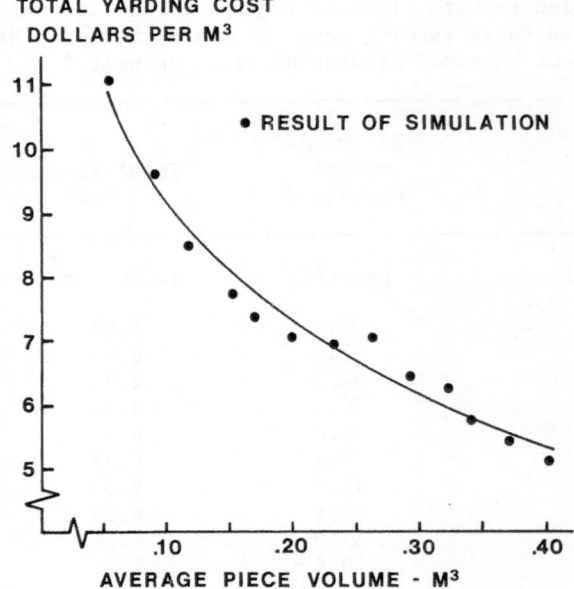

**TOTAL YARDING COST
DOLLARS PER M³**

• RESULT OF SIMULATION

AVERAGE PIECE VOLUME - M³

Figure 3. Simulation results showing the effect of average piece volume on total yarding cost

manual felling and limbing, yarding, loading, and hauling ranged from $17.85 to $25.00 per m^3 ($50.00 to $70.00 cunit^{-1}), for haul distances of 24 to 56 km (15 to 35 mi).[12] These costs exceeded the current prices of $12.00 to $15.00 per m^3 ($33.00 to $42.00 cunit^{-1}) generally paid for hardwood pulpwood or industrial fuelwood. Although prices of approximately $20.00 per m^3 ($55.00 cunit^{-1}) are paid for small-diameter roundwood delivered to manufacturers of wooden pallets[13] or processors of fuelwood,[14] these markets are very limited.

One alternative that could prove economically feasible would be to market fuelwood at roadside. The buyer would buck the bole-length pieces to the desired length and load them. This approach is now used where there is a high fuelwood demand for residential heating. Depending upon yarding conditions, eliminating the loading and hauling cost could reduce total costs to $8.90 to $14.30 per m^3 ($25.00 to $40.00 cunit^{-1}). Prices of $9.00 to $14.00 per 0.9 m^3 (1/3 cunit) pickup truckload would cover the cost of felling and yarding. When biomass removal is required for site preparation or regeneration, a portion of the harvesting cost might be borne by the landowner in lieu of alternative costs.

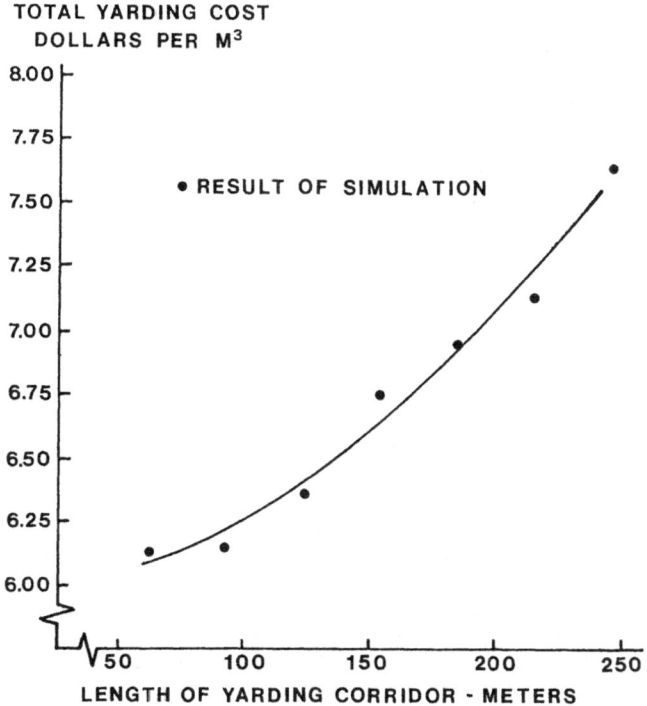

TOTAL YARDING COST
DOLLARS PER M³

• RESULT OF SIMULATION

LENGTH OF YARDING CORRIDOR - METERS

Figure 4. Simulation results showing the effect of corridor length on total yarding cost

CONCLUSION

Results from this study show that the cost of harvesting biomass with small cable yarder can be greatly reduced by using a two-person crew and yarding selectively with minimum piece-size limits. Results also show that average piece volume is the most important site-related variable affecting yarding cost and that significant variations in total yarding cost can be expected in response to changes in average piece size or yarding distance.

Because landing space on steep terrain sites and the production capacity of the small yarder are limited, biomass harvesting will often be limited to the utilization of bolewood. Difficult topography also requires costly manual felling and limbing which, coupled with loading and hauling costs, contributes to a total harvesting cost that is likely to exceed the value of the biomass removed. Marketing roundwood fuel at the logging site could be a more viable option. Environmental concerns may dictate the use of cable yarding. However, its application to low-volume biomass removals

141

will be limited by the distance to markets, market values, the demand for residential fuelwood, and alternative site preparation costs.

Although the results from field tests and simulations such as those summarized in this paper will not provide all the answers to biomass removal from steep terrain, they should give managers the data with which to develop economic harvesting policies.

REFERENCES

1. J. N. Kochenderfer and G. W. Wendel. 1978 Skyline Harvesting in Appalachia. Res. Pap. NE-400. USDA Forest Service, Broomall, PA. 9 p. 1978.
2. J. H. Patric. Some environmental effects of cable logging in the Appalachian forests. Gen. Tech. Rep. NE-55. USDA Forest Service, Broomall, PA. 8 p. 1980.
3. H. E. Matics. WESTVACO wood procurement cable logging program: the future situation. p. 17-23. In: Proceedings Appalachian cable logging symposium. June 12, 1982. Blacksburg, VA. Jefferson National Forest. Roanoke, VA. 1982.
4. UDSA Forest Service. Bitteroot Miniyarder. 8351-2504. USDA Forest Service Equipment Development Center, Missoula, MT. 1983.
5. S. L. Brown and J. A. Bergvall. In-woods testing and feasibility study of fuelwood recovery using a small-scale cable yarding system. State of Washington, Dept. of Natural Resources. Olympia, WA. 50 p. 1983.
6. F. W. Cubbage and A. H. Gorse, IV. Mountain logging with the Bitterroot Miniyarder. p. 81-91. In: P. A. Peters and J. Luchok, eds. Mountain Logging Symposium Proceedings, June 5-7, 1984 . Morgantown, WV. West Virginia Univ. Press, Morgantown. 1984.
7. J. E. Baumgras and P. A. Peters. A cost and production analysis of the Bitterroot Miniyarder on an Appalachian hardwood site. Res. Pap. NE-557. USDA Forest Service, Broomall, PA. 8 p. 1985.
8. C. B. LeDoux and D. A. Butler. Simulating cable thinning in young growth stands. Forest Science 27(4):745-757. 1981.
9. D. A. Butler and C. B. LeDoux. Reference manual for THIN: A cable yarding simulation model. Forestry Research Lab., Oreg. State Univ., Corvallis, OR. 35 p. 1983.
10. J. E. Baumgras. Cycle tie equations for the Bitterroot Miniyarder with a two person crew. Unpublished report on file at the USDA Forest Service. Northeastern Forest Experiment Station. 180 Canfield St. Morgantown, WV 26505. 1984.
11. C. B. LeDoux. Cable yarding residue after thinning young stands: A break-even simulation. Forest. Product Journal 34(9):35-40. 1984.
12. C. B. LeDoux. Stump-to-mill timber production cost equations for cable logging eastern hardwoods. Unpublished report on file at the USDA Forest Service, Northeastern Forest Experiment Station, 180 Canfield St. Morgantown, WV 26505. 1985.
13. Ohio Crop Report Service. Ohio Timber Prices. Columbus, OH. 2 p. 1984.
14. M. P. Folkema. Handbook on high-capacity production and marketing of fuelwood. FERIC Handbook No. 6. Forest Engineering Research Council of Canada. Quebec, Canada. 55 p. 1984.

COST SENSITIVITY ANALYSIS OF EUCALYPTUS

GRANDIS WOODY BIOMASS SYSTEMS

D. R. Dippon, D. L. Rockwood and C. W. Comer

Department of Forestry
University of Florida/IFAS
Gainesville, Florida 32611

ABSTRACT

A major decision variable in the short rotation woody biomass plantation process is the selection of that combination of planting density and
timing of stand initiation, coppices and final harvest that will minimize
the average dry tonne cost of production. The question of economic feasibility requires knowledge in advance of all costs and expected revenues.
This simple concept quickly becomes difficult to implement when planting and
site preparation methods along with the harvest technology are in development and not currently operational. Since large-scale woody biomass energy
generation plants are not in operation in Florida, the local market price of
the raw material is also unknown. Therefore, this analysis examines the
sensitivity of average delivered dry tonne costs to various factors of
production so that this raw material can be compared to other types of fuel
sources for energy conversion. The unprocessed cost of this energy source
ranges as low as $1.04 to as high as $2.60 gj^{-1} (gigajoule) which is less
than evaluations of other woody species by other authors.

Keywords. Economics, Eucalyptus grandis intensive culture, woody
biomass.

INTRODUCTION

The traditional sources of woody biomass for energy include waste fiber
from harvesting operations, by-products from primary forest products operations and the total above ground fiber removal of the woody biomass of

existing stands. The capture of raw material from such sources may be the least expensive currently, but the economics of these sources are heavily dependent on transportation costs and may be an unreliable supply in the long run. If woody biomass is to gain importance as an energy source, then resources must be allocated efficiently to insure a dependable supply of raw material. Short rotation intensive culture (SRIC) studies of woody species have focused on the evaluation of alternative species in various regions of North America. The SRIC concept may provide the reliable supplies, where needed, required to make woody energy a viable option for renewable energy.

Since 1978 Eucalyptus grandis Hill ex. Maid has been the focus of SRIC systems analysis at the University of Florida. Eucalyptus is without peer among presently evaluated species for productivity on drained muck soils in south Florida.[1] Growth rates and survival indicate that the species is well suited to SRIC systems. Percent seedling survival was similar for 1,600 and 10,000 trees ha^{-1} planting densities after one year but survival declined more rapidly for the denser planting by year two. Although individual seedling survival rates were lower for the denser planting, both basal area and volume growth were increased as planting density increased.

The ability of E. grandis to coppice, or to grow one or more new stems from a stool, increases its attractiveness as an energy source. Since Eucalyptus regenerates itself after each harvest operation, biomass accumulation can be generated over several growth cycles from the initial start-up cost of site preparation and planting. The effect is to reduce the average cost per dry Mg of woody biomass produced by foregoing future planting costs for two to three harvesting cycles. The number of coppice stems per stool is influenced by the initial planting density, genotype, stump height and climatic conditions immediately before and after harvest. Total volume growth of the coppice may be greater than initial seedling growth.[2]

This paper will discuss the impact that initial planting density and seedling and coppice rotation schedules have on the predicted average annual production. The average production cost per dry Mg is also evaluated for the aforementioned parameters along with various discount rates and other assumptions relating to the cost of operations. Finally, the estimated average cost of production for Eucalyptus grandis will be compared to other similar short rotation intensive culture (SRIC) systems elsewhere in the United States.

METHODS OF ANALYSIS

The ideal cultural practices required for SRIC are similar to, but more intensive than, the methods used for conventional Eucalyptus culture. The use of conventional vegetable crop site production practices on the drained organic soils of south Florida reduces competing weed competition to provide the seedlings with ideal growing conditions, sun, moisture, and nutrients. The schedule of the management practices recommended for an E. grandis SRIC system on this site are sequenced in Table 1. One report[2] suggests that a primary seedling growth, followed by two coppices, is a cycle system best suited for maintaining the site's biomass productivity before the original stool's coppicing vigor is reduced.

SRIC System Assumptions

Production and management assumptions are based on this and other research over the past years which provide a strong basis for cultural assumptions and rotation guidelines selected. The timing and complexity of SRIC operations varies by species and by broad soil type. The postulated series of operations in Table 1 is only for Eucalyptus on muck soils in south Florida. The drained muck soils encourage very rapid woody biomass accumulation, but for the most part this type of soil is currently under high-value agricultural cropping systems. Because of the high quality site, the yields represent the upper level expected from a SRIC system.

Since the sites are already under agricultural cropping methods, site clearing can be accomplished via disking operations. This prepares the site by reducing any herbaceous competition. However, we have assumed more intensive site preparation (chopping plus bedding) to better reflect likely stand establishment operations. Fertilizer is added in the form of triple-superphosphate to overcome future phosphorus deficiencies. The muck is rich in nutrients and no additional fertilization should be required for several rotations.

Planting densities of 1,600 and 10,000 trees ha^{-1} are evaluated with rotation growth period one- to four-years. Woody biomass production ranged from 12 to 23 dry Mg ha^{-1} per year for the respective spacings and rotation ages for the first growth cycle. Successive coppice regrowth are expected to match or even exceed the original production rates.

Table 1. Management scenario for biomass plantations of <u>Eucalyptus grandis</u> on drained muck soil

---

<u>Clear Site</u>: (May-June)
   -Double Chop

<u>Bed Site</u>:

<u>Apply Fertilizer</u>: (just before planting)
   50 kg ha^{-1} of P as triple-superphosphate

<u>Plant Site</u>: (June-August)
   1,600-10,000 trees ha^{-1}

                    (approx. 1-4 years)

<u>Harvest Stand</u>: (January-February)
   Yields of approximately 23 dry metric Mg ha^{-1} yr^{-1} for stands planted
at 10,000 trees ha^{-1} and yields of 12 dry metric Mg ha^{-1} yr^{-1} for stands
planted at 1,600 trees ha^{-1} have been reported.

                    (approx. 1-4 year)

<u>Harvest Stand</u>: (January-February)
   Yields approximately 10-20% higher than the original harvest are
expected.

                    (approx. 1-4 years)

<u>Harvest Stand</u>: (Early Summer)
   Yields are expected to be the same as the previous harvest.

The timing of operations substantially influences the annual average production. Planting the original stand requires that heavy equipment enter the site during the summer months (May - August). Since the sites are already engineered for drainage, soil water levels should not hinder these operations. Site preparation and planting during the summer will provide the seedlings time to establish under optimal growing conditions. Since south Florida is a subtropical environment other planting dates are feasible but the temperature/moisture combination of summer appears to be optimal for establishing the plantation.

The timing of the harvest/coppicing activity also has a major impact on the productivity realized from SRIC plantations. Early studies have demonstrated near complete failure of stand regeneration through coppicing/ harvest operations during the warm months of the year. Therefore, coppices should only occur during the winter months (January - February).

The restriction of coppice/harvest operations to certain periods of the year requires stockpiling of the harvested biomass. Whether the trees should be stored on the site, or chipped and stored in piles near the conversion facility, requires further study into their relative rates of decay and energy loss. Woody biomass will lose part of its energy potential as natural decay occurs. The optimal planting density and rotation length may be affected by the method chosen to stockpile the fiber.

Biomass Equations

Height and diameter at breast height (DBH) measurements were collected quarterly for six progenies and two planting densities (1,600 vs. 10,000 trees ha^{-1}) over a four year period. These measurements were used to derive volume equations used to estimate total plot volumes. Measurements for the six E. grandis progenies were pooled for estimating dry weight production ha^{-1}. Total stem dry weight ha^{-1} (Eq. 1.1), volume ha^{-1} (Eq. 1.2) and percent survival (Eq. 1.3) are listed below.

$$STEMDW = e^{(4.5862 - 29.9889/AGE + 11.7744 * SPACE/AGE)} \qquad (1.1)$$

$$STEMVOL = e^{(5.1963 - 29.9889/AGE + 11.7744 * SPACE/AGE)} \qquad (1.2)$$

$$SURV = 1.0/[1.0 + e^{(-3.0735 + 0.0536 * AGE + 0.0136 * AGE * SPACE)}] \qquad (1.3)$$

Where: STEMDW = dry Mg of woody biomass
STEMVOL = cubic meters of woody biomass
SURV = percentage of surviving stools
AGE = Age of plantation in months
SPACE = Dummy variable for initial planting density
0 for 1,600 trees ha^{-1}
1 for 10,000 trees ha^{-1}

147

Growth Responses

Table 2 shows that although the denser planting (10,000 trees ha^{-1}) demonstrates a poorer seedling survival rate throughout the four years (46% at four years vs. 62% for wider spacing), it consistently accumulates more volume and dry weight mass than the less dense (1,600 trees ha^{-1}) planting density. Even though the less dense planting accumulates less production per hectare, it appears that its growth could potentially equal or exceed the denser planting's net growth since the latter's mean annual increment (MAI) is declining at a more rapid rate. Both stand densities reach their respective peak MAI's before three years (Figure 1). Maximum MAI is obtained by the 18th month for the denser spacing with a production rate of 23.78 Mg ha^{-1} yr^{-1}. The less dense stand maximizes its MAI at 14.44 dry tonnes ha^{-1} 2.5 years after planting.

Table 2. <u>Eucalyptus</u> <u>grandis</u> survival, volume, and dry biomass production[*] by planting density through 48 months on drained muck soils

	Planting Density					
	10,000 Trees ha^{-1}			1,600 Trees ha^{-1}		
Age	Survival	Volume	Stem Biomass	Survival	Volume	Stem Biomass
(mos)	(%)	(m^3 ha^{-1})	(Mg ha^{-1})	(%)	(m^3 ha^{-1})	(Mg ha^{-1})
3	95	0.42	0.23	95	0.01	0.00
6	94	8.68	4.71	94	1.22	0.66
9	92	23.87	12.97	93	6.45	3.50
12	91	39.58	21.51	92	14.84	8.06
15	86	53.62	29.13	91	24.46	13.29
18	87	65.65	35.67	89	34.13	18.54
21	84	75.86	41.22	88	43.30	23.53
24	81	84.55	45.94	86	51.77	28.13
27	78	91.99	49.98	84	59.48	32.31
30	74	98.41	53.47	81	66.46	36.11
33	70	103.99	56.50	79	72.79	39.55
36	66	108.89	59.16	76	78.51	42.66
39	61	113.21	61.51	73	83.71	45.48
42	56	117.05	63.60	69	88.44	48.05
45	51	120.48	65.46	66	92.75	50.39
48	46	123.57	67.14	62	96.69	52.53

[*]Projections

Cost assumptions

Although the biological production function is of primary importance in studying a SRIC system, actual implementation will depend on the average delivered cost per dry Mg. This cost is a function of the operational costs incurred over the life and productivity of the system. An interactive simulation model called BIOCUT by Barron et al.[3] was used to calculate the discounted average costs.

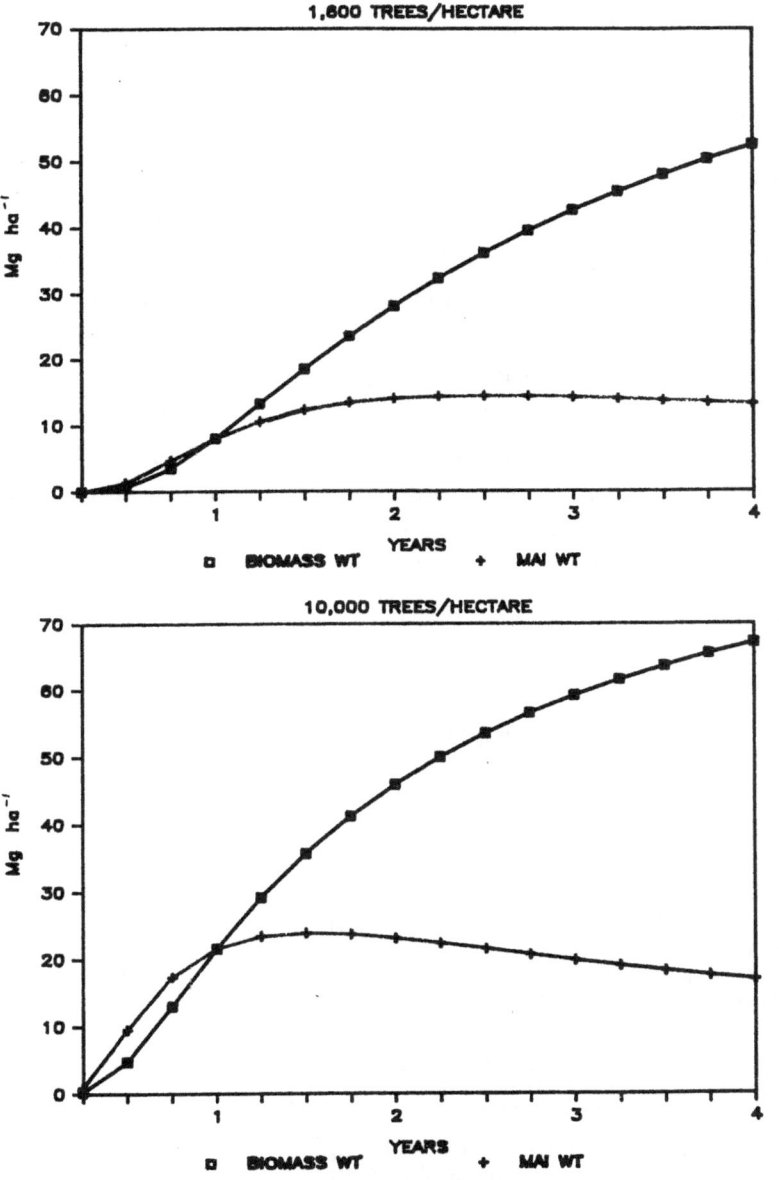

Figure 1. <u>Eucalyptus grandis</u> dry weight production and mean annual increment for two planting densities

The SRIC system was assumed to be comprised of 1,000 hectares with the entire area in plantations. The cost of such an operation is assumed to be divided into several components: regeneration activities at the start of each plantation and at coppices, initial and annual costs incurred, and the harvest/transportation cost incurred at the completion of each cycle. Table 3 lists the costs by activity. Tax effects are not considered since they are a form of transfer payments, and liabilities and credits vary by owner-ship and circumstance. Costs are assumed to increase at a rate equal to that of the overall inflation rate.

RESULTS

Examination of the data in Table 3 provides evidence that the site preparation and planting costs are major components of the computed average discounted cost per dry Mg, comprising nearly 85 percent of the costs in-curred with the 10,000 tree ha^{-1} plantation on a two-year cycle. To reduce the final impact of this cost component, various SRIC scenarios were evalu-ated. All the scenarios assumed two coppice cycles of equal length follow-ing an initial growing period for a total of three harvests per rotation (Table 4). In general, as the time period is increased, total production also increases; and for any time period examined the denser plantation pro-duces more total woody biomass than the less dense planting.

Table 3. Base case cost assumptions for an E. grandis SRIC woody biomass system

Initial/Annual Costs:		Plantation Cost:	
Area	1000 ha	Site Preparation	$177 ha^{-1}
Lease	$124 ha^{-1}	Seedling	$0.08 $tree^{-1}$
Start-up	$50 ha^{-1}	Planting	$0.04 $tree^{-1}$
Managerial	$30 ha^{-1}	Fertilization	$52 ha^{-1}

Harvest Costs:	
Fixed Costs	$150 ha^{-1}
Variable Costs	$6.60 Mg^{-1}
Transportation	$3.30 Mg^{-1}

Table 4. Eucalyptus grandis productivity for two planting densities and five cutting schedules

Periods (yr/yr/yr)	Density (trees ha^{-1})	Years per cycle	Total Volume Mg ha^{-1}
4/3/3	1,600	10	137.85
3/3/3	1,600	9	127.98
2/3/3	1,600	8	113.44
3/2/2	1,600	7	98.91
2/2/2	1,600	6	84.39
3/2/2	10,000	7	151.06
2/2/2	10,000	6	137.88
1/2/2	10,000	5	113.40
2/1/1	10,000	4	88.96
1/1/1	10,000	3	64.53

SRIC Production

In the previous table, coppice growth is assumed to be the same as seedling growth, but this assumption is probably incorrect. Therefore, production was examined for coppice growth projections of ±20 percent of seedling growth to determine the sensitivity of the average cost of dry tonne to the productivity of the coppice. Figure 2 demonstrates how these growth assumptions affect average annual production for the rotation and initial planting densities. Depending on the growth actually observed, the rotation which will maximize average annual production will entail either a five or six year, two-coppice, cutting cycle for the denser spacing. The less dense spacing is more variable with average annual production maximized by either a seven, eight, or nine year, two-coppice rotation.

SRIC Net Cost

So far, the discussion has only defined the optimal biological rotation while attempting to maintain high productivity rates. The woody biomass production data were combined with the cost data from Table 3 in the program BIOCUT to examine whether the optimal biological rotation is also producing the biomass at the least cost per Mg. Based upon our data and assumptions, it appears that costs are minimized (Figure 3) with the 5 and 8 year rotations (one rotation equals one planting cycle plus two coppice cycles) for

the dense and less dense stocking levels, respectively. Coppice productivity assumptions do not appear to affect the choice of the optimal rotation.

Figure 4 demonstrates the extent that changes in the assumed discount
rate will affect the average discount cost per Mg. The optimal rotation for

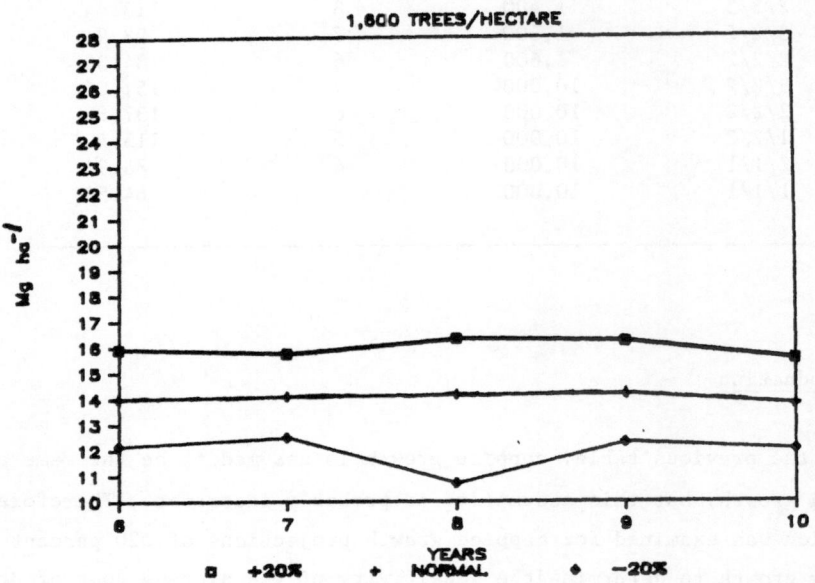

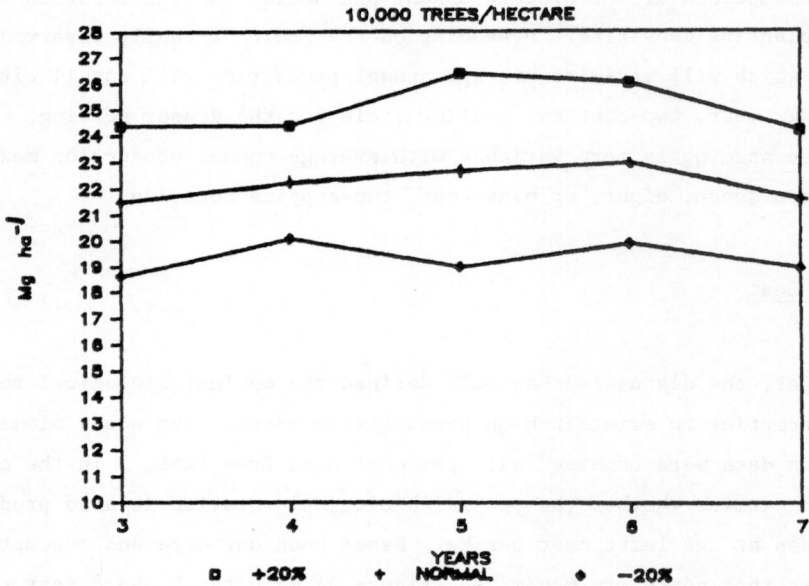

Figure 2. Mean annual increment for E. grandis for various planting
density, production and rotation combinations

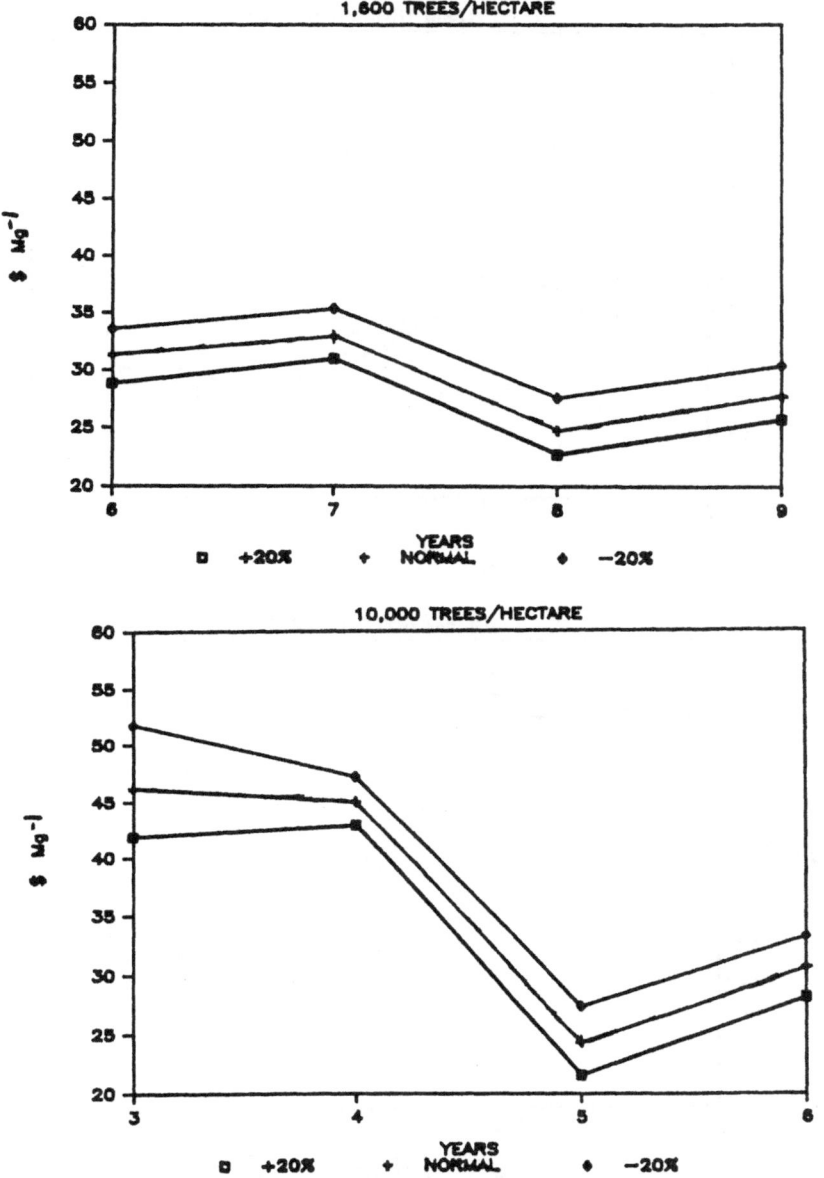

Figure 3. Net dry Mg (tonne) production costs for E. grandis for various
planting density, production and rotation combination

either planting density is unaffected by the discount rates examined. The
figure also incorporates a change in harvest and transportation cost assump-
tions.of \$12.00 Mg^{-1} and \$2.50 Mg^{-1}, respectively, and no fixed cost per
hectare. After reviewing several other permutations (Figures 2 and 3) our
results indicate a discounted average annual cost per Mg of between \$20 to

153

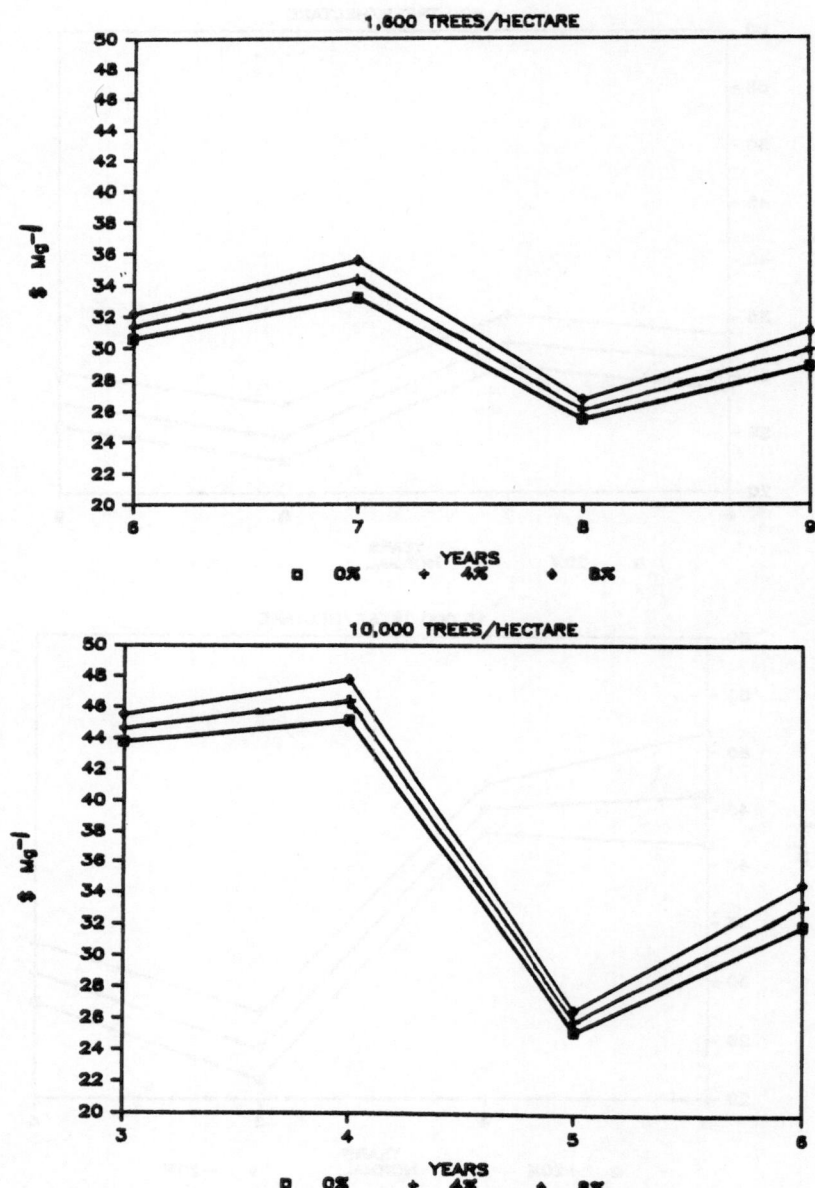

Figure 4. Net dry tonne production costs for E. grandis planting
density, discount rate and rotation combinations

$50. Assuming that there are 19.2 gj Mg^{-1} of E. grandis woody biomass[1] then
the unprocessed cost of this energy source ranges between $1.04 to $2.60
gj^{-1}.

Recently, Perlack[4] examined several SRIC woody biomass coppice systems
in terms of their average estimated cost per gigajoule. He determined that,

based on published data to date, Eucalyptus SRIC in Florida resulted in the lowest average discounted cost. Other species/regions combinations ranged from 107 (cottonwood in Kansas) to 232 (sand pine in Florida) percent above the Eucalyptus cost estimate. Although the Eucalyptus SRIC system on muck soils appears preferable, only 200,000 hectares were classified[1] as available in southern Florida. If energy systems are to be based on woody biomass in Florida, other sites and species will probably be incorporated in supplying biomass facilities be they electrical generating facilities, biogas systems, alcohol facilities or other uses.

CONCLUSIONS

The economics of producing potential energy for the least cost from a Eucalyptus SRIC energy plantation system appear to favor the denser stocking (10,000 vs. 1,600 trees ha^{-1}) with a five year rotation. Actually, the estimated cost Mg^{-1} is more sensitive to growth and coppicing cycle assumptions than it is to modifications in the assumed production functions (±20%) or real discount rates (0% to 8%). Given the optimal rotation length for both planting densities, the optimal planting density is dependent upon the growth function/discount factors assumed.

The discounted average dry biomass values cost figures presented in this paper only represent cost figures and production functions known to date. Eucalyptus seedling production on a large scale, with its corresponding cost economies of scale, does not currently exist. Harvesting and drying costs are major cost components of the woody biomass SRIC system, but the engineering of efficient equipment still remains in an early testing and design stage. Finally, we have production tests in place for the determination of coppice growth under several stocking/cutting cycle combinations.

Additional ongoing research is evaluating other genotypes stock for properties including faster initial growth and better coppice response in terms of growth rates and harvest success for the time of year and climatic conditions. If coppicing success continues to be limited to certain seasons of the year, then storage methodologies or alternative (off-season) sources of biomass need to be examined. Biomass can be produced at reasonable cost in southern Florida from a SRIC Eucalyptus system. A prototype operation with an electrical generator would determine the feasibility for supplying local markets with energy from a renewable woody biomass resource.

METRIC TO ENGLISH CONVERSIONS

1 gigajoule = 0.9478 million BTU   1 Mg = 1.10 ton

1 gj Mg^{-1} = 0.8598 million BTU ton^{-1} 1 kg ha^{-1} = .892 lb acre^{-1}

1 tree/ha = 0.405 tree acre^{-1} 1 ha^{-1} = .446 ton acre^{-1}

1 \$ Mg^{-1} = \$1.10 ton^{-1} 1 kj kg^{-1} = 2.320 BTU lb^{-1}

1 \$ gj^{-1} = \$1.055 million BTU   1 m^3 = 35.31 ft^3

1 ha   = 2.47 acre

ACKNOWLEDGEMENTS

Research reported here is supported by Oak Ridge National Laboratory under subcontract No. 19X-09050C and a cooperative program between the Institute of Food and Agricultural Sciences (IFAS) of the University of Florida and the Gas Research Institute, entitled "Methane from Biomass and Waste." The cooperation of the IFAS Belle Glade Agricultural Research and Education Center in provision and maintenance of the study site is gratefully acknowledged.

REFERENCES

1. D. L. Rockwood, C. W. Comer, L. F. Conde, D. R. Dippon, J. B. Huffman, H. Riekerk and S. Wang. Final Report: Energy and Chemicals from Woody Species in Florida. Oak Ridge National Laboratory. Oak Ridge, TN. ORNL/Sub/81-90501, 205 pp. 1983.
2. T. F. Geary, G. F. Meskimen and E. C. Franklin. Growing eucalyptus in Florida for industrial woody production. USDA For. Serv. Gen. Tech. Rpt. SE-23, 43 pp. 1983.
3. W. Barron, R. Perlack, P. Kroll, J. Cushman and J. W. Raney. FIRSTCUT: A Preliminary Assessment Model for Short-Rotation Intensive Silviculture: Model Description and User's Guide. Oak Ridge National Laboratory. Oak Ridge, TN. ORNL/TM-8566. 52 pp. 1983.
4. R. D. Perlack. The Economics of Short-Rotation Intensive Culture: A Cooperative Assessment. Oak Ridge National Laboratory. Oak Ridge, TN. (in press). 1985.

BIOMASS PROGRAMS OF THE SOUTHERN AGRICULTURAL ENERGY CENTER

J. L. Butler and J. H. Haynie

U.S. Department of Agriculture-ARS
Southern Agricultural Energy Center-SAA
Tifton, Georgia 31793

ABSTRACT

The biomass programs of the Southern Agricultural Energy Center receive primary funding from UDSA, with supplemental funding from DOE, and are directed toward the production, harvesting, processing, and converting of biomass into usable forms of energy. The goal is to develop methods and processes which will maximize the use of renewable energy for the production of food. Research is being conducted on the harvesting, storing, processing, and utilization of crop residues, animal manure, and herbaceous crops produced for energy and wood. The processes of conversion/utilization being studied include direct combustion, gasification and pyrolysis, anaerobic digestion, small scale alcohol production, and extraction of vegetable oils for diesel fuel substitutes. Liquid fuels are the premium fuels for agricultural production and it appears that peanut and rapeseed oil may be very useful, with only simple filtration and mixing with diesel fuel, in reducing the demand for diesel oil. Anaerobic digestion of animal waste can help alleviate the problem of manure disposal and also provide a gaseous energy form. Biomass energy represents an agricultural resource which is currently underutilized. In order for it to become a viable agricultural product, systems for its efficient production/utilization must be developed.

Keywords. Biomass, renewable-energy, vegetable-oils, methane, alcohol-fuels, wood.

INTRODUCTION

The Southern Agricultural Energy Center (SAEC) is one of two USDA
energy centers established as authorized in the Food and Agricultural Act of
1977. As established in December 1979, the goal of these centers is "to
discover, develop, and demonstrate technology which will permit agriculture
to be energy self-sufficient on a net basis by 1990 under conditions that
sustain productivity." The Science and Education Administration provided
input and formal support to these centers from three functional areas:
Agricultural Research (AR), Cooperative Research (CR), and Extension (EXT).
In addition, the two centers were designated to manage the programs funded
by "pass-through" money from USDOE to USDA.

As established, the SAEC is responsible for the on-farm collection,
storage and utilization of solar and wind energy and for the production,
harvesting, processing/converting and utilization of biomass. This encompasses both crop residues and crops which may be produced specifically for
energy. Research programs are conducted at Tifton, Georiga, with satellite
locations at Bushland, Texas, and Ames, Iowa, for wind energy research and
at Columbia, Missouri, for research on anaerobic digestion of manure for
methane production. Partial funding was supplied for three years to Belle
Glade, Florida, for research on utilization of the biomass residue from
sugarcane production.

At the peak of activity there were 65 projects, in addition to those at
Tifton and the satellite locations, conducting research on renewable energy.
Due primarily to the current excess supply of oil, the interest in and
support for renewable energy has diminished. Today, there are only 16 projects, plus those at Tifton and the three satellite locations, making up the
research program. Last year, new funding was received from Oak Ridge National Laboratoy (DOE) for vegetable oil/diesel fuel research.

VEGETABLE OIL

Research on vegetable oil/diesel fuel substitutes is currently being
conducted by the SAEC with emphasis on peanuts and winter rape. The oil
from these two crops is highly monounsaturated and can be used as a diesel
fuel substitute without refining, other than simple filtration. In order to
be injected satisfactorily, the viscosity of the oil must be reduced. This

may be done by mixing with diesel fuel, heating, transestification or by other means.

Peanut crops produce annual yields of 1.5 to 2.8 Mg Ha^{-1} and the seeds are about 48 percent oil. The research on peanuts is primarily aimed toward finding methods and/or processes to remove oil from the seed, suitable for on-farm or small cooperative operations. Because of the value of the residue as a protein substitute, the processes must leave the meal in a form which is suitable for livestock feed. In addition to the primary goals, some research on peanut oil utilization as a diesel fuel will be continued.

Winter rape, which is currently being grown in the United States only in the Pacific Northwest, could be grown in much of the country. Winter rapeseed oil is high in erucic acid (22:1) making it desirable as an industrial oil for high temperature and other industrial and marine lubrication problems, and as a mold release agent. This high erucic acid oil apparently is one of the better candidate oils for substituting for diesel fuel. The high erucic acid does limit the usefulness, by law in this country, of winter rapeseed oil for human consumption. The meal residue from the present varieties is high in glucosinolates, which limits its use as a protein supplement for monogastric animals.

The winter rape research program involves all aspects of the crop, from breeding through utilization. The breeding program at Moscow, Idaho, has developed lines which are: low in erucic acid and low in glucosinolates, high in erucic acid and low in glucosinolates, low in erucic acid and high in glucosinolates, and high in erucic acid and high in glucosinolates. From these lines, either a high quality food oil or a high quality industrial oil can be selected. For a protein supplement, the low glucosinolate meal should be selected. The high glucosinolate meal may have other agricultural uses. Both the high and low erucic acid oils appear equally suitable as candidates for diesel oil substitution.

The winter rapeseed research program also includes research on oil extraction, viscosity reduction, storage of both oil and meal and engine testing. In addition, variety trials are being conducted at nine locations from Virginia to California, and research is being conducted in Georgia on agronomic and engineering aspects of production, harvesting and storing of rapeseed in the southeast.

Fuel quality deterioration in both peanut and rapeseed oil as a function of storage conditions, extraction methods and vegetable oil composition is being researched. The chemical changes which occur during storage will be studied to determine the effect of these changes on fuel performance and life of lubricating oil.

ALCOHOL FUELS

Research at the SAEC on alcohol fuels is being conducted on sweet sorghum as the feedstock. This research seeks to determine the most effective methods to harvest, transport and extract the juice from the stalks and use of the residue (bagasse). To date, equipment to defoliate and harvest sweet sorghum by cutting the stalks into billets can be mechanically handled. Just prior to the expressing rolls, these billets are shredded to improve the juice yield. To evaluate the potential for storing, studies have been conducted on storing whole stalks. This method of storing appears to be impractical. Research is also being conducted on inoculation of the shredded stalks with yeast in an attempt to produce alcohol without extracting the juice first.

NON-WOODY BIOMASS

Methods and equipment to harvest, transport, store and utilize nonwoody biomass are being researched for crops such as Napiergrass and other high-yielded forage grasses, and residues such as stalks of corn, sorghum, sunflower, cotton, soybeans, straw and corncobs. Various forms of harvesting such as large round and conventional hay balers, forage harvesters, and modifications of combines to harvest cobs with the grass or cob savers are being evaluated. Because these residues remove from the soil fertilizer nutrients essential for plant growth and because some of them retard erosion, the effect of their removal on future productivity must be determined.

Because of low energy density of these biomass energy crops, methods of increasing the density, such as pelleting or cubing, are being evaluated to determine whether cost effective methods can be achieved. It is also much easier to regulate the rate of feeding of a pelleted material than of a bulky fibrous material. Thus, all the advantages of having a pelleted or cubed biomass unit must be weighed against the cost of producing that unit.

Direct combustion studies utilizing both a vortex furnace and a furnace capable of handling large round bales of non-woody biomass and logs are also being conducted. Because of the difficulty in getting complete combustion, the furnace capable of accepting the large units utilizes a heat exchanger so that the products of combustion do not go through the grain. It appears that this may also be required for any direct combustion furnace.

WOODY BIOMASS

Woody biomass research is being conducted to identify problems, and develop concepts, machines and systems to allow wood to be efficiently used to replace fossil fuels on the farm and to allow profitable sale off the farm. The research includes both the woody biomass from prunings or trimmings from trees, such as orchard trees and from farm woodlots. Although stick wood can, and is being utilized, the system is difficult to mechanize and automate, and consequently efficient operation of the conversion unit is not possible on a small scale. Also underway is research to develop concepts and equipment to efficiently gather, store and convert wood energy into more useful forms.

THERMOCHEMICAL CONVERSION OF BIOMASS

Biomass, in order to meet requirements, must be capable of supplying energy to a wide range of applications which are independent of feedstock. The thermal requirements may be met with direct combustion and research is being conducted to develop methods and equipment to burn biomass clearly and efficiently for space or water heating. For crop drying, direct combustion of biomass may impart undesirable compounds to the crop unless a heat exchanger is used. A heat exchanger represents additional cost and reduces the efficiency. Consequently, research is underway to develop gasifiers to convert the biomass energy into clean thermal energy. Gasifiers, which burn biomass under conditions of insufficient oxygen, produce a wide variety of gases, with carbon monoxide and hydrogen being the most important. This gas may be either burned to provide energy for heating or powering internal combustion engines.

Research in being conducted on updraft, downdraft and fluidized bed gasifiers to determine their potential for converting biomass into gases which may be used for a variety of needs.

METHANE PRODUCTION BY ANAEROBIC DIGESTION

Anaerobic digestion may be used to produce a gas (biogas) composed primarily of methane and carbon dioxide from a wide range of biomass feedstock. The methane content of this gas is generally in the 50-60% range with carbon dioxide making up most of the balance. This process appears to be ideally suited because the disposal of animal waste is both a necessity and an expense. Biogas production under both mesophylic (C.35°C) and psychrophylic (C.20°C) conditions is being researched with most of the effort on mesophylic production. Since the psychrophylic temperature is near the soil temperature (more than 1m deep) a minimum of insulation would be required and construction costs for digestor vessels could be reduced considerably. Other savings would be realized both in construction and operation if pumps and heating pipes were not required to maintain the higher temperatures.

Psychrophylic digestion research involves the identification of and microenvironmental conditions which favor the psychrophylic bacteria. If these conditions can be identified, it should be possible to make these bacteria more competitive throughout the psychrophylic range. Research on mesophylic biogas production includes geometry of the vessel, microenvironment, retention time, loading rate, solids content of feedstock, need for scrubbing the biogas, storage studies and utilization of biogas.

Utilization studies include cogeneration, in which the biogas is used in an internal combustion engine-generator (22kw) to produce electricity, with the waste heat from the engine being captured for other use (including processing grain into alcohol). The effluent from digestion is also a valuable fertilizer and research is underway as to its utilization and pollution potential for both surface runoff and subsurface water.

SUMMARY

The biomass research program of the SAEC is designed to use renewable resources to provide solid, liquid and gaseous fuels for a portion of the energy required by our agricultural production system. Successful completion of the program should result in the utilization of currently unused resources, reduce the atmospheric build-up of CO_2, reduce the dependence of agricultural production on petroleum fuels and provide the farmer with a new profitable commodity.

BIOMASS PRODUCTION FROM HERBACEOUS PLANTS

Glenn W. Burton

USDA, Agricultural Research Service
Coastal Plain Experiment Station
Tifton, Georgia 31793

ABSTRACT

Of the 358,000 named plant species in the world, only about 195,000 are herbaceous. Because of size and growth season, most have little potential as sources of biomass energy. Many with some potential are not adapted to the long, hot, humid summers and occasional winter freezes experienced in the southeastern United States. Most soils in this area are sandy, infertile, and droughty, requiring fertilization and deep-rooted, drought-tolerant plants for dependable high yields of biomass. Root crops and aquatic plants generally show less promise than the grasses. Sugar cane, Sacharrum offici- narum, cut once in 12 to 18 months and Napiergrass, Pennisetum purpureum, cut every 6 months top the grass perennials but are restricted largely to Florida and the tropics. Cutting more frequently drastically reduces their yields. Cut every 6 weeks to maximize annual dry matter yields, the hybrid bermudas, Cynodon spp., and Pensacola bahiagrass, Paspalum notatum, outyield other perennial grasses where frequent harvesting is required. Among the annual grasses, hybrids of sorghum, Sorghum bicolor, and pearl millet, Pennisetum americanum, head the list. Among higher yielding hybrids, which plant breeding can develop, will be pest-resistant, full season plants with greater stress tolerance and efficiency in the use of fertilizer and other growth factors. Herbaceous biomass will usually contain 1% or more of N and K and 0.2% of P, as well as other fertilizer elements that will be removed from the soil at harvest and, sooner or later, must be replaced.

Keywords. Grasses, root crops, aquatic plants, legumes.

The Plant Kingdom contains over 358,000 named species of plants.[1] Some 100,000 of these are fungi that lack chlorophyll and, hence, are incapable of fixing solar energy. This leaves over 250,000 green species that possess some potential as sources of biomass energy. Approximately 25,000 of these are algae, largely single celled organisms that form the scum on our ponds and are the initial food for most water animals. The mosses, liverworts, ferns, club mosses, and horsetails, totaling some 43,000 species, are mostly small, low growing plants that fail to yield enough to have biomass energy potential. If we omit the 650 species of gymnosperms (the pines, firs, etc.) and the several thousand woody species in the angiosperms, we have close to 195,000 named species of herbaceous plants in the plant kingdom capable of fixing solar energy. How many of these have potential as sources of economical biomass energy in the southeastern United States? Which have the greatest potential?

To answer these questions, we need to ask how many of the herbaceous species will grow in the southeastern United States? How many are adapted to this environment characterized by a long, hot summer, nearly 50 inches of rainfall annually, and scattered low winter temperatures and freezes that limit winter growth and kill many tropical species? Hot, humid summers, that foster plant diseases and droughts accentuated by the sandy soils, eliminate many species that might otherwise be adapted. With the exception of the black belt, river valleys, the Mississippi Delta, and limited muck areas, most of the soils are very low in the fertilizer elements required for plant growth.

The natural distribution of the plant kingdom's species will indicate which are likely to succeed in the Southeast. The many grasses and forbes that form a natural understory of southern forests with an open canopy are adapted but will produce less than 2 $Mg\ ha^{-1}$ ($Mg\ ha^{-1}$ x 0.446 = tons $acre^{-1}$) of dry matter per year without shade.[2] Under a 40% tree canopy, they will yield less than 0.5 $Mg\ ha^{-1}$. Most species well adapted to the northern half of the U.S. and Canada may be adapted to southern winters, but will not survive southern summers. Most tropical perennials will not survive the freezes experienced periodically in much of the South. Thus, poor adaptation eliminates a large percentage of the angiosperms that might efficiently produce biomass energy in the South.

Of the herbaceous angiosperms that are adapted to the South, many can be eliminated on the basis of their estimated annual yield. They are too small, or they utilize only a small part of the growing season.

Biomass energy sources must be dependable. Species that fail to yield well or survive unusual climatalogical or biological stresses can be eliminated without extensive testing.

What additional characteristics should the remaining herbaceous angiosperms possess? What tests should they pass?

If perennial, they must dependably survive the climatic extremes to be expected. Occasional low temperatures will determine the northern limit for many tropical species. They must also be able to tolerate the defoliation required as they are harvested.

Species vary greatly in the defoliation frequency that they can tolerate. Annual dry matter yields of heavily fertilized (672 kg ha^{-1} N + P and K) Coastal bermuda, Cynodon dactylon, can be maximized (20 Mg ha^{-1}) with a 6 to 12 week cutting frequency.[3] Cutting Coastal only once after 24 weeks of growth significantly reduced annual biomass yields 27%. On the other hand, cutting Napiergrass (elephantgrass), Pennisetum purpureum, twice a year instead of only once at Ona, Florida, reduced the total annual yield 70%.[4] In Puerto Rico for 3 years, Alexander cut sugar cane, Saccharum officinarum L., every 2, 4, 6, and 12 months and obtained average annual dry matter yields of 8, 22, 36, and 65 Mg ha^{-1}.[5] He also found that Napiergrass outyielded sugar cane when cut at 6 month intervals but yielded less than sugar cane when cut at 12 month intervals.

Ease of establishment, especially in annuals, will be very important. Seed propagation is usually a must. To qualify as good biomass species, seed yields must be good and the seed must be easily harvested. Establishment from seed must be dependable and the cost must be low. To reduce the weed potential of such species, their seeds should contain no hard-seeds that can lie in the ground many years and still germinate. Crotolaria spectabilis, a legume planted 40 years ago to add nitrogen to the soil, is a poisonous plant with poisonous hard seeds. Although never allowed to reseed since planted in 1940, volunteer plants from hard seeds still appear in my garden. Corn harvested with a combine from fields containing volunteer Crotolaria plants contained Crotolaria seed. Such grain killed many chickens to which it was fed until feed mills screened the Crotolaria seed from the corn.

The infertility of virgin Coastal Plain soils was well demonstrated in 1946 when 23 grasses and five legumes with good forage potential failed to produce 2.0 Mg ha^{-1} of dry matter per year on virgin Tifton loamy sand (one of the best) without fertilizer. When fertilized with 600 kg ha^{-1} of 4-8-6 fertilizer per ha, the three top grasses – weeping love, Eragrostis curvula; Coastal bermuda, Cynodon dactylon; and Pensacola bahia, Paspalum notatum, yielded 4.9; 2.0; and 1.5 Mg ha^{-1} of dry matter. The three top legumes, Indigofera hirsuta; Crotolaria lanceolata; and Lespedeza cuneata (Sericea) yielded 8.6; 5.9; and 0.3 Mg ha^{-1} with the same fertilization.

Obviously, the sandy soils of the South must be fertilized if they are to produce much biomass. The species used must be able to respond to fertilization. Hughes et al.[6] applied 50 kg ha^{-1} of N plus adequate P and K to the native grasses and forbes growing on southern sandy soils and only increased dry matter yields to 2.2 Mg ha^{-1}, demonstrating that the native vegetation is not an efficient source of biomass.

The highest yielding sorhgum, Sorghum bicolor, and pearl millet, Pennisetum americanum, hybrids, growing on soils that had been producing well fertilized, cultivated crops for many years, produced (with 130 kg ha^{-1} of N plus P and K) dry matter up to 25 Mg ha^{-1} at Tifton, Georgia, and 19 Mg ha^{-1} at Experiment, Georgia[7]. On unlimed, sandy soils, pearl millet hybrids will usually outyield sorghum. Adding lime will improve sorghum yields but usually fails to make them equal pearl millet yields. On heavy soils and on wet soils, sorghum will outyield pearl millet. Sorghum also ratoons better and grows faster than pearl millet at low temperatures.

O'Hair et al. measured dry root yields of a number of the most promising root crops that might be grown in Florida.[8] Topping the list, producing approximately 10 Mg ha^{-1} dry matter, were sweet potato, Ipomea batatas; fodder carrots, Daucus carota; fodder beets, Beta vulgaris; and cassava, Manihot esculenta. Although important in Brazil's biofuel program, cassava cannot be grown in most of the South because it will winterkill.

Recognizing that 8% of Florida's land area was occupied by bodies of water, lakes, canals, rivers, etc., Reddy et al. studied extensively the biomass potential of a number of aquatic plants that could be grown there.[9] They found emergent aquatic plants most productive and reported annual dry matter yields of 25, 8-23 and 16 Mg ha^{-1} for wild rice, _Zizania_ _aquatica_; cattails, _Typha_ _latifolia_; and soft rush, _Juncus_ _effusus_, respectively. Among the floating aquatic plants, they found water hyacinth, _Eichornia_ _crossipes_; penny wort, _Hydrocotyle_ _umbellata_; and water lettuce, _Pistia_ _stratiates_, the most productive with respective annual dry matter yields of 9-20, 5-10 and 3-5 Mg ha^{-1}. All of these species had the less efficient C_3 photosynthetic activity.

Many herbaceous species with forage potential also have biomass potential. They have also been studied more than most exotic herbaceous species. Prine and Mislevy[10] estimated dry matter yields of 27 field, forage, and fiber crops based on their 15-25 years of studying these crops in Florida as follows:

warm season grasses 6-24 Mg ha^{-1} yr
warm season legumes 2-7 Mg ha^{-1} yr
cool season grasses 2-9 Mg ha^{-1} yr
cool season legumes 2-7 Mg ha^{-1} yr
Florida 77 alfalfa, _Medicago_ _sativa_ 6-16 Mg ha^{-1} yr

Although kenaf yielded 12-14 Mg ha^{-1} yr of dry matter, its extreme susceptibility to nematode injury and its harvesting problems eliminated it as a suitable species for the production of biomass energy.

Prine and Mislevy[10] point out that shrub legumes such as leucaena, _Leucaena_ _leucocephala_, require several years to reach their maximum growth rate and that they are not winterhardy. Napiergrass is much more winterhardy than leucaena and goes into its maximum growth rate the first year it is planted.

Where they will survive as perennials throughout Florida and as far north as Tifton, Georgia, Napiergrass and sugar cane appear to be the highest yielding herbaceous species for biomass production presently available. Prine et al. agree and point out that, where freezing temperatures kill the tops to the ground, Napiergrass should yield more than sugar cane[4]. They report dry matter yields, when cut at the end of the growing season, to 48 to 55 Mg ha^{-1} yr, but suggest that these yields from small plots surrounded by sod would have been lower in solid stands.

At Tifton, Merkeron Napiergrass, over a 3-year period in solid stands surrounded by adequate Napiergrass borders, yielded about 30 Mg ha^{-1} yr of dry matter. Merkeron, a single clone, is the best of a number of F_1 hybrids between a tall Napiergrass plant and a leafy, short-internode plant only about half as tall. Both parents were resistant to the 'eyespot' disease caused by Helminthosporium sachari. Annual nitrogen applications of 134, 268, and 400 kg ha^{-1}, with adequate P and K made to replicated small plots, failed to show significant differences due to lateral movement of the fertilizer and the extensive root system of the grass. The average yield of 30 Mg ha^{-1} yr is considered the response to an average of about 268 kg ha^{-1} yr of N plus P and K. This forage contained 1% N, 0.2% P, and 1% K. Thus, the 300 kg ha^{-1} each of N and K and 60 kg ha^{-1} of P must be replaced eventually if such yields of forage are removed.

Napiergrass will outyield other herbaceous species considered if it is cut only once a year. Its ability to grow efficiently throughout the warm growing season, avoid lodging, and hold its leaves, even after they die, account for its high yields. Napiergrass is adapted to a wide range of soils and is able to extract more nutrients from a soil than most grasses. Merkeron Napiergrass planted in .9 m rows in 1978 but not fertilized since 1980 produced 8.0 Mg ha^{-1} of dry matter in 1984. The N, P, and K content of this dry matter was 0.4, 0.2, and 1.5%, respectively. Thus, Napiergrass not only yields more but produces more dry matter per unit of nitrogen and other plant nutrients than most herbaceous species. The border row of this test, with no applied fertilizer and competition with low growing weeds on one side, produced 45.0 Mg ha^{-1} if calculated on a 0.9 m row width, the area occupied by the base of plants. This calculated yield indicates that small plot yields of Napiergrass can be very misleading if they are not surrounded with an adequate border of the same species.

Like sugar cane, Napiergrass must be planted vegetatively. This poses no serious problem where it will survive the winters and produce biomass over a number of years. Its seeds are not suitable for commercial establishment. They are windborne and can make Napiergrass a weed in the tropics where the seeds can mature.

For biomass that can be harvested 2 to 4 times during the frost-free growing season without reducing annual yields, Pensacola bahiagrass and the hybrid bermudagrasses such as Coastal, Tifton 44, and Tifton 78 will be hard to beat. They are dependable perennials, with disease resistance, drought tolerance, and high dry matter yield potential producing up to 24 Mg ha^{-1} yr. They utilize fertilizer efficiently and their dry matter can be easily harvested and stored with hay making equipment.

Among the annuals, sorghum and pearl millet seem to offer the greatest promise. The best of the commercial hybrids on fertile soil, with 150 kg ha^{-1} yr of N plus other fertilizer nutrients as needed, can produce up to 25 Mg ha^{-1} yr of dry matter with two harvests per season. These must be planted each spring and will not use the entire growing season as efficiently as Napiergrass. They may experience establishment problems, will be more vulnerable to pest and environmental stresses, and will not resist lodging as well as Napiergrass. Harvesting and storage procedures for Napiergrass should also apply to sorghum and pearl millet hybrids.

For defoliation at 5 to 6 week intervals, alfalfa should lead the list of legumes. Annual dry matter yields of 15 or more Mg ha^{-1} yr will be possible and varieties such as Florida 77 may maintain good stands for several years. Costs of production will exceed those of the grasses listed above and the dry matter produced will be worth more as animal feed, except when damaged by poor hay making weather.

Finally, we may ask what the plant breeder can do to increase the yield and the efficiency of herbaceous species for biomass energy production. Given time and continued support, the plant breeder can help evaluate new species yet to be found. He can increase the dry matter yields and the efficiency of the best species now available. Here are two specific examples: In Africa where sorghum and pearl millet originated, there occur short-day sensitive landraces that will not initiate seed head primordia until the day length is 12 hours or less. Planted in the South, these landraces do not head until November and, instead of heading and ceasing to grow in August as current hybrids do, they continue active vegetative growth until frost. When cut only once, short-day hybrids carrying the genes of the short-day landraces will yield much more dry matter (up to 40+ kg ha^{-1} yr) than current hybrids.[11] We have developed such short day hybrids and need only to develop commercial seed production in the tropics to put them into commercial use.

In 1941, we crossed pearl millet with Napiergrass and obtained sterile hybrids, some of which yielded more than either parent.[12] In 1966, Powell and Burton suggested the commercial production of hybrid pearl millet x Napiergrass seed on cytoplasmic male sterile pearl millet Tift 23A, thus permitting seed propagation to replace the laborious and costly propagation of such hybrids.[13] More recently, Wayne Hanna has been producing and evaluating such hybrids that yield up to 36 Mg ha^{-1} yr and show much promise as perennials for south Florida and the tropics. As soon as commercial seed production can be developed in the tropics, these sterile hybrids will become another promising source of biomass for energy.

Plant breeding that increases yields can also modify the chemical composition of a plant.[3] Plant breeding has increased the digestibility of bermudagrass 12% and improved the performance of cattle consuming it some 30%.[14] It seems reasonable to assume, therefore, that plant breeding can modify herbaceous species so that microbes converting them to alcohol or some other energy source can increase the percentage produced.

Readers of this paper may well ask, "Can the annual yield of biomass for energy be increased by growing a winter annual with one of the summer growing species just considered?" Forty years of studying such double cropping systems permits us to answer, "Yes, but not much." Variations in winter weather, that are unpredictable, have made dry matter yields from double cropped winter annuals range from zero to about 4 to 6 $Mg\ ha^{-1}$. In at least 2 years out of 5, October plus November rainfall is inadequate to establish winter annual grasses and legumes in summer perennial sod crops at Tifton, Georgia. In at least half of the years, temperatures in one month will be too low for the best winter annuals to make noticeable growth. In the past two winters, many of the winter annual crops have been killed by the freeze and have given no return on the investment in seed, fertilizer, planting, etc. Most winter annuals make their maximum growth in the spring and, when growing in association with summer perennials, reduce their yields. The best use for winter annuals seems to be with summer annuals. Pearl millet or sorghum in the summer, followed by cereal rye or annual ryegrass, probably have the best potential in south Georgia. Crimson and arrowleaf clover are two of the most promising winter annual legumes for a double cropping system.

REFERENCES

1. G. W. Burton. Food resources in the plant kingdom. Soil and Crop Soc. of Fla. Proc. 28:222-232. 1968.
2. L. K. Halls, B. L. Southwell and F. E. Knox. Burning and grazing in Coastal Plain forests. Ga. Coastal Plain Exp. Sta. Bul. 51, pp. 26-27. 1952.
3. G. W. Burton, J. E. Jackson and R. R. Hart. Effects of cutting frequency and nitrogen on yield, in vitro digestibility and protein, fiber, and carotene content of Coastal bermudagrass. Agron. Jour. 55:500-502. 1963.
4. G. M. Prine, P. Mislevy, A. Shiralipour and P. Smith. Elephantgrass, energy crop for humid tropics. Proc. 4th ann. Solar and Biomass Energy Workshop. pp. 124-127. 1984.

5. A. G. Alexander. Tropical grasses as a renewable energy source. 3rd Ann. Solar and Biomass Workshop. pp. 209-212. 1983.

6. R. H. Hughes, J. B. Hilton and G. W. Burton. Improving forage on southern woodlands. Proc. 9th Int. Grassland Cong., pp. 1305-1307. 1965.

7. Georgia Agronomists. 1982.

8. S. K. O'Hair, S. J. Locascio, R. R. Forbes, J. M. White, D. R. Hensel, J. R. Shumaker and J. M. Dangler. Root crops and their biomass potential in Florida. Proc. Soil and Crop Sci. Soc. of Fla. 42:13-17. 1983.

9. K. R. Reddy, D. L. Sutton and G. Bowers. Freshwater aquatic plant biomass production in Florida. Proc. Soil and Crop Sci. Soc. of Fla. 42:28-40 (1983).

10. G. M. Prine and P. Mislevy. Grass and herbaceous plants for biomass. Proc. Soil and Crop Sci. Soc. of Fla. 42:8-12. 1983.

11. H. R. Sumner, R. E. Hellwig and G. E. Monroe. Methods to accelerate the drying rate of hybrid pearl millet and interspecific hybrids. Proc. 4th ann. Solar and Biomass energy Workshop, pp. 120-123 1984.

12. G. W. Burton. Hybrids between Napiergrass and cattail millet. Jour. Hered. 35:226-232. 1944.

13. J. B. Powell and G. W. Burton. A suggested commercial method of producing an interspecific hybrid forage in Pennisetum, Crop Sci. 6:378-379. 1966.

14. G. W. Burton. Registration of Coastcross-1 bermudagrass. Crop Sci. 12:125. 1972.

CRUCIFEROUS AND ROOT CROPS FOR YEAR-ROUND

BIOMASS PRODUCTION

S. K. O'Hair, J. M. Dangler, P. Everett, R. B. Forbes,
S. J. Locascio, S. M. Olson, J. R. Shumaker and J. M. White

Department of Vegetable Crops
University of Florida/IFAS
Gainesville, Florida 32611

ABSTRACT

Root and tuber crops have long been known for their excellent potential as energy crops. Their ability to concentrate starch and sugar in enlarged storage organs results in a plant that is readily convertible to bioenergy. Sweetpotato, a warm season crop, was planted in six Florida locations with different climates and soil types. Yields varied among locations with the highest dry matter productivity rate being 136 kg ha^{-1} day^{-1} by 160 days after planting (DAP) at Sanford. Cool season cruciferous crops were planted at three Florida locations. Among these crops, rape and turnip were the most promising. Dry matter productivity of 'Winfred forage' rape was 142 kg ha^{-1} day^{-1} at Quincy by 80 DAP. More than one planting per annum must be made using cruciferous and root crops, to achieve a year-round supply of fresh biomass. A succession of one sweetpotato crop and two turnip crops within one year yields 41 Mg ha^{-1} total biomass.

Keywords. Sweetpotato, turnip, rape, biomass, growth.

INTRODUCTION

Evaluation of plants including root crops for bioenergy production has been of interest for many years.[1] Once the evaluation of potential plants gained interest, it was soon realized that many could be used for more than fuel.[2,3] There are numerous chemicals that can be derived from plants, thus adding value to biomass production. In Florida, biomass production has been noted as having great potential.[4] This is attributed both to climate and land availability. Among the first crops to be evaluated were cassava, fodder carrot, fodderbeet, radish, rutabaga, and sweetpotato.[5,6]

The expected potential growth rate for biomass production in C3 plants, including the above mentioned crops, is 200 kg ha^{-1} day^{-1}.[7] Further, it was suggested that the potential growth rate differences among species and within species was small. Therefore it is necessary to select and evaluate species that are adapted to a specific environment, such as cool and warm weather.

Root crops have a special advantage in their ability to produce storage organs, which may enhance the photosynthesis rate during the enlargement stage.[8] Thus, their yield potential is outstanding. Additionally, storage organs provide a stabilizing effect whereby the optimal harvest time is flexible. Storage organ development also presents extended storage in the field without having appreciable losses.

Currently it is believed that the greatest yield increases will not be from improving the annual biomass production potential of short season annual cultivars, but rather by maximizing the number of crops produced on a given area of land.[9] Thus, yield per day rather than per crop is more important.

The optimum time for sweetpotato harvest has been widely studied. Chen and Yang[10] reported that roots continue to enlarge and yields increase in a near linear path up to 200 days after planting (DAP). Nearly all sweetpotato studies have concerned marketable root production. Therefore, little is known of total plant biomass production. Agata and Takeda[11] reported yields of 207 kg ha^{-1} day^{-1} in sweetpotato, which surpasses the projected maximum of van der Have.[14] Although sweetpotato yields in Florida have not been this high, it remains a most promising bioenergy crop.[12]

Since sweetpotato is considered to be a warm season crop, an alternate plant group was needed for securing a year-round supply of fresh biomass. The plants in the crucifer family fit this need as they flourish in cool weather, have a very wide range of adaptability, and have the ability to rapidly produce a crop. Most of what is known about the bioenergy potential of cool season crops is from forage trials. Rape and turnip biomass yields in Louisiana were as high as 10.3 Kg of dry forage ha^{-1} [13] Assuming that the growth of the crop was at least 100 days, this would amount to 103 kg ha^{-1} day^{-1}. This low value could be expected under cool short day conditions.

To better evaluate the potential of a multi-crop system in Florida, trials of sweetpotato and cruciferous crops were established at several locations. This paper reports the results from those trials.

MATERIALS AND METHODS

Sweetpotato trials were established at six locations in the spring of 1982 and 1983 to cover representative environmental conditions that can be found throughout Florida (Figure 1). In 1982, two cultivars, 'Morado' and 'Centennial', were planted using stem cuttings for the former and slips for the latter. In 1983, 'Centennial' and 'Jewel' were planted from slips. Planting densities were approximately 35,880 plants ha^{-1}. Fertilizer at the rate of 45 kg N, 90 kg P and 90 kg K ha^{-1} was applied once prior to planting and again approximately one month after planting. All plantings were designed as randomized complete blocks to allow for five monthly harvests to begin at 100 DAP and four replications. Irrigation was applied as needed. At each harvest, two plants were selected from each plot for vine fresh and dry weight measurements. Plant samples were oven dried to a constant weight at 80 C. Storage roots from 20 plants in each plot were harvested and fresh weight was measured. A root sub-sample of at least 300 g was taken to determine root dry weight percentage.

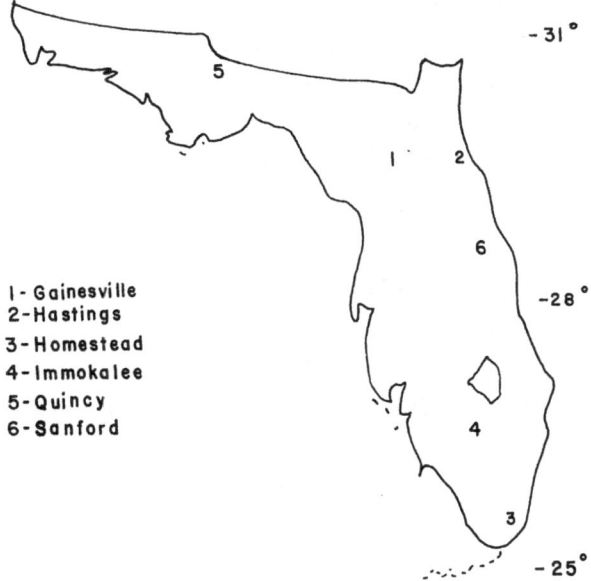

1- Gainesville
2- Hastings
3- Homestead
4- Immokalee
5- Quincy
6- Sanford

Figure 1. Biomass trial locations

175

Cruciferous crop trials were established between November and January at Homestead, Quincy, and Sanford to cover the extremes of the state (Figure 1). Fertilizer was applied at the same rate as previously stated for sweet-potato and trials were irrigated as needed. The planting and harvest schedules varied among the locations. At harvest, plants were divided into enlarged storage organs and plant tops. A sub-sample of at least 300 g from both parts was taken to estimate plant dry matter. Even though there were several entries in each trial, only the three highest yielding cultivars were used in this paper and in the development of the year-round production estimates.

All yields were converted to dry kg ha^{-1} day^{-1} to better equate yields from the various trials, since plant age at harvest varied with location and trial. This transformation also aided in determining the most efficient use of the land. Data were analyzed using general linear models and analysis of covariance with least squares regression.[14]

RESULTS AND DISCUSSION

In both 1982 and 1983, there were significant differences in sweet-potato root, top and total plant yields depending on the trial location (Table 1). Cultivar differences were observed only in 1982 when 'Morado' and 'Centennial' were compared. A significant location x cultivar interaction was recorded in both years for sweetpotato yields, with the exception of the plant top variable in 1982. Therefore, yields among locations were not combined. With the exception of the plant top variable in 1982, the linear and quadratic forms of plant age in DAP were significant in explaining variability. For this exception in 1982, only the linear form was significant. Thus, for total sweetpotato biomass production, there was a peak in productivity between the first and the last harvest.

'Morado' plant top production rate was highest by 100 DAP with as much as 80 kg ha^{-1} day^{-1} in the Hastings trial (Figure 2). By 220 DAP additional leaf production had nearly stopped. Similar trends were noted in 'Jewel' (Figure 3) and 'Centennial' (Figure 4). However, depending on the location where the latter two were grown, the drop in production did not occur until after the 130 DAP harvest. 'Morado' root production tended to be highest between 130 and 190 DAP, with productivity at Gainesville and Sanford being as high as 89 kg ha^{-1} day^{-1} between 100 and 190 DAP (Figure 2). At Homestead and Quincy, the peak in productivity appeared to be delayed by 30

176

Table 1. The effects of location, cultivar, year, and plant age on sweetpotato biomass yield.

	1982			1983	
Source of Variation	F Value Significance		Source of Variation	F Value Significance	
ROOT					
Location	***		Location	***	
Cultivar	***		Cultivar	NS	
Location x Cultivar Interaction	***		Location x Cultivar Interaction	*	
Plant age linear	***		Plant age linear	***	
Plant age quadratic	***		Plant age quadratic	***	
TOP					
Location	***		Location	***	
Cultivar	*		Cultivar	NS	
Location x Cultivar Interaction	NS		Location x Cultivar Interaction	NS	
Plant age linear	***		Plant age linear	**	
Plant age quadratic	NS		Plant age quadratic	**	
TOTAL PLANT					
Location	***		Location	***	
Cultivar	*		Cultivar	NS	
Location x Cultivar Interaction	*		Location x Cultivar Interaction	*	
Plant age linear	***		Plant age linear	***	
Plant age quadratic	**		Plant age quadratic	***	

F values were significant at the 5% (*), 1% (**), 0.1% (***) or not significant (NS).

days. 'Jewel' and 'Centennial' root productivity was highest at Sanford with a high of 119 and 110 kg ha^{-1} day^{-1} by 160 DAP, respectively (Figures 3, 4). In 'Jewel' a peak in productivity appeared to occur between 130 and 190 DAP at most locations (Figure 3). With the exception of the Sanford and Quincy trials, the peak in 'Centennial' root production appeared to occur between 100 and 160 DAP (Figure 4). This could be an indication that 'Centennial' has an earlier maturity than 'Jewel'.

Total plant productivity as high as 136 kg ha^{-1} day^{-1} was recorded for 'Centennial' at Sanford by 160 DAP (Figure 5). Total plant productivity for 'Morado' was 120 kg ha^{-1} day^{-1} at Hastings by 130 DAP (Figure 6). These

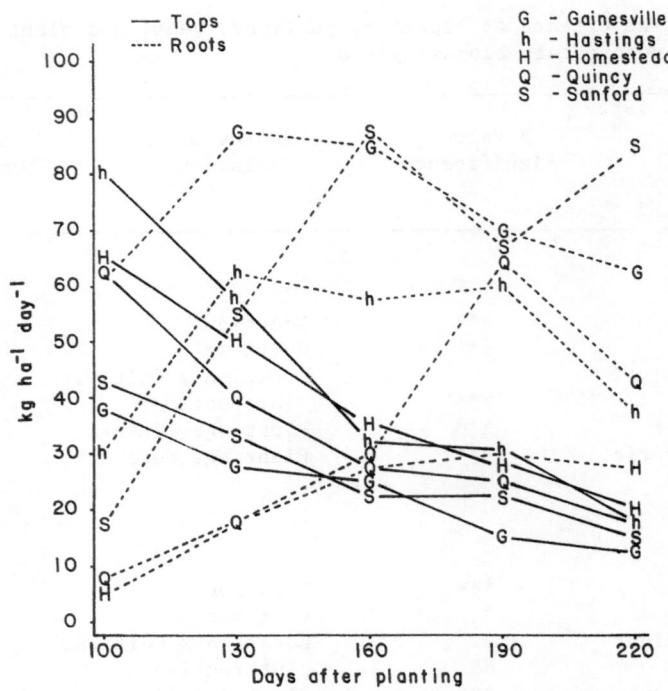

Figure 2. Growth rate of Morado sweetpotato tops and roots at five Florida
locations as affected by plant age

values were considerably less than the 200 kg ha^{-1} day^{-1} projected by van
der Have[14] and the 207 recorded in Japan.[11] 'Morado' productivity at
Gainesville and Hastings declined after the 130 DAP harvest, while it increased in the Quincy and Sanford trials. All of the increase was attributed to storage organ production, since leaf production rate was declining.
A decline was evident by 220 DAP among all locations with the exception of
Sanford. There was a more evident downward trend in productivity for
'Jewel' and 'Centennial' following the 130 DAP harvest (Figures 5, 7). The
peak in the Sanford trial appeared to be delayed by 30 days in both cultivars. This delay was probably correlated with the higher yield at Sanford.

'Winfred forage' rape produced 142 kg ha^{-1} day^{-1} by 80 DAP in the
Quincy trial (Figure 8). Dry matter production by 'White knight' turnip at
Sanford was 82 kg ha^{-1} day^{-1} by 83 DAP, while at Quincy production was 104
kg ha^{-1} day^{-1} by 80 DAP. 'Civasto' turnip, the best crop at Homestead,
produced 49 kg ha^{-1} day^{-1} by 100 DAP, while the maximum production rate for
this crop was 121 kg ha^{-1} day^{-1} by 60 DAP at Quincy. Low cruciferous crop
yields at Homestead could be attributed to warm winter temperatures that

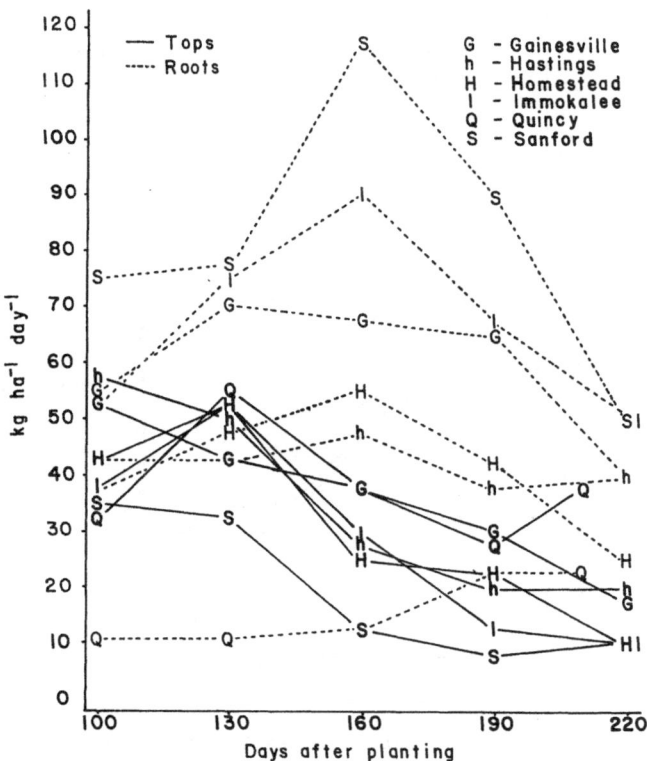

Figure 3. Growth rate of Jewel sweetpotato tops and roots at six Florida
locations as affected by plant age

were not favorable for good growth in these crops and an armyworm attack of
the foliage.

When year-round production is considered, it can be shown that as much
as 41 Mg ha^{-1} annum^{-1} can be produced (Table 2). This could be accomplished
as demonstrated at Sanford by producing one crop of 'Centennial' sweetpotato for 160 days in the warm season, followed by fall and spring plantings of 'White knight' turnip lasting 80 days each. Potential yields at
Homestead and Quincy appeared to be lower. Low sweetpotato yields at
Quincy were attributed to heavy summer rains that probably depleted the
soils of natural and supplemental fertilization. Since all projected yields
are lower than the estimated potential of 200 kg ha^{-1} day^{-1}, it appears as
though additional gains can be made. This may be done by identifying better
adapted cultivars and by altering the planting densities. Additional
fertilization may also increase the productivity; however, this must be
handled carefully, since energy inputs would increase with increased fertilizer application.

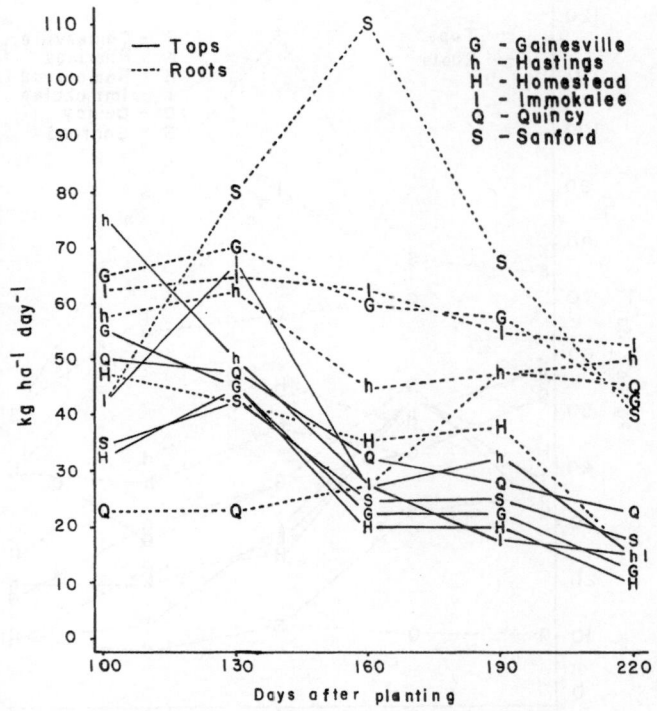

Figure 4. Growth rate of Centennial sweetpotato tops and roots at six Florida locations as affected by plant age

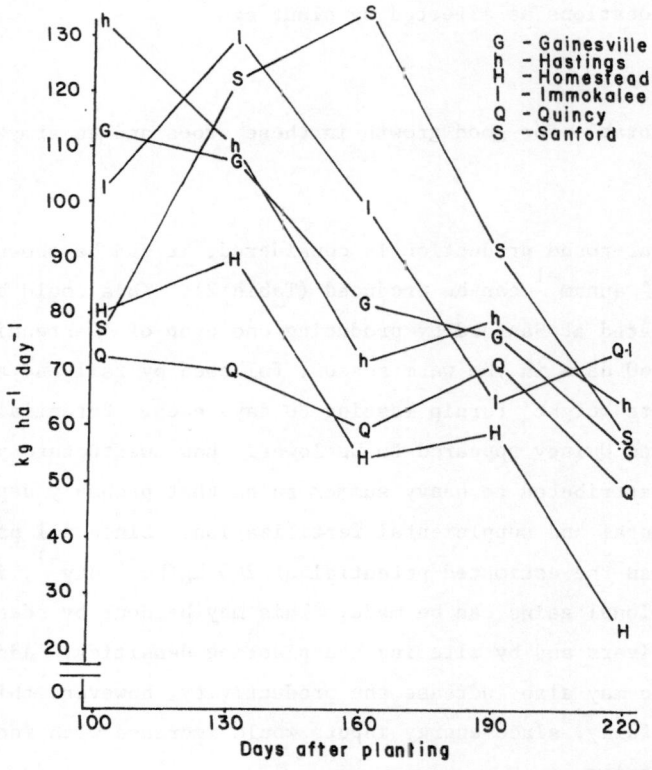

Figure 5. Centennial sweetpotato total biomass production rate at six Florida locations as affected by plant age

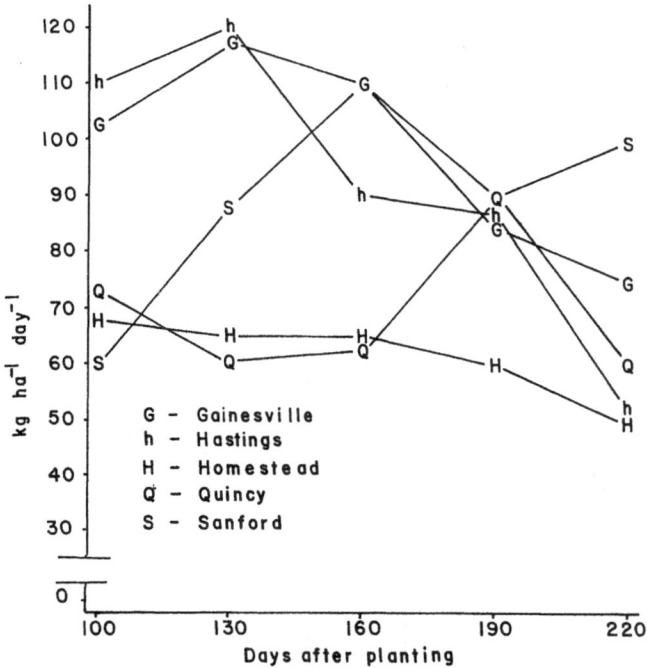

Figure 6. Morado sweetpotato total biomass production rate at five Florida locations as affected by plant age

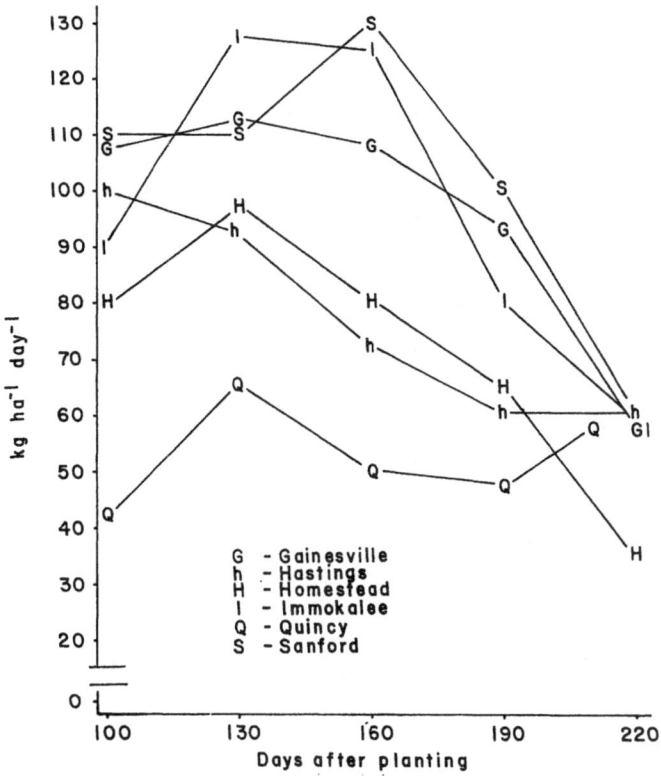

Figure 7. Jewel sweetpotato total biomass production rate at six Florida locations as affected by plant age

Table 2. Estimated biomass production based on a year-round production of 320 to 330 days.

	Homestead			Location Sanford			Quincy		
Crop	DAP	kg ha^{-1} day^{-1}	Mg ha^{-1}	Crop	DAP	kg ha^{-1} day^{-1}	Mg ha^{-1}	Crop	DAP	kg ha^{-1} day^{-1}	Mg ha^{-1}
Jewel sweetpotato	130.	98.0	12.7	Centennial sweetpotato	160	136.0	21.8	Centennial sweetpotato	160	61.0	9.8
Centennial sweetpotato	100	76.0	7.6	White knight turnip	80	120.0	9.6	Winfred forage rape	80	142.0	11.4
Civasto turnip	100	49.0	4.9	White knight turnip	80	120.0	9.6	Winfred forage rape	80	142.0	11.4
t ha^{-1} annum^{-1}		25.2				41.0				32.6	

DAP = days after planting.

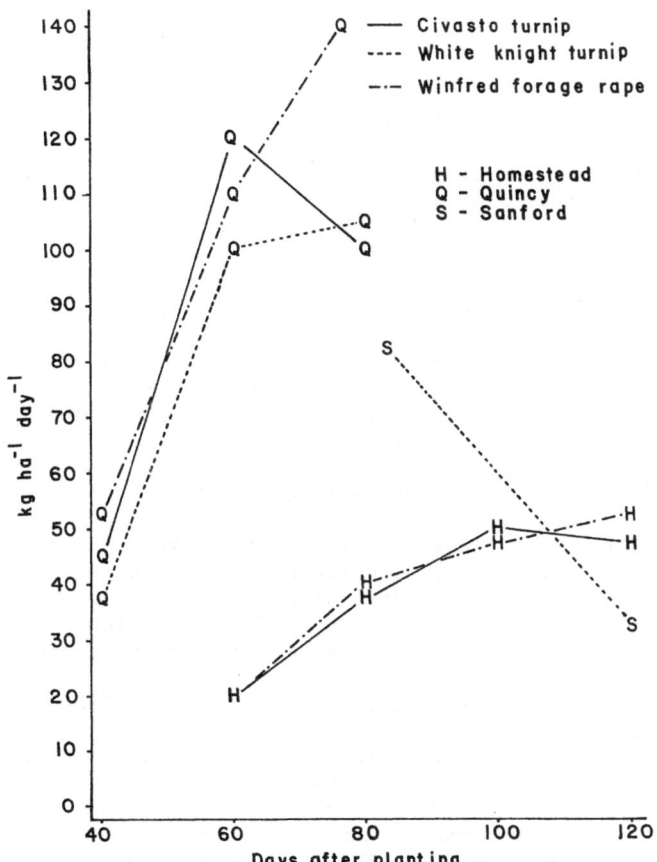

Figure 8. Cruciferous crop total biomass production rate at three Florida
locations as affected by plant age

ACKNOWLEDGEMENTS

Florida Agricultural Experiment Stations Journal Series No. 6591. This
paper reports results from a project that contributes to a cooperative
program between the Institute of Food and Agricultural Sciences of the
University of Florida and the Gas Research Institute, entitled "Methane from
Biomass and Waste". To convert values from kg ha^{-1} to lbs A^{-1} multiply the
value by 0.892. To convert values from Mg ha^{-1} to Mg A^{-1} multiply the value
by 0.446.

REFERENCES

1. J. G. da Silva, E. Serra, J. R. Moreira, J. C. Concalves and J. Goldemberg. Energy balance for ethyl alcohol production from crops. Science. 201:903–906. 1978.
2. E. S. Lipinsky. Chemicals from biomass: petrochemical substitution options. Science 212:1465–1471. 1981.
3. B. O. Palsson, S. Fathi-Afshar, D. F. Rudd and E. N. Lightfoot. Biomass as a source of chemical feedstocks: an economic evaluation. Science. 213:513–517. 1981.
4. W. H. Smith and M. L. Dowd. Biomass Production in Florida. Journal of Forestry. 79:508–511. 1981.
5. S. K. O'Hair, R. B. Forbes, S. J. Locascio, J. R. Rich and R. L. Stanley. Starch and glucose distribution within cassava roots as affected by cultivar and location. HortScience. 18:735–737. 1983a.
6. S. K. O'Hair, S. J. Locascio, R. R. Forbes, J. M. White, D. R. Hensel, J. R. Shumaker and J. M. Dangler. Root crops and their biomass potential in Florida. Proc. Soil Crop Sci. Soc. Fla. 42:13–17. 1983b.
7. D. J. van der Have. Plant breeding perspectives. In: J. Sneep and A. J. T. Hendriksen (Eds.) O. Holbeck (Coed.). Center for Agricultural Publishing and Documentation. Wageningen p. 53. 1979.
8. C. A. de Vries, J. D. Ferwerda and M. Flach. Choice of food crops in relation to actual and potential production in the tropics. Neth. J. Agr. Sci. 15:241–248. 1967.
9. L. T. Evans. The natural history of crop yield. Amer. Scientist 68:388–397. 1980.
10. L. H. Chen and C. C. Yang. Optimum starting date for the harvest of sweet potatoes. Trans. Amer. Soc. Agr. Eng. 23:284–287. 1980.
11. W. Agata and T. Takeda. Studies on matter production in sweet potato plants. 1. The characteristics of dry matter and yield production under field conditions. J. Fac. Agr. Kyushu Univ. 27:65–73. 1982.
12. J. M. Dangler, S. J. Locascio and L. H. Halsey. Sweet potato for biomass. Biomass 4:253–261. 1984.
13. J. M. Crane. National winter rapeseed variety trial 1980–81. Mimeo Report, Crop Sci. Dept., Oregon State Univ. 12 pp. 1981.
14. Anonymous. SAS users guide: statistics 1982 edition. SAS Institute Inc. Cary, North Carolina. pp 584. 1982.

BIOMASS PRODUCTION FROM TARO (<u>COLOCASIA</u> <u>ESCULENTA</u>) IN SUBTROPICAL WETLANDS

G. H. Snyder

Everglades Research and Education Center
University of Florida/IFAS
Belle Glade, Florida 33430

S. K. O'Hair

Tropical Research and Education Center
University of Florida/IFAS
Homestead, Florida 33031

ABSTRACT

In certain locations it may be more desirable to cultivate water tolerant crops in natural wetlands than to incur the capital and operating ex-
penses and the environmental consequences associated with providing sufficient drainage for cultivation of crops that are adversely affected by
periodic flooding. Taro (<u>Colocasia</u> <u>esculenta</u>) is being evaluated for biomass production under wetland conditions in southern Florida. Research has
centered on assessing biomass production of taro plant components under
various cultural conditions. Taro biomass from March plantings was comprised mainly of leaves until about July, after which time corms enlarged
and cormels formed. Close spacing (20 x 20 cm) suppressed cormel production
but increased overall per unit area biomass production (10 Mg ha^{-1}). Foliage harvests at two to four month intervals were better for sustained maximum production than monthly harvests. Taro cultivars were evaluated for
biomass production on mineral and organic soils. Production generally was
greater on organic soils. Cultivar 'Bun Long' produced the highest total
biomass yield on Histosol (20 Mg ha^{-1}). On the Spodosol, Islena had the
best yield (13 Mg ha^{-1}).

Keywords. Bioenergy, planting density, Histosol, aquatic plant, energy
crop.

INTRODUCTION

Many tropical and subtropical regions are poor in terms of conventional energy reserves, such as coal, oil and gas, yet they are relatively wealthy in terms of solar radiation, water, and extent of the frost free growing season. It is likely that these same environmental conditions were required for the accumulation of biomass that eventually metamorphosed into conventional energy sources. Therefore, it is reasonable to determine the extent to which humid tropical and subtropical regions can be used for the economic production of biofuels.

Wetlands are common features of humid regions, covering over 230 million hectares worldwide.[1] Generally these sites are among the last to be cultivated in a region, even though their potential for crop production often is recognized. In almost all cases, drainage is the first step in developing wetlands for agricultural use. In fact, the term "reclamation" essentially has come to be synonymous with drainage. However, many now recognize that wetlands play a useful and important role in the environment. They are sites for water storage, aquifer recharge, water purification, and provide habitat for many types of wildlife.[2] Permanent drainage largely eradicates the beneficial aspects of wetlands. In most cases drainage requires very large capital outlays at the outset, and high operating expense for maintenence of the system and fuel charges.

In Florida, many wetland sites have been reclaimed for agriculture. However, questions have been raised recently about the overall benefits of wetland drainage.[3,4,5] Money has been appropriated to "restore" the Kissimmee river valley. Proposals have surfaced for requiring on-farm impoundment of agricultural drainage.[2]

Utilization of flood-tolerant crop plants would allow production on wetland sites without imposing continuous drainage. Drainage still might be needed for certain cultural practices, although most practices at present can be conducted under flooded conditions. The ideal crop would be one that can tolerate flooding, but that does not absolutely require it. This would be particularly useful in a region like the Florida Everglades, which has fairly distinct wet and dry seasons.

Unfortunately, few aquatic or semi-aquatic plants are commercially cultivated. Rice (<u>Oryza</u> <u>sativa</u>) is the most widely grown aquatic crop. Taro (<u>Colocasia</u> <u>esculenta</u>) probably ranks second. In most locations, taro

has a number of advantages as a bioenergy relative to rice. Taro has fewer
pest problems, total biomass production probably is greater, and much of
this biomass is in the form of easily convertible materials. Several drawbacks to taro include an 8 to 10-month minimal growing season, the necessity
for vegetative propogation, and the paucity of production information.

Taro is propagated vegetatively from pieces of the central corm (an
enlarged stem), and from cormels, which arise from lateral buds off the
corm. In Hawaii it is propogated from "hulis," which are composed of about
15 cm of the lower petiole and approximately 1 cm of the upper corm. These
hulis are pressed into a flooded, prepared seedbed and permitted to root.
Plants generally are spaced 50 to 70 cm apart. Although taro survives with
little or no fertilization, it has been shown to be responsive to fertilizer.[6] There are thought to be over 1000 named cultivars of taro which have
been selected on the basis of such qualities as corm formation, taste,
texture, and color.

Although many uses for taro have been proposed,[7,8,9] virtually its only
use is for food, and then mostly on a limited basis. Published production
information has focused on food uses. The authors are unaware of any cultivation for biomass production. The objective of this paper is to present
three studies conducted to determine methods for cultivating taro as an
energy crop in south Florida. These studies were designed to 1) determine
the effect of plant spacing on corm and cormel biomass production, 2) determine the effect of harvest interval on foliage biomass production, and 3)
collect taro cultivars and observe their growth habits.

METHODS AND MATERIALS

Plant Spacing

"Lehua Maoli" taro (the main commercial cultivar in Hawaii) hulis were
planted in Pahokee muck (Euic, hyperthermic Lithic Medisaprist of the Histosol Order), on 20 x 20, 40 x 40 and 60 x 60 cm spacings in March 15, 1982.
Prior to planting P and K were soil incorporated at 25 and 160 kg ha^{-1},
respectively. There were three replications, each containing enough material to permit six bimonthly (two-month interval) harvests of 6 plants each,
surrounded by border rows. After roots formed, the field was maintained in
a flooded condition through the last harvest. Upon harvest, plants were

separated into leaf lamina, petioles, corms, cormels, roots, and unidentifiable senesced material for fresh and dry weight determination.

Foliage Biomass Production

Foliage from an established stand of C. esculenta var. "aquatilis" (a stoloniferous type) growing in flooded Pahokee muck was harvested on 1, 2, 3 and 4-month intervals beginning March 1983. There were three replications utilizing plots 1.5 meter square. The experiment was repeated in 1984 using separate plots in the same established stand. Foliage from the central 1 m^2 of each plot was separated into leaf lamina and petioles for fresh and dry weight determination. Fertilization in 1983 (kg ha^{-1}) was P-52 and K-162 in May, and N-100 and K-100 in August. In 1984, fertilization was P-100 and K-100 in May, and N-100 and K-100 in July.

Cultivar Collection and Evaluation

Taro cultivars were collected through the assistance of cooperators in Hawaii, Costa Rica, and the Windward Islands of the Caribbean. Cultivars were evaluated for biomass production at three locations in 1984, depending on planting material availability: 1) a flooded Histosol (first preference), 2) an upland Histosol, and 3) a flooded Spodosol. Four replicates of 9-plant plots surrounded by a "Lehua Maoli" border row were used at each location, when sufficient material was available. When material was very limited, non-replicated plots were established in the flooded Histosol. Cuttings were planted on 33 x 33 cm centers in late March and early April, and harvesting was conducted in mid-December. Pre-plant fertilization consisted of P and K at 290 and 470 kg ha^{-1} at all locations. Nitrogen was applied on the flooded Histosol site at 100 kg ha^{-1} on 26 July, and at the same rate on the Spodosol on 11 July and 16 August. Data were subjected to analysis of covariance[10] to adjust yields at a 9-plant plot, because in some cases individual plants died for reasons that seemed unrelated to the treatment (cultivars).

RESULTS AND DISCUSSION

Plant Spacing

Yields on a per plant basis (Table 1) were in sharp constrast with yields on a per unit area basis (Table 2). For example, corm dry weight per

Table 1. Number of leaves, cormels and dry weight of various "Lehua Maoli" cultivar taro plant parts,[1] per plant.

Harvest Date	Plant spacing[2]	Number of leaves	Number of cormels	Tops	Corms	Cormels	Roots	Total
M/D/Y	cm			\- \- \- \- \- g plant^{-1} \- \- \- \- \- \-				
5/17/82	20	2.4	0	4.0	2.2	0	1.7	7.9
	40	3.1	0	5.4	2.2	0	1.9	9.5
	60	3.1	0	5.9	2.3	0	2.2	10.4
Linear effect		NS[3]	NS	NS	NS	NS	*	NS
Quadratic effect		NS	NS	NS	NS	NS	NS	NS
7/12/82	20	4.6	–	19.5	9.4	0.3	5.1	34.3
	40	14.1	–	63.8	8.3	2.2	9.5	83.8
	60	12.7	–	64.6	8.7	2.1	11.7	87.1
Linear effect		*		*	NS	NS	*	*
Quadratic effect		*		*	NS	NS	NS	NS
9/13/82	20	3.1	–	13.4	24.4	0.4	5.2	43.4
	40	17.0	–	43.1	23.5	20.6	17.5	104.7
	60	30.0	–	92.9	35.4	46.1	32.8	207.2
Linear effect		*		**	*	**	**	**
Quadratic effect		NS		*	+	NS	NS	*
11/15/82	20	2.8	0.6	6.6	21.8	0.9	5.4	34.7
	40	14.2	7.2	25.8	30.9	34.3	23.6	114.6
	60	23.7	13.2	51.2	51.8	73.0	42.2	218.2
Linear effect		**	**	**	**	**	**	**
Quadratic effect		NS	NS	NS	NS	NS	NS	NS
1/17/83	20	3.9	1.6	7.4	17.0	4.5	5.3	34.2
	40	13.8	6.7	22.4	25.5	33.7	19.0	100.6
	60	27.3	15.2	50.9	45.2	102.3	49.0	247.4
Linear effect		**	**	**	**	**	**	**
Quadratic effect		NS	*	*	NS	*	NS	*
3/14/83	20	5.3	1.8	10.9	15.6	5.2	9.0	40.7
	40	15.6	7.7	24.5	21.4	31.8	24.4	102.1
	60	23.1	12.4	42.6	22.9	56.0	39.1	160.6
Linear effect		**	**	**	NS	**	**	**
Quadratic effect		NS	NS	NS	NS	NS	NS	NS

1. Tops include leaf lamina + petioles, roots include roots + unidentifiable senesced material.

2. In row and between row plant spacing.

3. **, *, + and NS represent statistical significance at $P<0.01$, $P<0.05$, $P<0.10$, and not significant, respectively.

Note: cm x 2.54 = inches, g x 0.0022 = pounds.

Table 2. Number of leaves, cormels, and dry weight of various "Lehua Maoli" cultivar taro plant parts[1], on an area basis.

Harvest Date	Plant spacing[2]	Number of leaves	Number of cormels	Tops	Corms	Cormels	Roots	Total
M/D/Y	cm	(1000 ha^{-1})		- - - - - - kg ha^{-1} - - - - - - - -				
5/17/82	20	611	0	1250	556	0	417	2223
	40	193	0	340	139	0	122	601
	60	86	0	166	65	0	60	291
Linear effect		**[3]	NS	**	**	NS	**	**
Quadratic effect		NS	NS	**	**	NS	**	**
7/12/82	20	1153	–	4861	2347	83	1278	8569
	40	879	–	3987	521	139	597	5244
	60	353	–	1794	242	57	324	2417
Linear effect		*		+	*	NS	**	**
Quadratic effect		NS		NS	NS	NS	NS	NS
9/13/82	20	764	–	3348	6098	111	1306	10863
	40	1063	–	2691	1469	1289	1098	6547
	60	833	–	2582	983	1279	912	5756
Linear effect		NS		+	**	*	*	**
Quadratic effect		**		NS	**	**	NS	*
11/15/82	20	708	153	1653	5445	222	1347	8667
	40	889	448	1615	1935	2143	1476	7169
	60	657	366	1425	1438	2028	1174	6065
Linear effect		NS	NS	NS	**	**	NS	**
Quadratic effect		NS	NS	NS	**	*	NS	NS
1/17/83	20	975	400	1847	4250	1125	1325	8547
	40	863	419	1396	1594	2105	1188	6283
	60	758	422	1415	1256	2843	1363	6877
Linear effect		NS	NS	NS	*	**	NS	NS
Quadratic effect		NS	NS	NS	+	NS	NS	*
3/14/84	20	1325	450	2710	3889	1292	2236	10127
	40	975	481	1532	1337	1987	1528	6384
	60	642	344	1182	637	1556	1087	4462
Linear effect		**	NS	*	*	NS	*	**
Quadratic effect		NS	NS	NS	NS	+	NS	NS

1. Tops include leaf lamina + petioles, roots include roots + unidentifiable senesced material.

2. In row and between row plant spacing.

3. **, *, + and NS represent statistical significance at P<0.01, P<0.05, P<0.10, and not significant, respectively.

Note: cm x 2.54 = inches, ha x 2.47 = acre, kg ha^{-1} x 0.89 = pounds acre^{-1}.

plant increased with increased plant spacing, but corm dry weight on an area basis decreased with increased plant spacing. With close spacing, corms were smaller. However, since there were more plants per unit area, close spacing nevertheless provided more total biomass. Taro traditionally has been grown at spacings of 50 x 50 cm or more, probably to encourage production of large corms. For biomass production, corm size is likely to be less important than total yield, so close spacing would be preferable. Cormel production was severly restricted by close spacing. For example, in the November harvest, less than 1 cormel per plant was observed for a 20 x 20 cm plant spacing, whereas there were over 13 cormels per plant for a 60 x 60 cm spacing. Total per unit area cormel yield also was greater at the 60 x 60 cm spacing, but the sum of corm + cormel dry matter yield was 5667 kg ha^{-1} for a 20 x 20 cm spacing, compared to 3466 for a 60 x 60 cm spacing. Other plant components (leaf lamina, petioles, and roots) responded to plant spacing in a manner similar to the corms, i.e., per plant weight increased with increasing plant spacing, but yield per unit area generally decreased with increasing plant spacing.

Maximum per unit area biomass production was achieved somewhat earlier with close plant spacing. For example, with a 20 x 20 cm spacing corm dry matter production was 2347, 6098, and 5445 kg ha^{-1} in July, September, and November, respectively, whereas with a 60 x 60 cm spacing it was 242, 983, and 1438, respectively for the same months. Considering total plant biomass production, for a 20 x 20 cm spacing statistically equivalent (Duncan's miltiple range test, $P<0.05$, statistical data not presented elsewhere) yields were obtained for all harvest dates except the first (May). For a 40 x 40 cm spacing, statistically equivalent yields were obtained from the third through the sixth harvest dates. But for the 60 x 60 cm plant spacing, the fifth harvest date (January) provided statistically greater total plant biomass yield than the other dates. Thus closer plant spacing favored earlier maximization of yield. This is an important consideration in Florida, where the growing season can be terminated by frost or freeze.

Data from the 1982 plant spacing study are used to examine the effect of time on the growth pattern of various plant components. These data, averaged over the 40 x 40 and 60 x 60 cm spacings (Figure 1), are in general agreement with those from a study conducted in 1981 using a 50 x 50 plant spacing.[11] As in the 1981 study, in 1982 root growth increased through November and remained constant thereafter. Top growth exceeded that of corms + cormels and roots for approximately 6 months after planting, as was observed in the previous year. It declined thereafter. Corm + cormel

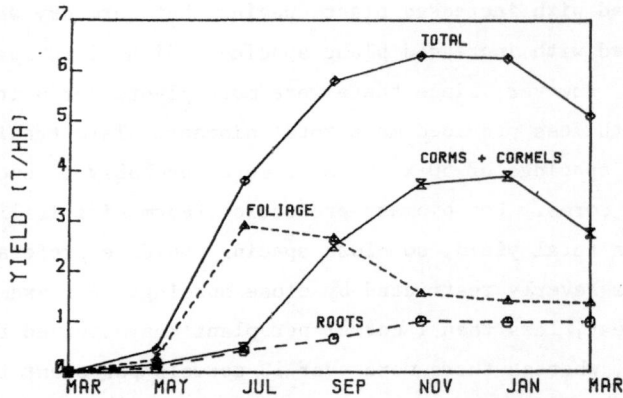

Figure 1. Production of 'Lehua Maoli' taro foliage (leaf lamina and peti-
oles), corms + cormels, roots, and total production. Planted at
EREC-Belle Glade, Florida, March 1982. Note: Mg ha^{-1} x 892 = lb
A^{-1}

production increased throughout the period from July through November.
Total biomass yield peaked during the November to January period. In 1981,
there was a severe freeze in January, and corm + cormel yield decreased
sharply from January to March. There was no comparable freeze in 1982, but
corm + cormel yield decreased by 31% during the January to March period,
nevertheless. This probably is a reflection of a reversal of the starch
storage process, whereby carbohydrates were mobilized to support new foliage
production. The implications for biomass production in south Florida are
obvious. March planted taro should be harvested from November through
January for maximum production, when planted on 50 x 50 cm spacings. Closer
spacing (20 x 20 cm) will permit earlier harvesting.

Foliage Biomass Production

Annual production with a 3-month harvest interval was significantly
greater than that with a 1-month interval (Table 3), but the 3-month interval was not different from 2 or 4 months. However, with monthly harvests,
there was a severe reduction in stand. This is shown by the general reduction in yield observed for each successive monthly harvest (Table 3), although part of this reduction, no doubt, was caused by the shorter days and
cooler temperatures that occurred at the end of the growing period. Another
measure of the apparent damage or stand depletion caused by monthly harvesting was obtained when whole plants from the 1983 study were harvested in

Table 3. Effect of harvest interval on dry weight foliage yield of 'aquatilis'[1] cultivar taro.

Harvest Month	Harvest interval (months)			
	1	2	3	4
	----------- (kg ha^{-1})- -------			
April	2380			
May	2250	4060		
June	2260		7020	
July	1870	4870		9630
August	1040			
September	1070	3181	5880	
October	790			
November	270	1650		5380
December	150		2510	
Total	12080 b[2]	13770 ab	15410 a	15000 ab

1. Yields averaged over a two-year period.
2. Values followed by the same letter are not significantly different by the Duncan's multiple range test ($P<0.05$).

Note: kg ha^{-1} x 0.89 = pounds acre^{-1}.

April 1984. Total corm + cormel dry matter yield was only 89 kg ha^{-1} when top growth was harvested monthly, whereas it was 1125, 564, and 2706 kg ha^{-1} for the 2, 3, and 4-month harvest intervals, respectively. It is not possible to determine why corm + cormel yield was lower for the 3-month harvest interval than for 2 or 4-months. However a contributing factor may have been that the 3-month plots were harvested only 10 days before a severe late December freeze in 1983, whereas the 2 and 4-month harvest interval plots were harvested 40 days before the freeze. The latter plots may have been better able to withstand the severe conditions of the freeze than the former. Corm + cormel fresh weight yield in April, 1985, following the 1984 foliage biomass study, was 250, 1236, 1882, and 2529 kg ha^{-1} for the 1, 2, 3, and 4-month harvest interval plots, respectively. These data, too, indicate that top growth harvest at a 1-month interval decreased tuber production relative to harvest at greater intervals, which probably accounts for the reduction in stand observed for the 1-month harvest interval.

In terms of total foliage yield during a growing season, there appears to be little difference among harvest intervals of 2-months or more. Therefore, the choice may center more on the economics of harvesting or the need for a steady supply of biomass feedstock. Harvesting economics may dictate that the longer harvest interval (4 months) is preferable, since the same

Table 4. Dry weight whole plant biomass production of 15 taro cultivars planted in three locations during the spring of 1984.

Cultivar	AREC-Belle Glade Flooded n*	AREC-Belle Glade Flooded Production	Histosol Unflooded n	Histosol Unflooded Production	AREC-Ft. Pierce Spodosol Flooded n	AREC-Ft. Pierce Spodosol Flooded Production
		Kg ha^{-1}		Kg ha^{-1}		Kg ha^{-1}
1 Lehua Maoli	4	5,180	4	6,452	4	5,505
2 Bun Long	4	19,652	4	20,348	4	6,072
3 Sawa Bastora**	4	7,140	4	7,710	3	5,201
4 Senorita**	4	4,889	1	6,580	–	–
5 Magaulusima	4	9,035	3	8,829	–	–
6 Sawa Puetata**	4	9,140	4	8,798	4	5,338
7 Keoni	4	15,117	4	10,887	3	5,571
8 Islena	4	9,006	4	10,598	4	13,081
9 Blanc	4	9,678	3	7,769	4	6,085
10 Souf**	4	8,803	4	6,125	4	6,896
11 Common	4	9,052	4	7,145	4	7,579
12 Noir**	4	10,264	4	7,439	4	5,563
13 Chi-sen**	1	6,293	–	–	–	–
14 "Costa Rica"**	1	6,293	–	–	–	–
15 Aquatilis**	4	7,601	4	9,717	4	5,237
LSD$_{0.05}$		294		489		358
Average (omitting 4,5,13,14)		10,058		9,363		6,557

* Number of replications; ** Stolons present.
Note: kg ha^{-1} x 0.89=1b A^{-1}

amount of biomass can be obtained with fewer harvests. Harvesting at shorter intervals would, of course, provide a more continuous source of biomass.

Cultivar Collection and Evaluation

Three fairly distinct growth habits have been observed among the cultivars collected. Some produce large corms, along with cormels closely attached to the corms. Among these types are the Hawaiian cultivars "Lehua Maoli" and "Bun Long." Other culitivars are very stoloniferous with abundant foliage, but produce relatively few corms and cormels. "Aquatilis," which is naturalized in canals, streams, and lakes throughout much of Florida and parts of Louisiana, is an example of this growth type. A third type produces large corms, but also has stolons. A cultivar obtained from Taiwan, "Chi-sen," has this growth pattern. Of the three habits, the latter may be most desirable for biomass production. For many purposes, the corms and cormels have advantages over the foliage, since their dry matter content is higher (ca. 25% as compared to ca. 5% for foliage), and they contain more starch. However, in harvesting the corms and cormels, the stand is destroyed. Since taro is propagated vegetatively, rather than by seed, replanting may be expensive. If a cultivar like "Chi-sen" is utilized, it may be possible to harvest in strips, leaving enough material in the field for revegatation by growth from stolons and plant parts left behind by the harvesting equipment.

A number of cultivars produced significantly higher total biomass production than "Lehua Maoli" (Table 4), which has been used in most of our tests. "Bun Long" produced the highest biomass (20 Mg ha^{-1}) on the Histosols. "Keoni," "Noir," and "Chi-sen" produced two to three times the biomass of "Lehua Maoli" on the flooded Histosol. Although "Islena" production was only slightly above average on the Histosols, it was two times the average on the Spodosol (13 Mg ha^{-1}). Average production (9 to 10 Mg ha^{-1}) on the Histosols did not appear to be greatly influenced by flooding. Lower average yields (6.5 Mg ha^{-1}) were obtained on the Spodosol.

The cultivar evaluation data demonstrate that large differences in growth type and biomass production occur among taro cultivars. It appears that additional collection and evaluation is merited, since only a very few of the many existing taro cultivars have been evaluated for biomass production.

ACKNOWLEDGEMENTS

This project contributes to a joint program between the Gas Research Institute and the Institute of Food and Agricultural Sciences, University of Florida. Assistance in collecting taro cultivars was received from Dr. Ramon de la Pena, University of Hawaii, and Mr. Herman Adams, Caribbean Agricultural Research and Development Institute. The authors also express appreciation to Ms. Myrene Hewitt for technical assistance provided.

REFERENCES

1. H. Angel and P. Wolseley. The water naturalist, In: Facts on File,
    Inc., New York. p. 54. 1982.
2. J. Browder, C. Littlejohn and D. Young. The South Florida Study.
    University of Florida Center for Wetlands. Gainesville. 1975.
3. R. Culpepper. Water war: battle rages along the Kissimmee. Florida
    Grower & Rancher. November. p. 8-9. 1983.
4. R. Loftis. Graham taps high-level panel to oversee saving of the
    Glades. Miami Herald (Palm Beach edition). p. 6D. Oct. 28, 1983.
5. A. M. Marshall. The future of Florida's salt and freshwater resources.
    p. 17-24, In: The environmental destruction of south Florida. W. R.
    McCluney (Ed.). University of Miami Press. Coral Gables, Florida.
    1971.
6. L. A. Sunell and J. Arditti. Physiology and Phytochemistry. In:
    Taro, J. K. Wang (Ed.), University of Hawaii Press, Honolulu. 1983.
7. J. R. Carpenter and W. E. Steinke. Animal feed. p. 269-300, In: Taro.
    J. K. Wang (Ed.), University of Hawaii Press, Honolulu. 1983.
8. G. J. L. Griffin and J. K. Wang. Industrial uses. p. 301-312, In:
    Taro. J. K. Wang (Ed.), University of Hawaii Press, Honolulu.
    1983.
9. J. H. Moy and W. K. Nip. Processed food. p. 261-268, In: Taro, J. K.
    Wang (Ed.). University of Hawaii Press, Honolulu. 1983.
10. SAS Institute, Inc. SAS User's Guide: Statistics, 1982 edition. SAS
    Institute Inc., Cary N. C. 1982.
11. S. K. O'Hair, G. H. Snyder and J. F. Morton. Wetland taro: a neglected
    crop for food, feed, and fuel. Proc. Fla. State Hort. Soc. 95:367-
    374. 1982.

STARCH AND MINERAL NUTRIENT ACCUMULATION BY

SWEET POTATO CULTIVARS

S. J. Locascio and J. M. Dangler

Department of Vegetable Crops
University of Florida/IFAS
Gainesville, Florida 32611

ABSTRACT

Four sweet potato cultivars [Ipomoea botatas L. (Lam)] were grown during 2 seasons to evaluate the influence of 8 times of harvest on root accumulation of dry weight, starch, and minerals. In both seasons, the yields of dry matter and starch were highest with 'GaTG-3' and 'GaTG-6' and lowest with 'White Star' and 'GATG-1'. The yields of all cultivars increased linearly with an increase in harvest time from about 3 to 6 months after transplanting. Root dry weights increased from 5.5 to 20.2 Mg ha^{-1} and starch weights from 3.0 to 11.1 Mg ha^{-1}. Root N concentrations were influenced by cultivars in 1 of the 2 seasons. In both seasons the root N concentrations were reduced from about 0.50 to 0.33% with an increase in growth period. The root K and Ca concentrations were not consistently influenced by time of harvest during the 2 seasons. It is apparent from these studies that one long growing season rather than 2 shorter consecutive seasons is required for maximum accumulation of starch by sweet potato.

Keywords. Ipomoea batatas, industrial sweet potato, biomass, starch, time of harvest.

INTRODUCTION

A number of crops have potential as substrate for the production of liquid and gaseous fuels.[1-10] Cassava and sweet potato are of particular interest because of their high potential for the production and storage of starch. Starch from cassava 'CMC 40' was estimated to be 5.1 Mg ha^{-1}.[11,12]

Sweet potato was found to be a more productive crop with a mean starch yield of 9.6 Mg ha^{-1} with some cultivars.[13] Sweet potatoes can be used to generate methane and CO_2 (0.3 and 0.2 m^3 kg^{-1} dry product, respectively, after 15 days of fermentation)[14] or used as fuel (4 kcal kg^{-1} dry product).[15]

The objective of this study was to evaluate the potential of 4 sweet potato cultivars for biomass production, especially starch and to establish the time required for maximum starch accumulation. Although starch accumulation as a function of time of harvest by sweet potato has been reported,[13] data was presented only for 2 harvest periods. More specific information which describes the trend in starch accumulation over long periods is required to determine with greater certainty the full potential of several cultivars that produce high starch concentration for industrial use. Root and plant top mineral nutrient concentrations are also reported because the concentration of mineral nutrients influences fermentation.

MATERIALS AND METHODS

Sweet potatoes were grown near Gainesville, Florida during 1981 on a Kanapaha fine sand and during 1982 on an Arredondo fine sand. Cultivars White Star, GaTG-1, GaTG-3, and GaTG-6 were arranged as main plots with 8 growth periods as subplots. Subplots were single rows 1.22 m wide and 4.58 m long and were replicated 2 times. Fertilizer was applied at 96-128-128-24 kg ha^{-1} N-P-K-FN 503 (micronutrient mix). One-half of the fertilizer was preplant incorporated into raised beds. Sweet potato slips were transplanted 0.30 m apart on April 20, 1981 and on April 13, 1982. Diphenamid (N,- N-dimethyl-2, 2-diphenylacetimide) was applied at 4.8 kg ha^{-1} for weed control. Four weeks after transplanting, the remainder of the fertilizer was applied to one side of the bed prior to mechanical cultivation. Sweet potatoes were harvested at 2 week intervals beginning on July 21 in 1981 and on July 22 in 1982. Two plant tops per plot were harvested for dry weight determination and mineral analysis. Plant tops were dried at 70°C and ground to pass a 1 mm screen. Subsamples (1 kg) from 10 roots per plot were cut into 1 cm slices and dried at 70°C.

Whole plant and root samples were analyzed for N, K and Ca. A 1.0 g sample of the tissue was ashed at 700°C for 5 hours. Total N concentrations were determined by the micro-Kjeldahl method,[16] Ca by atomic absorption spectrophotometry, and K by flame photometry emission.[17] Starch was solubilized and hydrolyzed as described by Dekker and Richards.[18] Starch concentrations of the root material were calculated from glucose concentrations obtained with a YSI Model 27 Industrial Analyzer (Yellow Springs Instrument Co., Inc., Ohio).

RESULTS

In 1981, the dry root yield of 'GaTG-3' (12.10 Mg ha^{-1}) was significantly greater than the yield of 'White Star' (8.79 Mg ha^{-1}) and 'GaTG-1' (8.52 Mg ha^{-1}) (Table 1). The main effect of cultivar on starch accumulation was similar to that for dry weight. The mean starch yield of 6.72 Mg ha^{-1} from 'GaTG-3' and 'GaTG-6' was approximately 40% higher than from 'GaTG-1' (4.58 Mg ha^{-1}). The ash content of dry root material ranged from 3.58% ('White Star') to 3.93% ('GaTG-3'). Differences among cultivars in root mineral concentrations except N were not significant.

Dry root and starch yields increased linearly from 92 to 190 days after planting (Figure 1). The increase in starch yield was approximately sixfold from 1.7 to 11.1 Mg ha^{-1} with an increase in growth period. Changes in root ash content were irregular (mean value 3.76%) (Table 2). The N concentration decreased linearly, the K concentration increased quadratically and the Ca content decreased irregularly with an increase in growth period.

Table 1. Main effects of sweet potato cultivar on root yield, ash, and mineral concentrations (1981 and 1982)

Sweet potato (cultivar)	Root					
	Dry wt. (Mg ha^{-1})	Starch[a] (Mg ha^{-1})	Ash wt. (%)	Mineral concn. (%)		
				N	K	Ca
1981						
White Star	8.79b[b]	5.54ab	3.58c	0.43b	1.36	0.10
GaTG-1	8.52b	4.58b	3.89ab	0.52a	1.50	0.11
GaTG-3	12.10a	6.87a	3.93a	0.48ab	1.46	0.12
GaTG-6	10.76ab	6.57a	3.67bc	0.47ab	1.38	0.11
1982						
White Star	12.44b	6.98b	3.77	0.36	0.71	0.10a
GaTG-1	13.76ab	7.37b	4.04	0.36	0.74	0.09ab
GaTG-3	15.78a	9.01a	3.56	0.35	0.68	0.11a
GaTG-6	14.80ab	8.37ab	3.50	0.35	0.66	0.07b

[a]Expressed as percent of solids
[b]Mean separation within columns by Duncan's multiple range test, 5%

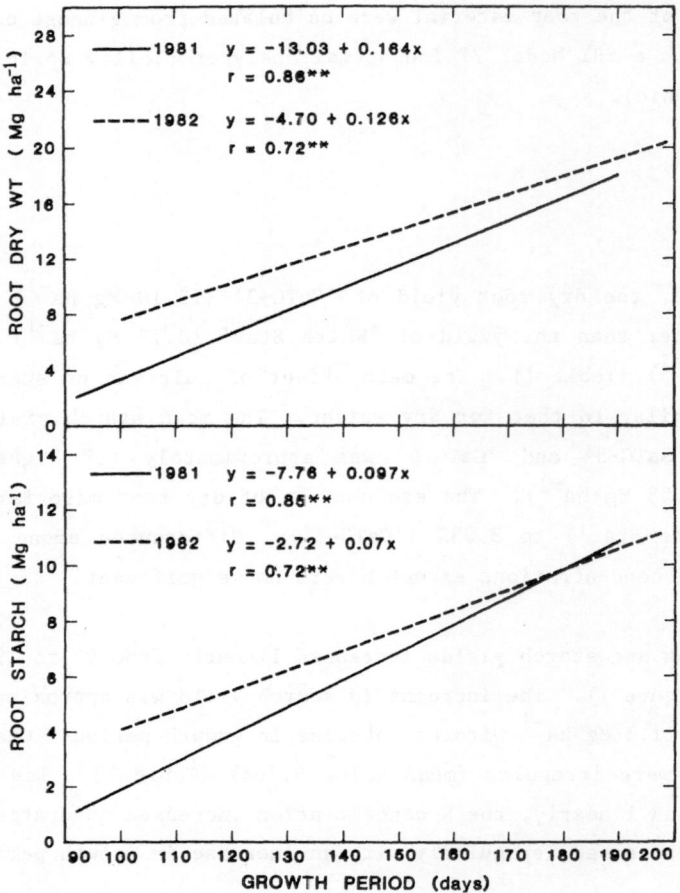

Figure 1.   Effect of growth period on sweet potato dry weight and starch
yields (1981 and 1982)

Plant top yield was not influenced by cultivar in 1981.   The mean dry
weight yield was 6.56 Mg ha^{-1} (Table 3).   The K concentration was influenced
by cultivars and ranged from 2.02% for 'GaTG-6' to 2.43% for 'GaTG-3'.   Dry
weight of plant tops increased linearly and quadratically between 92 to 190
days after planting (Table 4).   Maximum weight was obtained at 148 days
after transplanting.   Concentrations of N and K decreased and Ca increased
with an increase in growth period from 92 to 190 days.

In 1982, 'GaTG-3' produced the highest dry root (15.78 Mg ha^{-1}) and
starch (9.01 Mg ha^{-1}) yields (Table 1).   Calcium was the only mineral nutrient measured that was influenced by cultivar.   Dry root and starch yields
increased linearly between 100 and 198 days after planting (Figure 1).

Table 2. Main effects of sweet potato growth period on root ash and mineral concentrations (1981 and 1982)

Growth period (days)	Ash[a] (%)	Root mineral concn. (%)		
		N	K	Ca
		1981		
92	-.--	0.58	1.34	0.15
106	3.96	0.51	1.41	0.11
120	3.79	0.48	1.37	0.09
134	3.99	0.49	1.51	0.11
148	3.75	0.47	1.49	0.12
162	3.69	0.47	1.43	0.11
176	3.91	0.45	1.50	0.09
190	3.26	0.35	1.36	0.09
Signif.[b]	L**Q*	C*L**	Q*	L**Q*
		1982		
100	4.40	0.50	0.70	0.11
114	3.49	0.34	0.65	0.08
128	4.30	0.44	0.77	0.10
142	3.82	0.35	0.74	0.08
156	3.81	0.36	0.77	0.12
170	2.88	0.28	0.52	0.09
184	3.29	0.27	0.66	0.10
198	3.75	0.30	0.77	0.09
Signif.	L**Qu**	L**Qu**	C**	Qu**

[a] Expressed as percent of solids
[b] Significant effects were linear (L), quadratic (Q), cubic (C) or quartic (Qu) at the 5% (*) or 1% (**) levels

Starch yields increased about threefold from 4.2 to 11.1 Mg ha^{-1} from the first to the last harvest date. Root ash and N concentrations decreased but root K concentrations were irregular and Ca concentrations were not influenced by an increase in growth period (Table 2).

The dry weight yield of plant tops obtained in 1982 with 'GaTG-3' (8.07 Mg ha^{-1}) was significantly greater than that of 'White Star' (4.80 Mg ha^{-1}) and 'GaTG-1' (5.24 Mg ha^{-1}) but not 'GaTG-6' (5.91 Mg ha^{-1}) (Table 3). Mineral nutrient concentrations of plant tops were similar with all cultivars. Maximum accumulation of dry weight of plant tops occurred at 156 days after transplanting (Table 4). With further growth, plant dry weights were reduced.

Table 3. Main effects of sweet potato cultivar on plant top dry weight and mineral concentration

Sweet potato (cultivar)	Dry wt. (Mg ha^{-1})	Plant top		
		Mineral concn. (%)		
		N	K	Ca
		1981		
White Star	6.57	0.98	2.19b[a]	1.05
GaTG-1	6.47	1.09	2.26ab	1.06
GaTG-3	7.32	1.08	2.43a	1.12
GaTG-6	5.87	0.97	2.02b	1.09
		1982		
White Star	4.80b	1.13	1.06	0.88
GaTG-1	5.24b	1.08	1.14	0.84
GaTG-3	8.07a	0.98	0.99	0.85
GaTG-6	5.91ab	1.14	1.19	0.91

[a]Mean separation within columns by Duncan's multiple range test, 5%

DISCUSSION

In both years 'GaTG-3' (7.94 Mg ha^{-1}) and 'GaTG-6' (7.47 Mg ha^{-1}) produced the greatest starch yields. 'GaTG-3' has been reported to be a highly productive cultivar at this site (9.6 Mg ha^{-1} after 210-30 days).[13] No consistent differences in ash or mineral nutrient concentrations among cultivars were observed in the work reported here.

Since starch yields increased linearly within the last 100 days of growth, it appears that a growth period of 190-198 days or longer is essential to obtain maximum starch accumulation. Because of the large difference in the weight obtained at the first harvest (92-100 days) each year (1982, 1.74 Mg ha^{-1}; 1982, 4.15 Mg ha^{-1}) as well as the large difference in percent increase in starch yield obtained during the second one-half of the growth period each year (1981, 130% increase; 1982, 40% increase) it would appear that factors other than the length of growth significantly affected starch accumulation early in the growth period. In 1981 and 1982, rainfalls during the first 6 weeks of the growth period were 0.2 and 2.5 cm/wk, respectively. Nevertheless, mean starch yields after 190-8 days of growth were 11.1 Mg ha^{-1} in both seasons. Results of previous work[13] showed that when supplied with adequate water, the starch yield of 'GaTG-3' was high but did not increase significantly with a doubling of the length of the growth period

Table 4. Main effects of growth period on sweet potato plant top weight and mineral concentration

Growth period (days)	Dry wt. (Mg ha^{-1})	Plant top Mineral concn. (%) N	K	Ca
		1981		
92	1.38	1.43	2.62	1.05
106	4.71	1.26	2.16	1.06
120	5.30	1.18	2.58	1.13
134	6.78	1.06	2.61	0.94
148	9.19	0.91	2.28	1.08
162	7.82	0.77	1.93	1.09
176	8.39	0.77	1.94	1.13
190	8.90	0.85	1.69	1.14
Signif.[a]	L**Q**	L**Q*	L**Q**	
		1982		
100	6.28	1.55	1.74	0.87
114	4.58	1.32	0.63	0.84
128	6.07	1.17	1.41	1.11
142	6.51	1.04	1.00	0.80
156	9.13	0.90	1.08	0.76
170	4.86	0.98	0.93	1.06
184	5.22	0.89	1.10	0.82
198	5.38	0.80	0.88	0.72
Signif.	Qu*	L**Q**	L**Q*	L*Q*Qu**

[a]Significant effects were linear (L), quadratic (Q), cubic (C) or quartic (Qu) at the 5% (*) and 1% (**) levels

(105 days, 9.4 Mg ha^{-1}; 210 days, 10.4 Mg ha^{-1}). When rainfall was lower and no irrigation water was applied, starch yield almost doubled with a doubling in the length of the growth period (115 days, 4.9 Mg ha^{-1}; 230 days 8.7 Mg ha^{-1}). Results of these studies imply that when sweet potato is grown for biomass for an extended growth period, even when low rainfall limits early season starch accumulation, high yields can be obtained provided adequate rainfall occurs later in the season. When early season rainfall is adequate or when irrigation is provided, starch accumulation is rapid. Since most of the starch accumulation occurred in the second one-half of the 190-8 day growth period, it is unlikely that planting 2 successive crops would result in a greater starch yield than one crop grown for 190-8 days.

Root N concentration decreased with an increase in growth period during both years. The N concentration averaged less than 0.5% during the harvest periods. The N concentration of 'Maryland Gold', a table type sweet potato, supplied with one-half the added fertilizer N and approximately the same fertilizer P and K was reported to decrease from approximately 1.3 to 1.0% during the last 78 days of a 124 day growth period.[19] Root K concentrations decreased from approximately 2 to 1.5% during the same period[19] and were also higher than values reported here. Shoot N and K concentrations were also reported to be greater than those obtained in this study. Concentrations of both elements decreased as the growth period progressed in both studies.

'GaTG-3' was among the highest yielding cultivars in both years. With a mean starch yield from all cultivars of 11 Mg ha^{-1}, ethanol production can be estimated at 6446 liters ha^{-1} (0.586 kg ethanol kg^{-1} starch).[20] Sweet potato roots and tops also can be used to generate methane gas.[21] Ethanol production from 2 irrigated North Florida corn crops, which could be grown during a similar growth period, can be estimated at 7000 liters ha^{-1}.[11] Since starch yields from all cultivars increased linearly during the latter one-half of the 190-8 day growth period, the potential for ethanol from these high starch cultivars could be expected to be even greater when grown in regions with longer growing seasons.

ACKNOWLEGEMENTS

This paper is No. 6393 in the Journal Series of the Florida Agricultural Experiment Stations. This project is part of a joint program between the Institute of Food and Agricultural Sciences and the Gas Research Institute.

REFERENCES

1. Anon. Small-scale fuel alcohol production. U.S. Dept. of Agric. 1980.
2. Anon. Fuel from farms – a guide to small-scale ethanol production, Technical Information Center. U.S. Dept. of Energy, Oak Ridge, TN. 1980.
3. L. R. Brown. Food or fuel: New conpetition for the world's cropland. Worldwatch paper 35, Worldwatch Inst., Washington, DC. 1980.
4. A. L. Hammond. Alcohol: A Brazilizn answer to the energy crisis. Science 195:564–566. 1977.
5. J. F. Hills, S. S. Johnson, S. G. Akbar Abshahi and G. F. Peterson. Comparison of high-energy crops for alcohol production. Calif. Agric. 35 (11–12): 14–16. 1981.
6. D. L. Marzola and B. P. Bartholamew. Photosynthetic pathway and biomass energy production. Science 205: 555–558. 1979.

7. D. J. McCann and R. G. H. Prince. Agro-industrial systems for ethanol production. In: Proceedings of a conference held at the Sebel Town House. Sydney, Australia. Aug. 9-11, 1978. Institution of Chemical Engineers New South Wales Group. Ed. 1979.

8. R. K. Robinson and S. N. Kutianawala. Cassava: Its potential as an industrial crop. Wld Crops 31: 168-176. 1979.

9. M. Samways. Alcohol from cassava in Brazil. Wld Crops 31:181-186. 1979.

10. J. G. da Silva, G. E. Serra, J. R. Moreira, J. C. Concalves and J. Goldemberg. Energy balance for ethyl alcohol production from crops. Science 201: 903-6. 1978.

11. J. M. Dangler. Root and top yields of cassava and sweet potato as affect by cultivar, fertilizer rate, cowpea intercrop or time of harvest. MS Thesis. University of Florida, Gainesville, FL. 1982.

12. S. K. O'Hair, S. J. Locascio, R. R. Forbes, J. M. White, D. R. Hensel, J. R. Shumaker and J. M. Dangler. Root crops and their biomass potential in Florida. Soil and Crop Science Soc. Fla. 42: 13-17. 1983.

13. J. M. Dangler, S. J. Locascio and L. H. Halsey. Sweet potato for biomass. Biomass J. 4: 253-261. 1984.

14. Anon. Methane generation from human, animal and agricultural wastes. National Academy of Sciences. Washington, DC. 1977.

15. C. Chatfield and E. Adams. Proximate composition of fresh vegetables. U.S. Dept. Agr. Cir. 146. 1931.

16. P. L. Kirk. Kjeldahl method for total N, Anal. Chem. 22: 354-358. 1950.

17. L. M. Walsh. Ed. Instrumental methods for analysis of soil and plant tissue. Soil Science Society of Amer., Madison, WI. 1971.

18. R. F. G. Dekker 'and G. N. Richards. Determination of starch in plant material. J. Sci. Fd. & Agric. 22: 441-444. 1971.

19. L. E. Scott and W. L. Ogle. The mineral uptake by the sweet potato. Better Crops with Plant Food. 36: 12-16. 1952.

20. P. B. Jacobs and H. P. Newton. Motor fuels from farm products, Misc. Pub. 327. U.S. Dept. of Agric. 1938.

21. A. Shiralipour and P. H. Smith. Conversion of biomass into methane gas. Biomass J. 6: 85-92. 1984.

EFFECT OF PLANT POPULATION ON BIOMASS

PRODUCTION OF SIX WEED SPECIES

J. P. Gilreath

Gulf Coast Research and Education Center
University of Florida/IFAS
Bradenton, Florida 34203

ABSTRACT

The effect of plant population level on biomass production was determined for 6 common weed species which had shown promise as methane feedstock crops in preliminary screening studies. Biomass production of Amaranthus hybridus L. (smooth pigweed), Èupatorium capillifolium (Lam.) Small (dogfennel), and Sida rhombifolia L. (arrowleaf sida) was studied at four plant population levels (three levels for E. capillifolium) in August, 1982, and at four levels for all species in May, 1983. Yield (dry weight) of A. hybridus increased with increasing plant population up to 215,300 plants ha^{-1} with highest yields of 4.1 Mg ha^{-1} obtained in 1983. Dry matter production of E. capillifolium was not influenced by plant population in 1982; however, in 1983, yield increased to 13.4 Mg ha^{-1} with 215,300 plants ha^{-1}. Dry matter yield of S. rhombifolia was essentially the same at all population levels in 1982, while in 1983 yield increased to 16.7 Mg ha^{-1} with 215,300 plants ha^{-1}. In 1984, yields of Amaranthus australis (Gray), J. D. Sauer (giant pigweed), Ambrosia artemisiifolia L. (ragweed), and Setaria magna Griseb. (giant foxtail) were measured at three population levels. Dry weight of A. australis increased with increasing plant population to 17.5 Mg ha^{-1} with 26,900 plants ha^{-1}, the highest population level. Dry matter production by A. artemisiifolia was increased 49% to 14.3 Mg ha^{-1} by increasing the plant population 300% from 13,500 to 53,800 plants ha^{-1}. Increasing the plant population of S. magna from 26,900 to 107,600 plants ha^{-1} increased dry matter production 75% from 8.3 to 14.5 Mg ha^{-1}.

Keywords. Pigweed, dogfennel, sida, ragweed, giant foxtail.

INTRODUCTION

Production of green plants for use as sources of bioenergy has received considerable attention in the last few years. The solar energy stored in the form of plant tissue can be converted into numerous easily utilized energy sources, such as ethanol and methane. Conventional feed and fiber crops have historically been the mainstay of biomass research with the use of weedy plants being a rather recent concept.[1-6] For methane production, almost any organic material could serve as a feedstock; however, green plants are especially attractive due to the renewable nature of the raw material. Not all green plants have equal potential for use as methane feedstock as they vary considerably in chemical composition and production of biomass.

For any plant material to be attractive as a feedstock for methane production, cost of methane production must be competitive with the well head price of natural gas. Perhaps the easiest cost to control is the cost of producing the plant material. Production of conventional feed and fiber crops requires large inputs of fuel, fertilizer and pesticides, which are all costly, energy-dependent materials. Additionally, such crops frequently require irrigation and shifting of valuable arable land from production of feed and fiber to biomass. These requirements increase the cost factor, making conventional crops less attractive as feedstocks.

Utilization of nonconventional crops, such as trees, woody shrubs and weeds, is a more attractive alternative since their genetic base allows growth under a wider range of cultural and climatic conditions than conventional crops. In addition, most nonconventional crops have few serious known pests, are highly competitive, grow well in soils frequently unsuitable for crop production and respond well to small inputs of nitrogen.[5]

One important consideration in producing any plant species is the plant population per unit area required to produce the optimum yield. Research over the years has established guidelines for conventional crops, but none are available for nonconventional crops. This research was conducted to determine the effect of plant population on yield of six weed species.

MATERIALS AND METHODS

Studies were conducted in 1982, 1983, and 1984 at the Gulf Coast
Research and Education Center in Bradenton, Florida, located at 27° 30' N
latitude and 82° 30' W longitude in southwest peninsular Florida. Subsamples (5 Kg) of plant tissue were randomly selected from each population
level of each species at harvest and were subsequently dried to constant
weight at 60°C for dry matter determination.

1982 and 1983 Experiments

In August, 1982, Amaranthus hybridus (smooth pigweed) and Sida rhombifolia L. (arrowleaf sida) were planted at four population levels, while
Eupatorium capillifolium (Lam.) Small (dogfennel) was planted at 3 population levels. In May, 1983, four population levels for each species were
planted. The experimental area was prepared in 1982 by broadcasting 81 kg
ha^{-1} of nitrogen (N) and 93 kg ha^{-1} of potassium (K) and rototilling it into
the soil the day of planting. Land preparation in 1983 consisted of broadcasting 34, 15, and 28 kg ha^{-1} of N, P (phosphorus), and K, respectively,
and rototilling to incorporate the fertilizer. In 1982, plants were side-
dressed September 26 with an additional 34.7 kg ha^{-1} of N and 40 kg ha^{-1} of
K. During 1983, S. rhombifolia and E. capillifolium received supplemental
applications of 66 kg ha^{-1} N on July 29.

The soil at the experimental site was in Eau Gallie fine sand (sandy,
siliceous, Hypothermic, Alfic Haplaquod) with a pH ranging from 6.5 to 7.0
and approximately 1% organic matter. Five week-old, greenhouse-grown
seedlings of A. hybridus and S. rhombifolia and 5 week-old rooted cuttings
of E. capillifolium were transplanted at the appropriate population level
into plots 4 rows wide by 6.1 m long on August 27 (August 30 for E. capillifolium) in 1982 and May 26 in 1983. The populations studied for A. hybridus
and S. rhombifolia were 215,300, 107,600, 53,800 and 26,900 plants ha^{-1}. E.
capillifolium plants were transplanted at populations of 215,300, 107,600
and 53,800 plants ha^{-1} in the 1982 study and 215,300, 107,600, 53,800 and
26,900 plants ha^{-1} in 1983. Treatments (population levels) were assigned to
plots arranged in a ramdomized complete block design with each population
replicated 4 times during 1982 and 3 times in 1983. Unwanted weeds were
removed weekly by hoeing until growth of the desired plants physically
interfered. In 1982, A. hybridus and S. rhombifolia were harvested December
20, while E. capillifolium was harvested January 4, 1983. During the 1983
study, A. hybridus was harvested July 27, while E. capillifolium and S.

rhombifolia were harvested September 26 and 28, respectively. Yield data
were obtained from two center rows of each plot by cutting the plants to
within 10 cm of the soil surface. The first and last plants in each row
were not harvested so as to remove border effects.

1984 Experiments

Studies conducted in 1984 involved 3 weed species: Amaranthus
australis (Gray) J. D. Sauer (giant pigweed), Ambrosia artemisiifolia L.
(ragweed), and Setaria magna Griseb. (giant foxtail). Plants of each
species were grown in Eau Gallie fine sand (sandy, siliceous, Hyperthermic,
Alfic Haplaquod) (pH 6.8, 1% organic matter) with treatments (populations)
assigned to plots arranged in a randomized complete block design and replicated 4 times. Seedlings of A. australis, 30 day-old and 10 to 15 cm tall,
were transplanted into field plots on March 14, 1984. Plant populations of
6,700, 13,500 and 26,900 plants ha^{-1} were assigned to plots which consisted
of 4 to 6 rows (depending on between row spacing required to achieve the
desired population ha^{-1}) 12.2 m long. Each plot had a guard row on each
side to remove border effects. Replications were separated by 3 m of land
area. Fertilization consisted of 60, 26, and 50 kg ha^{-1} of N, P, and K,
respectively, applied in 2 equal applications on March 23, 9 days after
planting, and May 3. Plots were harvested June 12, 1984, 89 days after
transplanting, by cutting one row from the center of each plot, leaving one
plant at each end of the row to eliminate border effects. Plants were in
full bloom at this time.

Five to 8 cm tall, rooted cuttings of A. artemisiifolia were transplanted February 16, 1984. Plant populations of 13,500, 26,900 and 53,800
plants ha^{-1} were assigned to plots which were 3 rows wide by 9.1 m long with
1 guard row on each side of the experimental area. Replications were separated by 1.8 m of land. Fertilization consisted of 101, 44, and 84 kg ha^{-1}
of N, P, and K, respectively, applied in 2 equal applications on February
14, prior to planting, and on March 6. Approximately 50% of the plants had
initiated bloom buds on April 23 and were in the early stages of senescence
at harvest on June 12, 1984, 117 days after transplanting. One row per plot
was harvested for yield data.

Setaria magna was planted April 10, 1984, by transplanting 42 day-old
seedlings into 1.8 m by 6 m plots. Each plot consisted of 3 or 6 rows,
depending on plant population, with plots separated by 2 additional guard
rows. Plant populations studied were 26,900, 53,800, and 107,600 plants

Table 1. Effect of plant population on growth of _Amaranthus hybridus_, _Sida rhombifolia_ and _Eupatorium capillifolium_ after 115, 115 and 127 days[z], respectively. Bradenton, FL 1982

Number of plants ha^{-1}	Fresh weight		Oven-dry weight		% dry matter
	Mg ha^{-1}	g plant^{-1}	Mg ha^{-1}	g plant^{-1}	
Amaranthus hybridus					
26,900	4.2	134	1.0	30	22.7
35,800	6.7	111	1.6	27	24.1
107,600	9.2	78	2.0	17	22.0
215,300	9.9	43	2.4	10	23.9
Linear	*	**	*	*	NS
Quadratic	NS	*	NS	*	NS
R^2	0.51	0.68	0.54	0.63	0.13
Sida rhombifolia					
26,900	12.0	445	2.6	97.8	22.0
53,800	9.9	183	2.3	43.5	23.7
107,600	12.5	116	2.7	25.5	21.9
215,300	13.1	61	2.9	13.5	22.3
Linear	NS	**	NS	**	NS
Quadratic	NS	**	NS	**	NS
R^2	0.11	0.85	0.10	0.83	0.10
Eupatorium capillifolium					
53,800	9.1	169	2.7	50.9	30.1
107,600	10.5	97	3.5	32.7	33.7
215,300	11.9	55	3.5	16.4	29.8
Linear	NS	**	NS	*	NS
Quadratic	NS	NS	NS	NS	NS
R^2	0.34	0.69	0.29	0.62	0.28

[z]Days from transplanting 5 week-old plants
Significant at the 5% level (*), 1% level (**) or nonsignificant (NS)

ha^{-1}. Replications were separated by 1.5 m. Fertilization consisted of 81,35 and 56 kg ha^{-1} of N, P, and K, respectively, applied in two equal applications on April 10 and June 6, 1984. Plots were harvested July 18 and 19, 1984, 99 days after planting, by cutting plants in all treatment rows at a 15 cm stubble height. At this time plants were well into senescence.

RESULTS AND DISCUSSION

1982 Experiments

After 115 days of growth from date of transplanting, fresh and oven-dry weights of A. hybridus increased ca. 100% to 9.9 and 2.4 Mg ha^{-1}, respectively, as plant population increased eight fold from 26,900 to 215,300 plants ha^{-1}, while weight of individual plants decreased some 300% (Table 1). Plant tissue averaged 23.2% dry matter and did not vary with population. Fresh and dry matter production and percentage dry matter of S. rhombifolia were not influenced by plant population; however, individual plant weight decreased more than 700% when population level was increased from 26,900 to 215,300 plants ha^{-1} (Table 1). Growth of E. capillifolium followed the same trend as that of S. rhombifolia in that yield ha^{-1} did not change with increasing population, while individual plant weights decreased (Table 1). It is probable that this lack of response to population on a ha^{-1} basis is due to the late planting date for these species in 1982.

1983 Experiments

In 1983, additional growth parameters were measured in order to better characterize the effect of plant population on growth and biomass production by the previous three species (Table 2). Population did not significantly affect plant height, number of branches, number of leaves, leaf area per plant or biomass production per unit area (Table 2). As in 1982, fresh and dry weight of individual plants decreased with increased population. Percentage dry matter composition varied with population, with the highest percentage (26.2%) at the lowest population level (26,900) and the lowest percentage (22.2%) at 107,600 plants ha^{-1}. Fresh and oven-dry weights of S. rhombifolia increased from 14.0 and 4.1 to 57.0 and 16.7 Mg ha^{-1}, respectively, when the population level was increased from 26,900 to 215,300 plants ha^{-1} (Table 2). Along with this eight-fold increase in population was a slight increase in plant height, while fresh and dry weights of individual plants, number of leaves and average leaf area decreased as the plants became more crowded and competed for light, space, and nutrients. Percentage dry matter was 28.4%. Measurements made 123 days after transplanting E. capillifolium indicated a linear increase in fresh and dry matter production ha^{-1} with a corresponding decrease in per plant fresh and dry weight as a result of competition (Table 2). Highest yields of 34.4 and 13.4 Mg ha^{-1}, fresh and dry matter, respectively, were obtained with 215,300 plants ha^{-1}. Per plants fresh matter yields dropped from 430 g with 26,900

Table 2. Effect of plant population on growth of Amaranthus hybridus, Sida rhombifolia and Eupatorium capillifolium after 63, 125 and 123 days[z], respectively. Bradenton, FL 1983 (See footnotes on Table 1)

Number of plants ha[-1]	Fresh weight Mg ha[-1]	Fresh weight g plant[-1]	Oven dry weight Mg ha[-1]	Oven dry weight g plant[-1]	Height (cm)	Number of Branches	Number of Leaves	Leaf area (cm^2)	% dry matter
26,900	6.1	201	1.6	52	116	23	319	455	26.2
53,800	18.2	302	4.2	70	143	26	252	684	23.2
107,600	14.4	120	3.3	27	129	23	235	319	22.2
215,300	17.8	74	4.1	17	116	21	144	260	23.8
Linear	NS	*	NS	*	NS	NS	NS	NS	NS
Quadratic	NS	NS	NS	NS	NS	NS	NS	NS	*
R^2	0.20	0.34	0.18	0.39	0.14	0.18	0.12	0.15	0.47
Sida rhombifolia									
26,900	14.0	466	4.1	136	207	--	1614	17,108	29.3
53,800	29.3	486	7.4	122	211	--	948	10,240	25.0
107,600	44.6	370	13.4	111	221	--	997	12,480	30.0
215,300	57.0	237	16.7	69	231	--	333	4,568	29.3
Linear	**	**	**	**	**	--	**	**	NS
Quadratic	*	NS	**	NS	NS	--	NS	NS	NS
R^2	0.85	0.68	0.84	0.68	0.46		0.63	0.66	0.03
Eupatorium capillifolium									
26,900	13.0	430	4.8	160	213	--	--	--	37.1
53,800	15.3	253	5.9	97	215	--	--	--	38.3
107,600	28.5	236	10.8	89	221	--	--	--	37.9
215,300	34.4	143	13.4	56	227	--	--	--	23.0
Linear	**	**	**	**	NS	--	--	--	NS
Quadratic	NS	NS	NS	NS	NS	--	--	--	NS
R^2	0.75	0.63	0.76	0.63	0.14				0.09

Table 3. Effect of plant population on growth of <u>Amaranthus australis</u>, <u>Ambrosia artemisiifolia</u> and <u>Setaria magna</u> after 89, 117 and 134 days[z], respectively. Bradenton, FL 1984

Number of plants ha^{-1}	Fresh weight		Oven-dry weight		% dry matter
	Mg ha^{-1}	g plant^{-1}	Mg ha^{-1}	g plant^{-1}	
				<u>Amaranthus hybridus</u>	
6,700	36.8	4.9	8.9	1190	24.2
13,500	26.1	1.7	6.8	440	26.1
26,900	70.2	2.3	17.5	570	24.9
Linear	**	*	**	*	NS
Quadratic	**	**	**	**	NS
R^2	0.79	0.87	0.80	0.83	0.22
				<u>Ambrosia artemisiifolia</u>	
13,500	36.4	2.4	9.6	631	26.6
26,900	44.2	1.5	12.0	406	27.1
53,800	52.3	0.9	14.3	246	27.3
Linear	**	**	**	**	NS
Quadratic	NS	*	NS	*	NS
R^2	0.61	0.89	0.63	0.87	0.09
				<u>Setaria magna</u>	
26,900	37.6	1.2	8.3	265	22.1
53,800	49.7	0.8	11.5	186	23.2
107,600	63.2	0.5	14.5	114	22.9
Linear	**	**	**	**	NS
Quadratic	NS	NS	NS	NS	NS
R^2	0.68	0.78	0.69	0.76	0.17

[z]Days from transplanting 5 week old plants, except <u>S. magna</u> plants which were 42 days-old at transplanting

Significant at the 5% level (*), 1% level (**) or nonsignificant (NS)

plants ha^{-1} to 143 g at the highest population level. Percentage dry matter (38.1%) was not affected by plant population.

1984:

Biomass production of <u>A. australis</u>, <u>A. artemisiifolia</u> and <u>S. magna</u> increased with increased plant population in the 1984 study (Table 3). Increasing population of <u>A. australis</u> plants 400%, from 6,700 to 26,900 ha^{-1}, increased yields of fresh and dry matter ca. 2-fold from 36.8 and 8.9 to 70.2 and 17.5 Mg ha^{-1}, respectively. Coupled with the 200% increase in yield ha^{-1} was a corresponding decrease in individual plant weight. Plant fresh weight decreased from 4.9 to 2.3 kg, but percentage dry matter (25.1%) was unaffected by population level. Although competition between plants decreased the weight of individual plants and yield ha^{-1} when the population was doubled from 6,700 to 13,500, yield ha^{-1} increased significantly with an additional doubling of the population from 13,500 to 26,900 plants ha^{-1}. This increase was believed to be due to the greater number of plants per unit area and decreased weed competition as a result of shading.

Ambrosia artemisiifolia had a linear increase in yield with increased population with the highest yields of fresh and dry matter (52.3 and 14.3 Mg ha^{-1}, respectively) produced by 53,800 plants ha^{-1} (Table 3). Intra-species competition reduced individual plant fresh weight from 2.4 to 0.9 kg as population increased from 13,500 to 53,800 plants ha^{-1}. Percentage dry matter (27.0%) was not affected by plant population.

Production of S. magna, 134 days after transplanting, followed the same trend as A. artemisiifolia with fresh and dry matter yields increasing from 37.6 and 8.3 to 63.2 and 14.5 Mg ha^{-1}, respectively, as plant population increased four-fold from 26,900 to 107,600 plants ha^{-1} (Table 3). The effect of competition resulted in more than 200% reduction in average plant weight. The percentage dry matter in plant tissue was not affected by population and averaged 22.7%.

CONCLUSIONS

Although intra-species competition reduced individual plant weight for all species evaluated, increasing the plant population increased yield significantly. Amaranthus hybridus, S. rhombifolia, and E. capillifolium produced maximum dry matter yields of 4.1, 16.7, and 13.4 Mg ha^{-1}, respectively, when their respective populations were 215,300 plants ha^{-1}. Uniform populations higher than these would be difficult to establish in a row crop situation. If broadcast seeded rather than planted in rows, higher populations could be easily achieved; however, if handled as transplants, economic considerations would probably limit the plant population to those evaluated in this study.

Maximum dry matter production of 17.5, 14.3 and 14.5 Mg ha^{-1} by A. australis, A. artemisiifolia and S. magna, respectively, was obtained at populations of 26,900, 53,800 and 107,600 plants ha^{-1}, respectively.

Possibly these species could be planted economically at even higher population levels with a further increase in yield. Since $\underline{S}$. $\underline{magna}$ is a grass species, it could be broadcast seeded easily to achieve a higher population level. The effects of further increases in the populations and other cultural management treatments of these species should be investigated. Genetic manipulation is another area for potential improvement of these species as energy crops. Based on the data obtained in these and related studies, these species might be promising bioenergy crops for the future.

CONVERSION FACTORS

Number of plants A^{-1} = 0.405 (number of plants ha^{-1})
Ton A^{-1} = 0.446 (Mg ha^{-1})
lb = 2.205 (kg)
oz = 0.035 (g)
in = 0.3937 (cm)
in^2 = 0.155 (cm^2)

ACKNOWLEDGMENTS

Florida Agricultural Experiment Stations Journal Series No. 6206. This paper reports results from a project that contributes to a joint program between the Institute of Food and Agricultural Sciences of the University of Florida and the Gas Research Institute, Chicago, Ill., titled, "Methane from Biomass and Waste."

REFERENCES

1. W. C. Adamson. Weeds for oil, polyphenol, or hydrocarbon production in the southeast, Proc. 3rd Ann. Solar and Biomass Energy Workshop, Atlanta, GA, USA, 155. 1983.
2. W. C. Adamson. Dog-fennels for oil and polyphenol production, Proc. 4th Ann. Solar and Biomass Energy Workshop, Atlanta, GA, USA, 134. 1984.
3. F. Coppola and A. Brunori. The effect of the herbicide Gramixel on the heptane fraction of Euphorbia biomass, Biomass, 4, 59-68. 1984.
4. K. E. Foster and M. M. Karpiscak. Arid lands plants for fuel, Biomass, 3, 269-285. 1983.
5. J. P. Gilreath, W. D. Pitman and D. L. Rockwood. Production of nonconventional crops for methane feedstocks, Proc. of the 1983 Int. Gas Res. Conf., 340-350. 1983.
6. P. Vasudevan, G. S. Gujral and M. Madan. Saccharum munja Roxb., an underexploited weed, Biomass, 4, 143-149. 1984.

POTENTIAL SORGHUM BIOMASS PRODUCTION

IN NORTH FLORIDA

R. L. Stanley, Jr.

Department of Agronomy
University of Florida/IFAS
Quincy, Florida 32351

L. S. Dunavin

Department of Agronomy
University of Florida/IFAS
Jay, Florida 32565

ABSTRACT

Sorghums are well adapted to the North Florida area. In the past, sorghum has been evaluated primarily as a forage crop with limited evaluation of the older sweet (or so-called "syrup") types. Presently sorghum types are being evaluated in our biomass programs to determine maximum biomass production potential of the different types. Total biomass production of the forage (or "silage") types has been as high as 35.7 Mg ha^{-1} when harvested twice during a growing season. Grain content has varied widely, ranging from 25 to 50%. Production of the older syrup types has been as high as 45.1 Mg ha^{-1} for 2 harvests with grain content from 5 to 20%. When forage and syrup types were compared in the same experiments over a 3-year period, the syrup types produced more than the forage types. Recent entries from the Texas A&M research program have been evaluated for 2 years in North Florida. When harvested twice during a season, highest yield has been 31.6 Mg ha^{-1} for a 2-year average. When evaluated with the older syrup types none of the Texas entries have produced as much total biomass as the syrup types. In a row spacing and seeding rate study, seeding rate had a greater influence on production than row spacing. Several Texas entries and two of the older syrup types were planted at seed rates of 11.2 kg ha^{-1} (normal rate for syrup production). Total biomass production for all entries was greater at the higher rate.

Keywords. Sorghum types, seed rate, row spacing.

INTRODUCTION

Sweet sorghum [(<u>Sorghum</u> <u>bicolor</u> (L.) Moench.)] has been grown for many years in the southeastern United States and used to make syrup (Cowley and Smith, 1972). Its range of adaptation extends into the north Florida area. Cultivars with high sucrose content have been developed more recently for potential use as sugar crops. Recent increases in crude oil prices have resulted in research on sorghum as a potential source of sugars for ethanol fuel production.

The traditional use of sweet sorghum for human consumption has resulted in much research on cultural practices for syrup production. Cultural practices are well documented and summarized by Freeman et al., 1973. Grain and forage types of sorghum used for livestock feed are also well adapted to north Florida and cultural practices for production of livestock feed in Florida have been developed (Wright and Gorbet, 1984).

Broadhead and Freeman (1980) reported that narrow rows (52.5 cm) resulted in higher yields of gross and stripped stalks of sweet sorghum than conventional rows (105 cm). Stickler and Younis (1966) reported that grain yields in 50.8 cm rows were 11% greater than yields in 101.6 cm rows. Hiler and Isaacson (1984) reported a sorghum-methane system to be in the realm of economic feasibility, with the economics improved when it was feasible to harvest grain and vegetative material for separate purposes of food and energy.

The objective of the research reported here was to evaluate some of the different types of sorghum for maximum biomass production and to evaluate the effect of some cultural practices on total biomass production of the different types.

MATERIALS AND METHODS

This research was conducted at the Agricultural Research and Education Center near Jay, Florida (AREC, Jay) and at the North Florida Research and Education Center near Quincy, Florida (NFREC, Quincy).

AREC, Jay

Two sweet sorghums (MN1500 and M81E) and 2 forage types ('Red Top Kandy' and 'Titan R') were planted at a seeding rate of 11.2 kg ha^{-1} on

218

3-16-82, 5-10-83, and 4-6-84. Individual plots were 3 rows 4.9 m long and
.9 m wide. Treatments were replicated 3 times. In 1982 and 1983 the soil
was a Dothan fine sandy loam (Plinthic Paleudult) and in 1984 it was a Red
Bay sandy loam (Rhodic Paleudult). Preplant fertilization in 1982 and 1984
was 560 kg ha^{-1} of 8-24-24 and 280 kg ha^{-1} in 1983. When plants were about
3-4 weeks old each year, ammonium nitrate was applied at 224 kg ha^{-1}. A
second application at the same rate was applied after the first harvest
which had been made at the late dough stage of maturity.

The following sorghums were planted at 11.2 kg ha^{-1} seed on 5-12-83 and
on 4-6-84: (1) AT x 623 x Pickett-3, (2) AT x 623 x Rio, (3) A-Atlas x RT x
430, and (4) 'Rio'. Plot size, soil types and fertilization were the same
as previously described for each year. Each cultivar was harvested when it
reached the late dough stage. Ten cultivars from Texas A&M were planted at
11.2 and 3.4 seed on 5-12-83 with the same management practices as previously described.

Entry AT x 623 x Rio was planted on 6-1-83 at two seeding rates (11.2
and 3.4 kg ha^{-1}) in 2 row widths (.46 and .91 m). Plots were 2.4 m long,
with 5 rows making a plot at the .46 m spacing and 3 rows making a plot at
the .91 m spacing. Plots were replicated 3 times. Soil type was a Dothan
fine sandy loam (Plinthic Paleudult). Preplant fertilizer was 336 kg ha^{-1}
of 8-24-24, with ammonium nitrate at 224 kg ha^{-1} applied on 7-7-83. A
single harvest was made on 9-8-83 at the late dough stage.

For all experiments the center row was harvested, weighted, and a
sub-sample weighed and dried in a forced-air oven at 60 C to obtain dry
matter content.

NFREC, Quincy

In 1983 ten genotypes from Texas A&M and nine from the U.S. Sugar Crops
Field Station at Meridian, Mississippi were replicated four times in 3-row
plots 6 m long with rows spaced .9 m apart. Seeds were hand planted on
6-15-83 in hills 40.6 cm apart and hand thinned to 4 plants per hill after
emergence. Soil type was a Norfolk loamy fine sand (Plinthic Paleudult).
Preplant fertilizer was 560 kg ha^{-1} of 5-10-15, with 168 kg ha^{-1} ammonium
nitrate applied on 7-18-83. A 4.8 m section from the center row of each
plot was harvested at the hard dough stage for yield determination. In
1984, ten genotypes from Texas A&M (four of which were included in 1983 plus

six additional ones) were evaluated under the same practices as in 1983. Planting date in 1984 was 5-24-84.

RESULTS AND DISCUSSION

Table 1 shows total biomass production for 3 years of two of the highest producing sweet sorghum types and two of the highest producing forage types at the AREC, Jay, Florida. In 1982 the sweet types produced more total biomass at each harvest and for the entire season than the forage types. Production of the ratoon harvest relative to the first harvest was .80 and .76 for the sweet types, and .80 and 1.04 for the forage types. In 1983 the ratoon harvests were much lower for all cultivars. First harvest yields were greater for the two sweet types, while ratoon yields were greater for the forage types. Total yield for the season was lower for 'Red Top Kandy' than for the other three cultivars. In 1984, a significantly higher ratoon yield for 'Red Top Kandy' resulted in a season total greater than for the other forage type but not significantly different from the two sweet types. When total production is averaged for the 3 years, the two sweet types produced more than the forage types. Forage type 'Red Top

Table 1. Biomass production of two sweet and two forage sorghum cultivars at AREC, Jay, Fl

Year	Harvest	Sweet Type		Forage Type	
		MN 1500	M81E	Red Top Kandy	Titan R
		----------------Mg ha^{-1} Dry Matter-------------			
1982	First	21.5 a[1]	20.3 a	14.9 b	11.0 b
	Ratoon	17.2 a	15.4 a	12.0 b	11.4 b
	Season total	38.7 a	35.7 a	26.9 b	22.4 b
1983	First	24.8 a	25.0 a	18.0 b	19.0 b
	Ratoon	2.5 b	3.6 b	5.7 a	6.5 a
	Season total	27.3 a	28.6 a	23.7 b	25.5 b
1984	First	24.8 a	21.1 b	19.9 b	16.1 c
	Ratoon	7.0 c	9.0 bc	15.9 a	11.2 b
	Season total	31.8 ab	30.1 ab	35.8 a	27.3 b
3-yr. Average		31.6 a	31.5 a	28.8 b	25.1 c

[1]Means in a line (horizontal) followed by the same letter are not significantly different at the .05 probability level by Duncan's NMRT.

Kandy' produced more than the forage type 'Titan R'. Under favorable growing conditions, 'Red Top Kandy' has production potential as great as the two
syrup cultivars used in this experiment, as shown in 1984.

At the Agricultural Research and Education Center near Jay, Florida,
'Rio' was the lowest producing cultivar in 1983 at both the first harvest
and the ratoon harvest; consequently, total production for the season was
lower than for any of the other entries (Table 2). In 1984 'Rio' produced
more, however, than the other entries at the first harvest with season
totals no different among the entries. The ratoon harvest for all entries
in 1984 was greater relative to the first harvest than in 1983. Average
production for the two years was greater for AT x 623 x Pickett-3 than for
the other entries.

Yields at Quincy were much lower than at Jay both years (Table 3).
Planting dates were later at Quincy so that only a single harvest was made.
Plants were under considerable drouth stress at periods during the 1984
growing season. Rio was the highest producing cultivar in both years. For
the 2-year average A-Atlas x RT x 430 was lower in production than the
others. The spacing at Quincy was that which had been traditionally used
with the older syrup varieties for maximum syrup production; i.e. hills 40.6
cm apart with four plants per hill. As will be seen in some following data
this spacing may not be adequate for maximum biomass yield with the more
recent sugar types.

Table 2. Biomass production of four Texas A&M sweet sorghum entries for 2
years at AREC, Jay, FL.

Year	Harvest	AT x 623 x Pickett-3	Atlas x 623 x Rio	A-Atlas x RT x 430	Rio
		-----------------Mg ha^{-1} Dry Matter---------------			
1983	First	21.8 a[1]	16.5 b	17.0 b	12.7 c
	Ratoon	13.8 a	9.4 b	10.4 b	5.3 c
	Season total	35.6 a	25.9 b	27.4 b	18.0 c
1984	First	13.2 b	16.2 b	14.1 b	18.5 a
	Ratoon	14.4 a	12.3 b	10.6 b	11.4 b
	Season total	27.6 a	28.5 a	24.7 a	29.9 a
2-yr. Average		31.6 a	27.2 b	26.0 b	23.9 b

[1]Means in a line (horizontal) followed by the same letter are not
significantly different at the .05 probability level by Duncan's NMRT.

Table 3. Biomass production of four Texas A&M sweet sorghum entries at NFREC, Quincy, Fl

Entry	1983	1984	2-yr. Avg.
	------------------Mg ha^{-1} Dry Matter----------------		
AT x 623 x Pickett-3	20.2 b	11.2 c	15.7 a
AT x 623 x Rio	19.5 b	13.6 ab	16.5 a
A-Atlas x RT x 430	13.1 b	12.0 bc	12.5 b
Rio	22.4 a	14.2 a	18.3 a

[1]Yields in a column followed by same letter are not significantly different at the .05 probability level by Duncan's NMRT.

Table 4. Effect of seeding rate on biomass production of ten sorghums at AREC Jay, Fl

Entry	Seed rate	First harvest	Ratoon	Season total	Relative[1] yield
AT x 623 x Pickett-3	11.2	21.8	13.8	35.6	1.00
	3.4	7.5	7.2	14.7	0.41
A-Atlas x RT x 430	11.2	17.0	10.4	27.4	1.00
	3.4	9.4	10.6	20.0	0.73
AT x 623 x Rio	11.2	16.5	9.4	25.9	1.00
	3.4	10.6	9.8	20.4	0.78
AT x 623 x TMT x 430	11.2	13.8	7.3	21.1	1.00
	3.4	8.7	8.7	17.4	0.82
AT x 623 x 74 cs 5388	11.2	12.0	6.9	18.9	1.00
	3.4	9.4	6.8	16.2	0.85
Rio	11.2	12.7	5.3	18.0	1.00
	3.4	11.2	4.9	16.0	0.88
AT x 623 x SC0599-11E	11.2	10.1	5.4	15.5	1.00
	3.4	6.7	6.4	13.1	0.84
Wray	11.2	11.6	4.8	16.4	1.00
	3.4	8.5	6.2	14.7	0.89
AT x 623 x RT x 430	11.2	10.4	3.7	14.1	1.00
	3.4	6.4	3.7	10.1	0.71
Brown Midrib-12	11.2	6.8	2.5	9.3	1.00
	3.4	4.8	2.3	7.1	0.76
LSD .05		3.7	4.2	6.8	

[1]Relative yield = total yield @ 3.4 kg ha^{-1} divided by total yield @ 11.2 kg ha^{-1}

The effect of seeding rate on biomass production of ten sorghum entries
is shown in Table 4. Normal seeding rate for forage and grain sorghum types
is 11.2 kg ha^{-1}, whereas normal rate for syrup production is 3.4 kg ha^{-1}.
The two cultivars 'Rio' and 'Wray' are older syrup types, while the others
are sugar types from the Texas A&M research programs. The high seeding rate
gave the highest total yields for the season for all entries. At the first
harvest, yields were ususally much higher at the higher seeding rate. For
the ratoon harvest there was much less difference between the two rates,
with some entries producing more at the low seeding rate. At the high
seeding rate the first harvest was usually much greater than the ratoon
harvest, while at the low seeding rate the ratoon harvest more closely
approached the first harvest, being greater for some entries. The highest
producing entry produced only 41% as much total biomass at the low seeding
rate as it did at the high rate. The two syrup types ('Rio' and 'Wray') had
the highest relative producton at the low rate. Average production of all
entries at the low rate was 74% of the production at the high rate.

Table 5 shows the effect of row width and seeding rate on the total
biomass production of one of the higher producing entries (AT x 623 x Rio).
Yields are for a single harvest from a June planting that was harvested in
September. Regrowth was not sufficient for a ratoon harvest. Row width did
not affect biomass production at either of the seeding rates in the study;
however, yields were lower at the low seeding rate than at the high seeding
rate. Decreasing row width should provide a more favorable environment for
individual plants by allowing more space between plants in the row and the
effect should be more pronounced at higher seeding rates. However, the
narrow rows did not result in greater biomass production at either seeding
rate in this study. The data from Tables 4 and 5 indicate that for types

Table 5. Effect of row spacing and seeding rate on biomass production of
sorghum entry AT x 623 x Rio. AREC Jay, Fl., 1983.

Row width	Seeding rate	Dry matter
(m)		
0.46	11.2	25.6 a[1]
0.91	11.2	25.4 a
0.91	3.4	16.7 b
0.46	3.4	15.5 b

[1]Yield followed by same letter are not different at the .05 probability
level by Duncan's NMRT

223

similar to those used in these experiments, the normal seeding rate for syrup production (3.4 kg ha^{-1}) are not adequate for maximum biomass production of these types.

Table 6 shows the grain production as percent of the total biomass production for some of the different types that have been evaluated. For the years 1982-84 the two forage types 'Titan R' and 'Red Top Kandy' had more than doubled the percent grain of the 2 syrup types, 'MN1500' and 'M81E'. In 1982 there was not sufficient time for the ratoon crop to produce mature seed resulting in a low seed yield. In 1983 Titan R ratoon harvest was higher in percent grain than the first harvest, but total dry matter at the ratoon harvest was low (Table 1). For 1984 total dry matter yield at the ratoon harvest was relatively high for the forage types (Table 1) with grain content only slightly lower than for the first harvest (Table 6). The last four entries in Table 6 are some of the recent sugar types, and grain production of these four entries was higher than for the syrup types and approached that of the forage types. If high grain production is desired, the forage and sugar types would be more suitable than the syrup types.

Table 7 shows total biomass production at Quincy of sorghum cultivars from the research programs at Mississippi and Texas. The yields are for a single harvest from a June 15 planting in 1983. The higher producers were syrup types and the lower were the sugar types. A higher seeding rate for the lower producing types might have resulted in a higher relative production, as was shown in Table 4.

Table 6. Grain content as percent of total biomass for some sorghum entries. AREC Jay, Fl., 1982-1984

Cultivar	1982			1983			1984		
	First harvest	Ratoon harvest	Season total	First harvest	Ratoon harvest	Season total	First harvest	Ratoon harvest	Season total
					% Grain				
M81E	23	12	18	13	10	13	21	15	19
MN1500	8	4	6	3	0	3	15	4	13
Titan R	55	1	27	26	45	31	47	38	43
Red Top Kandy	49	5	29	27	9	23	37	27	33
AT x 623 x Pickett-3	--	--	--	12	--	--	27	17	22
A-Atlas x RT x 430	--	--	--	22	--	--	37	28	33
AT x 623 x Rio	--	--	--	35	--	--	33	38	35
Rio	--	--	--	18	--	--	16	21	18

Table 7. Total biomass production of some sorghum entries at NFREC Quincy, Fl., 1983

Entry	Dry Matter
	$(Mg\ ha^{-1})$
M81E	37.6
Mer 81-4	35.1
Mer 76-3	32.5
Mer 76-6	30.3
Mer 77-5	25.1
Theis	24.3
Rio	22.4
Mer 78-6	21.9
Mer 71-7	20.6
AT x 623 x P-3	20.2
AT x 623 x Rio	19.4
Wray	18.6
Dale	18.6
AT x 623 x TMT x 430	16.0
A-Atlas x RT x 430	13.0
AT x 623 x SC0599-11E	12.0
Brn Midrib 12 x AT x 623	10.8
AT x 623 x RT x 430	9.7
AT x 623 x 74CS5388	7.9
LSD$_{.05}$	5.26

SUMMARY AND CONCLUSIONS

Data presented indicate the production potential of some of the sorghum types in the North Florida area. The syrup types are generally higher in total biomass production, while the forage types have lower total production but higher grain content. Good ratoon yields can be obtained with all types from early plantings. Narrow rows have not increased total biomass production, but seeding rate recommended for forage types ($11.2\ kg\ ha^{-1}$) increased production over the rate recommended for syrup production ($3.4\ kg\ ha^{-1}$). Laboratory evaluation of samples should provide alcohol and/or methane production potential and give direction for further agronomic studies.

ACKNOWLEDGEMENT

This research is part of a joint program between the University of Florida, Institute of Food and Agricultural Sciences and the Gas Research

Institute. Dr. F. Miller and Mr. R. Monk, Plant and Soil Sciences Department, Texas A&M University provided sorghum varieties from their breeding program.

REFERENCES

1. D. M. Broadhead and K. C. Freeman. Stalk and sugar yield of sweet sorghum as affected by spacing. Agron. J. 72:523-524. 1980.
2. W. R. Cowley and B. A. Smith. Sweet sorghum as a potential sugar crop in south Texas. Proc. 15th Congress Int. Soc. Sugar Cane Technol. pp. 628-633. 1972.
3. K. C. Freeman, D. M. Broadhead and N. Zummo. Culture of sweet sorghum for syrup production. USDA-ARS Handbook No. 441. 1973.
4. E. A. Hiler and H. R. Isaacson. Sorghums for methane production. 1984 International Gas Research Conference pp. 605-612. 1984.
5. F. C. Stickler and M. A. Younis. Plant height as a factor affecting responses of sorghum to row width and stand density. Agron. J. 58:371-373. 1966.
6. D. L. Wright and D. W. Gorbet. Grain sorghum production. Fla. Agr. Expt. Sta. Circ. No. 506. 1984.

INSECT PEST MANAGEMENT OF SWEET SORGHUM IN SUGARCANE PRODUCTION

SYSTEMS OF LOUISIANA: PROBLEMS AND INTEGRATION

T. E. Reagan and J. L. Flynn

Department of Entomology
Louisiana State University Agricultural Center
Baton Rouge, Louisiana 70803

ABSTRACT

Sugarcane and sweet sorghum are being considered as potential biomass energy crops. Research in Louisiana has revealed a potential for increasing populations of the sugarcane borer, Diatraea saccharalis (F.), where sweet sorghum is grown near sugarcane. Early and late plantings of sweet sorghum (variety Wray) and corn with moderately resistant (CP 65-357) and susceptible (CP 61-37) sugarcane revealed a 1.4-fold increase in D. saccharalis pupal production in sorghum, and a 3-fold increase in corn as compared to the commercially grown CP 65-357. Preliminary studies on egg laying of D. saccharalis adults reared from corn, sorghum, and sugarcane showed 623, 583, and 473 eggs/pair, respectively.

A selection of six sweet sorghum fields adjacent to sugarcane (CP 65-357) for season-long monitoring of D. saccharalis during 1983 showed a 2-fold higher infestation in sorghum, and a resultant 21,835 vs. 5,440 adults emerged per ha ($P<0.05$), respectively. Factors which appear to be associated with these results include host plant, predation, and insecticide control efficacy differences between the crops. Future research is expected to reveal the feasibility of integrating cultural practices into a system to effectively manage this key insect pest and facilitate the compatible production of sweet sorghum and sugarcane.

Keywords. Pest management, insects, sugarcane, sorghum, biomass.

227

INTRODUCTION

The sugarcane borer (SCB), (<u>Diatraea</u> <u>saccharalis</u>, F.--Lepidoptera:
Pyralidae) is responsible for more than 90% of all sugarcane crop losses
ascribed to insect damage in Louisiana.[1] Sugar yield losses in fields where
SCB infestations are not held below the economic threshold (5% of plants
with live larvae in leaf sheaths) range up to 70% on susceptible commercial
varieties, and average 12-15% annually.[2] The principal cultivated hosts of
SCB in addition to sugarcane (<u>Saccharum</u> <u>officinarum</u>, L.) are sorghum (<u>Sor-
ghum</u> <u>bicolor</u>, L. Moench), corn (<u>Zea</u> <u>mays</u>, L.), and rice (<u>Oryza</u> <u>sativa</u>, L).[3]
The SCB has traditionally caused major yield losses in corn when grown in
sugarcane areas.[4] Recommendations encourage growing corn as far away from
sugarcane as possible.[5]

Primarily through research at Louisiana State University (LSU) during
the mid-1960's, integrated pest management systems for the SCB were developed as a direct response to heavy insecticide usage, pesticide related
ecosystem disruptions, and control failures caused by SCB insecticide resistance.[6] Though tactics which readily combat other sugarcane insect pests
are included in this system,[5] the emphasis is on management of SCB. The
system now stresses a balanced use of cultural, biological, and chemical
pest control strategies.[2] However, anticipated changes in cropping practices, particularly the production of sweet sorghum for biomass energy, may
be placing this highly successful pest management program in jeopardy.
These studies were undertaken to determine the potential role that sweet
sorghum and corn might play in the population dynamics of the sugarcane
borer as related to sugarcane production.

MATERIALS AND METHODS

During spring, 1981, a 6-replication field experiment was conducted to
compare SCB population development in two plantings (April 8, May 5) of corn
(Funk's 581), two (April 20, May 15) of sweet sorghum (Wray), and sugarcane
varieties moderately resistant (CP 65-357) and susceptible (CP 61-37) to SCB
(planted September 15, 1980). Individual plots were 6 rows 1.8 m (centers)
by 7.1 m (0.008 ha each) in a randomized complete block design. Sorghum and
corn were planted in two drills per row to simulate normal planting densities for these crops. Current cultural recommendations of the LSU Cooperative Extension Service for the various crops were followed. A ground application of chlordane (0.9 kg ai ha^{-1}) was made in early summer to suppress

predators and thus enhance SCB populations. Plant development determinations were made every 2-3 weeks.

At 5-7 day intervals after internodes first were visible in the sugarcane, data on the number and developmental stages of all larvae were determined by careful dissection of 20 plants randomly selected from the first 4 rows of each plot. SCB fecundity data were obtained from field collected pupae by returning them to the laboratory where they were segregated by sex (separately for each crop) and held in containers until adult emergence. Freshly emerged pairs of male and female moths were then placed in 0.5 l cartons containing moist vermiculite as well as wax paper upon which they could oviposit. After completion of egg laying (4-5 days), wax sheets were removed and the number of eggs and egg masses were counted.

The second study involved a 6-replication side-by-side comparison of sweet sorghum (Wray) and sugarcane (CP 65-357) fields (ca. 3 ha each) conducted on private farms in the Breaux Bridge/Henderson, Louisiana area during 1983. All sorghum fields were planted (2 drills on 1.8 m centers) between May 26 and 28. SCB infestation data was collected as previously described except that weekly infestation counts were made on 50 plants (systematically selected) in each sorghum and sugarcane field. Additionally, counts of plants harboring SCB predators including foraging imported fire ants (IFA), Solenopsis invicta, and carabid beetle larvae were made on all plants examined.

SCB damage determinations were made September 19th and 23rd by counting the number and position of internodes exhibiting entrance and exit holes on 50 randomly selected plants from each field. Yield data were taken from a particularly heavily infested sorghum field in which four 0.004 ha plots were sprayed with monocrotophos (0.84 kg ai ha^{-1}) for SCB control on August 29. Yield loss was determined via weight comparisons of 40 stripped (seed head removed) sorghum plants of medium diameter (4.3 ± SE 0.1 cm at the third basal internode) selected from each insecticide-treated and adjacent untreated plot.

Data for the 1981 study were subjected to analysis of variance with treatment mean separation by Duncan's multiple range test.[7] Three of the replications of sugarcane and sorghum in the 1983 study received an aerial application of azinphosmethyl (0.84 kg ai ha^{-1}) for suppression of SCB populations in early August. Since both crops were equally treated, data for this study were analyzed using a Paired-t test ($P<0.05$).

In the 1981 experiment, overall seasonal averages of percent infested plants were similar among all treatments (Table 1). However, treatment differences were detected during individual months. The late planted sorghum sustained lowest infestations in June and July ($P < 0.05$). Differences between early and late plantings of sorghum for these months strongly suggest a phenological influence as does the difference between the corn plantings in August. In the former case, the smaller size and younger age of the late planted sorghum (plants did not flower until early-mid August compared to mid July for the early planted sorghum) probably decreased the relative ovipositional attractiveness of this treatment. Likewise, the advanced state of senescence of the early planted corn by August probably reduced its attractiveness and/or host suitability for SCB.

Unlike the percent infested plant results, pupae production data displayed pronounced differences among crops (Table 1). Except for the corn plantings in June, planting dates apparently did not affect production of pupae. Pupa counts in corn were significantly higher ($P < 0.05$) than the other crops in June (early planting) and July (both plantings). August pupae production was similar for corn and sorghum, both of which were considerably higher than for sugarcane, especially CP 65-357. Average plot counts of pupae (number per ha) in corn remained above that of sugarcane throughout the season and that of sorghum until the last 2 sampling dates (Figure 1) (See also discussion of Table 2 for further explanation).

A breakdown of season average counts for the various SCB developmental stages are shown in Figure 2. First and second instar counts were markedly higher in sugarcane than in the other crops, probably due to sampling bias. Infestations of SCB are found throughout the plants in corn and sorghum while in sugarcane, young infesting stages (I-II instars) are generally confined to the upper leafsheaths. This undoubtedly hampered accurate detection of these early (I-II instar) larvae in corn and sorghum compared to sugarcane.

Third instar larval size and associated feeding damage allow for much easier detection and all 3 crops produced similar results for this stage. Average plant counts for the succeeding SCB stages demonstrated increasingly greater crop differences, possibly indicating late stage SCB mortality differences among the crops. This is demonstrated by the statistical comparison of differences between season average plant counts of third instars

Table 1. Monthly and seasonal comparisons of percent SCB infested plants and pupal production among two plantings each of corn and sweet sorghum and two varieties of sugarcane, St. Gabriel, La., 1981

Crop**	Mean % Infested plants*				Mean no. pupae ha^{-1}(*)			
	June	July	August	Season avg.	June	July	August	Season avg.
Corn (planting early)	11.2ab	40.2a	75.0b	46.2a	1420.8a	11,108a	16,878a	8387a
Corn (planting late)	9.1bc	40.5a	86.1a	51.2a	258.3b	10,288a	17,567a	7689ab
Sugarcane (CP 61-37)	14.8a	46.7a	75.6ab	53.1a	437.5b	6,854ab	10,500ab	5017bc
Sugarcane (CP 65-357)	9.2bc	40.0a	77.8ab	49.0a	0b	4,773b	5,833b	3017c
Sorghum (planted early)	5.7cd	38.5a	76.0ab	46.8a	0b	6,525ab	17,400a	6090abc
Sorghum (planted late)	1.9d	27.8b	78.6ab	51.0a	0b	4,125b	16,750a	5000bc

* Means in a column not followed by the same letter significantly differ (P<0.05; Duncan's (1951) miltiple range test)

** Planting dates sorghum: April 20, May 15; corn: April 8, May 5

231

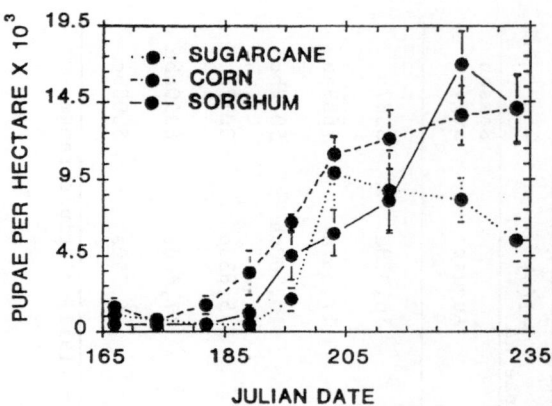

Figure 1. Seasonal production of D. saccharalis pupae (mean ± SE) in corn, sweet sorghum, and sugarcane. St. Gabriel, LA. 1981

(larvae generally establish in the stalk during this stage) and pupae (Table 2). The resultant apparent mortalities were highest for the moderately resistant sugarcane variety (CP 65-357) while these values were very low for corn. As expected, season total pupae production per hectare was highest for corn and lowest for CP 65-357. Overall, in this test, sorghum displayed a potential for SCB production comparable to the susceptible sugarcane variety CP 61-37.

Fecundity data were obtained from 15, 11 and 10 mated pairs of SCB moths reared from corn sorghum and sugarcane, respectively. Mean number of eggs (± SE) per pair from corn, sorghum and sugarcane were 623 ± 43, 538 ± 57 and 473 ± 72, respectively. Additional trials are needed to determine whether these trends represent actual differences. Plot yields in this experiment averaged ($\bar{x}$ ± SE) 73.3 ± 1.4 and 47.2 ± 4.3 Mg ha^{-1} (fresh weight) for sugarcane (varieties combined) and sorghum (both planting dates), respectively. Corn ear production averaged ($\bar{x}$ ± SE) 9,852.7 ± 580 kg ha^{-1} over both planting dates.

A phenological association of SCB and sorghum was also implicated in the 1983 study (Table 3). A dramatic rise in SCB infestation levels was noted in the sorghum fields in late July (J.D. 200) at which time sorghum plant heights had exceeded those of sugarcane. This differential prevailed throughout the remainder of the season. Overall season average percent infested plants were over 2 x higher in sorghum than in sugarcane (22.1% vs 10.9%, respectively, P<0.05). Average plant counts of SCB larvae in sorghum and sugarcane (0.48 and 0.17 larva/plant, respectively, P<0.05) indicated an even greater difference in SCB populations between the crops.

Plant examinations also revealed striking differences between the two crops in the populations of two important SCB predators, imported fire ants (IFA), Solenopsis invicta Buren, and carabid larvae. The percentage of sugarcane plants harboring IFA was over 4 x higher (P<0.01) than sorghum (17.9 ± SE 3.8% vs 4.0 ± SE 1.7%, respectively). Conversely, over 7 x as many (P<0.01) sorghum plants harbored carabid larvae as did sugarcane plants (9.2 ± SE 2.4% vs 1.2 ± SE 0.8%, respectively). IFA are considered the most important predator of SCB in Louisiana sugarcane[1] and their abundance in Louisiana sugarcane fields has been associated with the more stable semi-perennial nature of this crop.[8] IFA have been implicated as predators of carabid larvae as well[8,9] and this may at least partially account for the greater abundance of carabid larvae in the sorghum fields.

The intensity and pattern of SCB internode damage showed significant differences between sugarcane and sorghum (Table 4). Sorghum incurred nearly twice the bored internodes as did sugarcane. Sugarcane stalk damage increased from basal to upper internodes reflecting the successive increase in SCB population throughout the season. In sorghum, however, stalk damage was more evenly distributed over the plant. This corroborates field obser-vations in which SCB infesting stages could be found at any internode posi-tion infesting leaf sheaths or inside the sorghum stalk. This poses several important implications relative to pest management of sorghum. First, sorghum scouting would require more time and effort compared to that of sugarcane where the early infesting stages of SCB are more predictably confined to the upper leaf sheaths. In sugarcane, the lower leaves senesce, dry, and fall off (most commercial Louisiana varieties) while the rind becomes hardened, thus only the top portion of the plant (not more than 4-5 ligules from the whorl) can be successfully attacked by the borer. Such features as early rind hardness and leaf sheath appression are recognized plant resistance factors.[10] Additionally, this greater larval distribution on the sorghum plant may affect the efficacy of insecticidal SCB control strategies. Indeed, SCB suppression in sorghum fields receiving an insecti-cide application was much less pronounced than that of the adjacent sprayed sugarcane fields. Finally, yield loss considerations (sorghum as compared to sugarcane) could be an important implication of the more even distribu-tion of larvae since late infestations (referred to as top infestations in sugarcane) potentially could be more devastating in sorghum by indiscrimi-nately damaging middle and lower plant internodes.

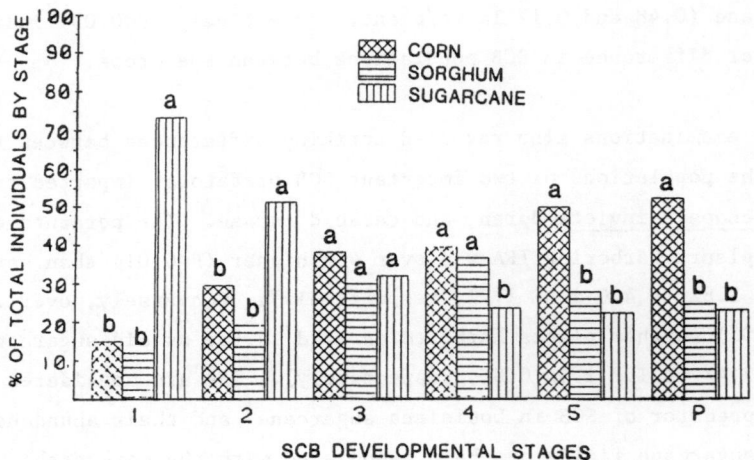

Figure 2. Relative proportion of sugarcane borers (SCB) found in sugarcane, corn, and sweet sorghum for all SCB development stages. P < 0.05 a vs. b.

Exit holes (indicating SCB moth departure) were much more numerous (P<0.01) in sorghum than in sugarcane, as was the ratio of exit holes to the number of bored internodes. This latter value indicated that 3 x more of the SCB larvae which had established within the sugarcane internode, completed development in sorghum. The overall higher SCB larval infestations combined with the higher success of completing development and higher plant population resulted in a 4 x greater (P<0.01) adult SCB production in sorghum.

A 10.1% reduction (P<0.01) in stalk weight was found in untreated (vs. insecticide treated) sorghum plants (small plot test at one of the survey fields).[11] The percentage of bored internodes ($\bar{x}$ ± SE) in the treated and untreated plots was 19.5 ± 4.6 and 48.6 ± 3.2 (P<0.01), respectively. This translates to an average yield loss of 0.35% per % bored internode.

In summary, the area-wide production of SCB pupae and adults in sweet sorghum is substantial, exceeded only by the superior within-season host, corn. Of particular interest is the extremely small amount of late-stage larval mortality in corn (usually only 10-20%), a crop which apparently offers an ideal haven after the insect bores into the plant.[12]

Table 2. Comparison of pupae production and the relative success of development completion of D. saccharalis larvae established on corn, sweet sorghum, and sugarcane, St. Gabriel, La., 1981. [1]

Crop	X̄ no. III [2] instars/plant	X̄ no. [2] pupae/plant	Apparent mortality [3] between stages III and pupae	Mean no. [4] pupae/ha
Corn	.175a	.145a	10.6b	77,877a
CP 61-37	.156a	.079b	43.7ab	40,167b
Sorghum	.137a	.068b	49.0ab	50,100b
CP 65-357	.138a	.048b	60.3a	29,050c

[1] Means in a column not followed by the same letter significantly differ ($P<0.05$; Duncan's (1951) multiple range test.

[2] Season averages of counts taken from 15 June to 20 August.

[3] Calculated from the percent difference between season averages of the number of III instar larvae/plant and the number of pupae/plant and provides a relative comparison of the proportion of established larvae that complete development.

[4] Calculated from the cumulative season totals of pupae/plant times the plant populations determined for each crop as follows: sugarcane 70,000 ha^{-1}; corn 62,000 ha^{-1}; sorghum 90,000 ha^{-1}.

Table 3. Comparison of seasonal progression of D. saccharalis infestations relative to growth of sugarcane (CP 65-357) and sweet sorghum. Breaux Bridge and Henderson, La., 1983.

Julian date	Plant development [1]		% SCB infested plants (±SE)		Mean no. larvae/plant	
	Sugarcane	Sorghum	Sugarcane	Sorghum	Sugarcane	Sorghum
182	44.5	30.1	2.0 ± 1.1	0.2 ± 0.1	0.02 ± .01	0.02 ± .02
200	84.5	98.1	6.1 ± 1.3	6.6 ± 1.0	0.06 ± .01	0.10 ± .02
210	109.8	170.6	6.7 ± 2.0	17.3 ± 2.3	0.01 ± .05	0.23 ± .03
217	112.8	(50% flowering)	7.2 ± 2.9	16.0 ± 4.6	0.09 ± .04	0.27 ± .12
227			6.9 ± 3.4	23.0 ± 8.1	0.10 ± .05	0.43 ± .18
235	192.8	248.4	12.8 ± 2.5	33.3 ± 6.5	0.23 ± .05	0.82 ± .30
			32.5 ± 12.2	57.9 ± 8.9	0.56 ± .24	1.50 ± .41

1/ Figures are the mean height (cm) from base of plant to the highest visible ligule determined from an average of 20 plants per field.

Table 4. Comparison of D. saccharalis moth emergence and plant damage between sugarcane (CP 65-357) and adjacent sweet sorghum fields. Breaux Bridge and Henderson, La., 1983. [1]

	Mean % D. saccharalis Bored Internodes				Mean Moth Emergence		
Crop	Bottom (internodes 1-4)	Middle (internodes 5-8)	Top (internodes 9 +)	Whole stalk	Mean no. exit holes plant [2]	Mean no. exit holes ha [3]	Ratio of exit holes to bored internodes [4]
Sorghum	19.9a	26.1a	21.4a	22.5a	.45a	21,834a	0.16a
Sugarcane	6.5b	12.9a	16.6a	11.4a	.08b	5,448b	0.05b

1/ Paired t-test (P<0.05)

2/ Exit holes represent external perforations in the stalk where the SCB moth has emerged and can be distinguished from larval entrance holes by their characteristic shape and size.

3/ Calculated using plant population estimates for each field.

4/ Gives a relative indication of the proportion of those larvae which bore into the stalk that successfully complete development.

237

The large and rapid buildup of 1983 SCB populations in sorghum was greater than would be predicted based on the 1981 study. We believe that several factors may be responsible for these results. First, less SCB predation by IFA, the key natural control agent in Louisiana sugarcane, occurred in sorghum than in sugarcane in the 1983 study. Ground predators were suppressed in 1981 so this was not a contributing factor in that year's study. Another factor could be the differential insecticidal control of SCB achieved in those replications treated in 1983. As mentioned, this may be associated with the greater SCB distribution on the sorghum plant (as compared to sugarcane). Finally, the large field plot sizes used in the 1983 study provided a more realistic situation for natural population development.

Although these initial findings would seem to cast aspersions on the future of sweet sorghum as a compatible energy or sugar producing crop for southern Louisiana, such interpretations may not be altogether bleak. Because manipulation of management and cultural practices has been successful in limiting sugarcane pest-related problems, such experiences may be useful and adopted to sweet sorghum production. These would include: (1) performing an SCB variety resistance search of sweet sorghum, (2) determining the effect of planting resistant sugarcane varieties adjacent to or in the vicinity of sorghum fields, (3) manipulating planting location to isolate sorghum fields, (4) screening for more efficacious insecticides for SCB control in sorghum, and (5) studying the role of planting density of SCB population development, as a few management practices that deserve consideration of future research.

REFERENCES

1. T. E. Reagan, G. Coburn and S. D. Hensley. Effects of mirex on the arthropod fauna of a Louisiana sugarcane field. Environ. Entomol. 1:588-91. 1972.
2. T. E. Reagan. A pest management system for sugarcane insects. Louisiana Agriculture. 24(2):12-4. 1981.
3. W. H. Long and S. D. Hensley. Insect pests of sugarcane. Ann. Rev. Entomol. 17:149-76. 1972.
4. E. H. Floyd, D. F. Clower and L. F. Mason. Effects of sugarcane borer infestations on the yield and grade of corn. J. Econ. Entomol. 53:935-7. 1960.
5. D. K. Pollet, T. E. Reagan and S. D. Hensley. Pest management of sugarcane insects. Louisiana Coop. Ext. Publ. 1982: 12p. 1978.
6. S. D. Hensley. Management of sugarcane borer populations in Louisiana, a decade of change. Entomophaga. 16:133-46. 1971.
7. D. B. Duncan. A significance test for differences between ranked treatments in an analysis of variance. Va. J. Sci. 2:171-89. 1951.

8. E. A. White. Effects of stubbling and weed control in sugarcane on the predation of the sugarcane borer, <u>Diatraea</u> <u>saccharalis</u> (F.). M. S. Thesis – Louisiana State University, Baton Rouge:216p. 1980.

9. D. W. Brown and R. A. Goyer. Effects of a predator complex on lepidopterous defoliators of soybeans. Environ. Entomol. 11:385-9. 1982.

10. F. A. Martin, C. A. Richard and S. D. Hensley. Host resistance of <u>Diatraea</u> <u>saccharalis</u> (F).: relationship of sugarcane internode hardness to larval damage. Environ. Entomol. 4:687-8. 1975.

11. J. L. Flynn, A. D. Ali and T. E. Reagan. Sugarcane borer control in sweet sorghum, small plot insecticide screening test. Insecticide and Acaricide Tests. 9:296. 1984a.

12. J. L. Flynn, T. E. Reagan and E. O. Ogunwolu. Establishment and damage of the sugarcane borer in corn as influenced by plant development. J. Econ. Entomol. 77:691-7. 1984b.

MARINE BIOMASS PRODUCTION

John H. Ryther

Division of Applied Biology
Harbor Branch Foundation, Inc.
Ft. Pierce, Florida 33450

ABSTRACT

Macroscopic marine algae (seaweeds) are, in many respects, ideal biomass species for conversion to fuel. In most parts of the world they are
not used for food, fiber or other commercial purposes, and they could be
grown extensively over large portions of the earth not otherwise used by
man. Few seaweeds are presently grown commercially for food or chemicals in
the Orient and Southeast Asia. Production practices are relatively low
yield, labor intensive enterprises that are economically viable only because
of the low cost of labor in the countries involved and the high local value
of the product. For biomass/energy purposes, seaweeds would have to be
grown at considerably higher yields and lower costs. The former has been
achieved in small, experimental systems, with annual production equivalent
to more than 100 dry Mg ha^{-1}, but only at prohibitive energy costs.

Both energy and labor costs, however, could be significantly reduced
through mechanization of the culture system, through recycling of nutrients
including carbon dioxide, and possibly through use of pelagic, floating
seaweeds.
Keywords. Biomass, seaweeds, macroalgae, energy farms.

INTRODUCTION

In many respects, macroscopic algae are ideal biomass species for
conversion to fuel. They are, with few exceptions (e.g., agar and other

241

gel-producing polysaccharides) not used for food, fiber, or other commercial purposes. Their growth rates in nature can be high and they can be cultivated in areas that are not useful for growing traditional food or fiber crops.

Seaweeds are among the most productive plants on earth, yielding in small experimental systems the equivalent of over 100 dry Mg ha^{-1} yr^{-1}.[1] They are fleshy plants with little supportive lignocellulosic type tissue and they readily converted through anaerobic digestion to biogas with a high proportion of methane.[2,3]

Seaweed cultivation for energy would appear to be particularly appropriate in mild, semi-tropical coastal waters where the climate allows year-round growth of many species. Florida, for example, has about 2,000 km of diverse coastline along both the Gulf of Mexico and the Atlantic Ocean. Off the west coast alone, there are some 50,000 square km of water less than 30 m deep, nearly 20,000 square km (about two million hectares or 5 million acres) less than 7 m deep. These shallow waters are characterized by having a smooth, hard sand bottom, clear seawater with good circulation, and normally moderate weather and sea state conditions. From ecological, engineering and logistic viewpoints, the region appears to be an ideal location for commercially cultivated marine plants.

Because seaweeds are not widely used, a well developed technology for their large-scale commercial cultivation does not exist. There are exceptions to be found, however, in Southeast Asia and the Orient, where a few species of seaweeds are grown for food or their chemicals. These are reviewed briefly below as a starting point for the conceptual design of more highly mechanized, less labor intensive seaweed energy farms of the future.

COMMERCIAL SEAWEED CULTURE IN SOUTHEAST ASIA AND THE ORIENT

In addition to several minor efforts scattered around the world, there are four larger commercial seaweed culture systems currently in operation. These are: (1) _Porphyra_ (nori) culture in Japan, (2) _Laminaria_ (kelp) culture in China, (3) _Eucheuma_ farming in the Philippines, and (4) _Gracilaria_ culture in Taiwan. The biology of the respective species and their cultivation technology have been described in detail in reviews by Bardach et al.[4], Hansen et al.[5], and Tseng[6] and have recently been summarized by Ryther.[7] The following brief resume has been exerpted from the latter publication.

Porphyra (nori) Culture in Japan

Several species of the red algae genus Porphyra have been grown as a highly-prized food since the seventeenth century in Japan, where it is commonly known as "nori."

The large sporophyte Porphyra plant (i.e. the edible stage) is an irregularly-shaped, flat, deep red blade that may grow 20 cm or more in length, depending upon the species and growing conditions. The Japanese crop, primarily P. tenera and P. yezoensis but including at least four other species, matures in late fall and may be harvested several times throughout the winter by cutting back the thallus without destroying its attachment.

In 1978, 60,000 hectares of sea surface were used to produce 21,150 Mg of nori with a value of 540 million U.S. dollars, by far the most economically important seaweed crop in the world.[6]

Originally, cultivation of Porphyra consisted of driving leafless tree branches into the bottom just above the mean water level along the open coastline in the fall of the year. The nonmobile nomospores of Porphyra would settle on the branches which could then be moved to more nutrient-rich habitats at the mouths of or within estuaries. The attached monospores subsequently developed into leafy sporophyte thalli, the edible stage and portion of the plant.

The branches have in recent times been replaced by man-made nets as spore-collected devices. Typically, these are made of synthetic twine 3-5 mm in diameter with 15 cm square mesh openings about one meter wide and ranging from 18 to 45 meters long. For grow-out of the attached spores, the nets are suspended from bamboo poles driven into the bottom in such a manner that the flat surface of the net is parallel to the water surface.

The productivity or yield of Porphyra is not usually given in the literature that describes its cultivation. A rough estimate[4] placed mean production at 0.75 dry Mg ha^{-1} yr^{-1}. A similar calculation may be made from the data on total area farmed and total annual production for 1978 reported by A. Miura as cited above by Tseng[6], i.e. 21,150 Mg produced over 60,000 hectares, for an average of 0.35 Mg ha^{-1} yr^{-1}. Both figures are probably conservative and higher yields are undoubtedly achieved, but apparently either the species and/or the culture method employed do not lend themselves to the high levels of production that would be required for purposes of conversion to energy.

Laminaria Culture in the People's Republic of China

In many parts of China, the inhabitants are subject to a chronic problem of goiter, a disease caused by iodine deficiency. The consumption of brown seaweeds rich in iodine is a prophylactic measure to prevent that disease and the small kelp, Laminaria japonica, therefore, became an important item in the Chinese diet.

As its name implies, Laminaria japonica is indigenous to the cold-water environment of Hokkaido, the northern island of Japan, from which some 3000 Mg yr^{-1} were formerly exported to China. Now it is grown in over 18,000 hectares of China's coastal waters with a production in 1979 or more than 275,000 dry Mg worth some 300 million U.S. dollars.

Chinese kelp culture begins along its northern coast in 15 large nurseries which are, in effect, large greenhouses (>5000 m^2) containing tanks through which fertilized, refrigerated (5-8°C) seawater is circulated.

In spring, the shallow (ca. 10 cm deep) tanks in each nursery are filled with 10,000 wooden frames, each roughly 40 x 60 cm around which 40 meters of rough, ca. 0.5 cm diameter string is wound. Mature sporophyte plants of Laminaria are briefly sun dried to stimulate the release of zoospores and are, then, spread over the wooden frames which are laid out flat in the nursery tanks. The zoospores are shed from the sporophyte plants and attached to the string frames ("spore curtains") within two hours. There, the complex life cycle of the Laminaria is completed. The spores develop into microscopic male and female gametophytes (the sexual form of the algae) which quickly mature to produce sperm and eggs, the motile sperm swims to and fertilizes the egg which germinates to produce the sporeling that eventually grows into the large mature, asexual sporophyte - the familiar, obvious seaweed plant.

All of the above stages in the life cycle, from the shedding of the zoospores to the development of the young sporelings, take place in the nursery during the period June-October. When the outside water temperature falls below 20°C, during mid- to late October in Tsingtao but late September to early October in the more northern Dalien area, the "spore curtains" are taken off their wooden frames and moved to the ocean, where they are suspended between parallel rows of large buoyed and moored ropes. At this point, the sporelings are 2-4 cm long and there are some 50,000 of them per 50-meter "spore curtain."

When they are small, the sporelings are tended daily, lifting each "spore curtain" from the water, meticulously brushing off the sediment and attached plants and animals from each plant, and immersing the entire curtain into a tub of concentrated liquid fertilizer. When they reach the size of about 10 cm, after 25-30 days in the ocean, the entire young crop is harvested, manually stripped off the strings to which they are attached, and bundles of four sporelings each are inserted into the weave of larger, 5 cm diameter, coarse, loosely woven ropes that are again tied across the parallel suspending lines. There they remain until they are harvested over a six-week period beginning in early June.

The plants are no longer individually tended after they are transplanted, but the crop is usually fertilized. Formerly, this was done by attaching to the ropes ceramic containers of fertilizer through which the nutrients could slowly diffuse. Now it is found more expedient and effective to broadcast or spray liquid fertilizer daily over the kelp beds.

In Tsingtao, where the growing season is some 230 days, the kelp reach a length of about three meters at the time of harvest. Yields average 12 dry Mg ha^{-1} year^{-1}. In the colder Dalien region, the season is perhaps one month longer. Because of that and/or for other reasons, the kelp plants there reach a length that may exceed 5 meters and yields of 20 dry Mg ha^{-1} year^{-1} are reported. The mature kelp plants are harvested in the late spring by manually hauling the lines with plants attached into a fleet of rowboats, each manned by three to four laborers.

Clearly, kelp culture in China, like its agriculture, is extremely labor intensive. There is no information available concerning the economics of this large industry, but such data would, in any event, be difficult to relate to western, capitalistic enterprise.

Eucheuma Culture in the Philippines

Eucheuma is a multiple-branched, fleshy red alga used for its contained polysaccharide, carrageenin. Originally harvested from wild stocks in Southeast Asia, such supplies were cut off by political upheavals in the 1950's and led to the development of cultivation methods for the species by the joint efforts of Marine Colloids, Inc. (Rockland, Maine), M.S. Doty (University of Hawaii), and the Philippines Bureau of Fisheries and Aquatic Resources.

Unlike Porphyra and Laminaria culture, which involves growing the seaweeds throughout all stages of their complex life cycles, Eucheuma is grown vegetatively by "planting" fragments of the fleshy thallus, allowing them to increase in size, and simply breaking or cutting off the new growth.

Initially, the Eucheuma was grown on nets closely resembling those used for Porphyra farming in Japan. These nets also are suspended off the bottom and parallel to the water surface by tying them to bamboo stakes. Eucheuma fragments are tied to the nylon mesh intersections.

Typically, a family Eucheuma farm consists of four modules each of 200 nets covering an area of about 2500 m^2, the total farm occupying one hectare.[9] More recently the net system has been replaced by the use of monolines, 10 m lengths of nylon monofilament staked at each end, 0.5 m apart, with Eucheuma propagules tied every ca 2.5 m along the line. The monoline method provides easier access to the plants and accomodates as many as 100,000 plants per hectare.[5]

As is also true of Porphyra culture, Eucheuma farming is a highly labor intensive operation, with constant tending of the plants, manual removal of epiphytes and predators, mending the nets and repairing the moorings, and harvesting the plants as they grow.[8] Siting of the farm is also critical, the alga requiring a vigorous exchange of seawater for rapid growth and favoring high-energy areas such as behind fringing reefs, but not so violent as to break the seaweed from the nets or otherwise damage the operation.

Parker[8] estimated annual production of Eucheuma at a pilot farm on Tapaan Island, Philippines, from six-months of harvest data, during 1971-1972, at 13 dry Mg ha^{-1} yr^{-1}. The annual depreciation cost of equipment and supplies, primarily nets, was placed at $364 (U.S.) for a four-module, one-hectare farm which Doty[9] states could be managed by one "enterprising family." Labor costs are not included in Parker's economic analysis of this cottage industry, nor is the size of Doty's enterprising family. If one estimates, from other information in Parker's report, an annual minimum wage for agricultural labor of $200 (U.S.), total cost of operation of a one-hectare Eucheuma farm employing four laborers would be $1164, making the cost of production $89 per dry Mg.

Gracilaria Culture in Taiwan

In 1962, first attempts were made to grow the red seaweed Gracilaria in southern Taiwan, using ponds that were originally constructed for fish culture. The Gracilaria is used locally or, after preliminary processing, is shipped to Japan for the extraction of the polysaccharide, agar.

Although Gracilaria normally grows attached to rocks or other substrata and undergoes the complex alternating life cycle between sexual and asexual reproduction that is typical of the red algae, certain species may grow unattached, in a drifting mode, on the bottoms of shallow ponds and estuaries. Such plants are usually sterile and grown entirely vegetatively, larger clumps breaking up into smaller fragments by the action of waves, currents and other forms of natural turbulence.

The old milkfish ponds where Gracilaria is now grown in Taiwan are usually rectangular, one to ten hectares in area and about one meter deep when filled to capacity. The pond bottoms are hard, sandy loam, soft mud bottom being considered undesirable for both growth and harvesting. The ponds are located adjacent to estuaries so that they may be filled and drained by tidal exchange assisted as needed by pumping. Water exchange is required to regulate salinity and to provide a new supply of nutrients from the normally enriched estuarine waters. Additional enrichment of the ponds with broadcast inorganic fertilizers or fermented pig manure is carried out irregularly between water exchanges, the farmer judging the need by the clarity of the water.

The best growth of the seaweed occurs in the temperature range of 20° – 25°C. Growth stops below about 12°C, but the plants can tolerate temperatures as low as 8°C. In southern Taiwan, where virtually all of the Gracilaria is grown, the normal water temperature range in the ponds is from about 10°C in winter to about 30°C in the summer. Pond depth is carefully regulated seasonally, partly to control temperature and partly, the intensity of sunlight that penetrates to the seaweed on the pond bottom. In summer, the ponds are maintained at 60–80 cm depth and in winter, at only 30 cm depth.

Several species of Gracilaria are cultured in Taiwan, often together in the same pond. The most popular appears to be that identified by the government biologists as G. confervoides. Cuttings or torn fragments of the seaweed, purchased from other farmers, are used for seed stock and are introduced to a new farm at a density of 3–5 kg wet wt m^{-2}. The plants are evenly spread over the pond bottom and grow there vegetatively throughout the year. When the population has roughly doubled in density and biomass, as estimated by eye, half the crop is harvested by a crew of 10–20 women, half of whom rake the seaweed into piled rows on the pond bottom, and the other half of whom dip-net the plants out of the water and into large bamboo baskets on wooden barges. The remaining half of the crop is then spread evenly over the pond bottom.

There are usually seven to eight harvests per year, each of one to three dry Mg ha^{-1}, mostly occurring from June through December. Little if any growth occurs during the late winter and very early spring and the stocks are sometimes held in deep, protected areas during the coldest part of the winter, using the same covered shelters as were originally designed for milkfish culture.

Some farmers harvest smaller crops more frequently during the growing season (i.e., every 10 days or so) but annual yields are approximately the same whatever the harvest routine. Yields range from 10 to 20 dry Mg and average about 14 Mg ha^{-1} yr^{-1}.

Shang[10] has provided a detailed economic analysis of a one-hectare Gracilaria farm in Taiwan in which the total cost of production of a crop of 10 dry tons was $1382 U.S., or $138 per Mg.

SUMMARY AND IMPLICATIONS OF ASIAN COMMERCIAL SEAWEED CULTURE PRACTICES

Excluding the Japanese Porphyra cultivation, where quality of the product is paramount and yield is of secondary importance, the three other major seaweed culture industries in the world, though involving different species and very different culture methods, produce remarkably similar average yields of 13–15 dry Mg ha^{-1} yr^{-1}. Considering the fact that these are young (15–30 yr.) and technologically rather simple, almost primitive industries, such yields are rather impressive. Many forms of modern agriculture, after centuries of evolution and improvement of both stocks and methods, produce biomass yields that are not significantly greater.[11]

Taking an average cost of production of about $100 U.S. dry Mg^{-1} and assuming the dry seaweeds to contain 60% volatile solids[2] capable of yielding 0.4 liter g vs^{-1} (6 SCF lb vs^{-1})[12] from their anaerobic digestion,

248

the cost of methane, not including its production from the seaweed, would be $12.63 10^3 SCF^{-1}, too high to be competitive with well-head natural gas prices by a factor of 3-4, but not an unreasonable starting place for a new industry that has not yet received the benefits of modern technology or genetic improvement.

MARINE BIOMASS RESEARCH IN THE UNITED STATES

There are marine biomass research projects in Japan, Sweden/Denmark, and in several places in the United States. Most have produced only very preliminary results to date and have not released definitive results. Some deal with unicellular marine algae, a subject which is beyond the scope of this report. Two projects will be discussed here: (1) production of the giant kelp, Macrocystes pyrifera, in California and (2) cultivation of several small, unattached species of seaweeds in Florida.

Giant Kelp Culture

The giant kelp, Macrocystes pyrifera, is the largest known alga and one of the world's largest plants, attaining a length in excess of 50 meters. It is one of the most important resources along the California coastline, not only because of its high commercial value (primarily for its polysaccaride alginic acid), but also because it is the dominant species and habitat of the nearshore ecosystem.

A decline in the natural kelp beds off California during the 1950's, due to heavy predation, pollution, and perhaps other factors, led to an extensive restocking and restoration program led by Dr. J. Wheeler North (Cal. Inst. of Technology). Because of the overwhelming success of these restoration efforts, the potential for cultivating Macrocystes in new areas was recognized.

In the mid-1970's, an ambitious Ocean Food and Energy Farm (OFEF) program was begun -- funded jointly by the Energy Research and Development Association, the National Science Foundation, the American Gas Association (AGA), the U.S. Navy, and various organizations in public and private sectors. The primary objective of the farm was to cultivate kelp as a source of energy. An ocean farm system was designed under the management of H. A. Wilcox of the Naval Undersea Center, San Diego.

The design consisted of an open-ocean farm covering 40,000 ha, 20 km on a side, located 160 km off the coast of southern California. After survey studies, three sites in southern California were recommended. The farm substrate, maintained at a depth of approximately 30 m, was to be made up of flexible triangular modules 300 m on a side, each covering about 4 ha. Each module would be held in place by diesel-powered propulsors. Nutrient-rich water was to be upwelled from a depth of about 100 m by wave-powered pumps. The upwelling pumps considered for the project included the Isaacs buoy propeller pump, a wave vane propeller pump, a modified Isaacs pump, and the Wilcox bellows pump.

Kelp plants, attached to the susstrate at a density of one plant per 100 m^2, would take about four years to mature; then the standing crop would be harvested by six times per year.

To test the technical and economic feasibility of the commercial-sized ocean farm, a research program was begun in 1976 jointly sponsored by ERDA (subsequently DOE) and AGA (subsequently Gas Research Institute—GRI) and managed by the General Electric Company. Scientific and engineering support was provided by the Institute of Gas Technology, the U.S. Department of Agriculture, and Global Marine Development, Inc. Under this program, a modular structure called the Test Farm, reminiscent of a single unit from Wilcox's sea farm, was installed at a site off Laguna Beach, California. The Test Farm consisted of a 3 m diameter buoy that stood upright in the water and was attached by a universal joint to an umbrella-shaped set of radial arms to which kelp plants were attached. Nutrient-rich water was pumped from depths of about 500 m, up through a 0.7 m diameter polyethylene pipe, using three pumps with capacities of 13 m^3 per minute, driven by 14 kw diesels. The Test Farm was deployed at sea in September, 1978, and thereafter was supplied with 103 adult kelp transplants.

Because of several technical problems, the initial plantings failed to survive and the test farm itself was ultimately lost. In the meantime, various analyses cast considerable doubt on the economic viability of such a system. Much of the doubt surrounding the argument resulted from uncertainty in the expected yields of the seaweed, which ranged from less than 10 to over 100 ash-free dry Mg ha^{-1} yr^{-1}.

That uncertainty led to the establishment of a small, near-shore test farm of <u>Macrocystis</u> at Goleta, California, by Neushal Mariculture, Inc. Two 0.2 ha test plots, one fertilized and one unfertilized, were stocked with juvenile kelp plants, anchored to the bottom, each at three different densities. Projected annual yields from four harvests during the year ranged from 8 (low density) to 62 (high density) dry Mg ha^{-1} yr^{-1}.[13] However, the high density plots, at a considerably greater density than that found in nature, were characterized by high mortality, presumably due to shading. As the dead plants were routinely replaced throughout the one-year experiment, the significance of the yields obtained under those conditions is difficult to assess. The lower yields obtained at natural kelp densities, where mortality was low, were comparable to those estimated for natural populations, while those from the intermediate-density plots, about 20 dry Mg ha^{-1} yr^{-1}, are comparable to the better yields from the Chinese rope culture of Laminaria.[6]

Small, Unattached Seaweed Culture in Florida

Gracilaria. Neish et. al.[14], first developed a technique for greatly accelerating the growth of seaweeds (in his case, the red alga, Chondrus crispus) by maintaining the plants in suspension through rotating the water in which they grow with a wooden paddle wheel. The present author and his colleagues, B. E. Lapointe, M. D. Hanisak, and others, employed a similar method for growing seaweeds in suspended culture, but with the use of vigorous aeration of the culture medium via on airline on the bottom of the culture tank. Using a series of small (50 l) semi-cylindrical culture tanks provided with flowing, enriched seawater and aeration, the above workers screened a large number of small seaweeds indigenous to central Florida for maximum production. The research was conducted at the mariculture facilities of the Harbor Branch Foundation (Ft. Pierce, FL) during 1976-1977, with support from the U.S. Energy Research and Development Association (now DOE).

Of The 50 or more species examined, the red alga Gracilaria tikvahiae, proved most satisfactory for several reasons, i.e.:

1. Rapid growth and high yield.

2. Continuous vegetative growth with no development of a sexual reproductive state and the attendant cessation of growth or disintegration of plant tissue.

3. Growth throughout the year (i.e. in Florida) and within the wide range of temperatures (ca. 12-35°C) and salinity (ca. 20-35%) that occurs in the Indian River and the experimental seawater supply.

4. Soft, fleshy thallus with little supportive ligno-cellulosic
    tissue, readily digested anaerobically at high conversion effi-
    ciency and with high methane content of the resulting biogas.

Following its selection from the screening process, Gracilaria was
grown under various environmental conditions and management strategies to
determine optimal operational parameters for its maximum sustained annual
yield. The latter varied seasonally from 12 (January) to 48 (July) grams
dry wt m^{-1} d^{-1}, with an annual mean of 34.8 g m^{-2} d^{-1}, equivalent to 127 dry
Mg ha^{-1} yr^{-1}.[1] This high yield was achieved by the following procedures:

1. Maintaining culture density at 2-4 kg wet wt m^{-2} by weekly harvest
    back to the starting density (2kg m^{-2}).

2. The exchange of 22 volumes per day of seawater enriched with 15 μm
    1^{-1} N (as NO_3^{-1}) and 4 μ m 1^{-1} P (as PO_4^{-3}).

3. Continuous aeration with sufficient velocity to maintain the sea-
    weed in suspension.

Further studies on the biology and nutrition of Gracilaria revealed
that growth under stagnant conditions became limited by high pH and the
accompanying unavailability of free CO_2. The benefit of seawater exchange
was, therefore, a new supply of CO_2, with yields being directly proportional
to seawater exchange rate. The same effect was achieved in stagnant water
by adding CO_2 gas.

Continuous enrichment of seawater with nitrogen and phosphorus led to
extreme problems of epiphytes (other, undesirable algae) overgrowing, smoth-
ering, and eventually killing the Gracilaria. That problem was solved by
pulse-feeding the Gracilaria, removing the plants from the culture tanks and
immersing them overnight in a concentrated nutrient solution. In the process of that treatment, algae are able to assimilate and store internally
enough nutrients to grow at their maximum rate in unenriched seawater for as
long as two weeks.[15] It was also found that the residue from the anaerobic
digestion of Gracilaria (and other algae) contain all the essential nutrients for growth of the plants and may be recycled as the medium for pulse
feeding the seaweeds.[16,17]

The role of and necessity for maintaining the algae in suspension
remains unclear, but intermittant aeration for about 25% of the time was as

effective as continuous aeration. Below that frequency yields declined.[18]
The movement of the plant is believed to result in better exposure of the
entire thallus to solar radiation, the breakdown of diffusion gradients of
nutrients, especially CO_2, or both. It may simply result in the removal of
sediment and epiphytes from the algal surface which may, in turn, affect
both light availability and CO_2 transfer.

Non-intensive Garcilaria culture in Florida with the algae lying on the
bottoms of ponds, with no aeration or other agitation, and with only a moderate exchange of seawater resulted in yields averaging less than 5 g m^{-2}
d^{-1}, almost an order of magnitude below the maximum and very similar to the
Gracilaria yields obtained from similar culture methods in Taiwan, as reported above (i.e. 10–20 Mg ha^{-1} yr^{-1}).

Ulva culture. Early studies with the green alga Ulva lactuca (sea
lettuce) were promising but frustrating because the alga periodically became
reproductive and disintegrated and, also, because it would not survive the
high summer temperatures of Florida. That problem has been resolved with
the discovery of a sterile, high temperature-tolerant strain of Ulva which
is now in culture. Annual yields of Ulva are somewhat lower than those of
Gracilaria, averaging about 20 g dry wt m^{-1} d^{-1} (73 Mg ha^{-1} yr^{-1}), but the
alga contains more soluble carbohydrates than Gracilaria and is more efficiently digested, producing more methane per unit of dry weight, which would
appear to compensate for the lower yields as a biomass source for
energy.[18,3]

Sargassum culture. A logistic problem exists in the large-scale cultivation of unattached, benthic seaweeds like Gracilaria and Ulva. Either
they must be grown in extremely shallow areas where they receive sufficient
sunlight and are accessible to manipulation, harvest, etc., or they must be
contained in baskets, tied to ropes, or otherwise held at or near the sea
surface, a practice that would be costly and difficult to manage on a large
scale. A floating species would resolve those problems and the two floating
brown algae, Sargassum natans and S. fluitans, are consequently a natural
target for investigation.

As natives of the central (Atlantic) gyre (i.e. the Sargasso Sea, which
derives its name from the plant), the two pelagic Sargassum species are
truely oceanic in distribution and do not tolerate brackish, low salinity
water or low temperature. The species are, also, never exposed to even
moderately high nutrient levels in their natural environment and they are

thought to grow very slowly. Such low yields would preclude the use of Sargassum as a biomass-for-energy species, but the plant has never been grown in culture and its actual yield potential is unknown.

Studies on the culture of Sargassum have, therefore, been initiated at a small field station in the Florida Keys, where access to a semi-tropical, oceanic seawater environment is available. Small cage cultures of the two pelagic species, suspended near shore in a free-flowing channel between the Keys, have resulted in growth equivalent to yields as high as 20 g dry wt $m^{-2} d^{-1}$ for periods of one to two weeks, but sustained growth or even survival has not yet been achieved.

In the high, often super-saturated carbonate environment of the shallow tropical marine environment, phosphorus is apparently scavenged by particulate carbonate and may often be the growth-limiting nutrient, in contrast to the nitrogen-limited situation in temperate coastal waters. In addition, the pelagic Sargassum species have a rich flora of epiphytic nitrogen-fixing bluegreen algae on their surfaces. As a result, nitrogen enrichment has little or no stimulatory effect upon the growth of Sargassum in that environment and might prove to be actually inhibitory to their growth. Phosphorus, on the other hand, and certain trace metals do appear to enhance growth of the plants and, also, to stimulate the nitrogen-fixing capacity of their epiphytic bluegreen algal flora. Thus, an entirely different nutritional strategy may be needed for Sargassum culture than for that of the more eutrophic species that have been previously grown. This research is partially supported by the joint program between the Gas Research Institute and the Institute of Food and Agricultural Sciences (IFAS), University of Florida.

CONCLUSION

Yields from the Asian commercial seaweed industry, while promising, would have to be increased by several-fold to make the industry economical as an energy source. Small-scale experimental studies have shown such increased yields.

Recycling of mineral nutrients (N, P, etc.) following digestion has been shown to be technically and biologically feasable. Such a practice would require the energy conversion process to occur at or very near the biomass production unit, to which the recycled nutrients could be pulse-fed

254

at appropriate intervals (e.g. weekly or biweekly) through a network of piping.

A nearshore marine biomass farm would inevitably experience some exchange of seawater through the action of tides, wind-driven currents and other forms of water motion. However, massive seaweed populations, at the density needed for maximum yield potential, would deflect water currents, particularly from the centers of large cultures, and it is doubtful that there would normally be sufficient water exchange to provide the carbon dioxide needed for peak production. Provision of gaseous CO_2, via a piped network similar to and perhaps the same as needed for other nutrients, would probably be more cost effective than artificially moving water through the system.

If the biomass were to be converted to biogas and the latter refined to pure methane on site, roughly half the original carbon content of the plants would be available as waste CO_2 gas for the production unit. The other half could similarly be made available if the methane were to be utilized (i.e. burned) on site for conversion to electricity or for some appropriate industrial application.

Periodic administration of the carbon dioxide gas, as short bursts under high compression for several minutes each hour of daylight, might, also provide the agitation necessary to turn the seaweed, dislodge sediment and/or epiphytes, or otherwise provide the function served by aeration in the experimental cultures.

Whether or not floating Sargassum proves to be the species of choice for a marine biomass farm will depend upon their inherent biological growth potential and, if sufficiently large, the ability to achieve that potential under economically reasonable culture conditions. That information should be available within the next year as a result of current investigations.

If pelagic Sargassum are not suitable biomass species, other non-attached algae like Gracilaria or Ulva still deserve consideration because of their proven high yields and other characteristics, at least for the large areas - hundreds of square miles - of shallow flats that exist throughout the American tropics and semi-tropics. In many respects, such as susceptibility to damage and disruption by winds and storms and obstruction of the surface to recreational activities, a shallow benthic population would be preferable to a floating one. Construction and management of such a system, in 3 m or less of water, should prove no more of an engineering challenge than for a hypothetical floating farm.

Finally, the large attached brown algae Macrocystis, Laminaria, certain species of Sargassum, and others should not be forgotten. Whether or not they can be grown at sufficiently high yields and low cost while attached to floating structures, as in the Chinese kelp culture mode, remains to be seen. An alternative may be their cultivation attached to the bottom, or to bottom structures, as with the Macrocystis test farm.

At this stage of the research, continued experimentation would appear to be warrented with all three types of seaweeds - floating and both unattached and attached benthic species. Basic research on their growth and nutrition and the genetic improvement of stocks is still needed, but, as ways are found to grow the plants, continuously and at high, sustainable yields, simultaneous studies must be made of the engineering design of systems capable of achieving those yields on a large scale and on an economical basis for energy application.

REFERENCES

1. B. E. Lapointe and J. H. Ryther. Some aspects of the growth and yield of Gracilaria tikvahiae in culture. Aquaculture 15:185-193. 1978.
2   M. D. Hanisak. Recycling the residues from anaerobic digesters as a nutrient source for seaweed growth. Bot. Mar. 24:57-61. 1981.
3. C. Habig and J. H. Ryther. Methane production from the anaerobic digestion of some marine macrophytes. Res. and Conserv. 8:271-279. 1983.
4. J. E. Bardach, J. H. Ryther and W. O. McLarney. Aquaculture. Wiley - Interscience, New York. 868 p. 1972.
5. J. E. Hansen, J. E. Packard and W. T. Doyle. Mariculture of Red Seaweeds. Cal. Sea Grant College Prog. Publ. T-CSGCP-002. 42 p. 1981.
6. C. K. Tseng. Commercial Cultivation. Chap 20 in the Biology of Seaweeds. C. S. Lobban and M. J. Wynne Eds. Bot. Monogr. 17. U. Cal. Press. p. 680-741. 1981.
7. J. H. Ryther. In press. Technology for the commercial production of macroalgae. Proc. National Mtg. on Biomass R. and D. for Energy Applications, Washington, D. C., 1-3 October, 1984.
8. H. S. Parker. The culture of the red algae genus Eucheuma in the Philippines. Aquaculture: 3:425-439. 1974.
9. M. S. Doty. Farming the red seaweed Eucheuma, for carrageenans. Micronesica 9:59-73. 1973.
10. Y. C. Shang. Economic Aspects of Gracilaria Culture in Taiwan. Aquaculture 8:1-7. 1976.
11. J. P. Cooper. Ed. Photosynthesis and productivity in different environments. Cambridge Univ. Press, London. 715 p. 1975.

12. K. F. Fannin, V. J. Srivastava and D. P. Chynoweth. Unconventional anaerobic digester designs for improving methane yields from sea kelp. Symp. Papers Energy from Biomass and Wastes IV, Lake Buena Vista, Fl., January 25-29. 1982. Institute of Gas Technology, Chicago. 373-399 p. 1982.

13. M. Neushal and B. W. W. Harger. Kelp biomass production. Yield, genetics and planting technology. Annual Report to Gas Research Institute 1/83 - 8/84. GRI - 85/0012. 1985.

14. A. C. Neish, P. F. Shacklack, C. H. Fox and F. J. Simpson. The cultivation of Chondrus crispus: Factors affecting growth under greenhouse conditions. Can. J. Bot. 55:2263-2271. 1977.

15. J. H. Ryther, N. Corwin, T. A. DeBusk and L. D. Williams. Nitrogen uptake and storage by the red alga Gracilaria tikvahiae (McLachlan, 1979). Aquaculture 26:107-115. 1981/82.

16. J. H. Ryther and M. D. Hanisak. Anaerobic digestion and nutrient recycling of small benthic or floating seaweeds. Proc. Symp. Energy from Biomass and Wastes V. Inst. Gas Tech. Jan., 1981. Lake Buena Vista, Florida. 109 p. 1981.

17. C. Habic, D. A. Andrews and J. H. Ryther. Nitrogen recycling and methane production using Gracilaria tikvahiae: A closed system approach. Res. and Conserv. 10:303-313. 1984.

18. J. H. Ryther. Cultivation of macroscopic marine algae. Subcontrast report to Solar Energy Research Institute SERI/STR - 231-1820 under subcontract XR-9-8133-1, November, 1982. 33 p. 1982.

HARVESTING SYSTEMS FOR AQUATIC BIOMASS

Larry O. Bagnall

Department of Agricultural Engineering
University of Florida/IFAS
Gainesville, Florida 32611

ABSTRACT

Aquatic weeds are among the most productive plants on earth, but aquatic biomass harvesting systems are the major missing elements in development of economically feasible aquatic biomass utilization. Existing aquatic plant harvesting systems harvest less than 50 wet Mg h^{-1} (55 wet T h^{-1}) [2.5 dry Mg h^{-1} (2.8 T h^{-1})], are expensive to install, are excessively labor-intensive, and deliver a product that must be further processed before converting to a useful energy form. No single harvesting system is likely to be suitable for all aquatic plant species. Variability in plant size and growth habit, water chemistry, and site extent, depth and irregularity impose a range of navigational and gathering design requirements. Beyond these are the institutional and environmental constraints involving living organisms in non-pristine water subject to competing uses.

The elements of aquatic harvesting systems are cutting, gathering, moving, elevating, draining, chopping, and transporting the plants. These elements can be accomplished by a variety of mechanisms and combined in many sequences in a successful system design. Commercial systems designed for weed control are currently available but must be modified for adequate performance in biomass systems. Development of improved systems depends on advanced knowledge of the biomass characteristics and perception of an adequate systems market.

Keywords. Biomass, energy crop, harvesting system, harvester.

INTRODUCTION

If the growth of highly productive aquatic weeds could be managed and uses found, they could become crops. Efforts are underway around the world to find uses for aquatic plants and to develop equipment, processes and systems to convert the "weeds" to "crops."[1] Feasibility of commercial production of compost, animal feed, and paper from aquatic plants has been investigated.[1,2,3]

Most recent research and commercial interest in utilization of aquatic plants in this country has been for production of methane.[4,5,6,7] The major economic and technical impediment to commercial production of methane from water hyacinth has been the system for harvesting the plants and preparing them for digestion.

SYSTEM CONSTRAINTS

Aquatic biomass harvesting systems must operate within the constraints imposed by the characteristics of the plants, the physical environment, the end product, and human institutions.

Plant Characteristics

Aquatic plants are categorized as floating, emersed or emergent, and submersed plants. Floating plants are supported and nurtured by the water and are not normally attached to the hydrosoil. Emergent plants are rooted in the hydrosoil and have a significant portion projecting above the water surface. Submersed plants are rooted in the hydrosoil, may extend to the surface of the water, but do not have any appreciable portion of the plant projecting above the surface. Each category and species has specialized harvesting requirements.

Water hyacinth (<u>Eichhornia crassipes</u>) is the most promising floating aquatic biomass species.[4,5,6,7,8] Typical standing crops of water hyacinth range in density from 200 to 300 Mg ha^{-1} (90 to 150 T A^{-1}), depending on water quality.[7] Dry matter productivity is as high as 35 Mg ha^{-1} yr^{-1} [16 T (A^{-1} yr)].[8] Water hyacinth has been heavily promoted for wastewater treatment, widely used in pilot wastewater treatment systems and is an excellent methane digestion feedstock.[4,5,8] Floating plants appear to be easy to harvest because they do not need to be cut from the hydrosoil and can be moved readily over the surface of the water.

The most productive and widespread emergent aquatic plant is cattail (Typha spp.). A less common, but promising, species is pennywort (Hydrocotyle umbellata). Dry matter productivity of both is about 8 Mg ha^{-1} yr^{-1} [4 T A^{-1} yr^{-1}].[1] Both species have been cultured in wastewater treatment systems where they were managed by harvesting. Cattails are promising sources of methane and ethanol.[9,10] The emergent plants must be cut or uprooted to be harvested. Cattail beds can, in some cases, be drained and the plants harvested as terrestrial plants. If the starchy rhizomes are to be harvested, the plants must be uprooted or follow-up root harvest conducted. Pennywort is semi-floating, vine-like, and may be only loosely attached to the hydrosoil. Cutting the floating mat is more difficult and important than cutting from the hydrosoil.

Hydrilla (Hydrilla verticillata) and Eurasian watermilfoil (Myriophyllum spicatum) are the most widespread and possibly the most productive submersed plants. Both usually grow in monoculture, crowding out any competitors and filling the water column with biomass, with some concentration near the surface. Typical stand density is 18 to 31 Mg ha^{-1} (8 to 14 T A^{-1}) and annual dry matter productivity is on the order of 3 Mg ha^{-1} yr^{-1} [1 T (A^{-1} yr^{-1})].[11] Hydrilla digests erratically.[7] The submersed plants must be cut from the hydrosoil. The cut part is marginally bouyant and will slowly rise to the surface and stay there. The cut plants can be moved in the water, but tend to form large, intractible balls. The vinelike stems tend to wrap on parts of the harvesting equipment with which they come in contact.

Environmental Constraints

Aquatic plants grow in a wide range of aquatic environments which affect harvester design. Access, draft and obstructions affect the design of the system and the economic feasibility of harvesting. Marshes, high banks, incompatible development and long hauling distances restrict physical and economic access to otherwise good aquatic biomass production sites. Long distance transportation cost of low dry matter density aquatic biomass is prohibitive.

Adequate uniform draft is required for a simple system to operate effectively. High-flotation wheeled vehicles can be used in drained or

extremely shallow sites and floating equipment can harvest where water is
deep enough, but variations over a wide range on the same site require very
specialized, expensive equipment. Floating plants grow in water of any
depth, but are most vigorous in depths greater than 0.2 m (0.7 ft), where
roots are not soil-bound. Emergent plants grow in water less than 0.5 m
(1.6 ft) deep. Submersed plants grow in water from 0.2 to 6 m (0.7 to 20
ft) deep.

Obstructions, such as sand bars, stumps, posts, piers and floating logs
damage machines and their operators. The effort required to navigate around
obstructions severely reduces the economic viability of harvesting.

Most proposed systems for production of aquatic biomass are associated
with use of the plants for wastewater treatment in artificial ponds. The
problems of access, draft and obstructions are controlled in the design of
the ponds.

End Product Requirements

The harvested biomass must have physical, chemical and biological
characteristics suitable for conversion to methane and, to some extent,
these are controlled by the harvesting system. The material must be finely
divided enough to be handled easily, preferably as a fluid or semi-fluid,
and to react rapidly in the bioconversion process. The harvesting system
may affect the chemical and biological characteristics by selectively harvesting the plants at the most appropriate stage of maturity.

Institutional Constraints

The institutional constraints on design and operation of harvesting
systems, apart from the production system, are dispersion of plant fragments
and other pollutants, and interference with competing uses of the water
body. The benefits are greater and the detrimental effects less when the
biomass is produced on wastewater in artificial ponds.

SYSTEM DESIGN

Design of an aquatic biomass harvesting system includes determining the
functions to be performed, the sequence in which they will be performed and

262

the mechanisms with which they will be performed. The best combinations of functions, sequence and mechanisms have not been determined.

System Functions and Components

The functions performed by aquatic biomass harvesting systems are cutting, gathering, over-water transporting, elevating, draining, chopping, and overland transporting. Not all of the functions are required on all plant species in all systems, and some functions are combined in some systems.

Aquatic plants are usually cut with a reciprocating or rotary cutter, but can be separated from the hydrosoil and other plants by pulling, tearing, uprooting or dislodging. The reciprocating cutterbar is similar to an agricultural mower. Reciprocating speed is lower and stub guards or stationary sickle sections are used to prevent wrapping of plant material on the stationary guard points. Usually a horizontal blade and two vertical blades are used, with the vertical blades located at the ends of the horizontal blade. In one wide harvesting system, no vertical cutters were used, because the edge effects were thought to be negligible. Harvesters for floating plants don't need a horizontal cutterbar.

Plants are gathered into a large mat, feed stream or windrow by accumulation of smaller mats, clusters and individual plants. Usually the gatherer is some type of rake, often mounted on the front of a pusher boat. In some cases, a pickup conveyor is assisted by rotary or linear lateral gatherers or by a feeding reel mounted above the conveyor.

Gathered plants can be moved to the transfer site either before or after elevation from the water. Because intact plants are buoyant, many operators prefer to move them in the water, using less expensive equipment than would be required if they were elevated before movement. However, the combination of applied force, mass, buoyancy and drag often causes the plants to swirl around or roll under the rake, especially if it has inadequate lateral and vertical containment, the mat being attempted is too large, or the speed is too high. Typically, the length of mat cannot exceed half the width of the mat, the vertical containment must extend to the depth of the floating mass, and speed should not exceed 0.5 m s^{-1} (1 mph).

The plants may be elevated from the water in batches by construction equipment or continuously by conveyor or pump. Batch elevation is slow and expensive for biomass harvesting. Pumps are effective but the energy required to move the entraining water is excessive. Chain-and-flight and flat-wire-belt conveyors are the types most commonly used to elevate aquatic plants. Plants roll or slide down conveyors sloped more than 30°. Water being moved upward by the flights spills over and out, washing the plants from the conveyor unless they are restrained by forward motion of the harvester or by a feeding reel. In mobile systems, the forward speed of pickup conveyors is limited to about 1.6 km h^{-1} (1 mph); at higher speeds, water and large plant fragments flow around the matted plants on the conveyor. A tined drag conveyor design raises the elevating mechanism above the water and restrains the plants from above, forcing them into and up a sloped trough, eliminating the bow wave and allowing slopes up to 90°. Theoretical energy required to elevate is 3 W h Mg^{-1}m^{-1} [0.0012 hp h T^{-1} ft^{-1}]. The best observed energy requirement was 34 W h Mg^{-1}m^{-1} [0.013 hp h T^{-1} ft^{-1}].[3]

Plants are usually drained as they are elevated and the elevating medium is selected to have good drainage characteristics. In pump elevating systems, the drainage function is performed separately on a screen with adequate provision for return of the surplus water to the pond.

Harvested plants should be chopped before digestion and if they are chopped immediately after harvesting, the entire subsequent handling system can be simplified and reduced in size. Harvested plant bulk density is about 170 kg m^{-3} (11 lbm ft^{-3}).[12,13] Chopping increases the bulk density to as much as 670 kg m^{-3} (42 lbm ft^{-3}) and, at least as importantly, increases the fluidity, so that the chopped material can be readily handled and processed.[14] A cylinder-shearbar chopper requires 0.5 J g^{-1} (0.18 hp h T^{-1}) for the chopping mechanism and can chop 615 Mg h^{-1} per meter (200 T h^{-1} per foot) of mechanism width.[14] The feeding mechanism requires approximately an additional 0.5 J g^{-1} (0.18 hp h T^{-1}).[3]

Overland transportation of aquatic biomass to a utilization site has usually been by truck. Because of the low solids content (5%) and low density of the plants, even when chopped, the trucks are not used efficiently and transportation cost is high and capacity low. If the utilization plant can be placed near the harvest site, transportation by light conveyor can be the most efficient method. Properly prepared water hyacinth can be pumped efficiently by a progressing-cavity pump.

System Configuration

The functions of aquatic biomass harvesting can be combined in various ways in mobile, immobile and hybrid systems, as suggested in Figure 1. All three configurations contains mobile and immobile elements, the distinction between the configurations is based on whether the plants are elevated from the water before or after over-water transportation.

In a mobile system, the plants are cut, gathered, elevated, chopped, and transported over-water by a mobile machine or system of machines. The advantages of this system are that the plants are harvested by a compact machine or system, the volume and intractability are reduced immediately, and the over-water transportation is maneuverable so that harvest can be more selective. The disadvantages are that a complex, expensive, mobile machine or system must be devised to work in a difficult environment and that the over-water transportation capacity may be extremely limiting. Often the systems will include multiple separable barges for over-water transportation to reduce cost and increase capacity. Capital and labor costs of these systems are usually high and capacity low.

In an immobile harvesting system, the plants are cut, gathered and transported over-water by relatively light-weight, low-cost equipment utilizing the bouyancy of the plants, then elevated, chopped and injected into overland transportation by shore-based or floating machinery at a fixed site. The advantages of this configuration are that over-water transportation does not require complex, expensive equipment, relative to the quantity of biomass moved and that the system can be devised to be less capital and

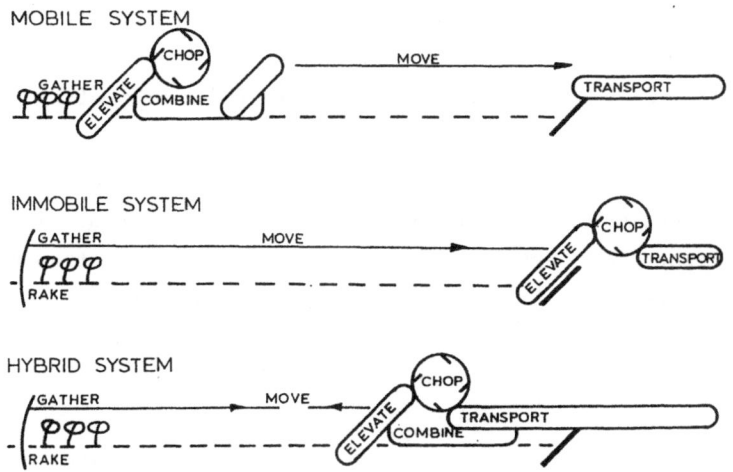

Figure 1. Alternative aquatic plant biomass harvesting systems

265

labor intensive. The disadvantages are lack of manuverability and flexibility, irregular loading of the high-cost elevating/chopping component, and difficulty at the elevating interface caused by the high density and disorientation of the compacted mat.

A hybrid system is one in which the plants are gathered into a small area by equipment similar to that in an immobile system, then elevated, drained, chopped and injected into overland transportation by a tethered mobile harvester. The advantages of this system are that the difficulty of transition from the moving to the stationary components is reduced, the unloading delay of the gathering equipment is reduced, and the processing unit is presented with a continuous supply of plants to be processed.

EXAMPLE SYSTEMS

Design of aquatic plant harvesting equipment began in the late 19th century with the development of machinery to control water hyacinth in Louisiana and Florida.[15,16] Examples of recent commercial and developmental harvesters and systems show the similarities and diversity of current development and how they fit into the configuration patterns.

Mobile Systems

Harvesting systems for submersed and emergent plants have tended to be of the mobile configuration; primarily because the volume of material to be carried was relatively low and a relatively complex mechanism for cutting and gathering was required to be mobile, and the addition of the elevating function offset the difficulty of moving cut submersed plants in the water.

The most widely used commercial submersed plant harvesting system is Aquamarine's Aqua-trio. It consists of a harvester which has an integrated cutter-pickup and an 18 m^3 (650 ft^3) live-bed hold, a transport which is essentially identical to the harvester except that it has no pickup, and a chain-and-flight shore conveyor to elevate plants from the transport to a truck. The system is capable of harvesting about 0.4 ha h^{-1} (1 A h^{-1}). The harvester can be used independently at reduced system capacity. Aquamarine-/Altosar also makes smaller and larger harvesters.

Limnos, Ltd., of Canada, made a two-stage submersed plant harvesting system. The cutter was separate from the pickup-mill. It cut a 6 m (20 ft) swath at 5 to 6 km h^{-1} (3 to 4 mph). The cut plants floated to the surface, where the lateral gathering wheels of the harvester windrowed the swath and fed it into the shallow-running pickup conveyor. The windrow was elevated into a large hammer-mill which pulped the plants and discharged the pulp into a detachable barge. The barge transported the pulped plants to the shore and pumped them to disposal, transportion or processing. The Limnos design was based largely on research and development at the University of Wisconsin, where the lateral-gathering, on-board chopping and bulk handling concepts were first demonstrated.[17]

The water hyacinth crimper-harvester was based on a mechanism, similar to a forage crimper, which elevated the plants from the water and coarsely chopped them in a single, high-speed operation.[18] The mechanism consisted of a set of 1.8 m (6 ft) wide intermeshing fluted rolls, the lower, 200 mm (8 in) diameter roll just above the water, and the upper, 300 mm (12 in) diameter roll floating 45° above and ahead of it. The rolls were driven at a peripheral speed of 5 m s^{-1} (1000 ft min^{-1}). The mechanism was mounted on a paddlewheel-propelled barge. Efficiency and capacity was poor but, by extensive modification, could be improved. Mean capacity of the harvester was 5.1 Mg h^{-1} (5.6 T h^{-1}). Average gross pickup unit energy was 3 J g^{-1} (1 hp h T^{-1}); net energy was 0.8 J g^{-1} (0.3 hp h T^{-1}).[18]

A harvester-chopper, or combine harvester, shown in Figure 2, consisting of a gathering screw and a cylinder/shearbar chopper has been built and is being tested. A 0.9 m (3 ft) wide x 0.6 m (2 ft) diameter center-feeding screw with a large, buoyant core and undershot feeding paddles gathered and elevated water hyacinth into the feed rolls of a 300 mm (12 in) chopper. A chain-and-flight drag conveyor was installed under the chopper to carry the chopped material to storage or further transportation.[18]

Seiga, of Denmark, has developed a harvesting system around a large-tired amphibious vehicle which is used, among other things, for cutting and transporting reeds.

Immobile Systems

Harvesting systems for floating plants have tended to be of the immobile configuration because the buoyancy and mobility of the plants has been obvious and mechanisms to move the plants to fixed transfer sites were developed early.

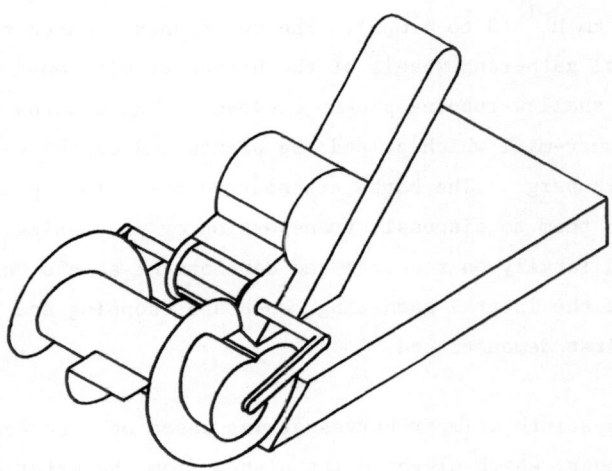

Figure 2. Combined gatherer chopper for water hyacinth biomass

Several stationary conveyor harvesting systems have been developed, mostly for water hyacinth. Aquamarine built a harvester for the Disney World wastewater project in which a narrow flat-wire-belt carried the plants under a flail chopper. The National Space Technology Laboratory used a heavy stationary conveyor and a stationary chopper, feeding them with pusher boats.

Sarasota Weed and Feed developed a sequence of harvesters, the most recent of which is a truck-mounted 3 m (10 ft) wide unit. A similar preceding model, equiped with a flail chopper over a 1 m (3 ft) wide cross-conveyor which increased the density and improved the handling characteristics of the plant material, was capable of harvesting 40 Mg h^{-1} (44 T h^{-1}).[19] Plants were fed to the harvester by an airboat and reel and removed by truck. The Florida Department of Natural Resources built a large harvester loosely based on this design.

The elevating and reducing functions of the crimper-harvester were separated in the harvester-crimpers. The plants were gathered and elevated by a reel-assisted 30°, 1.2 m (4 ft) wide chain-and-flight conveyor, then reduced by 1.2 m (4 ft) wide x 150 mm (6 in) diameter closed rolls with 25 mm (1 in) meshing flights. The crimped plants were transfered laterally by a conveying screw to a narrow chain-and-flight conveyor which elevated them to transportation, processing or storage. Capacity of the first model was 1.1 Mg h^{-1} (1.2 T h^{-1}) and of the second was 2.5 Mg h^{-1} (2.8 T h^{-1}).[18]

The immobile system installed at the IFAS facility at Zellwood (Figures 3 and 4) consists of a cable-driven rake which gathers and moves the plants in the water, and a take-out conveyor which moves the gathered mat laterally, elevates it, and feeds it into a chopper. The rake shown in Figure 3 was a 6 m (20 ft) wide light weight, pontoon-supported tubular frame with tines projecting 150 mm (6 in) below the water and 300 mm (12 in) above. The rake pivoted about a horizontal axis about 400 mm (16 in) above the water so that it cleared the plants on its return strike. The laterally-folding rake shown in Figure 4 has been installed to improve passage through the mat on the haulback and to provide greater depth to trap mats on the pull stroke. The laterally-folding configuration will be more appropriate to large systems than the pivoting-lifting configuration. The take-out is a 6 m (20 ft) wide drag conveyor with 0.6 m (2 ft) cantilevered, tined flights. Anticipated capacity of this system, as installed in a 0.4 ha (1 A) pond, is 20 Mg h^{-1} (22 T h^{-1}).

Based on the average dry matter productivity and a constant daily supply of biomass, system capacity should be 2.3 wet Mg $ha^{-1}d^{-1}$ [1 wet T $(A^{-1}d^{-1})$].[8] Average standing crop wet density is 217 Mg ha^{-1} (95 T A^{-1}), but the crop will pack in the rake to a density of at least 300 Mg ha^{-1} (130 T A^{-1}). Assuming an eight-hour working day, plants uniformly distributed in a semi-circular pool, and a speed limitation of 0.5 m s^{-1} (100 ft min^{-1}), the system size requirements are shown in Figure 5.

Amasek has developed a harvesting system consisting of a "water tractor" and an immobile harvester with a multi-stage chopper. The "water tractor" runs on a pair of large, cleated drums, the rear one of which is steerable. A rake with side projections is mounted on the front of the "tractor" and gathers and pushes water hyacinth to the feeding mechanism of the harvester. The feeding mechanism is a long-stoke reciprocating rake with tines that swing to pass over the hyacinth as the mechanism is extended, then lock behind the hyacinth as it is drawn in. The feeder feeds the hyacinth under a spirally cleated drum which partially crushes them and carries them rearward and upward between itself and a counter-rotating "kicker" drum, feeding them into a "hog", which coarsely shears them. The large fragments are fed laterally to an elevator which takes them to the top feed port of a recutter-equipped cylinder/shearbar chopper, which reduces particle size to a few millimeters. The finely-chopped material is directed to a progressing-cavity pump, which pumps it to storage, transportation or conversion.

Figure 3. Immobile water hyacinth rake-take-out conveyor harvesting system
installed on pond R1 at Zellwood

Figure 4. Modified Immobile Water Hyacinth Rake-Take-Out Conveyor
Harvesting System

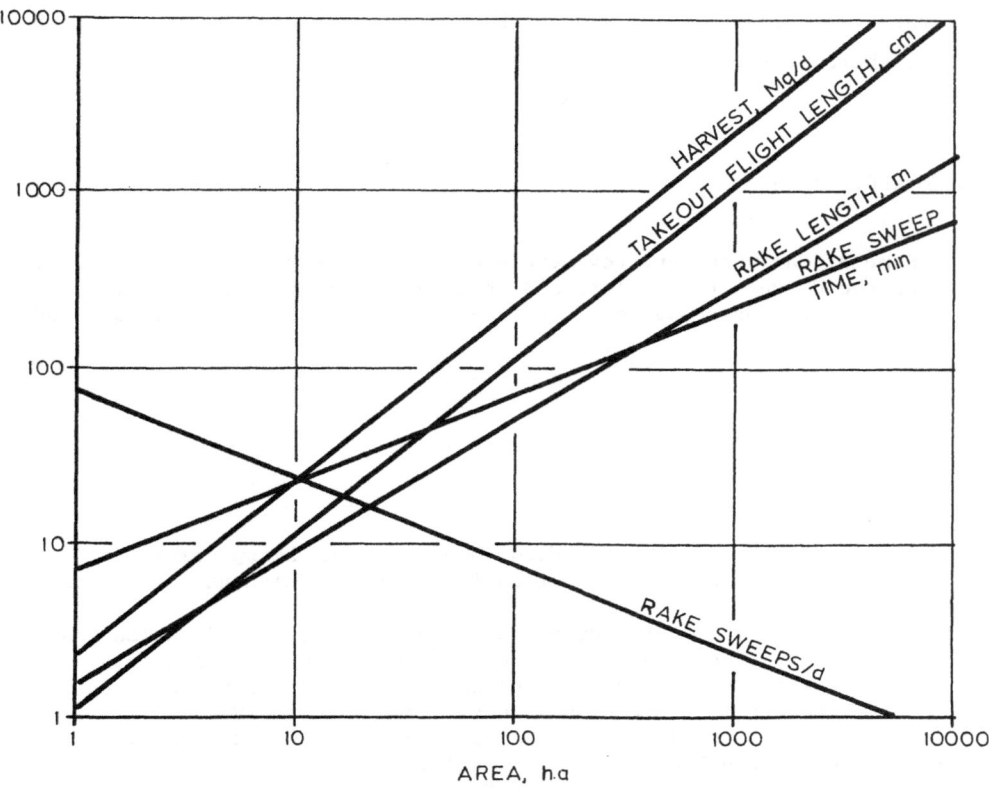

Figure 5. Effect of harvest area on rake-take-out design and operation
parameters

Hybrid Systems

No hybrid systems are currently in operation, with the possible exception of the Amasek system, but some of the mobile and immobile systems have
been operated in hybrid configuration when they were being used to harvest
plant types for which they were not designed. We expect that some existing
components will be adapted to hybrid systems in the near future.

The proposed hybrid system will use the same rake as the immobile
system for wide-ranging gathering and transportation of the water hyacinth
to a small harvesting pool. The combined gatherer-chopper will sweep the
pool, tethered to the transportation system, probably by a pipeline. The
rake can then be cycled more rapidly, not having to dwell at the take-out
conveyor, and the larger pool and greater range of the combine should keep
the transportation system more uniformly loaded. Neither the combine nor

the rake will need to be manned, as both can be operated by one operator from a shore station.

ACKNOWLEDGEMENTS

Some of the work reported herein was supported by the joint program between the University of Florida's Institute of Food and Agricultural Sciences and the Gas Research Institute titled "Methane from Biomass and Waste."

REFERENCES

1. Anonymous. Making Aquatic Weeds Useful: Some Perspectives for Developing Countries. National Academy of Sciences. Washington. 1976.
2. L. O. Bagnall, T. deS. Furman, J. F. Hentges Jr., W. J. Nolan and R. L. Shirley. Feed and fiber from effluent-grown water hyacinth, in: Wastewater Use in the Production of Feed and Fiber - Proceedings, Environmental Protection Agency. Washington. 1974.
3. L. O. Bagnall and J. F. Hentges, Jr. Processing and conservation of water hyacinth and hydrilla for livestock feeding. In: Aquatic Plants, Lake Management, and Ecosystem Consequences of Lake Harvesting. J. E. Breck, R. T. Prentki and O. L. Loucks, eds., University of Wisconsin, Madison. 1979.
4. B. C. Wolverton and R. C. McDonald. Upgrading facultative wastewater lagoons with vascular aquatic plants. J. Water Pollut. Control Fed. 51:305-313. 1979.
5. D. P. Chynoweth, D. A. Dolenc, B. Schwegler and K. R. Reddy. Wastewater reclamation and methane production using water hyacinth and anaerobic digestion. Institute of Gas Technology. Chicago. 1983.
6. L. O. Bagnall. Water hyacinth energy aquaculture. ASAE paper 80-5042, American Society of Agricultural Engineers. St. Joseph. 1980.
7. L. O. Bagnall. Performance of aquatic plant biogas systems. Aquatic Plant Management Society. Las Vegas. 1982.
8. K. R. Reddy, W. F. DeBusk and L. O. Bagnall. Water hyacinth biomass production in eutrophic lake water, in: Proceedings - 1984 International Gas Research Conference. Government Institutes, Rockville. 1984.
9. N. J. Andrews and D. C. Pratt. The potential of cattails (Typha spp.) as an Energy Source; productivity in managed stands, J. Minn. Acad. Sci. 44:5-8. 1978.
10. K. Ladenburg. Ethanol production from cattails. ASAE paper 82-3600, American Society of Agricultural Engineers. St. Joseph. 1982.
11. W. T. Haller. Hydrilla: a new and rapidly spreading aquatic weed problem. Florida Agricultural Experiment Station, Gainesville. 1976.
12. L. O. Bagnall. Bulk mechanical properties of water hyacinth. J. Aquat. Plant Manage. 20:49-53. 1982.
13. L. O. Bagnall. Bulk mechanical properties of hydrilla, J. Aquat. Plant Manage. 18:23-26. 1980.
14. J. S. Stewart III. Energy and Flow Requirements for Chopping Waterhyacinths. ME thesis, University of Florida. Gainesville. 1972.

15. W. E. Wunderlich. History of water hyacinths control in Louisiana, <u>Hyacinth Control</u> J. 1:14-18. 1963.
16. A. Tabita and J. W. Woods. History of hyacinth control in Florida. <u>Hyacinth Control</u> J. 1:19-22. 1963.
17. D. F. Livermore and R. G. Koegel. Mechanical harvesting of aquatic plants: an assessment of the state of the art, in: Aquatic Plants, Lake Management and Ecosystem Consequences of Lake Harvesting. J. E. Breck, R. T. Prentki and O. L. Loucks, eds., University of Wisconsin, Madison. 1979.
18. L. O. Bagnall. Development of a water hyacinth biomass combine. ASAE paper 83-5038. American Society of Agricultural Engineers. St. Joseph. 1983.
19. C. L. Phillippy and J. M. Perryman. Mechanical harvesting of water hyacinth (<u>Eichhornia</u> <u>crassipes</u>) in Gant Lake Canal, Sumter County, Florida. Florida Game and Fresh Water Fish Commission. Tallahassee. 1972.

MANAGEMENT STRATEGIES FOR WATER HYACINTH PRODUCTION IN A

NUTRIENT-LIMITED SYSTEM

W. F. DeBusk, K. R. Reddy and J. C. Tucker

Central Florida Research and Education Center
University of Florida/IFAS
Sanford, Florida 32771

ABSTRACT

Experimental water hyacinth [Eichhornia crassipes (Mart.) Solms] cultures were established, in mesocosm flow-through channels and microcosm
tanks containing water from Lake Apopka, at the University of Florida (IFAS)
research farm in Zellwood. The baseline dry biomass yield for water hyacinth grown in Lake Apopka water was 67 Mg ha^{-1} yr^{-1}. Foliar application
twice per week of 20 kg N ha^{-1} and 3 kg P ha^{-1} increased growth rate by as
much as 68%. Microcosm studies revealed that 62-74% of foliar applied
urea-N was assimilated by plants, with losses probably due to NH_3 volatilization. Projection of current data suggests that a water hyacinth system
covering 21% of Lake Apopka could produce a methane yield of 2.07 X 109 MJ
yr^{-1}, while causing a net reduction of nutrients in the lake.

Keywords. Water hyacinth, production, nutrients, aquatic plants,
biomass.

INTRODUCTION

The water hyacinth [Eichhornia crassipes (Mart.) Solms] is a floating
aquatic macrophyte found throughout tropical and subtropical areas of the
world. Recent studies have demonstrated the feasibility of using water
hyacinth as feedstock for anaerobic digestors producing methane gas for
fuel.[1,2]   Water hyacinth is extremely productive and readily digested,
making it a desirable choice as a biomass crop. Among several aquatic
macrophyte species screened as biomass crops in Florida, water hyacinth

275

showed the highest annual yield, with the highest growth rate during the active growing season from May through October.[3,4] Water hyacinth has also proven to be an efficient biological filter for wastewater and other nutrient-rich waters.[5,6,7,8]

A large-scale water hyacinth production system has been proposed for Lake Apopka, near Orlando, Florida, by Reynolds, Smith, and Hills of Jacksonville, in a feasibility study conducted for the joint program of the Gas Research Institute and the University of Florida's Institute of Food and Agricultural Sciences (IFAS).[9] The proposed system would produce methane for the Lake Apopka Natural Gas District. Lake Apopka is the largest lake in Central Florida, and is the last in a chain of lakes making up the Oklawaha River Basin. Over the course of many years, this lake has become highly eutrophic. This change is largely a result of the influx of nutrient-enriched waters from food processing, agricultural and municipal activities.[10]

In order to determine the feasibility of utilizing water hyacinth for biomass production and nutrient removal in Lake Apopka, experiments were designed with the following objectives: (1) to determine the base-line biomass yield of water hyacinth cultured in Lake Apopka water, and (2) to develop management strategies for optimizing conditions for plant growth.

METHODS

Site Description

This study was conducted at the University of Florida's Zellwood research farm, a branch of the Central Florida Research and Education Center (IFAS) in Sanford. This farm is part of the Zellwood muck land vegetable farming area, adjacent to Lake Apopka. This research builds upon the base of data accumulated on aquatic macrophyte growth and water quality renovation during the past decade.[6,11] Aquaculture research facilities include retention reservoirs, mesocosm-scale channels and microcosm tanks. All experimental facilities can be supplied with either drainage water from surrounding farmland or water from Lake Apopka.

276

Mesocosm study

Two concrete channels, each 61 m x 6.1 m x 0.6 m deep, were stocked with water hyacinths at a fresh weight density of 15 kg m^{-2} (150 Mg ha^{-1}). These flow-through systems received water from Lake Apopka at an average rate of 95 l min^{-1}, resulting in a retention time of 1.5 days. Both channels were divided by floating barriers into 6 cells of equal area, effectively providing replicate growth cultures. A 1 m^2 PVC frame with an attached basket of Vexar plastic mesh was placed into each cell. Water hyacinths were placed into the baskets at the same density as those outside the baskets. Plant growth was determined by weighing the growth baskets and contents, after draining excess water, and monitoring biomass weight changes over time. As in all subsequent experiments, plant subsamples were weighed, dried at 70 C, then reweighed, for converting yields to a dry weight basis. Plants in both channels were harvested, bi-weekly during the active growing season and monthly during the winter, back to their original density (15 kg m^{-2}), in order to maintain the optimum density range for plant growth. At the time of harvest, plant density in the growth baskets was also readjusted to 15 kg m^{-2}. Samples of lake water inflow and channel outflows were collected twice per week for analysis of N and P.

Plants in channel 1 were sprayed 3 times during the summer months with nitrogen and phosphorus fertilizer. Each spray treatment contained N and P at the rate of 10 and 3 kg ha^{-1}, respectively. A spray volume of 1 m^3 ha^{-1} was used in order to minimize leaf run-off while keeping the ionic concentration sufficiently low to prevent leaf burning. Plants in channel 2 received no foliar spray during the study period.

Microcosm studies

In order to determine which nutrient(s) might be growth-limiting in Lake Apopka, and to subsequently optimize foliar application of nutrients, a microcosm study incorporating various foliar spray treatments was implemented. Water hyacinths were grown in small outdoor tanks (0.62 m^2 water surface area) containing water from Lake Apopka, at a density of 10 kg m^{-2} (100 Mg ha^{-1}). Six foliar spray treatments, with 3 replications each, were imposed on the water hyacinths; N only (as urea), P only (as KH_2PO_4), micronutrients only (Nutrispray, a commercial minor element spray; Sunniland, Chase & Co., Sanford, FL), N + P, N + P + micronutrients, and no spray (control). Plants were sprayed at 0, 2, 4 and 6 weeks, with application rates, in kg ha^{-1}, of: N = 28.1, P = 6.0, Fe = 4.0, Cu = 0.2, Zn = 1.5, B =

0.04, Mo = 0.02, S = 3.0. In order to quantify plant assimilation of foliar applied nutrients, the final nutrient spray application was prepared using urea with labeled N (^{15}N, a stable isotope) as a tracer.

Water in the tanks was changed 3 times per week throughout the duration of the 8-week study period. Plants were weighed every 2 weeks to measure growth rates. After 4 weeks, cultures were harvested back to 10 kg m^{-2}. At 4 weeks and again at the conclusion of the study, plant samples were taken from each tank for analysis. Plant tissue was analyzed for total N, ^{15}N, and P. Assimilation of ^{15}N was calculated from the total N concentration and the ratio of ^{15}N to total N, which was determined by mass spectroscopy.

A follow-up study was designed to optimize the rate of application of foliar N + P spray to water hyacinth grown in Lake Apopka water. Water hyacinths were stocked at a density of 10 kg m^{-2} in small outdoor tanks and harvested back to initial density after 4 weeks. Lake water was changed 3 times per week. Six levels of foliar N and P fertilization were implemented by varying the number of applications during the 8 week study period: 0 (control), 1, 2, 4, 8, and 16 times. Each application contained 20 kg N ha^{-1} and 3 kg P ha^{-1}. Plants were weighed every 2 weeks in order to monitor growth rates.

RESULTS

The baseline biomass yield (1984-85) for water hyacinth grown in Lake Apopka water in the flow-through channels was about 67 Mg ha^{-1} yr^{-1}, with average monthly yield ranging from 5.2 to 25.6 g m^{-2} day^{-1} (Figures 1 and 2). Differences between yields in channel 1 and channel 2 were not significant, indicating that the foliar applications of nutrients had no effect on plant growth. It is possible that, at the time of foliar application, nutrients were not limiting in the channels because of internal nutrient cycling. The high degree of variability in plant growth rate may have been influenced by several factors, including solar radiation and nutrient availability.

Total N concentration in channel 1 outflow varied considerably (Figure 3), primarily due to large variations in lake water N concentration (channel inflow). Short-term changes in nutrient levels in Lake Apopka may be attributed to fluctuations in algal populations, which temporarily immobilize nutrients in the water column. Typical N removal rates in channel 1

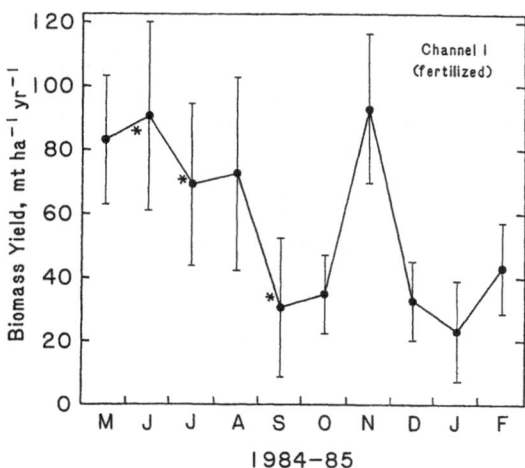

Figure 1. Average monthly biomass yield in flow-through mesocosm water
hyacinth system (channel 1) receiving foliar applications of N +
P (*) at rates of 10 and 3 kg ha^{-1}, respectively. Data points
represent means of 6 replicate cells ± standard deviations

were on the order of 50%. Principal N removal pathways probably included
plant uptake, nitrification-denitrification, and microbial assimilation in
the plant root zone.[12,13] Phosphorus removal in channel 1 followed the same
trend as N removal (Figure 4). Inflow P concentration varied considerably,
resulting in corresponding variations in outflow P levels. Average P
removal was about 60%. Phosphorus removal was due primarily to plant
uptake, microbial assimilation and sediment adsorption.[12,13]

Results of microcosm studies showed strong evidence that both N and P
are growth-limiting nutrients in Lake Apopka, since the application of N + P
to water hyacinths increased net productivity by an average of 37%, compared
with a 6% increase resulting from application of N only (Table 1). Net
productivity (biomass yield) of plants treated with N + P spray averaged
24.2 g m^{-2} day^{-1} (88.3 Mg ha^{-1} yr^{-1}) while the average yield of control
cultures was 17.7 g m^{-2} day^{-1} (64.6 Mg ha^{-1} yr^{-1}). The addition of micronutrients tended to suppress growth, possibly due to toxicity or the high
ionic strength of the spray, although no visible burning of the leaves was
noted.

A mass balance for ^{15}N revealed that between 61.5 and 73.6% of the
foliar applied N was assimilated into plant tissue (Table 2). The bulk of N

Table 1. Effect of foliar application of N, P, and micronutrients on growth of water hyacinth in Lake Apopka water. Cultures were maintained for 8 weeks. Plants were sprayed at 0, 2, 4 and 6 weeks. Data represent means of 3 replicate cultures. Values with the same letter suffix are not significant at P = 0.05

Foliar spray treatment	Net productivity (dw)	% increase over control
	$g\ m^{-2}\ day^{-1}$	
Control	17.7^a	---
N only	18.8^a	6.2
P only	18.0^a	1.7
Micronutrients only	13.8^b	-22.0
N + P	24.2^c	36.7
N + P + micronutrients	23.8^c	34.5

Table 2. Recovery of ^{15}N-urea in foliar application of N, P and micronutrients by water hyacinths cultured in Lake Apopka water. Application rates in kg ha^{-1} were N = 28.1, P = 6.0, Fe = 4.0, Cu = 0.2, Zn = 1.5, B = 0.04, Mo = 0.02, S = 3.0. Data points represent means of 3 replications

Treatment	Standing crop	Plant N	Total N	Total ^{15}N	^{15}N recovery
	$g\ (dw)\ m^{-2}$	$-\%N-$	$mg\ N\ m^{-2}$	$mg\ ^{15}N\ m^{-2}$	% added ^{15}N
Control	1175	1.16	13624	----	----
N only	1197	1.47	17600	1728	61.5
N + P	1448	1.35	19551	1814	64.5
N + P + micronutr.	1380	1.23	16872	2070	73.6

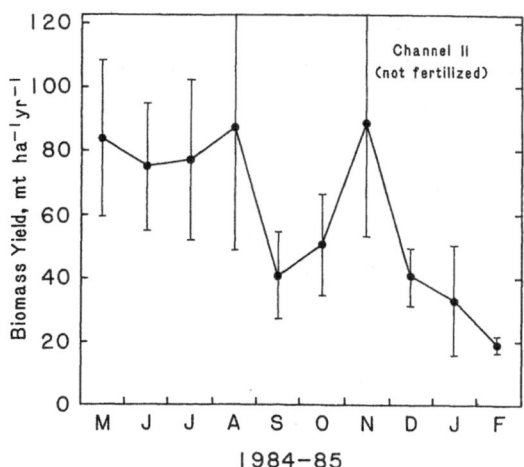

Figure 2. Average monthly biomass yield in flow-through mesocosm water
hyacinth system (channel 2) receiving no foliar application of
nutrients. Data points represent means of 6 replicate cells ±
standard deviations

loss was probably due to volatilization of NH_3, since no significant increase was observed in N or P levels at the water surface immediately
following spraying.

In a follow up study, water hyacinth biomass yields in Lake Apopka
water increased as the rate of nutrient application was increased (Table 3).
Plant growth increased slightly up to a frequency of 8 applications (once
per week). The foliar spray application rate of twice per week resulted in
a growth rate of 28.1 g m^{-2} day^{-1} (102.6 Mg ha^{-1} yr^{-1}), which was significantly higher than all other treatments.

Water hyacinth biomass production data have been generated since 1979
for plants grown in either agricultural drainage water or eutrophic lake
water (Table 4). Average annual biomass yield during that period ranged
from 39.7 to 67.5 Mg ha^{-1}. Variability was attributed primarily to changes
in weather conditions, specifically, the number of freezes during the winter
months and total annual solar radiation influx. Yields were generally
higher for water hyacinths cultured in drainage water, which is somewhat
richer in available nutrients than lake water. Drainage water contains
relatively high concentrations of inorganic N and P, whereas most of the
total N and P in Lake Apopka water is fixed in algal biomass and, therefore,
is only available for plant uptake subsequent to mineralization of dead
algal cells.

Table 3. Effect of foliar N and P application rate on growth and growth-related parameters of water hyacinth in Lake Apopka water. Plants were grown in microcosm tanks (0.62 m^2 water surface area). Study period was 8 weeks; plant density readjusted to initial density of 10 kg m^{-2} after 4 weeks. Data points are means of 3 replicate cultures. Values with the same letter suffix are not significant at P = 0.05

Number of foliar applications	Total applied		Net productivity (dw)	% increase over control
	N	P		
	--kg ha^{-1}--		g m^{-2} day^{-1}	
0 (control)	0	0	16.7[a]	----
1	20	3	19.7[ab]	18.0
2	40	6	20.7[bc]	24.0
4	80	12	21.9[bc]	31.1
8	160	24	23.2[c]	38.9
16	320	48	28.1[d]	68.3

Table 4. Biomass yield of water hyacinth in agricultural drainage water or eutrophic lake water.

Year	Source of water	Biomass yield (dw)
		Mg ha^{-1} yr^{-1}
1979	Agricultural drainage	67.5
1980	Agricultural drainage	50.0
1981	Agricultural drainage	65.2
1982	Lake Apopka	63.0
1983	Lake Apopka	39.7
1984	Lake Apopka	63.9

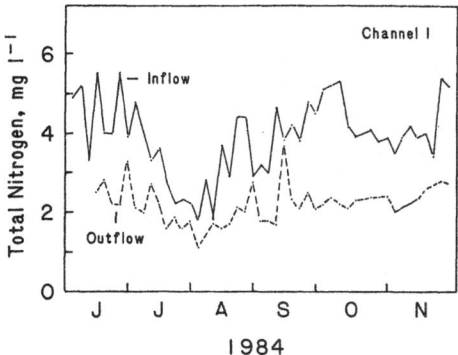

Figure 3. Concentration of total N in channel 1 inflow (Lake Apopka water)
and outflow

DISCUSSION

Results of the mesocosm channel study indicated a baseline yield of
about 67 Mg ha^{-1} yr^{-1} (projection of 1984-85 data) for water hyacinth grown
in Lake Apopka water. Obtaining higher yields involves manipulation of
factors which limit plant growth. For water hyacinth, principal growth-
limiting factors include light availability, nutrient availability and plant
density. The latter may affect growth by governing light and nutrient
availability, i.e. extremely high plant density results in competition for
both light and nutrients. The optimum fresh weight plant density range for
maximum water hyacinth yield is about 150 - 400 Mg ha^{-1} when grown under
high nutrient conditions.[3,14] In nutrient limited waters, however, optimum
density range is restricted to about 150-250 Mg ha^{-1} (Reddy and DeBusk,
Unpublished Results, University of Florida, CFREC, Sanford, FL 1983). By
using the lower end of this range as a starting density and through bi-
weekly (summer) or monthly (winter) harvest, cultures were maintained within
this optimum range.

The primary growth-limiting factor for water hyacinth in Lake Apopka
water is nutrient availability. This was well illustrated by the positive
response of water hyacinth to foliar application of N and P. A large
portion of the total nutrient mass in the lake system is fixed in algal
biomass, or is stored in the highly organic lake sediment.[15] Availability
to plants of the existing nutrient pool in the lake could be increased
through proper management of a large-scale water hyacinth production system.
Management strategies include maintaining the system at a sufficiently high

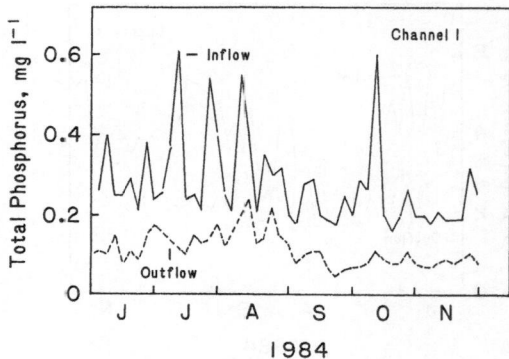

Figure 4. Concentration of total P in channel 1 inflow (Lake Apopka water) and outflow

plant density level to restrict algal growth, thus increasing nutrient availability. Depletion of nutrients in the water by plant uptake causes the formation of a concentration gradient at the sediment-water interface, resulting in a net flux of nutrients from sediment to water.

Nutrient availability in the lake may not be sufficient for optimum yields. Nutrient additions in the form of foliar spray can significantly increase the growth rate of water hyacinths; however, data from microcosm studies suggest that this may involve substantial amounts and high frequency of fertilization.

Projection of available data indicates that a net reduction of nutrients would occur in Lake Apopka if 21% of the lake surface area (2649 ha) were dedicated to water hyacinth production. Assuming an average biomass yield of 67 Mg ha^{-1} yr^{-1}, about 670 and 134 kg ha^{-1} yr^{-1} of N and P, respectively, would be removed by plant uptake. The potential methane gas yield from a water hyacinth crop of this size would be sufficient to supply the needs of the 680 square mile Lake Apopka Natural Gas District. At a rate of 7.8 x 10^5 MJ ha^{-1} yr^{-1} produced from water hyacinth, a total methane yield of about 2.07 x 10^9 MJ yr^{-1} could be realized. This exceeds the current annual natural gas consumption in the district of 1.26 x 10^9 MJ yr^{-1}.

Data generated by the meso- and microcosm water hyacinth production systems suggest that, using current non-intensive management schemes, significant amounts of natural gas may be produced from a similar, large-scale system on Lake Apopka, while at the same time, reducing nutrient concentrations in the water. Further studies must be employed to refine certain management techniques, such as foliar application of fertilizer and harvesting, so that biomass yield may be increased to a level of greater economic benefit.

ACKNOWLEDGEMENTS

Florida Agricultural Experiment Stations Journal Series No. 6497. This paper reports results from a project that contributes to a cooperative program between the Institute of Food and Agricultural Sciences (IFAS) of the University of Florida and the Gas Research Institute (GRI), titled, "Methane from Biomass and Waste."

REFERENCES

1. B. C. Wolverton and R. C. McDonald. Energy from vascular plant wastewater treatment systems. Econ. Bot. 35:224-232. 1981.
2. A. Shiralipour and P. H. Smith. Conversion of biomass into methane. Biomass 6:85-92. 1984.
3. K. R. Reddy and W. F. DeBusk. Growth characteristics of aquatic macrophytes cultured in nutrient enriched water. I. Water hyacinth, water lettuce, and pennywort. Econ. Bot. 38:225-235. 1984.
4. K. R. Reddy and W. F. DeBusk. Growth characteristics of aquatic macrophytes cultured in nutrient enriched water. II. Azolla, duckweed, and salvinia. Econ. Bot. 39:200-208. 1985.
5. D. A. Cornwell, J. Zoltek, Jr., C. D. Patrinely, T. deS. Furman and J. I. Kim. Nutrient removal by waterhyacinth. J. Water Pollution Control Federation 49:57-65. 1977.
6. K. R. Reddy, K. L. Campbell, D. A. Graetz and K. M. Portier. Use of biological filters for agricultural drainage water treatment. J. Environmental Quality. 11:591-595. 1982.
7. K. R. Reddy, F. M. Hueston and T. McKim. Biomass production and nutrient removal potential of water hyacinth cultured in sewage effluent. J. Solar Engineering (in press). 1985.
8. J. L. Yount and R. Crossman. Eutrophication control by plant harvesting. J. Water Pollution Control Federation 42:173-183. 1970.
9. C. A. Warren, D. E. Bruderly, M. Angelieri, L. J. Bilello, S. Bucalo, G. W. Finger, R. Hart, S. W. Hinton, J.R. Newman and J. W. Vinzant. The methane from biomass and waste program. Task 1. Evaluation of the Lake Apopka Natural Gas District Report. Gas Research Institute. Contract No. 50080-323-0423. Chicago, IL.
10. U. S. EPA. Environmental impact statement. Lake Apopka restoration project. Lake and Orange Counties. EPA 904/4-79-043. 1979.
11. L. R. Sinclair and R. B. Forbes. Nutrient removal from drainage waters with systems containing aquatic macrophytes. Trans. ASAE 23:1189-1194. 1980.
12. H. H. Rogers and D. E. Davis. Nutrient removal by water hyacinth. Weed Science 20:423-428. 1972.
13. K. R. Reddy. Fate of nitrogen and phosphorus in a waste-water retention reservoir containing aquatic macrophytes. J. Environ. Qual. 12:137-141. 1983.

14. T. A. DeBusk, J. H. Ryther, M. D. Hanisak and L. D. Williams. Effects of seasonality and plant density on the productivity of some freshwater macrophytes. Aquat. Bot. 10:133-142. 1981.

15. P. L. Brezonik, C. D. Pollman, T. L. Crisman, J. N. Allison and J. L. Fox. Limnological studies on Lake Apopka and the Oklawaha chain of lakes: I. Water quality in 1977. Rept. No. ENV-07-78-01. Dept. of Env. Eng. Sci., Univ. of FL, Gainesville, FL. 1978.

GRAVEL BED HYDROPONICS FOR WASTEWATER

RENOVATION AND BIOMASS PRODUCTION

Linda Lea Handley, Lee S. Casey, Jose L. Lopez, Juan M.
Sutija, Hussein I. Abdel-Shafy and Susan B. Colley

Drinking Water Research Center
Florida International University
Miami, Florida 33199

ABSTRACT

This paper presents an overview and some preliminary data for a relatively new technology for biomass production and wastewater renovation. The
two main goals of our GBH (gravel bed hydroponics) research are: (1) optimizing both crop production and wastewater renovation and (2) construction
of an empirical model for system performance. Economic analyses must await
improvement of the GBH method.

We are using two C_4 tropical grasses which have shown some of the
highest net productivities known for vascular plants: paragrass (Brachiaria
mutica (Forsk.) (Stapf) and Napiergrass (Pennisetum purpureum Schumach). In
the past, under heavy effluent irrigation, paragrass has produced as much as
1.93×10^5 kg ha^{-1} yr^{-1} dry biomass. In the GBH, Napiergrass has occa-
sionally produced even greater harvest equivalents. These grasses have a
large potential for nutrient uptake and appear to adequately clean wastewaters. Additionally, they can accumulate large concentrations of crude
protein with low nitrate-nitrogen contents. This makes them good candidates
for animal feed. They are being tested for use as chemical feedstocks.

These grasses will probably be most useful in tropical latitudes. In
South Florida we can test their performance against seasonal weather
factors while still having little risk of frost.

This paper begins by discussing the rationale for biomass production
based on wastewater reuse. The limits to biomass production are set by the

amounts of available resources, by crop selection, and by site selection. The availability of wastewater nutrients and water may enhance the economy of biomass production.

Keywords. GBH, limiting resources, crop productivity, Napiergrass, Paragrass.

INTRODUCTION

The nutrient film technique (NFT) consists of growing crops in a channel through which a thin film of nutrient solution flows. The GBH (gravel bed hydroponics) differs in only one respect. A layer of gravel encloses the crop roots. Our NFT and GBH systems use domestic wastewater for the thin-film nutrient solution.

We are attempting to derive an empirical model for these systems. This model will describe crop productivity and wastewater renovation as functions of seasonal weather variables. Measured variables include temperature, relative humidity, sunlight, and evapotranspiration for two species of C_4, panicoid grasses: paragrass (Brachiara mutica) and Napiergrass (Pennisetum purpureum). Mass balances will be calculated for water, nitrogen, and phosphorus. Crop quality will be assessed as contents of crude protein, carbon, nitrogen, phosphorus, and potassium. The need for such models was pointed out by O'Brien and Stowell et al.[1,2]

This paper briefly discusses three topics: (1) the relevance of wastewater use to biomass production, (2) the historical development of these systems, and (3) some preliminary results of our work with these systems in South Florida. At this writing, our results are not rigorously quantitative, merely indicative. We do not yet have long-term continuous measurements.

DISCUSSION

Relationship of Wastewater GBH and NFT to Biomass Production

For biomass to compete economically as an energy source, its production must be economical. More specifically, the value of the end product must substantially exceed the cost of the inputs.[3] Two major strategies have been practised: (1) crop selection and breeding and (2) improvement of agronomic practices for given sites and crops.

Given these strategies alone, the potential for improving crop production and economy reaches an impassible ceiling determined by species capacity. This capacity may be expressed as resource use efficiency (amount of dry biomass produced per unit of resource supplied). Different taxa have different resource use efficiencies. Many highly efficient plants (e.g., CAM plants) are not greatly productive.[4,5] The growth curve plateaus at relatively low amounts of the resource with which the plant is most efficient. Beyond this point, addition of more resource does not increase growth, and efficiency of production declines.

Other taxa effectively use resources in greater amounts before the growth plateau is reached. However, a crop's inherent efficiency of resource use (relative to the amounts supplied) cannot be appreciably changed by breeding improvements.[6] The relative allocations of fixed carbon may be altered (e.g., stem and leaves versus seeds or root:shoot ratios[7,8]). The total amount of resource effectively used may be increased and the plants better adapted to a particular site.[7] But these changes only create the need for greater inputs (e.g., fertilizer and water) and, hence, expense.

In no case has it been demonstrated that selective breeding has changed the overall resource use efficiency of crop plants. Several workers have shown that below the plateau of the growth curve, plants adjust their physiological responses to optimize resource use efficiency for a given level of inputs.[9,6,10] Given the nature of plant life, resource use efficiencies are already as optimized as they can be.

Therefore, breeding improvements can only shift the resource saturation point, increasing the need for resources, and increasing the cost of production. Given the above, the solution to decreasing production costs, while concurrently increasing production, lies elsewhere.

We suggest that one possible solution lies in the following approach: (1) crop selection, (2) site selection, and (3) the use of wastewater as a fertilizer and water source. A recent review paper[11] reinforces this by pointing out that increased crop yields in North America have been largely due to only two factors (1) selection of crops genetically better site-adapted so that they can effectively use greater fertilizer and water inputs and (2) supply of greater inputs. Where inputs have been restricted because of cost, breeding improvements have not improved yields.

Of the physiological plant types, C_4 plants demonstrate, as a group, the greatest net productivities.[12] In efficiencies of light and water use, they are thought to fall between the C_3 and CAM plants.[8,13] However, to achieve maximum growth, they not only require commensurately large inputs of water, nitrogen and phosphorus, they also need large amounts of heat and light.[14,12] This implies selection of warm, sunny sites, e.g., the lowland tropics and subtropics.

Where soil and hydrologic conditions permit, wastewater irrigation may be most appropriate. Where they do not permit, a thin film hydroponics method can be used to good advantage. The next section discusses the historical development of two such methods: the nutrient film technique (NFT) and gravel bed hydroponics (GBH).

Discussion of NFT & GBH Development

The Nutrient Film Technique (NFT) and the GBH (gravel bed hydroponics) are closely related systems for feeding controlled amounts of nutrient solution to crops grown in water-tight channels. They are both thin-film, flow-through hydroponics systems. The NFT contains no rooting medium; the GBH contains gravel in which the crops are rooted. There are advantages to both systems. The NFT was found by Bouzon[15] to perform much like a plug-flow reactor with respect to BOD removals from wastewater. The GBH is thought to perform more like a trickling filter.

Both systems share the advantages that (1) they are site independent of soil and hydrologic conditions; (2) they are high-rate wastewater renovation methods; and (3) they can eliminate the problems of root aeration found with traditional hydroponics. The GBH appears to mix and distribute the waste-waters better, provide more microbial habitat, and provide more physical support for crops.

Flow-through hydroponics were designed with the growth of food and horticultural crops in mind. In particular, the NFT has become widespread as a glass house cropping method. Graves[16] recently discussed operational aspects of commercial NFT operations, including crop nutritional aspects.

It was noted incidentally that nutrients and pollutants were removed from the hydroponic solution. The emphasis was on crop growth and optimal crop growth conditions. This should still be the primary focus. When crop growth is optimized, nutrient stripping and pollutant removals are optimized as a consequence.

The Nutrient Film Technique was originally developed by A. J. Cooper[17,18] at the Glasshouse Research Institute in England. Cooper found that dilute nutrient solutions could be used. Sewage effluents are not usually as concentrated in nutrients as are traditional hydroponic solutions. Therefore, Cooper's success with dilute solutions suggested that the NFT might be useful in wastewater renovation. This line of inquiry was pursued by a group at Portsmouth Polytechnic in England, working at a wastewater treatment facility called Budd's Farm.[19]

They found, however, that mixing of the solution was sometimes poor, and that field crops needed a physical matrix for support.[20] They solved these problems by adding a layer of gravel to the system and calling it gravel bed hydroponics (GBH).

This brought the development of the NFT process almost full circle so that the resulting GBH resembled the systems developed and investigated in North Europe by Seidel and De Jong,[21] among others. In these early systems, bulrushes were grown in deep gravel-filled trenches for the sole purpose of wastewater renovation.

The workers at Budd's Farm have experimented with a variety of temperate-climate crops. Some of these were tomatoes, courgettes (a kind of squash), sugar beet, broad beans, and strawberries.[19] They are now working mainly with sugar beet. Small-scale trials have been successful over several years. They are prepared to use the technique developed with sugar beets on a commercial scale. Bone et al[22] reported that sugar beet yields were comparable to those found in conventional agriculture (e.g., 7.7×10^4 kg ha^{-1} yr^{-1} fresh weight). Quality was also comparable.

The effluent at Budd's Farm is secondary municipal wastewater. The GBH provides tertiary treatment. Micronutrients, such as iron and manganese, are added. There are usually too little micronutrients in secondary effluents for healthy crop growth.[20,23] The pH is also routinely adjusted at Budd's Farm. The high pH normally found in effluents is thought to interfere with plant uptake of metal ions from solution.[16,24] With wastewater irrigation, micronutrients are supplied by the soil. The NFT and GBH contain no soil.

Winfield and Bone[19] gave typical N, P, K values for the Portsmouth effluent of 20-25, 5-12, and 10-18 mg dm^{-3}. GBH of sugar beet and tomato removed about 50% of the nitrate-N, more than 90% of the ammonium-N, 20 and 50% of the P and 38 and 89% of the K.

Two major glass house NFT experiments were done in the United States. These were done concurrently by W. J. Jewel and his colleagues at Cornell University in New York and by John Bouzon and his group at Cold Regions Research and Engineering Laboratories (CRREL) in New Hampshire. These were feasibility trials and lasted for less than two years. Nevertheless, considerable information was obtained concerning the performance of the NFT system in cold climates, using glass houses and a variety of crops.

Rather remarkable removals were reported from New York where the effluents averaged only 9°C. An air temperature of nine to ten degrees centigrade is usually considered the lower temperature limit for crop growth.[25] Even Napiergrass, a subtropical C_4 plant, grew well and gave good wastewater renovation results.

Jewell et al[23,26] chose reed canary grass as their major test crop. It is a temperate-climate, C_3 grass which withstands cold weather and moderate flooding. It has been used extensively in effluent irrigation trials in the northern US.[3] In the Cornell trials, it tolerated up to 10 cm day^{-1} of effluent. The Cornell group reported a net productivity for this grass of 4.69×10^4 kg ha^{-1} yr^{-1} dry weight. The reported value for Napiergrass was 1.5×10^5 kg ha^{-1} yr^{-1} dry weight.

They also tried 17 other plants to various extents. The ones which flourished were cattails, watercress, bulrush, strawflowers, Japanese millet, roses, marigolds, wheat, and Phragmites. The Cornell group also examined evapotranspiration (ET) for possible clean water recovery. In some of the units, ET was as great as 100,000 dm^3 ha^{-1} day^{-1} ($= 10$ mm day^{-1}).

There is, in general, a positive and linear relationship between transpiration and biomass,[27] although the slope and intercept of such a curve varies with taxa and environmental factors. At high biomass, the evaporation component becomes negligible. The large ET rates observed by the

Cornell group were probably due to allowing these tall-growing grasses to achieve a large standing crop biomass.

The above implies a compromise between ET and nutrient uptake. As the plants become more mature, they assimilate nutrients at a slower rate. However, harvesting them to force increased nutrient uptake would depress ET. This is a trade-off which can be manipulated to fit the purposes of a system, i.e., ET recovery or nutrient stripping and biomass harvest.

The Cornell group envisioned an eventual three-stage NFT: Stage 1 being a rapid-flow roughing section to remove TSS and BOD as well as some nutrients, Stage 2 being a nutrient stripping unit for production of biomass or food crops, and Stage 3 being nutrient-limited plant production for final polishing of the effluent. We see our tropical grass-GBH system as corresponding functionally to Stages 1 and 2.

Bouzon's group at CRREL[15,28] also used reedcanary grass in an NFT. While Jewell's group used mainly synthetic sewage, the CRREL group used primary effluent which averaged 11°C. Effluent was applied at three flow rates (379, 757, and 1,514 dm^3 day^{-1}). The grass was harvested six times. Their data indicate a mean calculated net productivity of 1.5×10^4 kg ha^{-1} yr^{-1} dry weight. However, productivity varied over the course of the experiment from 0.9 to 2.1×10^4 kg ha^{-1} yr^{-1}. N, P, and K in the harvested grass averaged 5.46, 0.68, and 3.83% of dry weight. No nutrient deficiencies were reported when using primary effluent.

The grass was reported to be of high quality, being 81% digestible and containing 35% crude protein, by dry weight. Nitrate-N contents were not reported. Since nitrate-N can be toxic in animal feeds,[29] this could be an important factor.

Bouzon[28] was able to model BOD removals as a first order, plug-flow reactor process. We have been unable to substantiate this model. Nutrient removals are more complicated to describe. Nitrogen species change along the length of the channel, as shown by data from both of the U.S. projects. Also the availability of N and P decreases along the length of the channel. The net effect may be to depress the crop's uptake capacity (hence, growth) with increasing tray length.

The three projects discussed above demonstrated the potential of the NFT and GBH for treating various types of wastewater while growing substan-

tial crops, either in glass houses or as field crops. The mechanical technology is not difficult. Two critical elements are still lacking: predict-
ability of system processes and system optimization.[1,2]

Critical system processes are crop yield and quality, nutrient stripping capacity, TSS and BOD removals. All of these processes depend upon the
interaction of the crops with environmental variables. The next task is to
correlate system performance for several crops with weather factors and
nutrient loadings. It is to this task we have addressed our work in South
Florida.

Florida NFT/GBH: Preliminary Results and Discussion

The Florida NFT (now GBH) project is located at Florida City, south of
Miami, on the site of the Gateway Wastewater Treatment Plant. The NFT
system originally consisted of 18 trays of grasses. The trays were plastic-
lined plywood, 1 m wide, 12 m long, and 20 cm deep. The slope was 5%.
Para-, Napier-, and torpedograss were each given three flow rates of effluent (2.73, 3.65 and 8.18 m^3 day^{-1} = 1, 3, and 5 gpm) and at each flow rate
two types of effluent (secondary and raw). Torpedo grass did not thrive and
was eliminated.

Our first project period consisted of two main phases: pilot plant
construction and a problem identification and shake-down phase. During the
second phase, sampling of wastewaters consisted of two types. Initially,
weekly grab samples were taken for testing wastewater quality. Later, four
intensive sampling and spiking experiments were done in which water quality
was tested hourly from dawn to dusk. The intensive samplings were done
under cool weather conditions, probably yielding minimal performance data.

Because we are trying to define optimum conditions for a minimum-
technology system, we use out-of-doors crop growth, hardy tropical grasses,
and minimal wastewater pre-treatment.

Because of funding calendars and deadlines, most of our data during the
first project period were taken during the cold season, when the grasses
grew the least.

Several crops were evaluated before it was determined that para- and
Napiergrass were the most suitable. We also tried _Phragmites_, _Cyperus_
esculentus (yellow nutsedge), _Hydrocotyl_ sp. (pennywort), _Commelina_ _diffusa_

(dayflower), and _Panicum_ _repens_ (torpedo grass) with only lackluster success.

Napiergrass (Pennisetum purpureum Schumach)

Napiergrass is a tropical, C_4 plant, closely related to sugarcane. Its appearance is grossly similar to that of sugarcane, having large woody stems. Larcher[12] credits this grass with the highest known productivities in natural stands. In South Florida it is a weed of roadsides and disturbed areas. Vicente-Chandler et al[30] described Napiergrass as a good forage and discussed its cultivation and nutritional characteristics. For systematic and habitat descriptions, see Rotar[31] and Whitney et al.[32]

Workers with the Gas Research Institute (GRI) have found high productivities. They chose Napiergrass as a candidate for biomass and biogas production.[33,34] If Napiergrass could be produced with the low cost inputs of the NFT or GBH, then high biomass yields might be obtained at very low cost, while simultaneously renovating wastewaters.

Paragrass (Brachiaria mutica (Forsk.) Stapf)

Despite being a C_4 grass and tropical close relative of sugarcane, its appearance and growth habit are unlike that of sugarcane. It is mat forming and herbaceous. However, like sugarcane, it is a wetland plant, which Napiergrass is not. Paragrass is tolerant of anaerobic root conditions and will survive under a variety of nutrient regimes.

Handley and Ekern[35] found it to be an excellent grass for effluent irrigation in Hawaii. It stripped large quantities of N and P. It also produced large amounts of biomass (up to a maximum of 1.93×10^5 kg ha^{-1} yr^{-1} dry weight), contained up to 13% crude protein, and less than 0.1% nitrate-N.

The chief characteristics of paragrass have been discussed by several authors.[30,32,35,36,37] ,

Napiergrass is the more difficult of the two grasses to cultivate. It is initially slower growing, taking several weeks longer than paragrass to establish a dense root mat. When young, it is more easily harmed by anaerobic root conditions and cannot tolerate heavy solids loadings. It is subject to windthrow unless deeply rooted in gravel. It is more difficult to harvest because of its large, woody stems. Because it is a stool-forming grass, it does not form a dense continuous root mat as does the paragrass. Its reported large productivities, however, made it appear worthwhile to attempt to optimize its growth in the GBH. We have since found it unsuitable for outdoor GBH use and have eliminated it from our project.

Para- and Napiergrasses grew best given raw sewage. Secondary effluent did not contain enough nutrients, nor was the nitrogen in the correct form. In tests conducted at our site, these grasses preferentially used ammonium-N. It has long been known that many plants can be grouped by their effectiveness in using nitrate or ammonium.[12,38] All of the secondary effluent nitrogen was in the nitrate form. Both grasses assimilated nitrate-N from the ammonium-rich raw sewage. In the secondary effluent trays, where ammonium was lacking, almost no nitrate-N was assimilated by the plants.

There were also insufficient micronutrients in the secondary effluent for crop growth. Chlorosis was evident. We have lately found that simple clarification removes effluent mironutrients. From this we hypothesize that the micronutrient metals were not in solution, but absorbed onto particles, which were removed by the clarifier unit. It has long been known that the digestion process of secondary sewage treatment removes micronutrients from sewage.[20,23] We are not aware of any literature sources reporting this removal from the simple, physical process of solids sedimentation. This may be a special case, caused by the high pH (ca. 7.2 - 7.6) of South Florida waters. For the moment, we are applying daily foliar sprays while we experiment with means of replacing micronutrients in the wastewater after the clarifier treatment, as well as lowering wastewater pH.

A pattern was observed with regard to flow rates. We have no verified explanation for it. Crop growth and nutrient removals were greatest for both grasses, at flow rates of 2.7 and 13.6 m^3 day^{-1} and poorest at 8.2 m^3 day^{-1}. (These rates correspond to 1, 5, and 3 gpm over 12 hours each day). Phosphorus removals were also positively related to air temperature, both daily changes and mean monthly temperature. Several parameters did not change in the NFT trays. Those showing no statistically significant change were: amount of dissolved oxygen, pH, conductivity, and chlorinity.

Hydraulic retention times were difficult to assess. Plant roots tended to retain the dye, which then leached for several hours. It was difficult to find a clear dye peak. However, Table 1 contains our best estimates of

Table 1. Hydraulic retention times (HRT) in NFT trays at three flow rates and planted with two grasses

Grass	Flow Rate (m^3 day^{-1})	HRT (minutes)
Napiergrass	2.73	90
	3.65	40
	8.18	15
Paragrass	2.73	80
	3.65	66
	8.18	20

retention times under different flow rates. These flow rates were estimated for the NFT, not the GBH.

After an initial baseline harvest, the crops were harvested twice. The greatest equivalent dry weight harvest amounts from optimal growth periods were 153 kg ha^{-1} day^{-1} for paragrass and 1,047 kg ha^{-1} day^{-1} Napiergrass.

We found that using an NFT system created the following problems for wastewater renovation. The wastewaters tended to carve channels in the root mats of the plants, bypassing the root matrix, its nutrient stripping and filtering effects. This also made the quality of the finished wastewaters highly variable. In addition, the grasses tended over time to pull away slightly from the bottom of the tray (i.e., contact between roots and tray was no longer continuous), and much of the wastewater simply flowed under the root mat, again circumventing renovation. Because of these problems, we have emulated the Budd's Farm-Portsmouth Polytechnic project and converted our system to GBH by adding a layer of locally available crushed coral gravel.

The polyethylene tray liners deteriorated and were easily punctured. We have now fiberglassed the trays and eliminated the plastic liners.

At first we tried to monitor the removals of all species of nitrogen. We found that nitrogen species changed as the wastewater flowed through the trays, as also reported by Wolverton et al[39] for a similar system. Although this was not specifically reported by Jewell et al[23,26] and Bouzon and Palazzo[15] for their NFT work, it can be inferred from their data. Urea was hydrolyzed to ammonia, and some ammonia may have been oxidized to nitrate.

In addition, we found that the crops preferentially assimilated ammonium. Because of these complications, we are now monitoring only total nitrogen, in and out of the GBH trays.

Flow rates through the GBH are too great for directly estimating an accurate water balance. The maximum expected evapotranspiration (ET) of 8-10 mm day^{-1} would be masked in the experimental error of measuring the present application rate of 2.73 m^3 day^{-1} (= 227 mm day^{-1} or 1 gpm for 12 hours each day.) We have, therefore, constructed a flow-through lysimeter for each grass. ET in the lysimeters will be related to ET in the trays by regressing ET against harvest biomass.[1]

We are simultaneously maintaining a class A evaporation pan and a new type of evaporimeter, devised by Dr. Paul C. Ekern of the University of Hawaii. Testing in Hawaii showed this evaporimeter to more closely track the actual ET of grass sod than does a class A pan.

Our new experimental design is simpler. We have now three replicate trays of paragrass. There is also a control tray containing only gravel. All trays are given clarified, primary effluent at the rate of 2.73 m^3 day^{-1} (1 gpm) for 12 hours each day. The effluent is also acidified with indus-trial grade sulfuric acid to pH = 6.5 in the clarifier.

The acidification step was added to the system in an attempt to correct micronutrient deficiencies, especially iron deficiency. As shown in Table 2, preliminary data indicate that acidification does solubilize iron and that the crops are apparently using this iron. The dissolved iron content of the effluent is 8.7 times greater after acidification; it drops slightly

Table 2. Dissolved iron content of effluent at four points in the GBH-paragrass system. Means in mg L^{-1}

Sample Point	Iron Concentration
Screened raw wastewater	0.0503
Acidified and clarified wastewater	0.4397
Influent to GBH crop tray	0.3363
Effluent from GBH crop tray	0.1078

between the clarifier and the head of the GBH crop trays; and it is reduced in the crop trays by a factor of three. In addition, the crops appear to be much more vigorous.

Weather factors are now continuously recorded. These are temperature, relative humidity, and solar radiation (total, net and PAR). Rainfall is taken with a U.S. standard rain guage.

All previous NFT/GBH studies used grab samples to estimate wastewater removals, as we did during the first project period. However, weather factors change over the day. We have, therefore, started taking daylight composite samples to monitor the net performance of the system. Flow rates are negligible at night. Grab samples are still taken to establish the daily variations in removals and the variations between replicates. Composite samplers have only recently been put into operation. As grab sampling proved to be unreliable as an indicator of nutrient stripping, none of these data are reported here.

Table 3 contains data taken during the first few months of our new GBH-paragrass experiment with composite sampling. These data are preliminary, but they do serve to indicate the great range and potential of wastewater removals possible with this system. The data were taken during late winter and early spring when the weather was variable. The range of values presented here reflect the range of responses to daily weather conditions. Weather conditions, during this period, ranged from cold, cloudy days (small removals) to warm, sunny days (high removals). Although this data set clearly demonstrates the relationship between weather and system performance, we do not yet have enough data to quantify the relationship.

Table 3. GBH removals. Composite samples taken 25 February to 28 April, 1985. Means and ranges in percent of parameter removed

GBH Tray Type:	Paragrass		Control*	
Parameter	Mean	Range	Mean	Range
BOD	48	8-91	52	5-58
Total N	12	0-92	13	0-41
P**	21	0-61	12	0-68

* The control tray contains only gravel, no crop; ** ortho-phosphate-P

CONCLUSIONS

Appropriate crop selection and site selection offer a viable approach
to reducing the expense of biomass production. The most promising plants
for maximum production are C_4 crops in warm-climate sites. Use of domestic
wastewater completes the suite by providing low-cost water and fertilizer
inputs. When the GBH method is further developed from a technical standpoint, an economic analysis will be needed.

Many crop-wastewater systems are similar: NFT, GBH, and Seidel's
bulrushes in deep gravel trenches. Yet none of these has been fully developed so that system performance is both optimized and predictable. The GBH
project attempts this with C_4 crops in a subtropical climate.

Initial results are encouraging for the GBH system using acidified
primary effluent with para grass.

ACKNOWLEDGMENTS

This work is supported by U.S. Geological Survey Contract number 4-FG-
93-00070 and by National Science Foundation Grant number CEE-831589. We
thank Prof. John A. Raven for reviewing the manuscript and making helpful
comments.

REFERENCES

1. W. J. O'Brien. Use of aquatic macrophytes for wastewater treatment.
   Proc. Amer. Soc. Civil Engg. J. Environ. Engg. Div. 107(EE4):681-
   698. 1981.
2. R. Stowell, R. Ludwig, J. Colt and G. Tchobanoglous. Concepts in
   aquatic treatment system design. Proc. Amer. Soc. Civil Engg. J.
   Environ. Engg. Div. 107 (EE5):919-940. 1981.
3. D. R. Linden, C. E. Clapp and J. R. Gilley. Effects of scheduling
   municipal wastewater effluent irrigation of reed canary grass on
   nitrogen renovation and grass production. J. Environ. Qual. 10:507-
   510. 1981.
4. D. L. Marzola and D. P. Bartholomew. Photosynthetic pathway and biomass
   energy production. Sci. 205:555-559. 1979.
5. H. T. Odum and R. C. Pinkerton. Time's speed regulator: the maximum
   power output in physical and biological systems. Amer. Sci. 43:331-
   343. 1955.
6. J. R. Evans and J. R. Seeman. Differences between wheat genotypes in
   specific activity of ribulose-1,5-bisphosphate carboxylase and the
   relationship to photosynthesis. Plant Physiol. 74:759-765. 1984.
7. R. M. Gifford, J. H. Thorne, W. D. Hitz and R. T. Giaquinta. Crop
   productivity and photoassimilate partitioning. Sci. 225:801-807.
   1984.

8. R. H. Brown. A difference in N use efficiency in C_3 and C_4 plants and its implications in adaptation and evolution. Crop Sci. 18:93-98. 1978.

9. F. S. Chapin, III. The mineral nutrition of wild plants. Ann. Rev. Ecol. Syst. 11:233-260. 1980.

10. A. D. Hanson and W. D. Hitz. Metabolic responses of mesophytes to plant water deficits. Ann. Rev. Plant Physiol. 33:163-203. 1982.

11. J. S. Boyer. Plant productivity and environment. Science 218:443-448. 1982.

12. W. Larcher. Physiological Plant Ecology. Springer Verlag. New York. 303 p. 1980.

13. J. R. Ehleringer. Photosynthesis and photorespiration: biochemistry, physiology, and ecological implications. HortSci. 14:217-22. 1979.

14. R. M. Gifford. A comparison of potential photosynthesis, productivity and yield of plant species with differing photosynthetic metabolism. Aust. J. Plant Physiol. 1:107-117. 1974.

15. J. R. Bouzon and A. J. Palazzo. Preliminary Assessment of the Nutrient Film Technique for Wastewater Treatment. U.S. Army Corps Engineers, Cold Regions Research and Engineering Laboratory Spec. Rpt. 82-4. 15 p. 1982.

16. C. J. Graves. The nutrient film technique. Chap. 1 IN J. Janick (ed.) Horticultural Reviews. Vol. 5. AVI Publ. Westport, Conn. 415 p. 1983.

17. A. J. Cooper. The Nutrient Film Technique of Growing Crops. Grower Books. London. 1976.

18. A. J. Cooper. The ABC of NFT. Grower Books. London. 1979.

19. B. A. Winfield and D. A. Bone. The sewage farm reborn. J. Inst. Public Health Engineers. 9:185-186 + 198. 1981.

20. S. Lucas. Making the most of waste. Water. Sept.: 26-44. 1981.

21. J. Tourbier and R. W. Pierson (eds.) Biological Control of Water Pollution. Univ. Penn. Press. Philadelphia. 1976.

22. D. Bone, B. Loveridge and B. Winfield. Sugar from sewage. British Sugar Beet Review. 49:42-44. 1981.

23. W. J. Jewell, J. J. Madras, W. W. Clarkson, H. DeLancey-Pompe and R. M. Kabrick. Wastewater Treatment with Plants in Nutrient Films. Final Rpt. to U.S. EPA. 1983a.

24. H. D. Foth. Fundamentals of Soil Science. 6th Edit. John Wiley and Sons. N.Y. 436 p. 1978.

25. M. P. Russelle, W. W. Wilhelm, R. A. Olsen and J. F. Power. Growth analysis based on degree day. Crop Sci. 24:28-32. 1984.

26. W. J. Jewell, J. J. Madras, W. W Clarkson, H. DeLancey-Pompe and R. M. Kabrick. Wastewater treatment in nutrient films. Project Summary. EPA-600/S2-83-067. 1983b.

27. R. J. Arkley. Relationships between plant growth and transpiration. Hilgardia 34:559-584. 1963.

28. J. R. Bouzon. A wastewater treatment and reuse process for cold regions. IN Cold Regions Environ. Engg. Conf. 18-20 May, Fairbanks, AK. 1983.

29. R. F. Crawford and W. K. Kennedy. Nitrates in forage crops. World Farming 5:8-24. 1963.

30. J. Vicente-Chandler, F. Abruna, R. Caro-Costas, J. Figarella, S. Silva and R. W. Pearson. Intensive Grassland Management in the Humid Tropics of Puerto Rico. Bull 233. Univ. Puerto Rico, Coll. Agric. Sci., Rio Piedras. 1974.

31. P. P. Rotar. Grasses of Hawaii. Univ. Hawaii Press. Honolulu. 355 p. 1968.

32. L. D. Whitney, E. Y. Hosaka and J. C. Ripperton. Grasses of the Hawaiian Ranges. Bull. 82 Hawaiian Agric. Exp. Sta. Univ. Hawaii, Honolulu. 148 p. 1939.

33. G. M Prine and P. Mislevy. Grass and herbaceous plants for biomass. Soil and Crop Sci. Soc. Florida 42:8-12. 1983.

34. W. H. Smith. Methane from Biomass and Waste. Ann. Rpt. Gas Res. Inst., Chicago, Ill. 165p. 1983.

35. L. L. Handley and P. C. Ekern. Effluent irrigation of para grass: water, nitrogen and biomass budgets. Water Resources Bull. 20:669-677. 1984.

36. L. L. Smith, (nee' Handley). Development of emergent vegetation in a tropical marsh (kawainui, Oahu) Part I. Newsletter Hawaiian Bot. Soc. 16 (3/4):37-47. 1977.

37. L. L. Smith, (nee' Handley). Development of emergent vegetation in a tropical marsh (Kawainui, Oahu) Part II. Newsletter Hawaiian Bot. Soc. 17 (1/2):3-27. 1978.

38. G. R. Stewart, J. A. Lee, T. O. Orebamjo and D. C. Havill. Ecological aspects of nitrogen metabolism. p.41-17 IN R. L. Bielski, A. R. Ferguson and M. M. Cresswell (eds.) Mechanisms of Regulation of Plant Growth. Bull. 12. The Royal Soc. New Zealand. 1974.

39. B. C. Wolverton, R. C. McDonald and W. R. Duffer. Microorganisms and higher plants for wastewater treatment. J. Environ. Qual. 12:236-242. 1983.

ALGINATE LYASES OF VARYING SUBSTRATE SPECIFICITIES

FROM MARINE BACTERIA

Tony Romeo, John C. Bromley and James F. Preston, III

Department of Microbiology and Cell Science
University of Florida/IFAS
Gainesville, Florida   32611

ABSTRACT

Alginate lyase activities of several marine bacterial isolates associated with species of marine brown algae, genus Sargassum, have been characterized with respect to substrate specificities and modes of substrate depolymerization. The intracellular activities of oxidative isolates, genus Alteromonas, show preferences for poly D-mannuronate (poly M) over poly L-guluronate (poly G) and native alginate. The extracellular activities degrade alginate more rapidly than either poly G or poly M. The analogous enzymes from the fermentative organisms include intracellular activities which are slightly more active on poly M than on alginate or poly G. Extracellular preparations from 2 of the isolates have little or no activity on poly G and are only slightly more active on poly M than alginate. One of the fermentative isolates had extracellular activity greater toward alginate and poly G than poly M. In general all extracellular activities showed a greater decrease in viscosity of alginate relative to glycan cleavage than did the intracellular extracts, indicating higher endoeliminase activities in the extracellular fractions. The fermentative isolates which produce alginate lyases of varied specificities may prove valuable in converting marine brown algal biomass to methane. Specific alginase enzymes of the fermentative as well as the oxidative organisms should allow development of defined methods for producing protoplasts from the marine algae of the family Phaeophyceae.

Keywords. Alginate, alginate lyase, marine bacteria, Alteromonas, marine algae, Sargassum.

<u>Abbreviations</u>: poly G, poly α-1,4-L-guluronate; poly M, poly β-1,4-D-mannuronate; TBA, 2-thiobarbituric acid; PESI, Provasoli's enriched seawater with 0.27 mg/liter KI; Δ, an unsaturated carbon - carbon bond.

INTRODUCTION

The Phaeophyta are marine brown algae which inhabit oceanic coastal waters throughout the world. Although several species have been commercially exploited for their anionic carbohydrates, in particular the alginates, the bulk of the world's supply remains untapped. The giant kelp <u>Macrocystis pyrifera</u> has been evaluated for its bioconversion to methane and shown to provide methane yields that are competitive with other biomass and waste sources.[1] Studies are in progress to determine the feasibility of farming <u>Macrocystis pyrifera</u>[2] and <u>Laminaria saccharina</u>[3] for the production of feedstocks for methane generation. Species of the genus <u>Sargassum</u> represent an alternative source which includes benthic species common to the colder waters and pelagic as well as benthic species found along the coast of Florida and other subtropical and warmer temperate waters. Trawling collections of the pelagic species, <u>S. natans</u> and <u>S. fluitans</u>, have placed their estimated biomass in the Sargasso sea alone at 4 to 40 million metric tons.[4] The high carbohydrate content of these algae, and their present lack of commercial exploitation, makes them attractive as a potential source of biomass for conversion to methane.

The most abundant carbohydrate polymer of phaeophyta is alginate, a linear polyuronide which contributes to the structural and ion exchange properties of the cell wall.[5] Enzymes capable of degrading alginate are almost exclusively lyases which depolymerize by eliminative cleavage of the 1 → 4 glycosidic bonds.[6] Alginate lyases have been shown to be produced by, and in some cases partially characterized from marine invertebrates,[7-13] fungi,[14] marine[15-24] and terrestrial bacteria,[25-27] and marine brown algae.[28] The arrangement of the two constituent uronic acids of alginate, β-D-mannuronic and α-L-guluronic, into homopolymeric (poly M and poly G) and heteropolymeric (poly MG) blocks[29,30] determines the chemical properties of alginate, and necessitates the use of lyases of varied substrate specificities to achieve maximal depolymerization. Some generalizations regarding the substrate specificities of enzymes have been noted.[20,25] Mollusks tend to produce high levels of poly M lyase activity, marine pseudomonads produce poly G lyases, and several bacterial isolates with the properties of the genus <u>Alteromonas</u> produce both types of enzymes.[20,24] Requirements for the complete degradation of alginate should include the combined activities of endolytic and exolytic eliminases capable of degrading poly G and poly M regions of alginate. Endolytic enzymes can attack native alginate to produce oligomeric products which would be further degraded by exolytic enzymes to provide monomers for complete dissimilation.

Both exolytic and endolytic alginases have recently been identified in several bacterial isolates obtained from actively growing Sargassum tissue, of which the oxidative species have been tentatively assigned to the genus Alteromonas with the facultative species having metabolic properties most closely conforming to those expected for the genus Photobacterium.[24] This study compares substrate specificities and modes of cleavage of intracellular and extracellular preparations from aerobic and facultative marine bacteria isolated from Sargassum tissues. Some of these activities have been evaluated for their ability to degrade alginate associated with the actively growing Sargassum tissues with the objectives of preparing viable protoplasts for manipulating candidate species and for the bioconversion of alginate to methane.[31]

CARBOHYDRATES OF PHAEOPHYTA

The phaeophyta produce large amounts of carbohydrate, e.g. almost 50% of the dry weight in Macrocystis pyrifera.[1] The functions of the major carbohydrates are, in general, related to roles in maintenance of structural integrity or in supplying short or long term energy reserves for the plants. Figure 1 depicts the chemical structures of the three types of polymers which are localized extracellularly and presumably help to maintain wall properties.

The major wall component is alginate, a linear $1 \rightarrow 4$ polymer of β-D-mannuronic and α-L-guluronic acid which generally comprises 10 to 25% of the dry weight of brown algae (Table 1). A considerable body of information has been obtained regarding the fine structure of alginate and its effects on solution properties. The two uronic acids are arranged into homo- and heteropolymeric blocks of dp (degree of polymerization) around 20 for the homopolymers, which are interspersed in native alginate.[29,30] The configuration of the poly G sequences has been shown to resemble buckled chains; the poly M regions form flat ribbons (Figure 2).[32] Only poly G binds to calcium and other divalent cations with high affinity, thus rendering purified poly G insoluble or causing the native polymer to form a gel.[33-36]

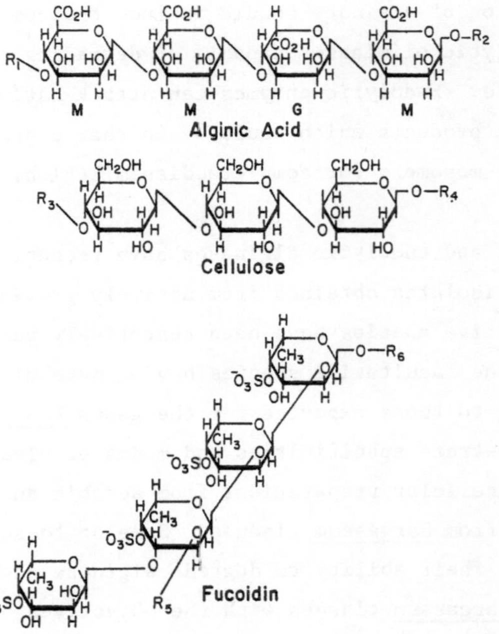

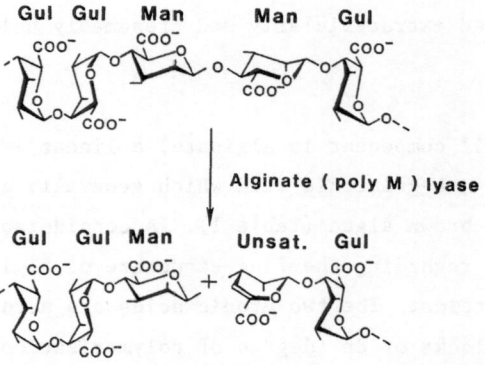

Figure 1. Chemical structures of the three polymers which comprise the bulk of marine brown algal cell walls (Phaeophyta). Unit saccharides, glycan bonds, and modifications of the saccharides are indicated. Configurational aspects of structures are not implied.

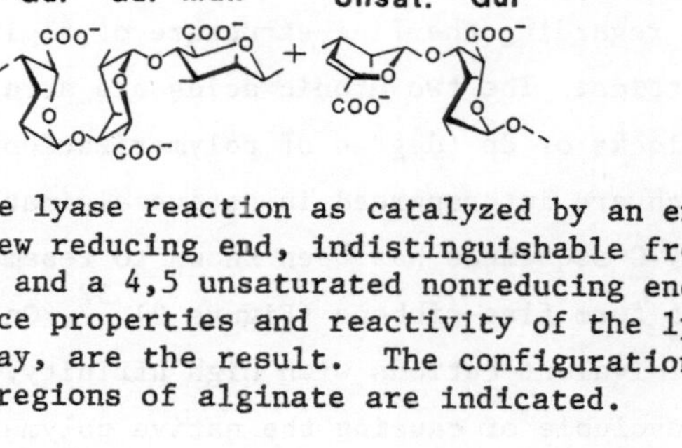

Figure 2. The alginate lyase reaction as catalyzed by an endo- poly M lyase. A new reducing end, indistinguishable from that formed by hydrolysis, and a 4,5 unsaturated nonreducing end which confers uv absorbance properties and reactivity of the lyase products in the TBA assay, are the result. The configurations of the poly G and poly M regions of alginate are indicated.

Table 1. Alginate levels in Phaeophyta species

Species	Alginic acid % Dry Wt.
S. fluitans	18.8[a]
S. natans	19.9[a]
S. filipendula	13.3 - 23.5[b]
S. polyceratium	20.3[a]
S. vulgare	17.9[a]
L. cloustoni	14 - 22[c]
M. pyrifera	14.1[d]

[a,b,c,d] Values from references 64, 65, 66, and 1, respectively.

The gel is formed due to calcium mediated cooperative interchain interactions of poly G regions, which has been likened to an "egg box" in configuration,[37,38,39] and extension of soluble, hydrated, poly MG and poly M "ribbons" away from the insoluble complexes. Some of the brown algae have been shown to form tough, firm, holdfast tissues using alginate with high levels of guluronic acid, and in the same plant produce flexible apical tissues containing high levels of mannuronic acid.[40,41]

Fucoidin or fucoidan is a sulfated fucose polymer, which is found in levels from less than 1% of the dry tissue weight for species such as Macrocystis pyrifera[1] to 24% for species such as Fucus spirilis and Pelvetia canaliculata which are extensively exposed to air during growth.[5] It consists of α,1 → 2 linked L-fucose with sulfate esterified primarily at position 4 (Figure 1). The molecule is probably branched at positions 3 and 4 and as isolated, may contain other saccharides, including galactose, xylose, and uronic acids, and metal ions as well.[5] Fucoidin is believed to reside in extracellular mucilaginous material and non-fibrillar portions of the wall, and as a result of its hygroscopic properties may protect the brown algae against dehydration.[5,42,43]

Cellulose is a β,1→4 unbranched glucan which is an important structural component of higher plant cell walls, and is consistently found in the cell wall of brown algae, in levels which range from about 1 to 10% of the dry weight of the plant.[5] Studies on the histology of the walls of Fucus have identified crystalline components which consist at least in part of cellulose.[44] Studies on the zygotes of Fucus indicate that the shape is maintained to a large extent by cellulose, as procedures which extract the other components, including alginate, leave an intact sack-like structure which retains the original form of the zygote.[45]

ENZYME CATALYSED ALGINATE DEGRADATION

Depolymerization of alginate occurs chemically by acid hydrolysis, base catalysed β-elimination, and by a free radical mediated mechanism.[46-48] Enzymes which catalyse degradation of alginate, with a single exception,[49,50] have been found to be eliminases, or alginate lyases. The elimination reaction as catalyzed by alginate lyases is depicted in Figure 2.[6] The reaction mechanism relies upon the existance of an electron withdrawing carboxyl group, an extractable α-proton (on C-5), and an O-linked uronide (on C-4) which is the leaving group.[51] The result, in the case of endolytic cleavage (internal to the polymer), is the generation of a new reducing end which is identical to that produced by hydrolysis, and a Δ4,5 non-reducing end. All alginate lyase enzymes which have been examined exhibit differential activity on purified poly G versus poly M, although a bacterial enzyme active on poly G was reported to cleave the G → M linkage.[26]

Assays for bond scission by alginate lyases measure either generation of new reducing or nonreducing unsaturated ends. Unsaturated terminal residues of oligomeric products absorb light at 230 nm,[6] although the monomer product, 4 deoxy-L-erythro-5-hexoseulose uronic acid, does not do so appreciably. A specific and sensitive assay for both unsaturated monomer and nonreducing end groups is based upon the spectrophotometric determination at 548 nm of the chromogen formed upon reaction of thiobarbituric acid (TBA) with periodate treated products.[6]

ISOLATION, PROPERTIES, AND GROWTH OF ALGINATE LYASE PRODUCING BACTERIA

Viable tissues of Sargassum natans and Sargassum fluitans provided the source for isolation of alginate lyase producing bacteria, except in the

case of isolate FM, which was obtained from decaying <u>Sargassum</u>. The algae
were transported from the Atlantic Coast of Florida to Gainesville, Florida
where identifications were confirmed, voucher specimens saved, and epiphytic
bacteria isolated.

<u>Sargassum</u> tissue (1 g quantities) was subjected to mild sonication in
sterile sea water (Instant Ocean from Aquarium Systems, Mentor, Ohio), and
dilutions of the sea water which contained bacteria removed from the tissue
were plated onto solid alginate medium (2% agar, 1% sodium alginate in PESI,
Provasoli's enriched seawater, supplemented with 0.27 g/L iodine).[53,54]
Colonies which exhibited substantial clearing of the calcium alginate haze
of the medium after several days of growth were selected for further
studies. Purity of the cultures has been established by subculturing on
solid alginate medium, growth in liquid alginate medium (0.1% sodium alginate, PESI containing 1.0 mM calcium and 5.5 mM magnesium), and growth on a
rich solid medium (2% agar, 1% glucose, 0.8% nutrient broth, 1% yeast
extract in PESI).

Some of the morphological and physiological properties of the bacteria
have been described[24] and along with other properties are shown in Table 2.
All isolates are gram negative polarly flagellated rods. All produce
clearing zones on alginate agar and release products indicative of lyase
mediated degradation of alginate (see the above description of the TBA assay
for alginate lyase). Of seven organisms isolated, four are oxidative and
three fermentative. All oxidative organisms are oxidase positive and all
fermentative isolates oxidase negative. All fermentative isolates, but none
of the oxidative bacteria, produce acid in liquid glucose medium (Table 2).
None of the isolates showed evidence of gas production on glucose or alginate containing media. Morphological, physiological, and DNA base composition data allowed the assignment of aerobes to the genus <u>Alteromonas</u>.[24] The
fermentative isolates, although morphologically and physiologically similar
to bacteria of the genus <u>Photobacterium</u>, are excluded from this genus by
their DNA base composition, and have not been assigned to any existing
genus.

For enzyme isolation, bacteria were grown in Fernbach flasks containing
1 L of liquid alginate medium at room temperature (22°C) with rapid gyrotory
shaking and were harvested at late exponential phase.

Table 2. Morphological and biochemical properties of alginase secreting bacteria associated with _Sargassum_ species

Isolate	Morphology[a]	Oxidase[b] Reaction	Glucose[c] +O_2	Glucose[c] -O_2	Alginate[d] +O_2	Alginate[d] -O_2	pH[e]
SO12382 FM	short rod	+	+	-	+	-	5.84
SFFB080483 A	0.9 - 1.1 x 1.6 - 2.2	-	+	+	+	+	4.69
SNFB080483 B	0.5 - 0.6 x 2.1 - 2.6	+	+	-	+	-	6.61
SNFB080483 C	0.7 - 0.9 x 1.6 - 2.0	+	+	-	+	-	6.91
SNFB080483 D	0.4 - 0.6 x 1.4 - 2.1	-	+	+	+	+	5.27
SFFB080483 F	0.5 x 1.2 - 2.0	+	+	-	+	-	6.73
SFFB080483 G	0.5 x 1.1 - 1.3	±	+	+	+	+	4.98

a Morphologies and dimensions in μm determined by measurements from scanning electron micrographs with the exception of isolate FM, which was analyzed with the optical microscope. All isolates were gram negative.

b Oxidase reactions were carried out according to methods described in reference 24.

c Glucose medium formulation consisted of 1% glucose, 0.8% nutrient broth, 1% yeast extract in PESI, A positive reaction indicates growth relative to controls which had no added glucose.

d Alginate medium formulation consisted of 0.1% sodium alginate in PESI.

e pH determinations were carried out directly on cultures grown on glucose medium incubated under aerobic conditions for 4 days; uninoculated glucose medium had a pH of 5.63.

INTRACELLULAR AND EXTRACELLULAR ALGINATE LYASES: SUBSTRATE SPECIFICITIES
AND CLEAVAGE PATTERNS

Enzyme Isolation

Bacterial cells were removed from culture medium by centrifugation,
frozen in liquid nitrogen, and stored at -70°C. For analysis of extracellular enzymes the spent medium was concentrated by tangential flow filtration
using a Millipore Pellicon cassette system with a polysulfone (PTGC) membrane which allowed retention of molecules larger than 10,000 Da, and
dialyzed against distilled, deionized water.

For intracellular preparations cells were thawed, suspended in 4
volumes of ice cold 0.1 M sodium phosphate buffered at pH 7.5, and disrupted
with a French pressure cell at 16,000 lb in^{-2}. Unbroken cells and cell
debris were removed by centrifugation, and acidic polymers were rendered
insoluble in the supernatant solution by adding 5% streptomycin sulfate
dropwise with stirring to a beaker at 0°C to give a final concentration of
2%. After stirring the mixture for 10 minutes at 0°C the resulting precipitate was removed by centrifugation and the supernatant solution containing
alginate lyase was treated with solid ammonium sulfate (to 65% saturation).
The protein precipitate was pelleted by centrifugation at 10,000 x g, 10
min, 4°C, redissolved in pH 7.5 phosphate buffer, and dialyzed against
distilled deionized water or sodium phosphate buffered at a desired pH.

Preparation of Substrates

Sodium alginate was purchased from Fisher Scientific Company as a
purified grade originally isolated from Macrocystis. Prior to use in
viscometric determinations alginate was centrifuged at 100,000 x g for 5
h.[55] Poly M and poly G were obtained from HCl hydrolyzed alginate, following the methods developed by Haug et al. [30] Preparations of poly G and poly
M were further fractionated on Sephadex G-50 with 0.5 M NaCl as eluent, and
selected fractions analyzed by reducing sugar and total carbohydrate assays,[52,56,57] and ^{1}H and ^{13}C NMR,[58,59] to assess uniformity of size and
purity of substrates.

Substrate Specificities of Intracellular and Extracellular Preparations

Alginate lyase was quantified by the TBA assay.[6,60] Substrate mixtures
contained either 0.1% sodium alginate, poly G, poly M, or no carbohydrate

(controls for endogenous substrate), in 0.05 M KCl, buffered with 0.03 M sodium phosphate from pH 5 to 8, or 0.05 M sodium acetate at pH 4.

With three exceptions, activities of the enzyme preparations with each substrate was maximal at pH 8. Extracellular preparations from facultative organisms A and D were most active on poly G at pH 7 and extracellular activity of isolate G was highest with poly M at pH 7. Figure 3 compares intracellular and extracellular activities from isolate A, on poly G, poly M, and alginate under several pH conditions.

Comparison of levels of activities of intracellular and extracellular preparations on poly G, poly M, and alginate at pH 8.0 are shown in Table 3. Intracellular preparations were, in all cases, most active on poly M, and generally slightly higher on alginate than poly G. Extracellular preparations generally were highly active on alginate. Fermentative isolates A and D showed little or no activity toward poly G extracellularly, and fermenta-

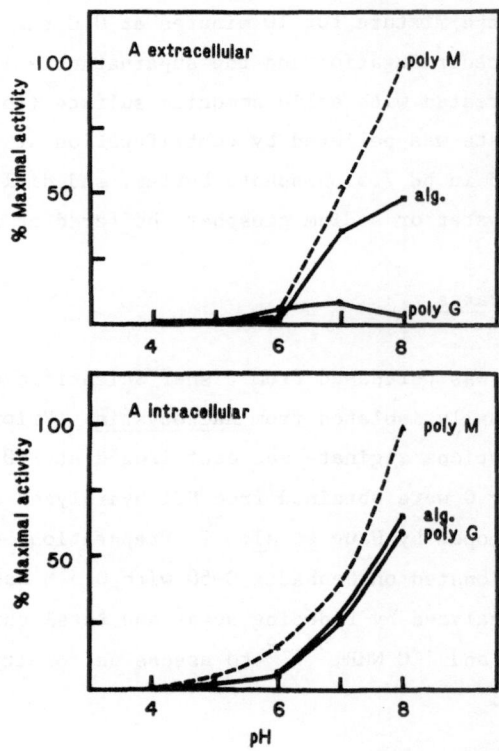

Figure 3. Activities of intracellular and extracellular preparations from isolate A, under various pH conditions, toward poly G, poly M, and alginate. Values have been normalized to the condition which allowed the maximal number of bonds to be cleaved, as determined using the TBA assay.

tive isolate G showed little activity on poly M. Levels of extracellular activities of the oxidative isolates, FM, B, and C, were comparable on poly G and poly M.

Patterns of Substrate Cleavage

Measurement of relative level of endo- and exoeliminase activities in intracellular versus extracellular preparations was carried out at pH 8.0 by following the decrease in viscosity during depolymerization by capillary viscometry,[61] and measuring rate of glycan bond cleavage by the TBA assay. Plots of the reciprocal of specific viscosity, i.e., specific fluidity, ϕsp, versus periodic acid generated TBA reactive products produced straight lines with slopes proportional to the relative level of endolytic activity. In the organisms examined, the slopes were greatest, and therefore the endolytic activities highest, in the extracellular fractions. The oxidative isolate FM, in particular, shows striking partitioning of exo- and endolytic activities. The comparisons of these slopes are given in Table 3.

Digestion of Sargassum filipendula Tissue by Intracellular and Extracellular Alginate Lyases

Active apical tissue with no visible epiphytic growth was excised and subjected to mild sonication, weighed, finely chopped with a scalpel, and incubated with enzyme preparations from facultative isolate A under conditions described in Table 4. The total number of unsaturated nonreducing termini produced by the intracellular extract was at each time point greater than that produced by the extracellular extract. However, when samples of each reaction mixture (minus tissue) were removed, mixed with a combination of intracellular and extracellular enzymes, and allowed to incubate further, the material released from the tissue by the extracellular preparation was observed to be accessible to further depolymerization, whereas the material released by the intracellular preparation was not. The most plausible explanation for this is that the extracellular preparation, which was shown in Table 3 to be highly endolytic, depolymerized the alginate of the tissue to yield oligomers which were substrates for the enzymes in the second incubation. The intracellular enzymes were also capable of releasing and depolymerizing alginate from the tissue, but released less total mass of alginate. The material released by the intracellular preparation was in a more highly depolymerized state than that released by the extracellular preparation, and could not be further degraded by the enzyme mixture in the second incubation.

DISCUSSION

This paper and our previous work[24] document observations that fermentative as well as oxidative marine bacteria are capable of producing alginate lyases of varied substrate specificities. Marine bacterial poly G lyases were previously described by several investigators[15,18,23] and oxidative bacteria from _Fucus_, which produced both poly G and poly M lyases' were isolated by Doubet and Quatrano.[20,21] Intracellular extracts from our isolates were high in poly M lyase, making them similar to the isolates from _Fucus_. Extracellular preparations, except from isolates A and D, were most active on native alginate. Levels of poly G and poly M lyases were comparable in the extracellular preparations from all oxidative isolates that were examined. On the other hand fermentative isolate G produced much more poly G lyase than poly M lyase, and isolates A and D produced poly M lyase in large excess over poly G lyase. These last two organisms are in contrast with the isolates from _Fucus_ which tended to produce higher levels of poly G lyase in the extracellular fractions. An organism such as isolate A, which produces little if any extracellular poly G lyase but makes intracellular poly G lyase may depend upon other bacteria in the environment to secrete endo poly G lyases for the further depolymerization of alginate.

The observation that endolytic activities are higher in the extracellular fractions as compared to the intracellular fraction presumably reflects a requirement for degradation of large native alginate molecules to allow entry into the bacterial cells. By retaining exoeliminases either in

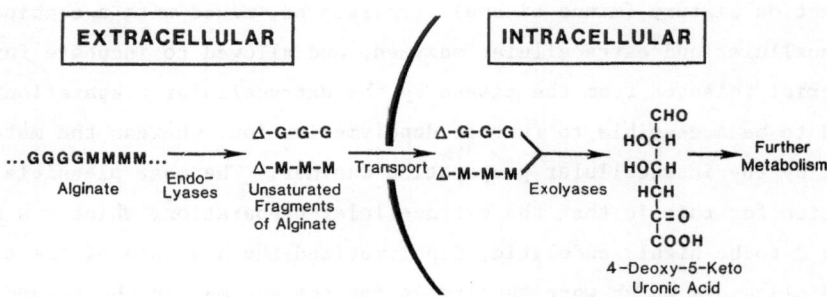

Figure 4. Proposed model for metabolism of alginate by bacteria colonizing tissues of Phaeophyta. Alginate polymers are released from the wall by endolytic depolymerization to smaller oligomers by lyases secreted by the bacteria. These oligomeric products are of various sizes, and also include poly GM regions of alginate which are not shown. Exolytic enzymes continue depolymerization of oligomers which enter the cell, with the eventual production of the unsaturated monomer compound, and its use as an energy source.

Table 3. Substrate specificities and modes of cleavage of alginate by bacterial lyases at pH 8.0

Isolate[c]	Intracellular[a]				Extracellular[b]			
	Poly G	Poly M	Alginate	ϕ sp[e] A^{548}	Poly G	Poly M	Alginate	ϕ sp[e] A^{548}
FM	0.190	0.527	0.235	1.2	0.300	0.569	0.647	20.0
A	0.689	1.140	0.764	8.6	0.000	0.370	0.225	12.0
B	0.026	0.091	0.032	–	0.276	0.287	0.367	–
C	0.015	0.042	0.018	–	0.423	0.373	0.648	–
D	0.825	0.917	0.880	–	0.087	1.058	0.981	–
G	1.287	2.445	1.200	7.3	2.230	0.347	3.135	10.5

Isolate[d]	Poly G	Poly M	Alginate		Poly G	Poly M	Alginate	
FM	0.80	2.24	1.00		0.46	0.87	1.00	
A	0.90	1.46	1.00		0.00	1.64	1.00	
B	0.81	2.84	1.00		0.75	0.78	1.00	
C	0.83	2.33	1.00		0.65	0.58	1.00	
D	0.94	1.04	1.00		0.09	1.08	1.00	
G	1.07	2.04	1.00		0.71	0.11	1.00	

[a] Intracellular activities were obtained after disrupting bacteria in a French pressure cell followed by partial purification to remove anionic polymers. This fraction may include activities bound to the cell surface as well as those which are truly intracellular.

[b] Extracellular activities were measured in the medium after concentration and dialysis but without further purification.

[c] Values in the upper panel are μ moles of unsaturated product formed per min per g (wet weight) of cells. All activities presented were calculated after subtracting activities observed in the absence of added substrate.

[d] Values in the lower panel are calculated as ratios of lyase activity dependent upon added substrate divided by the activity dependent upon native alginate.

[e] Slopes of straight lines obtained by plotting specific fluidity, ϕsp, against the results of TBA assays, A^{548}, measured during depolymerization of alginate, indicate relative level of endolytic vs. exolytic cleavage.

Table 4. Degradation of <u>S. filipendula</u> tissue by intracellular and extra-cellular alginate lyase preparations from isolate A[a]

Time (h)	Direct Assay[b]		+Enz. Mixture[c]	
	n.r. ends u moles		n.r. ends u moles	
	Intra.	Extra.	Intra.	Extra.
0	0.01	0.007	–	–
2	0.255	0.095	0.272	0.318
6	0.617	0.325	0.722	1.010
10	0.816	0.440	0.840	1.368
24	1.267	0.649	–	–

[a] Two 25 mg apical portions of sonicated <u>S. filipendula</u> tissue, containing an estimated 2.5 μ moles of uronic acid residues each, were finely chopped and incubated with enzyme preparations in PESI lacking added Ca^{++} and Mg^{++}, and containing 1.2 mM EDTA. The activities of both enzyme solutions at the start of the experiment were 0.0386 μ mole/ min per ml with alginate as a substrate.

[b] Samples of sea water media were removed at indicated times and assayed for products using the TBA assay, which measures nonreducing (n.r.) unsaturated terminal residues.

[c] Samples of the seawater solutions (excluding tissue) were removed at the indicated times and added to a mixture of intra- and extracellular algi-nate lyase preparations containing activities of 0.0193 μ mole/min per ml from each source, incubated further for 12 h and assayed for products of the lyase reactions as above.

the cytoplasm or bound to the cell surface these organisms avoid producing large amounts of metabolizable monomers in the external environment which through diffusion and/or utilization by other organisms would be lost to the bacteria producing the enzymes.

A model for bacterial utilization of alginate based on this study and the work of others is shown in Figure 4. Native alginate is endolytically depolymerized to fragments possessing unsaturated nonreducing ends. These fragments are internalized and exolytically degraded, possibly undergoing degradation during the entry process. The metabolic pathway for degradation of the monomer product by a pseudomonad has been described.[62]

Under conditions of anaerobic digestion, the metabolites of the monomer should be readily converted to methane by a consortium of bacteria. Shiralipour et al. have observed that improved yields of methane can be obtained from digestion of Sargassum tissues using microflora associated with Sargassum.[63] The rate and/or extent of methane production might be further improved by supplementary inocula of organisms such as the facultative isolates of this study, which produce alginate lyases with a spectrum of substrate specificities and modes of cleavage, or by addition of alginate lyase enzymes to the fermentor.

Purification of alginate lyase enzymes, as noted,[19,26] will provide specific reagents for dissection of brown algal cell walls and modification of alginate structure. We are developing methods of producing protoplasts of Sargassum to permit genetic manipulation of somatic cells.[31] The extracellular endolytic poly M lyase from isolate A has recently been purified (Romeo, T., and Preston, J.F., III, unpublished), and along with other specific enzymes including poly G lyases, cellulases, and fucoidinases, should assist in the generation of protoplasts in reproducibly high yields.

ACKNOWLEDGEMENTS

We wish to thank Dr. John E. Gander, Cynthia Jackson, and Sandra Bonetti for NMR spectroscopy of substrates, and Humerto J. Jiminez, who isolated bacterium SO12382FM. We are also grateful to Donna Huseman for preparation of some of the figures, and Ms. Tonie Henry and Ms. Tricia Williams for typing this manuscript. This work was supported by the Gas Research Institute, and the Institute of Food and Agricultural Sciences, University of Florida, CRIS project No. MCS2170 and represents Journal Series No. 6450 of the University of Florida Institute of Food and Agricultural Sciences Experiment Station.

REFERENCES

1. D. P. Chynoweth, G. Sambhurath and D. L. Klass. Anaerobic Digestion of Kelp. In: Biomass Conservation Processes for Energy and Fuels. O. R. Zaborsky and S. Soter, (Ed.), Plenum Publishing Corporation, New York. 1980.
2. M. Neushul. The Domestication of the Giant Kelp, Macrocystis, as a Marine Plant Biomass Producer. In: The Marine Plant Biomass of the Pacific Northwest Coast. p. 163-181. D.W. Krauss, (Ed.), Oregon State University Press. 1977.

3. B. H. Brinkhuis, E. C. Mariani, V. A. Breda and M. M. Brady-Campbell.
    Cultivation of _Laminaria saccharina_ in the New York Marine Biomass
    Program. Hydrobiologia. 116/117:266-271. 1984.

4. E. Parr. Quantitative Observations on the Pelagic _Sargassum_ Vegetation
    of the Western North Atlantic. Bull. Bingham Oceanogr. Coll. Vol.
    VI: 1-94. 1939.

5. E. Percival and R. H. McDowell. Chemistry and Enzymology of Marine
    Algal Polysaccharides. Academic Press, New York. 1967.

6. J. Preiss and G. Ashwell. Alginic Acid Metabolism in Bacteria. I.
    Enzymatic Formation of Unsaturated Oligosaccharides and 4-Deoxy-L-
    Erythro-5-Hexoseulose Uronic Acid. J. Biol. Chem. 237:309-316.
    1962.

7. V. V. Favorov and V. E. Vaskovsky. Alginases of Marine Invertebrates.
    Comp. Biochem. Physiol. 38B:689-696. 1971.

8. L. A. Elyakova and V. V. Favorov. Isolation and Certain Properties of
    Alinate Lyase VI from the Mollusk _Littorina_ Sp. Biochim. Biophys.
    Acta. 358:341-354. 1974.

9. V. V. Favorov, E. I. Vozhova, V. A. Denisenko and L. A. Elyakova. A
    Study of the Reaction Catalyzed by Alginate Lyase VI from the Sea
    Mollusc, _Littorina_ sp. Biochim. Biophys. Acta. 569:259-266. 1979.

10. T. Muramatsu, S. Hirose and M. Katayose. Isolation and Properties of
    Alginate Lyase from the Mid-Gut Gland of Wreath Shell _Turbo cornu-
    tus_. Agric. Biol. Chem. 41:1939-1946. 1977

11. T. Muramatsu. Additional Evidence for Substrate Specificities of
    Alginate Lyase Isozymes from _Turbo cornutus_. Agric. Biol. Chem.
    48:811-813. 1984.

12. H. I. Nakada and P. C. Sweeny. Alginic Acid Degradation by Eliminases
    from Abalone Hepatopancreas. J. Biol. Chem. 242:845-851. 1967.

13. L. F. Jacober, L. Rice and A. G. Rand, Jr. Characterization of the
    Carbohydrate Degrading Enzymes in the Surf Clam Crystalline Style.
    J. Food Sci. 45:381-385. 1980.

14. M. Wainwright and V. Sherbrock-Cox. Factors Influencing Alginate
    Degradation by the Marine Fungi: _Dendryphiella salina_ and _D.
    arenaria_. Botanica Marina. XXIV:489-491. 1981.

15. Y. Kashiwabara, S. Hiroshi and K. Nisizawa. Alginate Lyases of
    Pseudomonads. J. Biochem. 66:503-512. 1969.

16. S. Fujibayashi, H. Habe and K. Nisizawa. Heterogeneity of Alginate in
    Special Reference to the Enzymatic Degradation. J. Biochem. 67:37-
    45. 1970.

17. K. H. Min, S. F. Sasaki, Y. Kashiwabara, U. Makoto and K. Nisizawa.
    Fine Structure of SMG Alginate in the Light of Its Degradation by
    Alginate Lyases of _Pseudomonas_ sp. J. Biochem. 81:555-562. 1977.

18. I. W. Davidson, I. W. Sutherland and C. J. Lawson. Purification and
    Properties of an Alginate Lyase from a Marine Bacterium. Biochem.
    J. 159:707-713. 1976.

19. R. S. Quatrano and B. A. Caldwell. Isolation of a Unique Marine
    Bacterium Capable of Growth on a Wide Variety of Polysaccharides
    from Macroalgae. Appl. Environ. Microbiol. 36:979-981. 1978.

20. R. S. Doubet and R. S. Quatrano. Isolation of Marine Bacteria Capable
    of Producing Specific Lyases for Alginate Degradation. Appl.
    Environ. Microbiol. 44:754-756. 1982.

21. R. S. Doubet and R. S. Quatrano. Properties of Alginate Lyases from
    Marine Bacteria. Appl. Environ. Microbiol. 47:699-703. 1984.

22. T. L. Pitt and L. L. Raisbeck. Degradation of the Mucoid
    Polysaccharide of _Pseudomonas aeruginosa_ by _Beneckea pelagia_. J.
    Appl. Bacteriol. 45:297-300. 1978.

23. I. W. Sutherland and G. A. Keen. Alginases from _Beneckea pelagia_ and
    Pseudomonas spp. J. Appl. Biochem. 3:48-57. 1981.

24. J. F. Preston III, T. Romeo, J. C. Bromley, R. W. Robinson and H. C.
    Aldrich. Alginate Lyase-Secreting Bacteria Associated with the
    Algal Genus _Sargassum_. pp. 727-740. In: Developments in Industrial

Microbiology, Vol. 26. L.A. Underkofler (Ed.) Society for Industrial Microbiology Press. Arlington, VA. 1985.

25. J. B. Boyd and J. R. Turvey. Isolation of a Poly-L-Guluronate Lyase from Klebsiella aerogenes. Carbohyd. Res. 57:163-171. 1977.

26. J. B. Boyd and J. R. Turvey. Structural Studies of Alginic Acid, Using a Bacterial Poly-L-Guluronate Lyase. Carbohyd. Res. 66:187-194. 1978.

27. J. B. Hansen, R. S. Doubet and J. Ram. Alginase Enzyme Production by Bacillus circulans. Appl. Environ. Microbiol. 47:704-709. 1984.

28. J. Madgwick, A. Haug and B. Larsen. Alginate Lyase in the Brown Alga Laminaria digitata. Acta Chem. Scand. 27:711-712. 1973.

29. A. Haug, B. Larsen and O. Smidsrod. A Study of the Constitution of Alginic Acid by Partial Acid Hydrolysis. Acta Chem. Scand. 20:183-190. 1966.

30. A. Haug, B. Larsen and O. Smidsrod. Studies on the Sequence of Uronic Acid Residues in Alginic Acid. Acta Chem. Scand. 21:691-704. 1967.

31. J. F. Preston III, T. Romeo, A. Gibor and M. Polne-Fuller. Investigations of Sargassum species for Bioconversion to Methane: Mannitol Levels, Temperature Requirements, and Protoplast Formation. In Press. Proceedings of the 1984 International Gas Research Conference.

32. E. D. T. Atkins, W. Mackie, K. D. Parker and E. E. Smolko. Crystalline Structures of Poly-D-mannuronic and Poly-L-guluronic acids. J. Polymer Sci. Part B: Polymer Letters 9:311-316. 1971.

33. O. Smidsrod and A. Haug. The Effect of Divalent Metals on the Properties of Alginate Solutions. I. Calcium Ions. Acta Chem. Scand. 19:329-340. 1965.

34. A. Haug and Smidsrod. The Effect of Divalent Metals on the Properties of Alginate Solutions. II. Comparison of Different Metal Ions. Acta Chem. Scand. 19:341-351. 1965.

35. R. Kohn, I. Furda, A. Haug and O. Smidsrod. Binding of Calcium and Potassium Ions to Some Polyuronides and Monouronates. Acta Chem. Scand. 22:3098-3102. 1968.

36. A. Haug and O. Smidsrod. Fractionation of Alginates by Precipitation with Calcium and Magnesium Ions. Acta Chem. Scand. 19:1221-1226. 1965.

37. G. T. Grant, E. R. Morris, D. A. Rees, P. J. C. Smith and D. Thom. Biological Interactions Between Polysaccharides and Divalent Cations. The Egg-Box Model. FEBS Lett. 32:195-198. 1973.

38. E. R. Morris, D. A. Rees, D. Thom and J. Boyd. Chiroptical and Stoichiometric Evidence of a Specific, Primary Dimerization Process in Alginate Gelation. Carbohyd. Res. 66:145-154. 1978.

39. D. A. Rees, E. R. Morris, D. Thom and J. K. Madden. Shapes and Interactions of Carbohydrate Chains. pp. 195-290. In: The Polysaccharides, Vol. 1, G.O. Aspinall, (Ed.), Acad. Press, Inc., New York. 1982.

40. A. Haug, B. Larsen and O. Smidsrod. Uronic Acid Sequence in Alginate from Different Sources. Carbohyd. Res. 32:217-225. 1974.

41. I. L. Andresen, A. O. Skipnes and O. Smidsrod. Some Biological Functions of Matrix Components in Benthic Algae in Relation to Their Chemistry and the Composition of Seawater. p. 362-381. In: Cellulose Chemistry and Technology. J.C. Arthur, Jr., (Ed.), American Chemical Society, Washington, D.C. 1977.

42. M. E. McCully. Histological Studies on the Genus Fucus. I. Light Microscopy of the Mature Vegetative Plant. Protoplasma 62:205-230. 1966.

43. L. V. Evans, M. Simpson and Callow. Sulfated Polysaccharide Synthesis in Brown Algae. Planta (Berl.) 110:237-252. 1973.

44. M. E. McCully. The Histological Localization of the Structural Polysaccharides of Seaweeds. New York Acad. Sci. 175:702-711. 1970.

45. R. S. Quatrano and P. T. Stevens.   Cell Wall Assembly in Fucus
    Zygotes. Plant Physiol. 58:224–231. 1976.
46. A. Haug, B. Larsen and O. Smidsrod.   The Degradation of Alginates at
    Different pH Values. Acta Chem. Scand. 17:1466–1468. 1963.
47. O. Smidsrod, A. Haug and B. Larsen.   Degradation of Alginate in the
    Presence of Reducing Compounds. Acta Chem. Scand. 17:2628–2637.
    1963.
48. O. Smidsrod, A. Haug and B. Larsen.   Kinetic Studies on the
    Degradation of Alginic Acid by Hydrogen Peroxide in the Presence of
    Iron Salts. Acta Chem. Scand. 19:143–152. 1965.
49. R. A. Stevens and R. E. Levin.   Viscometric Assay of Bacterial
    Alginase. Appl. Environ. Microbiol. 31:896–899. 1976.
50. R. A. Stevens and R. E. Levin.   Purification and Properties of an
    Alginase from Alginovibrio aquatilis. Appl. Environ. Microbiol.
    33:1156–1161. 1976.
51. J. Kiss. Eliminative Degradation of Carbohydrates Containing Uronic
    Acid Residues. Advan. Carbohyd. Chem. 29:229–303. 1974.
52. N. Nelson.   A Photometric Adaptation of the Somogyi Method for the
    Determination of Glucose. J. Biol. Chem. 153:375–380. 1944.
53. L. Provasoli.   Media and Prospects for Cultivation of Marine Algae.
    p. 63–75. In: Cultures and Collections of Algae. A. Watunabe, and
    A. Hattori (Eds.). Japan Society Plant Physiology, Tokyo. 1968.
54. M. Polne-Fuller, N. Saga and A. Gibor.   Algal Cell, Callus, and Tissue
    Cultures and Selection of Algal Strains. Proceedings of the Algal
    Biomass Symposium, Golden, Colorado (in press). 1984.
55. O. Smidsrod and A. Haug.   A Light Scattering Study of Alginate. Acta
    Chem. Scand. 22:797–810. 1968.
56. M. Dubois, K. A. Gilles, J. K. Hamilton, P. A. Rebers and F. Smith.
    Colorimetric Method for Determination of Sugars and Related Sub-
    stances. Anal. Chem. 28:350–356. 1956.
57. A. Haug and B. Larsen.   Quantitative Determination of the Uronic Acid
    Composition of Alginates. Acta Chem. Scand. 16:1908–1918. 1962.
58. H. Grasdalen, B. Larsen and O. Smidsrod.   A P.M.R. Study of the
    Composition and Sequence of Uronate residues in Alginates. Carbo-
    hyd. Res. 68:23–31. 1979.
59. H. Grasdalen, B. Larsen and O. Smidsrod.   ^{13}C-N.M.R. Studies of
    Monomeric composition and Sequence in Alginate. Carbohyd. Res.
    89:179–191. 1981.
60. A. Weissbach and Hurwitz.   The Formation of 2-Keto-3-Deoxyheptonic
    Acid in Extracts of Escherichia coli B. I. Identification. J.
    Biol. Chem. 234:705–709. 1959.
61. J. E. McKie and J. F. Brandts.   High Precision Capillary Viscometry.
    Methods in Enzymology 126:257–288. 1982.
62. J. Preiss and G. Ashwell. Alginic Acid Metabolism in Bacteria. II.
    The Enzymatic Reduction of 4-Deoxy-L-erythro-5-hexoseulose Uronic
    Acid to 2-Keto-3-deoxy-D-gluconic Acid. J. Biol. Chem.
    237:317–321. 1962.
63. A. Shiralipour, P. H. Smith and J. E. Moore.   Annual Report, Gas
    Research Institute, IFAS Program. 1984.
64. N. E. Aponte de Otaola, M. Diaz-Piferrer and H. Graham.   Seasonal
    Variation and Anatomical Distribution of Alginic Acid in Sargassum
    Spp. Found Along the Coasts of Puerto Rico. J. Agr. Univ. Puerto
    Rico. 67:469–475. 1983.
65. F. W. Davis. Algin from Sargassum. Science 111:150. 1950.
66. W. A. P. Black.   Seasonal Variation in Weight and Chemical Composition
    of the Common British Laminaraceae. J. Mar. Biolog. Assoc. U.K.
    29:45–72. 1950.

THERMOCHEMICAL PROCESSES FOR BIOENERGY PRODUCTION

Ed J. Soltes

Department of Forest Science
Texas A & M University
College Station, Texas   77843-2135

ABSTRACT

Biomass, as produced in its various forms, is generally poorly suited to energy production without pretreatment to change its physical and chemical nature. Thermochemical conversion of biomass refers to the alteration of this nature through heating. Depending on conditions used (e.g., temperature reached, time at temperature, oxygen availability) biomass can be altered very slightly, or be completely changed along a continuum of process conditions permitting a variety of outcomes. At lower temperatures and oxygen access (pyrolysis), biomass materials are cracked and partially oxidized to tars, chars and gases. Pyrolysis process conditions can be defined in which either char (carbonization) or tar (tarification) can be favored. Carbonization has importance in a number of petroleum-poor countries, while tarification is still being researched (described as a route to liquid fuels).

At intermediate temperatures, biomass can be selectively processed into gaseous products (gasification) with sufficient calorific value to be used as gaseous fuels. Biomass gasification has found use in the retrofit of natural gas furnaces and gasoline engines. At the other extreme of the process continuum at elevated temperatures and in the presence of at least a stoichiometric amount of oxygen (combustion), biomass can be directly processed for energy. In direct combustion, the primary tar, char and gaseous products of biomass thermal cracking are oxidized completely in exothermic conditions to heat and oxidized gases.

Keywords. Biomass processing, bioenergy, thermochemical conversion, carbonization, tarification, gasification, combustion.

INTRODUCTION

Harvest and process residues in forestry and agricultural operations can serve on-site energy requirements, but the biomass materials are generally poorly suited for direct use. Modern agricultural and process machinery seldom use solid fuels, therefore pretreatments are often required to change their chemical and physical nature. Similar considerations apply to biomass crops and off-farm energy uses. Thermochemical processes for biomass are basically conversion processes that alter the chemical and physical nature of biomass to permit use.

Thermochemical processing is flexible. Depending on conditions used (primarily the temperature reached, the oxygen to fuel ratio, and residence time at temperature) biomass can be altered very slightly, or be completely changed. These three variables, for example, can define conditions for pyrolysis, gasification and combustion but there is often little distinction between these processes, and in fact there is a continuum of process conditions. Selection of treatment conditions permits a variety of outcomes of importance to the production of bioenergy products. The products differ in the proportions of solid, gaseous or liquid forms. Useful in discussion of thermochemical processing is to first consider direct and complete combustion of biomass, and then follow the thermochemical processing continuum through various stages of incomplete combustion.[1]

DIRECT COMBUSTION

The objective in direct and complete combustion is heat. Air is supplied in excess to afford combustion of biomass in the production of fully oxidized, permanent gases. Resulting solids are mainly ash and highly resistant carbon compounds. Liquid products are non-existent, or result as a trace of condensates. The exothermic reactions involved in this process produce heat. Since wood and agricultural materials have been burned since before recorded time to provide heat, one might assume that there are few, if any, remaining problems in direct combustion of biomass materials. This is not so.

In order to understand the problems involved in burning biomass, we must first understand the combustion process. The most important combustible elements in any wood or agricultural material are carbon and hydrogen which, on combination with oxygen, form carbon dioxide, water, and heat. In most cases, air is used as a source of oxygen, and the associated nitrogen in the air must also enter the furnace. In order to insure complete combustion, somewhat more than the minimum theoretical, or stoichiometric, air must be supplied. Further, solid fuels, such as wood and coal, normally require larger amounts of excess air than do liquid or gaseous fuels because of problems in air-fuel mixing. The excess air may run from 15 to 150 percent. At 50% excess air, about 11% of the heat produced in the combustion of hogged fuel can be lost to stack gases.[2]

Water is still an even more serious problem in the combustion of green biomass materials. For example, the wet basis moisture content of green southern pine wood and bark is generally about 50%, and about 13% of the total available energy is required just for the evaporation of the water. Furthermore, more air must normally be introduced through the fuel pile to facilitate evaporation. Since water vapor at 250 degrees C occupies nearly 2000 times the volume of liquid water, the net result is an increase in stack gas volume of some 11% for wood at 50% M.C. This increase in gas velocities can compound problems in particulate carryover and losses of heat to stack gases. For efficient combustion, time is required to complete the combustion process before particles are blown out of the stack, high temperatures are needed to initiate and promote the oxidation reactions, and turbulence is necessary to improve air-fuel contact.

For green biomass, water creates additional problems. Per unit of mass, green wood has a gross heating value about one-fourth that of fuel oil (due not only to the moisture content but also to the oxygen content of dry biomass). On a volumetric basis, however, green wood chips have about one-tenth the heating value of fuel oil. Since fuel is shipped and metered by volume, green chips are then ten times more costly than fuel oil to transport by truck on a dollar per unit of energy basis. For this reason, there is development activity in the densification and drying of biomass materials destined for fuel usage, at the point of production prior to transportation.

There are yet additional problems in high ash production when agricultural materials containing large amounts of chemicals are combusted. In fluidized-bed combustion of such biomass (for example, cotton gin trash and sorghum bagasse) serious slagging and fouling problems have been observed, and metal coupons placed in the hot exhaust gas stream experienced high rates of corrosion and erosion.[3] These problems can be traced to the high content of low fusion temperature ash in the feeds to combustion, and the formation of volatile inorganic compounds.

Despite the problems associated with direct combustion of biomass, it is currently the most direct and most widely used method for converting biomass into energy. Direct combustion, however, is not suited to all energy needs. Biomass, like the 200-year supply of coal possessed by the US, occurs in the solid form -- the form in which only 7% of the US energy usages consume. Solid biomass fuels cannot be used directly in higher efficiency liquid or gaseous furnaces and engines. Problems with biomass combustion are then related to its direct use as a fuel. Since there are physical feed problems, as well as lower energy content and lower efficiency in use of wood in combustion, direct substitution involves high costs for burners which have to be changed and energy installations which must be enlarged. When wood must be used in its solid form, relatively quick fixes to these problems can be, for example, drying for increased efficiency and higher steam production rates, or densifying for cost containment in transportation and storage. However, the more appropriate solution to this problem is to convert wood into physical fuel forms that can be used in existing high efficiency equipment. In other words -- change the fuel, not the furnace. In this context remember that, with the exception of direct combustion for which the objective is heat, thermochemical processes are fuel conversion processes that convert biomass into more useful fuels.

To understand these fuel conversion processes, it is instructive to look at the combustion of biomass as involving a series of conversion steps. In the presence of heat and air, biomass is first converted to several primary products -- combustible gases, liquid tar and solid char. These primary products are themselves fuels, and if given sufficient air and temperature will eventually undergo secondary conversion towards complete combustion. However, conditions can be determined, and processes using these conditions can be developed, which deliberately favor incomplete combustion such that the primary fuel products do not undergo secondary reactions, and can be recovered. Further, conditions can be identified which favor either the gaseous, liquid, or solid fuel product of biomass primary conversion, and these conditions and processes are what we normally refer to as gasification, pyrolysis or carbonization, respectively. Several problems in the generation of energy from biomass via direct combustion are then avoided in prior conversion of biomass, via gasification or pyrolysis, into better fuels.

GASIFICATION

At elevated temperatures just short of those required for combustion, but in the presence of a limited amount of oxygen or air, biomass will be converted primarily into a mixture of carbon monoxide, hydrogen and volatile hydrocarbons. This process is called gasification and the objective is gaseous fuel. To maximize the production of gaseous fuel product, and to minimize that of the other two primary conversion products tar and char, sufficient residence time is required at conversion temperatures such that tar and char undergo secondary reactions towards gaseous product.

Gasification is also not without problems. Unless extraordinary steps are taken to provide for conditions which result in complete gasification, some tar and/or char is always produced. These are undesirable and often troublesome products. Problems can be compounded in the feeding of high-ash biomass materials. To a large extent, the solid natures of char, and ash if it is also produced, are such that handling and disposition problems of these materials can be relatively easily solved, especially if ash fusion temperatures are avoided. Tar, however, is a sticky semi-liquid with corrosive properties and creates troublesome handling problems, not to mention gumming and corrosion problems when entrained in the gaseous fuel product and combusted in engines.[4]

Although tars can be collected by various disposal techniques such as scrubbing and recovery techniques such as condensation, there is currently no use for the tar except reentry into the gasification reactor. Inexpensive gasifiers, such as those based on updraft, gravitating bed principles, are easy to build and offer acceptable conversion efficiencies, but these produce tars in excessive quantities, especially if wet wood is gasified. Downdraft gasifiers reduce the problem but their use is limited to small scale applications because the design has scaling problems. Rather than having to cope with tar recycle, the accepted solution to the tar problem normally involves added sophistication and expense to reactor design to eliminate, or at least minimize, tar production.

325

Staged Gasification/Combustion

Staged gasification/combustion systems are another solution to tar problems. Biomass fuel is first gasified using about 70% stoichiometric air required for complete combustion, but here the gasification reactor is close-coupled with a combustion furnace. Gaseous products of gasification, while hot and containing vapors and tars in a volatile state, come in contact with another source of air for secondary combustion. The sensible heat of the product gases and the subsequent heat produced from combustion of the tars and vapors in the gas provide high conversion efficiency in the production of heat.

Energy Uses for Biomass Combustion and Gasification

Wood Residues. There are many isolated cases of successful implementation of wood combustion and gasification. Gaseous fuel is especially attractive for stationary engines developing shaft power. Wood residues at forest products mill sites are being gasified to provide clean fuel gas for wood drying, for boiler use in the production of needed process steam, and to provide high percentages of the energy requirements of dual-fueled gas/diesel engines in applications requiring shaft power. Wood residues are also being combusted or gasified to fuel turbines for electrical power generation, both for public utilities, and in cogeneration modes to provide electrical power and process steam requirements for the forest products industry. Staged gasification/combustion systems hold promise for retrofitting existing or new oil- or gas-fired systems while using solid biomass fuel feeds -- attractive features for process energy needs of the forest products industry.

These processes, however, are not yet economically competitive for use on broad scales. However, as such technologies develop, as fossil fuel pricing dictates the use of less expensive fuels, as industries are requested to cut back on fossil fuel usage through government dictate, and as the forest products industry continues its trends toward materials and energy self-sufficiency, more of these processes may become desirable.[5]

Agricultural Residues. Agriculture is a large consumer of energy. Energy pricing pressures and low prices paid to farmers for their product are forcing farmers to adopt less energy-intensive cultural practices and to consider growing their own fuels and/or growing crops for their commodity value as energy. Energy requirements for agriculture are additionally

326

decentralized, that is, needs are for reliable, small-scale biomass energy conversion systems. Although traditional resource recovery systems (such as for example used to process municipal waste) derive large benefits from economics of scale, biomass conversion systems in agriculture are required to be competitive in the range of some 1 to 100 Mg per day. In Texas alone, it is inconceivable that the 73,000 internal combustion engines and some 67,000 electrical motors used state-wide to pump irrigation water would be displaced by an equal number of biomass-fed engines.[3] Answers to this problem are not yet apparent.

Other applications for the combustion and gasification of agricultural residues relate to satisfying the energy requirements of various agricultural processes, such as the drying of grain, or the ginning of cotton. These would still be relatively small compared with, for example, an electrical generation plant sized for a forest products operation. The challenge then for energy from and for agriculture is the identification and development of technologies which are at the same time efficient, small-scale, safe and reliable. If such systems could be built, uses would extend beyond agricultural needs.

Larger Scale Gasification for Chemicals and Fuels Synthesis

The carbon monoxide/hydrogen product of wood gasification, similar to that obtained in the gasification of coal, can be used for more than gaseous fuel. When oxygen is used in biomass gasification, the product gas is free of nitrogen and can be used as synthesis gas. Methanol can be snythesized from this gas mixture and used directly as a liquid vehicular fuel, or be converted via the Mobil process into gasoline hydrocarbons.[6,7] Alternately, the gas product can be converted via the Fischer-Tropsch process into diesel hydrocarbon fuels or, following reforming, gasoline.[8,9]

Wood gasification has been compared with coal gasification.[10,11] It appears that wood may offer several technical advantages over coal in gasification, because of lower oxygen or steam requirements, and lower shift and desulfurization requirements. Coal contains little oxygen and requires an oxygen-donor, usually large amounts of water, for successful production of gaseous fuel.

Despite these and several other technical advantages, and despite growing use of wood gasification systems, especially by the forest products industry in efforts to replace natural gas fuels with available wood process

residue fuels, large-scale wood gasification systems are not now economically possible. The relatively high transportation costs for wood vs coal
limit the effective harvesting radius around a conversion plant, thus limiting wood conversion plant size. Especially for chemicals, where wood would
be in direct competition with coal, the quantities of wood needed to approach the gas production scales contemplated for coal gasification may not
be available.[5]

PYROLYSIS

If thermochemical conversion of biomass is conducted at temperatures
still lower than those used for gasification (generally below 600 degrees
C), and in the absence of air or under air-starve conditions, all three
primary products -- gas, tar and char -- can be recovered. Using suitable
process conditions, either the char or tar product can be favored. Process
design guidelines can be identified which favor one type of product. For
example, higher heating rates generally favor tar production at the expense
of char and gas.

Pyrolysis for Carbonization

The most common form of biomass pyrolysis is carbonization, that is, to
produce char. Char is superior as a fuel compared to wood or agricultural
residues, and its production can be afforded in very simple systems. In
primitive carbonization systems, the gaseous and tar products are partially
combusted to provide heat as the driving force for the conversion process
but to some extent are allowed to escape into the atmosphere. Even under
such conditions, chars can be produced in 20% mass yield, but because of the
higher calorific value of char compared with biomass, are in reality produced in some 30% energy yield.

Chars are easily produced, are more energy-dense than their parent
biomass forms, and have considerable advantages. These include being essentially smokeless fuels, and are thus heavily used as fuels for home cooking
and heating purposes in developing countries without fossil energy and the
resources to purchase fossil fuels. In other countries with abundant biomass resources, notably Brazil, charcoal is produced in relatively sophisticated processes as a primary fuel for industrial energy needs and for steel
manufacture, as well as in relatively primitive processes for cottage industries and home use, especially in remote rural areas. In more highly indus-

trialized societies, using still more sophistication, carbonization can be
an efficient process for the conversion of biomass into a more useful solid
fuel form. In the US, however, carbonization is relegated primarily to the
production of charcoal briquets for outdoor home cooking, even though wood
charcoal has some interesting and useful properties which make it a unique
raw material for the production of metallurgical and chemical products.

Pyrolysis for Tarification

Pyrolysis has had problems in definition. It is a process term which
has been used for both char and tar processes. Char is produced in carbonization processes under conditions of long residence time (hours to weeks) at
lower temperatures (generally not greater than 400 degrees C) while re-
stricting air supply. Tar product is favored under conditions of short
residence time (fractions of a second to minutes) at higher temperatures
(generally greater than 400 degrees C) under oxygen deficient or inert
atmospheres. The meanings of gasification and carbonization, both product
oriented terms, are clear. There is no good term for processes which favor
tar production. The term tarification vis-a-vis the terms carbonization and
gasification might clarify terminology in this area.

The key to commercial implementation of tarification is the identification of economically competitive technology for the production of higher-
valued tar products. We have written on the chemical composition of biomass
thermochemical conversion tars, on the potentials for biomass pyrolysis as a
tarification reaction, and the identification of some higher-valued uses for
tars.[1,12,13,14]

Tars have undesirable physical properties. Reported attempts at tar
utilization usually relate to direct combustion, although tars are poor
fuels. Besides being viscous at ambient temperatures, not completely volatile, and exhibiting high oxygen content, tars are also gummy, corrosive,
carcinogenic, subject to chemical changes in storage, and do not mix with
conventional fuels. If tars are to be used, then they must then be reprocessed to improve a number of properties.

The chemical compositions of tars are very complex.[13,15] High temperature reactions in the absence of oxygen or under air-starve conditions are
not specific, and the availability of sufficient energy for alternative
pathways results in a series of complex concurrent and consecutive reactions
which provide a wide spectrum of pyrolytic products in small yields. Util-

329

ity of the tar in any chemical or fuel sense must recognize this chemical complexity, and it has been suggested that the tar either be fractionated into simpler mixtures, or be reprocessed into more useful mixtures.

Tars produced from the gasification or pyrolysis of several biomass residues are, however, similar in chemical composition. The composition of a tar derived from the pyrolysis of a pine sawdust and bark feed is similar to another produced in the gasification of pecan shells or from corn cobs. This suggests that different tars can be processed to produce similar products. Hydroprocessing (hydrogen and catalyst at elevated pressure) of these tars does produce a mixture of hydrocarbons with compositions which exhibit similarities to those found in both gasoline and diesel fuels. These have been fractionated into gasoline and diesel fuels which have been run in engines.[14]

CONCLUSIONS

Thermochemical conversion processes can be used on biomass materials to produce not only useable heat, but a variety of useful, energy-dense gaseous, liquid and solid fuels. Thermochemical conversion in the form of combustion can assist us in meeting the energy needs of our forestry and agricultural operations, as well as newer bioenergy industries. Thermochemical conversion in the form of gasification can assist in adapting bulky biomass fuels to efficient furnaces designed for gaseous fuel usage. Thermochemical conversion in the form of pyrolysis assists the needy without access to fossil fuels by providing charcoal fuels, and may yet provide new approaches to chemicals and hydrocarbons for material and vehicular engine fuel needs. Thermochemical processes can allow the use of non-marketed biomass residues and biomass crops by converting them into useful energy products.

REFERENCES

1. E. J. Soltes. Thermochemical routes to chemicals, fuels and energy from forestry and agricultural residues. In: Biomass Utilization. W. A. Cote, Jr., Ed. Plenum Press, New York. 1983.
2. E. J. Soltes and A. T. Wiley. Energy self-sufficiency for the Texas forest products industry. Annual Meeting. Texas Chapter. Society of American Foresters. Lufkin TX. April. 1978.
3. E. J. Soltes, W. A. LePori and T. C. Pollock. Fluidized-bed energy technology for biomass conversion. Biotech. Bioeng. Symp. 12:15. 1982.

4. W. A. LePori and E. J. Soltes. Thermochemical conversion to energy and fuels. In: Biomass Energy. E. A. Hiler and B. A. Stout, Eds. TEES Monograph. Texas A&M University, College Station TX. 1985.

5. E. J. Soltes, J. G. Massey and W. K. Murphey. Implications of selected wood use scenarios for the production of energy and industrial materials. Biotech. Bioeng. Symp. 11:3. 1982.

6. N. Y. Chen and W. E. Garwood. Some catalytic properties of ZSM-5, a new shape selective zeolite. J. Catal. 52:453. 1978.

7. P. B. Weisz, W. O. Haag and P. G. Rodewald. Catalytic production of high-grade fuel from biomass by shape-selective catalysts, Science 206:57. 1979.

8. J. L. Kuester. Liquid hydrocarbon fuels from biomass. In: Biomass as a Nonfossil Fuel Source. D. L. Klass, Ed. American Chemical Society Symp. Series No. 144:163. 1981.

9. J. L. Kuester. Catalytic conversion of biomass-derived synthesis gas to diesel fuel in a slurry reactor. In: Proceedings of the Third Annual Solar and Biomass Workshop. USDA, Atlanta, GA. 108. 1983.

10. W. H. Klausmeier. Applications for Biomass Gasification. Volume 1, Argonne National Laboratories. Argonne, IL. 1982.

11. T. B. Reed. Biomass Gasification Principles and Technology. Noyes Data Corp. Park Ridge, NJ. 1983.

12. E. J. Soltes and S-C. K. Lin. Vehicular fuels and oxychemicals from biomass thermochemical tars. Biotech. Bioeng. Symp. 13:53. 1983.

13. Y-H. E. Sheu, C. V. Philip, R. G. Anthony and E. J. Soltes. Separation of functionalities in pyrolytic tar by gel permeation chromatography - gas chromatography. J. Chromatog. Sci. 22:497. 1984.

14. E. J. Soltes and S-C. K. Lin. Hydroprocessing of biomass tars for liquid engine fuels. In: Progress in Biomass Conversion. Volume V, D. A. Tillman and E. C. Jahn, Eds. Academic Press. New York. 1984.

15. Y-H. E. Sheu. Kinetic Studies of Upgrading Pine Pyrolytic Oil by Hydrotreatment. Ph.D. Dissertation. Texas A&M University. College Station, TX. August 1985.

BIOMASS DERIVED LEVULINIC ACID DERIVATIVES AND

THEIR USE AS LIQUID FUEL EXTENDERS

John J. Thomas and Ronald G. Barile

Medical Research Institute and Chemical Engineering Program
Florida Institute of Technology
Melbourne, Florida 32901

ABSTRACT

Ethanol is now the only biomass-derived gasoline additive being mass-marketed today. Although ethanol has some attraction as an antiknock agent, it has serious drawbacks as a gasoline fuel extender because of its low energy value. This low energy value limits its practical concentration in gasoline to around 10% in present-day engines. There are, however, other potential biomass derived fuel additives. One generally overlooked scheme involves the utilization of levulinic acid (LA) derivatives. LA is easily prepared from hexoses and hexose polymers such as cellulose. This general procedure has been known since the 1830's and various commercial products have been derived from LA. This presentation summarizes optimization of the levulinic acid production process.

Keywords. Levulinic acid, fuel extenders, cellulose, biomass, hexose.

INTRODUCTION

When one performs an acid hydrolysis under forcing conditions on a six carbon sugar, the product is a material called levulinic acid (LA).[1] For example, cellulose is first hydrolyzed to glucose by acid. Then acid dehydration can occur to form hydroxymethylfurfural which decomposes to levulinic acid and formic acid.

H–C=O

Cellulose $\longrightarrow$ $(HOCH)_4$ $\xrightarrow[\text{pressure}]{\text{H+}\ \ \text{heat,}}$ $HOCH_2$ ⟨furan ring⟩ CHO

CH$_2$OH

glucose   hydroxymethyl-
                                                           furfural (HMF)

$HMF \longrightarrow CH_3\overset{O}{\underset{\|}{C}}CH_2CH_2\overset{O}{\underset{\|}{C}}OH$ + $HC\overset{O}{\underset{\|}{}}-OH$

levulinic acid   formic acid

The six carbon sugar feedstock can be formed by the hydrolysis of the cellulose in sawdust, wood chips, or waste paper. It can also be glucose, fructose (corn sugar), sucrose (cane sugar), or a component of starch. Possible cheap and economical sources of levulinic acid are shown in Table 1. Levulinic acid (LA) has been known since 1830 and the technology for producing it has been thoroughly studied. At the present time, however, no large scale commercial use has been found for the material. However, levulinic acid has been used to make resins, plasticizers, textiles, animal feed, and in the preparation of diphenolic acid which is an intermediate for surface coatings.[2] Of most immediate interest is the conversion of LA to alpha-angelicalactone (AL).[1,3]

$CH_3\overset{O}{\underset{\|}{C}}CH_2CH_2\overset{O}{\underset{\|}{C}}-OH \xrightarrow[\text{distillation}]{\text{156°C}\ \ \text{100 kPa}} CH_3 - C \underset{O}{\overset{HC---CH_2}{\diagdown\diagup}} C{=}O + H_2O$

levulinic acid (LA)   alpha-angelicalactone (AL)

(1.00 kPa = 0.295 in. of Hg)

Alpha-angelicalactone is soluble in gasoline and seems to be a suitable fuel extender. However, it can also be converted into the following materials using known technology and cheap materials.

CH_3 ⟨tetrahydrofuran ring⟩   (1)

alpha-methyltetrahydrofuran

Table 1. Biomass feedstocks for levulinic acid production

1. Waste plant material: - hardwood or beech bark[4]
2. Fiberboard industry wastewater[5]
3. Bagasse pity, bagasse, molasses[6,7]
4. Post-fermentation liquor[8]
5. Furfural still residues[9]
6. Aqueous oakwood extracts (from manufacture of tanning agents)[10,11]
7. Rice hulls[12]
8. Oats residues[7]
9. Wood sugar slops[13]
10. Fir sawdust[2]
11. Naptha[14]
12. Corncop furfural residues[15]
13. Cotton balls, rice straw, soybean skin, soybean oil residue, corn husks[16]
14. Cotton stems[17]
15. Cottonseed hulls[18]
16. Molasses[19]
17. Starch[20]
18. Potatoes, sweet potatoes, lactose[21]
19. Wastewood, pulping residues[22,11,23]
20. Sunflower seed husks[24]
21. Tapioca meal[25]

$$CH_3\overset{O}{\overset{\|}{C}}CH_2CH_2\overset{O}{\overset{\|}{C}}\text{-OR} \qquad (2)$$

levulinate esters

$$CH_3 \text{---} \underset{OR}{\overset{H_2C \text{------} CH_2}{\underset{\diagdown}{\overset{|}{C}}}} \underset{O}{\overset{|}{C}} = 0 \qquad (3)$$

pseudo-levulinyls

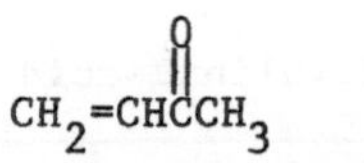

(4)

methylvinylketone

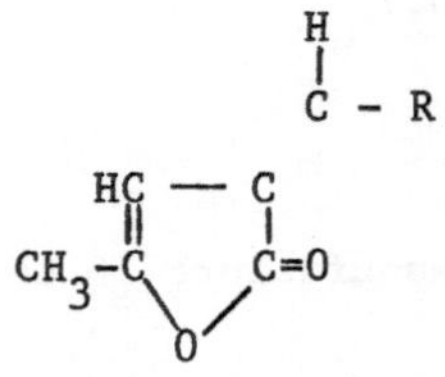

(5)

aldehyde derivatives

Alpha-methyltetrahydrofuran has been prepared on a commercial scale and used as a solvent and a chemical intermediate. Esters of levulinic acid have been used as plasticizers and in solvent refining oil. Pseudo-esters have been used as paint removers, solvents, and lacquer. Methylvinylkeytone has been used as commercial starting material for plastics, and can be derived from AL using heat or light.[25,1] Aldehyde derivatives of AL have been used as muscular depressants, anthelmintics, and eelworm hatching agents. Alpha-methyltetrahydrofuran is particularly attractive as a fuel extender since it is a cyclic ether with high mass-air/mass-fuel ratios (12.4) and a structure resembling the accepted anti-knock agent methyl t-butylether.[31] Alpha-methyltetrahydrofuran has been prepared from AL in the following manner:[26,27,28]

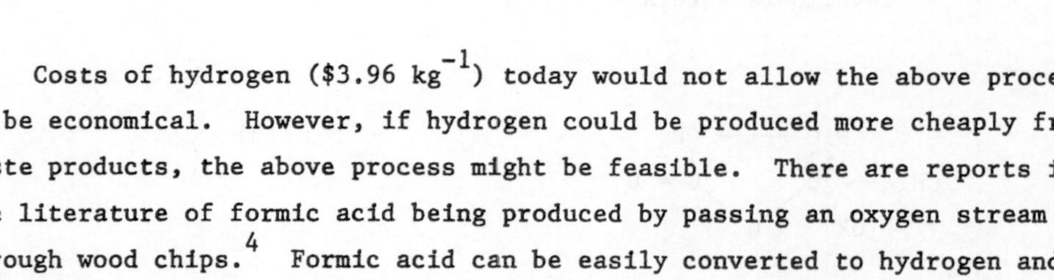

Costs of hydrogen (\$3.96 kg^{-1}) today would not allow the above process to be economical. However, if hydrogen could be produced more cheaply from waste products, the above process might be feasible. There are reports in the literature of formic acid being produced by passing an oxygen stream through wood chips.[4] Formic acid can be easily converted to hydrogen and carbon dioxide using a platinum catalyst at room temperature. Hydrogen produced similarly to this method, perhaps with air in place of oxygen, may be more economical.[29]

Levulinic acid itself has 24,000 kJ l^{-1} (86,000 BTU gal^{-1}) and has been described as an anti-knock agent, although it is not very soluble in gasoline.[30] The theoretical weight yield of levulinic acid from a hexose is 64.5% but the literature shows that only two-thirds of the theoretical yield can be obtained. Thus, a weight yield of 35-45% is expected, based on data given by Harris.[32] Conversion of LA to AL occurs at around 90% or better. Optimal yields of glucose formed by hydrolyzing cellulose is around 70-90%, depending on hydrolysis method.

As mentioned previously, for every mole of levulinic acid produced, a mole of formic acid is also produced. A significant market exists for formic acid (95%) at a current price of $1.12 kg^{-1}.[33] This situation enhances significantly the economic attractiveness of production of angelicalactone.

BIOMASS HYDROLYSIS

The batch reactor used is a simple autoclave reactor (Figure 1). The feed streams into the reactor are the biomass, and the mineral acid (such as HCl or H_2SO_4). Glass tanks serve as the reaction vessels for the autoclave. High density plastic (PVDF) serves as the lids tanks since tests conducted show that PVDF plastic is compatible with the reagents at working temperatures and pressures. Controls on the autoclave monitor pressure, temperature, and time of reaction. A steam generator is used as the heat source. Initial experiments were performed to see what cellulosic feedstocks gave the best levulinic acid yields under varying conditions.

Experiments were conducted to correlate effects of various reaction parameters on molar yields of levulinic acid. The parameters found to most influence yields include the type of biomass and mineral acid employed, the reaction time and temperature, and the presence of salts in the digestion medium.

Results of the hydrolysis reactions are summarized in Table 2. Experiments 9 through 12H were similar in that each was performed under the same conditions of temperature, reaction time, feedstock and acid type. The only difference for each was the heating time required to reach the digestion temperature, and the time allowed for cooling to room temperature after the digestion period. In these experiments 3% (v/v) HCl was used to digest shredded newspaper. The digestion time for each was three hours. No salts

Figure 1. Autoclave for production of crude levulinic acid

338

Table 2. Hydrolysis reaction condition and production in several experiments using various biomass materials

Experiment No.	9	10	11	12H	12S	13C	13N	14	15	16	17
Biomass Type	SNP	SNP	SNP	SNP	SNP	SCP	SNP	SCP	SCP	SD	SNP
Biomass Wt. (kg)	4.0	4.0	4.0	2.0	2.0	2.0	2.0	4.0	4.0	15.2	2.1
Acid Type	3% HCl	3% HCl	3% HCl	3% HCl	3% H_2SO_4[a]	3% H_2SO_4[a]	3% H_2SO_4[a]	3% H_2SO_4[a]	3% H_2SO_4[a]	3% H_2SO_4[a]	[b]
Acid Vol. (l)	36	36	36	18	18	17	17	34	34	33	40
Rxn Time (hr)	3	3	3	3	3	3	3	3	3	5	24
Heat Up Time (min)	140	40	20	10	10	20	20	30	25	30	4
Cooling Time (min)	70	100	25	20	20	25	20	30	25	25	85
Rxn Temp. (°C)	135	135	135	135	135	135	135	135	135	134	135
Final pH	0.30	0.30	0.30	0.30	0.50	0.38	0.58	0.50	0.50	N.D.	0.10
Residue Wt. (kg)	2.04	2.15	2.15	1.04	0.69	2.18	1.21	2.61	3.15	4.05	2.33
Soln Vol (l)	34	33	33	13	7	16	7	14	30	0.5	25
w/v% LA in Soln.	1.032	0.595	0.789	0.805	1.628	0.975	1.437	0.919	0.787	N.D.	1.65
w/w% LA in Solid	0.005	0.606	0.658	0.053	12.799	0.543	14.53	0.129	0.787	1.53	8.83
w/w% Yield LA	8.77	5.23	6.86	5.26	10.11	8.39	13.82	3.30	6.18	0.41	19.20
Molar % Yield LA	15.31[e]	9.14[e]	11.98[e]	9.18[e]	17.66[e]	14.65[e]	24.10[e]	5.76[d,e]	10.79[d,e]	Trace	33.50[c,e]

KEY: SNP = shredded newspaper; SCP = shredded computer paper; SD = saw dust; N. D. = no data; a = added 6% (w/v) sodium chloride; b = used recycle exp. no. 9 thru 11, and 1 vol% HCl; c = adjusted for initial LA content in solution; d = low yield due to loss from cracked vats; e = assumed cellulose concentration of 80%.

were added to the acid solutions. Molar yields of levulinic acid ranged from 9.1 to 15.3%, with a mean of 11.4 ± 2.9%. The different yields could not be directly correlated to heating and cooling times. However, the experiment which produced the highest yield (15.3%) also had the longest equilibration times (140 minutes to achieve required digestion temperature and 70 minutes of cooling time before removal from the autoclave).

Similar experiments had previously been performed (5, 6A, and 6B.) These reactions utilized 5% (v/v) HCl, shredded newspaper and no salts. The reaction temperatures were similar, but the reaction times were twice as long (six hours versus three hours). The average molar yield was 21.0 ± 0.7%; almost double that for the reactions using shorter reaction periods and lower acid concentrations.

Experiments 12H and 12S compare effects of different acid types on molar yields. Three percent (v/v) HCl was used for 12H, and 3% (v/v) H_2SO_4 was used for 12S. The reactions were performed under identical conditions. The vessel containing sulfuric acid produced a yield of levulinic acid twice that for the vessel containing hydrochloric acid (17.7% as opposed to 9.2%).

Experiments 13C and 13N compare results using different feedstocks. Computer paper was used for 13C and newspaper was used for 13N. In each case, the digestion liquor contained sulfuric acid and 6% (w/v) sodium chloride. Again, the reactions were performed under the same conditions. The vessel containing newspaper produced a higher yield of levulinic acid than the vessel containing computer paper (24.1% as opposed to 14.6%). It also was observed that compared to the yield from 12S using no salts, the yield from 13N using salts was higher by a factor of almost 27%. The cellulose content of newspaper and computer paper is approximately 80% and the molar yields are based on this figure. Newspaper is very high in lignin content while computer paper is low in lignin but high in fillers such as $CaCO_3$, rosin, and TiO_2.[34] These latter materials may interfere with the hydrolysis reactions.

During experiments 14 and 15, computer paper was digested using a solution of 3% sulfuric acid with 6% sodium chloride. The molar yield of levulinic acid was low for each case, 5.76 and 10.79%, respectively. The low yield was partially due to loss of liquid when the glass reaction vessels cracked.

In experiment 16, sawdust was used as a feedstock. The result of the digestion was absorption of the mineral acids by the biomass which left only moist solids by the end of the reactions. The partially digested sawdust was extracted with water, followed by methyl isobutylketone (MIBK). The combined extracts provided only traces of levulinic acid.

In experiment 17, shredded newspaper was digested using solutions from previous autoclave runs (9 through 11), which were digested for three hours. This reaction mixture was allowed to digest for a period of 24 hours. Levulinic acid was produced at a molar yield of 33.5%. This value was adjusted for the amount of levulinic acid initially in solution.

HPLC data for the newspaper digestion solution in 17 is provided in Figure 2. The chromatograph depicts peaks which correspond to individual components of the sample; (a) depicts a series of levulinic acid standards used to identify and quantify the levulinic acid component of the hydrolyzed sample, (b) depicts a chromatograph of the 9 through 12 mixture before the 24 hour reaction, and (c) depicts a chromatograph of this mixture after digestion for 24 hours.

In Figure 2(b), the peak area for levulinic acid is relatively small compared to the areas of the other peaks in the sample. In Figure 2(c), the LA peak area is much greater in comparison with the other peaks. This suggests that the other peaks may be intermediates that further react to form levulinic acid, or that these compounds are destroyed or volatized during the longer reaction period. It was determined by spiking with hydroxymethyl furfural that HMF did not correspond to one of these peaks. Quantitation of these peak areas indicated the molar yield of levulinic acid to be 13% for the 3 hour digested solution, and 33.5% for the 24 hour solution.

Results from the analysis of the hydrolysis reaction products indicate that levulinic acid is produced in significant yield. The presence of acetic acid, formic acid, and hydroxymethylfurfural (HMF) in the digested solutions has not been substantiated. Spikes of these compounds in the digested solutions indicate that they are readily detected using the analytical conditions employed. Thus it is reasonable to assume that if these compounds are produced during digestion they may undergo changes that inhibit their detection, such as acid-catalyzed polymerization.

HPLC Assays

HPLC assays were developed to identify the quantify the compounds produced during the hydrolysis reactions.

The following procedures were used in the sampling, analysis and quantification of the hydrolysis reaction products:

After being autoclaved, the digestion solution was allowed to cool to room temperature. The resulting sludge was mixed well and a one liter, representative sample was obtained. One hundred ml of the sludge was homogenized and weighed to $\pm$ 0.2 g. The solution was filtered and the residue was dried in a vacuum oven at 80°C for a few hours and weighed to $\pm$ 0.05 g. The solids were extracted with distilled water and the extract was filtered with a 0.2 μm nylon membrane and assayed without further dilution. The filtrate from the 100 ml sample was diluted with "HPLC Grade" water, filtered, and assayed with the extract.

The instruments used for the HPLC analysis of the extract and filtrate include an Altex Model 100A solvent metering pump, an Altex Model 1601 solvent programmer, and Altex Model 400 solvent mixer, a Rheodyne Model 7066 column selector with a valve injection port, and an LDC model 1107 Refracto-monitor.

The following analytic conditions were used to quantify and identify the products from the digestive, hydrolysis reactions:

ANALYTICAL COLUMN:		250 x 4.6 mm i.d., 10 μm, Polypore H
GUARD COLUMN	:	50 x 4.6 mm i.d., 10 μm, RP-C18 ODS
MOBILE PHASE	:	0.0025 N sulfuric acid
FLOW RATE	:	0.2 ml min^{-1}
PRESSURE	:	2068 kP (300 psi)
DETECTION	:	Refractive Index
RANGE	:	16 A.U.F.S.

The Polypore H analytical column is packed with 10 μm, resin based, spherical particles of a porous polymer in the hydrogen form. It provides separation of organic acids, carbohydrates and alcohols using ion exclusion, partition and ligand exchange mechanisms. The column has been successfully used to simultaneously resolve mixtures of levulinic acid, formic acid,

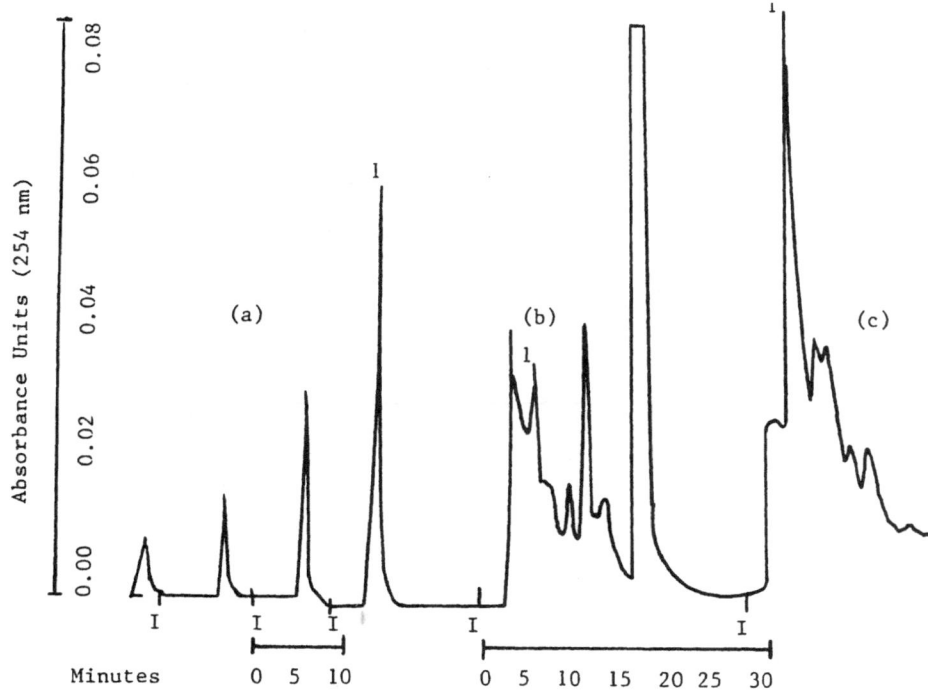

Figure 2. Liquid chromatographic analysis of (a) four Levulinic acid
injections at four concnetrations, (b) 17b a 3 hour digested
solution, and (c) 17a a 24 hour digested solution. Analytic
conditions: Column: Polypore H, 250 x 4.6mm, 10 μm; Mobile
phase: 0.005 N H_2SO_4; Flow rate: 0.5 ml/min; Temperature: 25°C;
Detection: UV 254nm; Peak #1 = Levulinic acid; I = injection

acetic acid, hexose and hydroxymethylfurfural. Detection of these compounds
has been successfully performed using refractive index.[35,36,37]

Standard curves are prepared using granular, anhydrous dextrose, 90%
formic acid, 99.8% glacial acetic acid, crystalline levulinic acid and
crystalline hydroxymethylfurfural. Figure 3 depicts a chromatograph for a
standard mixture of these compounds using the analytical conditions above.
Identification of each compound by specific retention time and the individual detection limits for each are provided in Table 3.

Extraction of the digested biomass solutions with MIBK results in the
formation of an aqueous and organic phase. Each of these phases contain
differing amounts of LA and MIBK. HPLC assays have been developed that

343

Table 3. HPLC Assay of Standard Mixture of Expected Compounds

Peak No.	Ret. Time (min)	Identification	Detection Limit
1	7.8	Dextrose	40 ppm
2	11.4	Formic Acid	50 ppm
3	13.2	Acetic Acid	65 ppm
4	18.0	Levulinic Acid	50 ppm
5	35.4	Hydroxymethyl-furfural	40 ppm

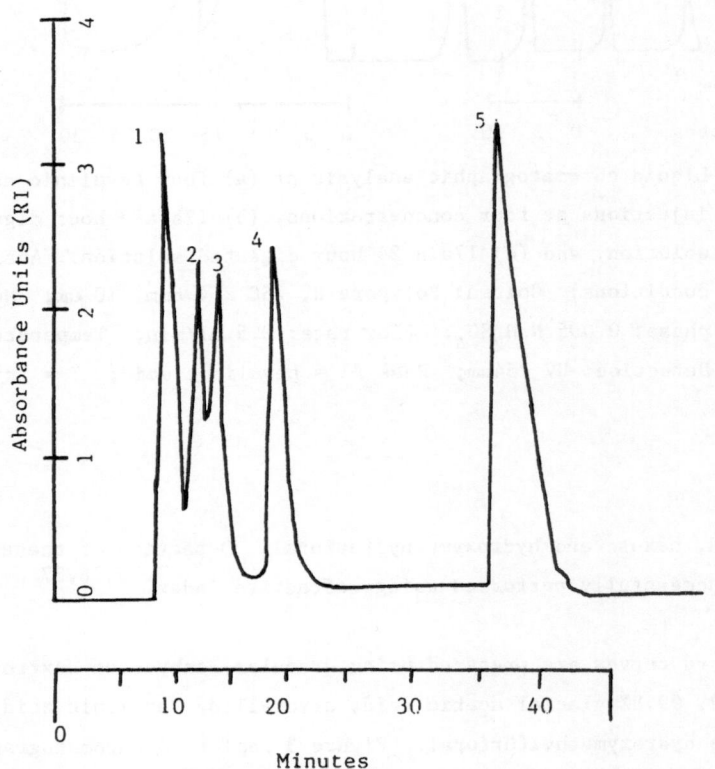

Figure 3. Liquid chromatograph of dextrose. Formic acid. Acetic acid.
Levulinic acid and hydroxymethyl furfural standard misture.
Columns: Polymore H, 250 x 4.6mm, 10 um; Mobile phase: 0.0025 N
H_2SO_4; Flow rate: 0.02 ml/min; Temperature: 25°C; Detection:
Refractive index; Sensitivity: 16 A.U.F.S.; Peaks: all 10 ppt -
1 = Dextrose, 2 = Formic acid, 3 = Acetic acid, 4 = Levulinic
acid, 5 = Hydroxymethyl furfural

quantify the amounts of LA and MIBK that are contained in these solutions. The analytic conditions for the Assay follows:

COLUMN	:	250 x 4.6 mm i.d., RP-C18 ODS
MOBILE PHASE	:	100% water to 50% Acetonitrile:Water
GRADIENT	:	10 minutes at exponent = 5
FLOW RATE	:	1.5 ml min^{-1}
DETECTION	:	UV at 280 nm
RANGE	:	0.04 A.U.F.S.

The reverse phase, C-18 packing material of the column provides excellent separation of the LA and MIBK when utilizing the above analytic conditions. The assay procedure has been used to quantify the LA and MIBK in both the aqueous and organic phases when appropriate dilutions with acetonitrile were made.

Standard curves were first prepared using crystalline LA and high purity MIBK. The aqueous and organic phase solutions phase solutions were diluted as needed with water and acetonitrile, respectively. The samples were then filtered with a 0.2 μm nylon membrane and were analyzed. Figure 4 depicts the chromatographs of the LA and MIBK standards, and of the aqueous and organic extracts using the analytic conditions outlined. Resolution of the LA and MIBK in the samples was easily accomplished using the guidelines stated above. A compensating, polar planimeter was used to obtain values of the peak areas. Quantification of the concentrations was then performed by correlating the peak areas to the standard curves.

CONCLUSION

Results of these hydrolysis reactions show that levulinic acid yields are primarily influenced by the type of feedstock and mineral acid used, the amount of time allowed for the reaction, and concentration of salts in the digestion mixture. Optimum yields were obtained using newspaper, as opposed to computer paper or sawdust, as a feedstock. Sulfuric acid provided greater yields than did hydrochloric acid. Reaction periods longer than three hours are needed to insure more complete hydrolysis of the biomass feedstock when pilot plant size reactions are performed. The addition of salts to the digestion solution tends to enhance levulinic acid yields.

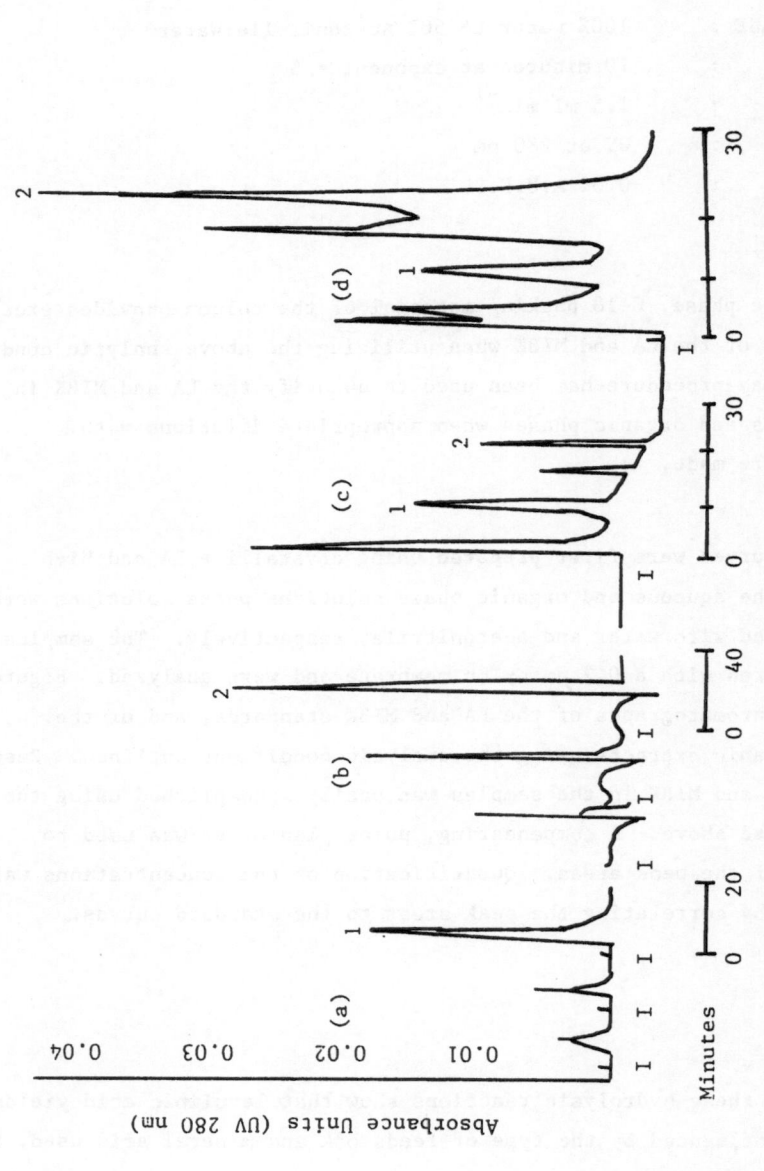

Figure 4. Liquid chromatograph of (a) levulinic acid standards, (b) MIBK standards, (c) aqueous phase of extraction liquor, and (d) organic phase of extraction liquor. Analytic condition: Column: ODS RP-C18, 250 x 4.6mm i.d., 10um; Mobile phase: 100% H_2O to 50% H_2O: acetonitrile; Gradient: 15 minutes at exponent = .5; Flow rate: 1.5 ml/min; Detection: UV @ 280 nm; Sensitivity: 0.04 A.U.F.S.; Peaks: 1 = Levulinic acid, 2 = Methyl isobutylketone

REFERENCES

1. R. H. Leonard. Levulinic Acid as a Basic Chemical Raw Material. Ind. Eng. Chem. 48:1331-1334. 1956.
2. C. P. Haworth and W. L. Shilling. Levulinic Acid from Hexose-Containing Material. US 3,258,481. 1966.
3. R. H. Leonard. Conversion of Levulinic Acid into Alpha-Angelicalactone. US 2,809,203 Chem. Abst. 52:2891b. 1958.
4. Z. Kin, H. Kowalczyk, L. Gorski, R. Klajenski, B. Tonzewski, J. Jaworski and E. Przybylak. Simultaneous Preparation of Furfural, Levulinic Acid, and Humic-Nitrogen Fertilizer from Waste Plant Material. Pol. 99,185 Chem. Abst. 88:3643c. 1978.
5. J. Pajak and M. Kryczko. Treatment of Fiberboard Industry Wastewaters. Pol. 99,879 Chem. Abst. 91:180991j. 1979.
6. C. I. Nee and W. J. Yse. Furfural and Levulinic Acid Prepared Concomitantly from Bagasse Pith. Taiwan Sugar. 22(2):49-53 Chem. Abst. 83:117532e. 1975.
7. E. Ramos-Rodriguez. Jointly Producing Furfural and Levulinic Acid from Bagasse and Other Lignocellulosic Materials. US 3,701,789 Chem. Abst. 78:16017g. 1973.
8. N. P. Mel'nikov, B. M. Levitin and L. A. Sergeeva. Levulinic Acid. USSR 463,657 Chem. Abst. 83:42838y. 1975.
9. L. A. Badovskaya, V. G. Kul'nevich, L. I. Firsova, M. L. Kurzin and V. V. Chudaev. Conversion of Still Residues from Furfural Manufacture by their Oxidation with Hydrogen Peroxide to Decarboxylic Acids and Keto Acids. Tr. Krasnodar. Politekh. Inst. 29:107-8 Chem. Abst. 76:115075q. 1972.
10. S. Prosinski, Z. Adamski and A. Kwasniewski. Analysis of Chemical Components of a Hydrolyzate Obtained from Oak Extraction Chips after Distilling of Furfural. Rocz. Wyzsz. Szk. Roln. Poznaniu. 52:89-98 Chem. Abst. 77:903162. 1971.
11. Z. Khin and J. Stawecka. Use of Waste Lignocellulose for the Production of Levulinic Acid. Przegl. Papier. 24(9):303-6 Chem. Abst. 70:5302e. 1968.
12. Y. Sumiki. Levulinic Acid. Japan 176,438 Chem. Abst. 45:7589i. 1948.
13. E. Faerber. Recovery of Products, such as Furfural and Levulinic Acid and its Esters, from Slops from the Wood-Sugar Process or the like. US 2,293,724 Chem. Abst. 37:10402. 1943.
14. Y. Kikuchi and K. Ikematsu. Separation and Recovery of Levulinic Acid from Naptha Liquid-Phase Oxidation Waste Liquid. Japan Kokai 74: 51,220 Chem. Abst. 81:119997t. 1974.
15. A. P. Dunlop and P. A. Wells. Levulinic Acid US 2,813,900 Chem. Abst. 52:9199b. 1957.
16. Y. Sumiki and A. Kojima. Preparation of Levulinic Acid and its Utilization. I. Levulinic Acid from Agricultural Produce Waste. J. Agr. Chem. Soc. Japan 20:651-2 Chem. Abst. 42:5422. 1944.
17. V. S. Minina, A. E. Sarukhanova and K. U. Usmanov. Preparation of Furfural and Levulinic Acid by Hydrolysis of Pressed Cotton Stems. Fix.i Khim. Prirodn. i Sintetich. Polimerov, Akad. Nauk Uz. SSR, Inst. Khim. Polimerov. 1:87-93 Chem. Abst. 60:835h. 1962.
18. K. Akmamedov, V. S. Minina and K. U. Usmanov. Cottonseed Coverings as a Valuable Raw Material for the Hydrolysis Industry. Fiz. i Khim. Prirodn. i Sintetich. Polimerov, Akad. Nauk Uz. SSR, Inst. Khim. Polimerov 1:78-86 Chem Abst. 60:757f. 1962.
19. C. K. Rao, C. C. Reddy, G. S. Sidhu, I. K. Kachler and S. H. Zaheer. Isolation of Levulinic Acid from Molasses. Indian 70,171 Chem. Abst. 56:2623h. 1959.
20. C. H. G. Hands and F. R. Whitt. The Preparation of Levulinic Acid on a Semitechnical Scale. J. Soc. Chem. Ind. 66:415-16. 1947.

21. T. Takahashi. Studies on Decomposition of Carbohydrates by Strong Mineral Acids. I. Determination of Decomposition Products. J. Agr. Chem. Soc Japan. 20:553-6 Chem. Abst. 42:8166f. 1944.

22. A. J. Wiley, J. F. Harris, J. F. Salman and E. K. Locke. Wood Industries as a Source of Carbohydrates Ind. Eng. Chem. 47:1397-1405. 1955.

23. W. I. Shilling. Levulinic Acid from Wood Residues. Tappi 48,(10):105A-8A Chem. Abst. 63:18455. 1965.

24. L. L. Shil'nikova. Complex Processing of Plant Raw Materials with the Production of Furfural and Levulinic Acid. Khim. Pererab. Drev. 8:7-9 Chem. Abst. 67:118315t. 1967.

25. O. Chapman and C. McIntosh. Photochemical Decarboxylation of Unsaturated Lactones and Carbonates. J. Chem. Soc. D, (8):383-4 Chem. Abst. 75:35566q. 1971.

26. Y. Hachihama and M. Imoto. Synthesis of High-Molecular Compounds from Furfural. II. A Synthesis of 1,4-pentanediol from Furfural. J. Soc. Chem. Ind. Japan 45:19-22 Chem. Abst. 44:8859d. 1942.

27. I. Hayashi, E. Negoro and Y. Hachihama. Levulinic Acid and its Derivatives. II. Preparation and Reduction of Gamma-Valerolactone by Catalytic Hydrogenation. J. Chem. Soc. Japan 57:67-9 Chem. Abst. 49:1154d. 1954.

28. J. H. Helberger, S. Ulubay and H. Civelekoglu. Simple Procedure for Preparing Alpha Angelicalactone; Fission of Oxygen heterocycles on Hydrogenation. Ann. 561:215-20 Chem. Abst. 43:4640. 1949.

29. L. J. Fieser and M. Fieser. Advanced Organic Chemistry. Reinhold Publishing Co., New York. 1962.

30. G. W. Eckert and H. V. Hess. Motor Fuels Containing Oxo-Carboxylic Acids. US 3,074,787. 1963.

31. B. Torck, A. Convers and A. Chauvel. Methanol for motor fuel via the ethers route, Chemical Engineering Progress. 78:38-45. 1982.

32. J. Harris. Acid hydrolysis and dehydration reactions for utilizing plant carbohydrates. Appl. Polym. Symp. 28:131-144. 1975.

33. Chemical Marketing Reporter. March 25, 1985.

34. B. Moore. The effect of changing technology on the use of rosin in the paper industry. Naval stores Review. 94(1):14-16. 1984.

35. R. Andersson, B. Hedlund. HPLC analysis of organic acids in lactic acid fermented vegetables. Z. Lebbensn. Unters Forsch. 176:440-443. 1983.

36. R. McVeeters, R. Thompson and H. Fleming. Liquid chromatographic analysis of sugar acids and ethanol in lactic acid vegetable fermentation. 67(4):710-714. 1984.

37. W. Abeydeera. HPLC determination of sugars on cation exchange resins. Proc. Aus. Soc. Sugar Cane Technologists. 171-186. 1983.

UTILIZATION OF BIOMASS FUEL FOR PRODUCTION

OF ELECTRIC POWER

William S. Bulpitt and James L. Walsh, Jr.

Mechanical Systems Division
Georgia Tech Research Institute
Atlanta, Georgia   30332

ABSTRACT

The main focus of this paper will be a presentation of results of a
project recently completed by Georgia Tech in the Republic of the Philippines. Georgia Tech was retained by the Asian Development Bank to review
the progress of the "Dendro Thermal Power Production" project underway in
the Philippines in the 1-3.5 MW size range, being built by French and British equipment suppliers. Current plans call for an additional eight plants
to be built by the French and British, three 5 MW plants to be built by the
U.S., and nine 3 MW plants to be built by the Japanese. Eventually, there
may be 70 wood-fired power plants in the Philippines contributing some 200
MW to the national power system.

In general, the overall conclusion of the twenty-two member project
team was that the dendro thermal concept is viable, but that many mistakes
have been made in the Philippines project which need not be repeated by the
implementing agency (the National Electrification Administration, or NEA) or
other nations seeking to develop biomass energy.

Keywords. Biomass, cogeneration, power plants, wood energy, Philippines, dendro thermal, gasification.

INTRODUCTION

During the 1800s wood provided an important component of the fuel
supply for industrial plants in the United States. As exploration for

fossil fuels became more developed, however, wood was replaced first by coal and later by fuel oil and natural gas. By 1970, the only place where wood could be found as an industrial fuel was in pulp and paper mills. In these mills, wood residue (resulting from the pulping process) was considered a bothersome byproduct more than a low-cost fuel, and combustion of the wood in boilers was more of a disposal exercise than a steam production exercise. During the mid-1970's, however, when the price of fuel oil jumped from $2 to $12 per 159 liters ($2 to $12 per barrel) in the U.S., industrial plants and utility plants both began to look at alternate fuels for their processes.

This paper will give a brief overview of the use of biomass (principally wood) fuels for power generation in the mid-1980s. The discussion will consider systems operating in the United States and in the Philippines.

THE FUEL RESOURCE

In the United States, the regions where wood (and consequently wood fuel) is most readily available include the Southeast, the Northeast, and the Pacific Northwest (including Alaska). Other less important concentrations of the resource are available in various western states. In addition, various forms of agricultural residue are available in farming areas.

The least expensive and most widespread form of wood fuel is mill residue. This material varies in size from sawdust through chips and may include boards, blocks, and slabs. Its non-uniform nature and high (50%) moisture content make handling, storage, and burning of this fuel costly and difficult. Its low price, currently $6.60 to $11 Mg^{-1} ($6 to $10 T^{-1}) delivered, makes it an attractive fuel for industrial use in the range of 6.8×10^3 to 90.8×10^3 kg hr^{-1} (15,000 to 200,000 lb hr^{-1}) of steam. While this residue is widely available, it may have competing uses such as a feedstock for particle board and other forest products. It is not usually wise for a facility not associated with the forest products industry (such as a free standing power generation plant) to depend solely on mill residue for fuel. One large industrial user may tie up the majority of mill waste available in a given locality, and long-term supplies may be in question.

Another widely used form of wood fuel is whole tree chips. This fuel is the product of in-field chipping operations which utilize large machines chipping entire trees. Thus the fuel mix can include bark, limbs, needles and leaves in addition to chips. This material is generally more expensive

350

($14.4 to $22 Mg^{-1} or $13 to $20 t^{-1}) than sawmill residue but may be readily available through long-term contracts with loggers or wood brokers.
This form of wood has been the primary source of fuel for some of the larger
wood energy plants built recently in the United States.

As already mentioned, agricultural residue fuels may be available in
certain locations, but due to the dispersed nature of this material (peanut
hulls, walnut shells, cotton stalks, etc.) it is generally more difficult to
obtain sizeable quantities for use as fuel in larger power plants.

The last source of wood fuels currently in use or under consideration
is fuel from energy plantations. These plantations use fast-growing tree
species such as sycamore or eucalyptus which regenerate naturally from an
existing stump. These energy plantations can have short rotations from
three to seven years and may be particularly appropriate in tropical regions
with high rainfall and long growing seasons. In the United States, research
on this concept is being carried out by a number of organizations including
the University of Florida and the Oak Ridge National Laboratory.[1,2]

Power Plants in the U.S.

Since the mid-1970's, a number of wood-fired power plants have been
built which have as their primary purpose the production of electricity. A
representative sample of these plants is shown in Table 1. As can be seen
from the table, there is a wide range in size represented, and each project
represents a very site-specific situation. Burlington Electric, for
example, is located in one of the northernmost sections of Vermont, a long
distance from coal suppliers. Thus the price of coal delivered to Burlington tends to be relatively high, and feasibility studies showed two important things about the local wood supply: a) there is more than enough wood
available on a long-term basis to supply the plant, and b) the price of the
wood is low enough to make the plant economically feasible.

Other plants shown on the table have really sprung up in response to
the federal PURPA (Public Utility Regulatory Policies Act), legislation
which has opened the door to some power producers to receive a "fair" price
for their electricity. These arrangements have been particularly successful
in the regions of New England, Florida and California. These areas have
traditionally relied on oil-fired power plants which have become very expensive to run and operate. They have also been the areas with the toughest
environmental regulations, making conversion to coal difficult.

Table 1. Representative wood-fired power plants installed since 1970

Plant Name	Location	Size	Fuel	Start-UP Date
Biomass Power Corp.	Monticello, FL	7.5 MW	Wood chips, peanut hulls	1984
Greensboro Lumber	Greensboro, GA	7.5 MW	Sawmill residue	1979
Burlington Electric	Burlington, VT	50 MW	Total tree chips	1984
Kettle Falls	Kettle Falls, WA	40 MW	Bark, sawmill residue	1983
French Island	Lacrosse, WI	15 MW	Sawmill residue	1981
Bio-Energy Corp.	West Hopkinton, NH	12 MW	Sawmill residue	1984
Dixon Lumber	Eufala, AL	2.5 MW	Sawmill residue	1982
Molokai Electric	Hawaii	4 MW	Straw, chips, pineapple waste	1982
Cogen of Tenn., Inc.	Red Boiling Springs, TN	12 MW	Sawdust	1984
Ultra-Power, Inc.	Burney, CA	11 MW	Forest residue	1984

*References 4, 5, 6.

In other states it has been difficult for plants to obtain what might be considered a fair price for their electricity, and fewer plants have been built. This has been the case in Georgia and Alabama, for example.

The parameters that must be examined most closely when considering a small wood-fired power production plant include the following: a) wood (or other fuel) availability and cost, b) construction cost of the plant and c) buyback rate from the connected utility. Many recent installations have incorporated used equipment which has kept the overall price of the installations down. Using totally new equipment, it would not be surprising to see a cost of $1500/installed kW for a wood-fired power plant in the 5-15 MW range, while some of the recent plants have cost in the neighborhood of $600-900/installed kW utilizing rebuilt steam turbines, generators, and switchgear and a "no frills" approach to boiler and woodyard design. The future proliferation of these types of installations is uncertain, and much will depend on interpretation and enforcement of the PURPA regulations.

Power Plants in Developing Countries

The price of oil and other fuels has had a tremendous impact on industrial plants and power generating facilities in developed nations, but this impact has been more profound in less developed countries, particularly those with no domestic fuel sources. These countries tend to have power plants that were constructed at the lowest prices possible and have suffered from irregular maintenance programs. They also tend to be oil fired because that fuel was traditionally available and among the easiest fuels to deal with.

Fluctuations in the world oil prices have thus had great effects on the power producing facilities in these less developed nations, but only a few have launched the ambitious programs needed to deal with these problems. One such nation is Brazil, which has embarked on a massive program to produce alcohol fuels for transportation and other uses. Another such nation has been the Philippines. It has been estimated that up to 40% of the export earnings of the Philippines has gone to oil-producing nations for the purchase of fuel oil for transportation and power production. In 1979, the Philippines embarked on an ambitious program to do something about these problems, and the main thrust of the program included the construction of a network of wood-fired power plants and small hydroelectric plants. The Georgia Tech Research Institute has recently completed an evaluation of the wood-fired power plant system, and some of the results will be discussed here.

Power Plants in the Philippines

The wood-fired power plant system in the Philippines has been called the "Dendro Thermal Development Program" and has as its ultimate goal the construction of 70 wood-fired power plants in various locations throughout the country, producing some 200 MW of electric power. Fuel will be grown in energy plantations of at least 1100 hectares (2500 acres) in size. The main tree species will be leucaena, known locally as "ipil-ipil," a species which can exhibit very high growth rates under the proper conditions.[8] Plans call for a rotation interval of four years, and the original plans called for all of the wood to be provided by the plantation adjacent to the power plant.

353

As of mid-1984, equipment had been ordered by the Government of the Philippines for 17 power plants. This equipment will be provided by British and French suppliers. In addition, preliminary designs have been generated for nine power plants to be supplied by the Japanese, and negotiations are underway between the Philippines and the United States Agency for International Development (USAID) for three power plants of U.S. manufacture. Figure 1 shows the location of the plants.

Of the 17 power plants ordered, nine are currently in operation or near startup. These include two 3.4 MW steam plants of French design, five 3.1 MW steam plants of British design, and two 1 MW gasifier-engine systems of French design.

The technology employed in the power plants is quite conventional, although the wood gasifier plants tend to be rather unique. The French power plants operate at a pressure of about 4.14 MPa (600 psi) and utilize two stages of hammermills to crush the wood fuel. The wood is supplied to the plant as small logs with a maximum diameter of 10 cm (4 in) from the plantation. A flue gas dryer is employed to lower the moisture content of the wood fuel before it enters the furnace. Startup problems with the French plants have centered on the drying and fuel preparation systems. A three-stage steam turbine is used, giving an overall net plant thermal efficiency of about 16%. A typical French power plant layout is shown in Figure 2.

The British power plant design is simpler than the French, utilizing a 2.07 MPa (300 psi) boiler but a seven-stage steam turbine. A typical plant layout is shown in Figure 3. The net plant thermal efficiency for the British units is also about 16%. The fuel handling system for the British units is quite simple and utilizes logs with a 10 cm (4 in) maximum diameter as the fuel form. This concept has proved to be a problem where the wood available from the plantation (in the form of small diameter logs) has not been adequate and outside wood supplies of irregular shapes have been used. As the plantation development progresses, these problems should become less significant.

The French gasifier plants rely on two downdraft gasifier systems each connected to two dual fuel diesel engines. These systems were not started up as of mid-1984, so the performance of the systems is unknown at this time. Protection of the engines requires extensive gas cleanup, and so the cleanup systems tend to be rather complicated, which may cause long-term

354

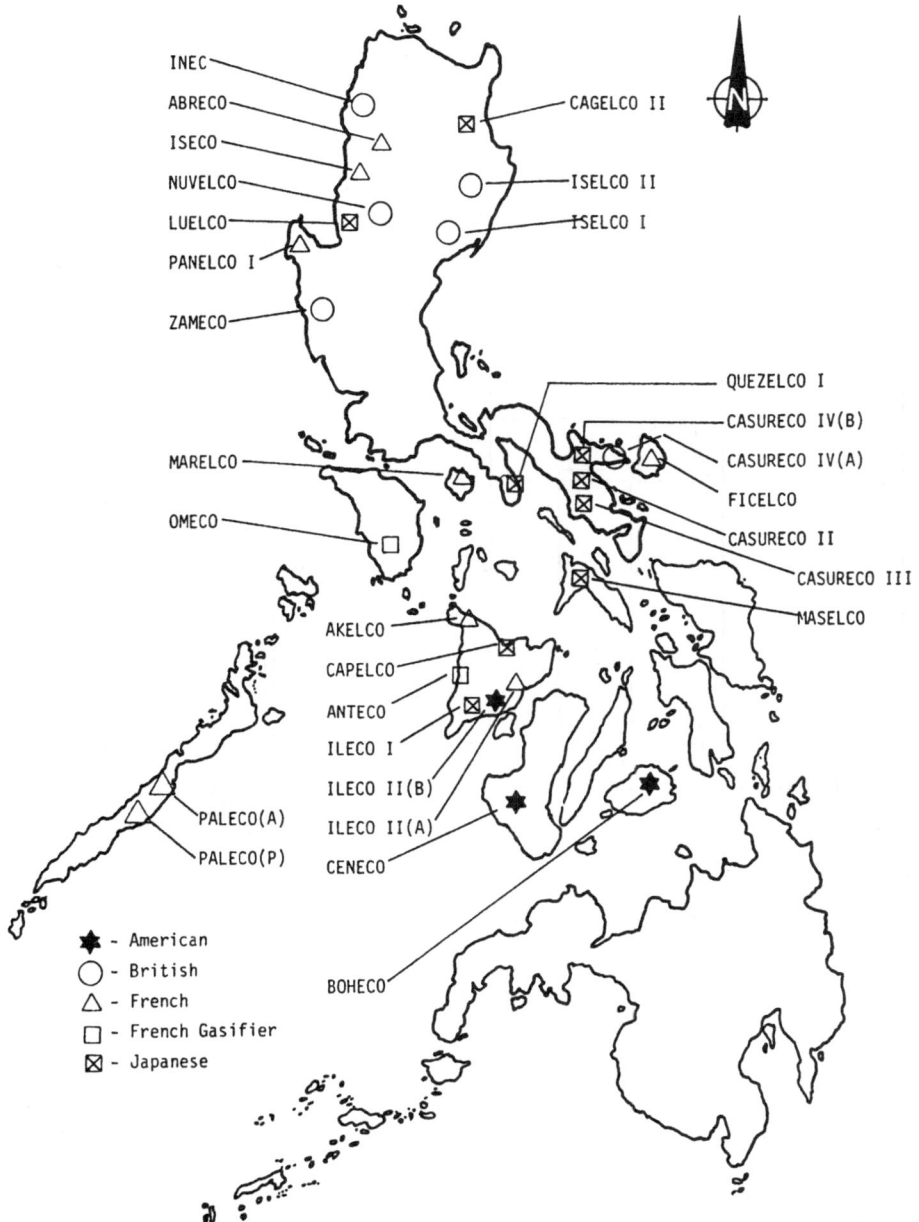

Figure 1. Location of power plants

maintenance and operating problems. The layout of a typical gasifier plant is shown in Figure 4.

In general, however, the problems encountered with startup of these power plants are relatively simple. In fact, the logistics of constructing the facilities in very remote areas probably present the greatest problems.

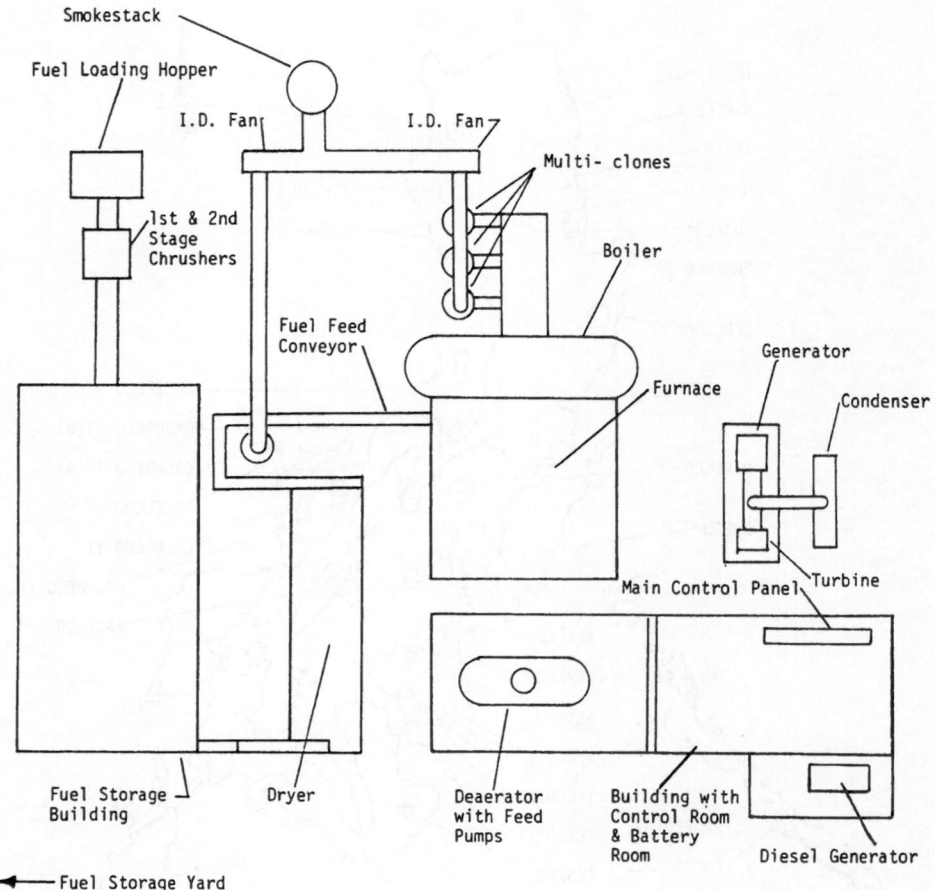

Figure 2. Alsthom–atlantique plant layout

Given a little experience, the plant operators should be able to deal with
the day-to-day operations adequately.

One problem that has become evident is that with such a diversity of
power plant designs, the NEA will have some difficulty in training operators, providing spare parts, and maintaining the plants. Thus there should
be some effort in the future to standardize a design, and it is recommended
that this design utilize fuel chipping, centralized controls, communication
links with the National Power Corporation grid, and cooling towers rather
than steam flow cooling.

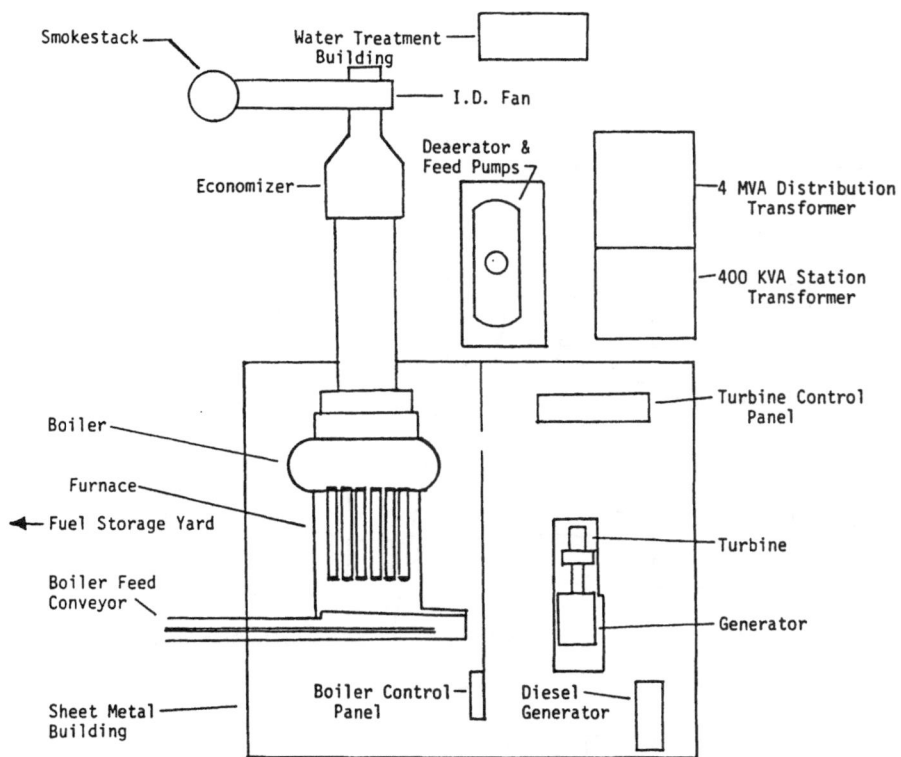

Figure 3. Balfour beatty plant layout

For the existing and future plants, a detailed electrical transmission and distribution analysis must be conducted, including plans for interconnection, to avoid costly problems experienced at some early plants.

At some of the power plants there has not been adequate attention paid to operator training. In developing countries, there is often a lack of qualified personnel, and every effort must be made to develop local talent if the projects are to succeed.

With regard to environmental issues, the power plants currently have little or no facilities for pollution control. Considering the remote locations, this is probably not a major issue at present but must be considered if these plants proliferate in the future. One particular aspect that must be addressed is the matter of liquid effluent from the gasifier scrubbers.

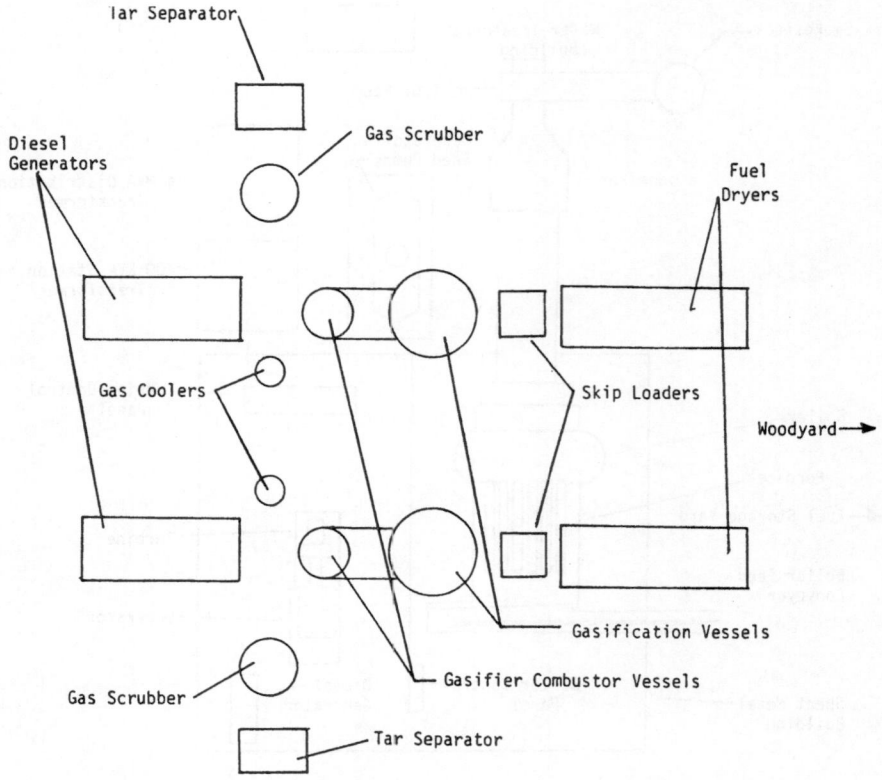

Figure 4. Moteurs duvant plant layout

Energy Plantations

The energy plantations supplying wood fuel for the power plants are approximately 1100 hectares (2500 acres) in size and are divided into 10 to 12 modules with about 10 tree farms in each module. All of the plantations were originally planted with the leucaena or "ipil-ipil" tree which will reproduce itself directly from a stump in about three years after harvesting. The trees are harvested by the farmers and transported by land to the pick-up point for the plantation cable system for transportation to the power plant. This cable system is similar to a ski lift and connects all of the modules to the power plant.

The primary problem with the current plantations is the selection of ipil-ipil as the only species to be planted. Of all the sites visited by the evaluation team, only Panelco I had a reasonably well developed plantation. Many of the sites had a power plant nearing completion, but the plantation was non-existent. Ipil-ipil would simply not grow in all situa-

358

tions. It is currently projected that the plantations will produce about 15 dry Mg ha^{-1} yr^{-1} (16.7 dry t A^{-1} yr^{-1}). Thus a 1100 hectare plantation can supply only 60% of the energy needed for a 3 MW plant.[7]

Several solutions to this problem have been implemented or proposed. There is not a single solution to the problem as soil and climate conditions vary at different sites. An experimental nursery has been started at INEC and three other sites where a number of species are being evaluated. In addition, the forestry team has recommended a number of additional species as candidates to replace the ipil-ipil. It has also been recommended that a senior forester be added to the NEA staff in Manila. While these solutions are being implemented, there is sufficient wood available on the local market for plant operation; but, as expected, the cost of this wood is higher, usually by 30 to 50 percent.

The use of the cable system for harvesting also presented problems. The primary problem was an acceptable system for feeding wood to the cable pick-up point in each module. These pick-up points are usually located at the top of a hill, and methodology for transporting the wood had not been established at most sites. Additionally, the "packaged" cable concept was not considered to be a sound idea as the design of most wood transport systems requires a site specification analysis.

As part of the evaluation, a revised harvesting plan was generated for four of the sites. These plans considered a number of alternatives to the current cable system. The plans are general concepts, and detailed engineering and analyses are still required.

Institutional Aspects

Implementation of the Dendro Thermal Program has been through the National Electrification Administration (NEA), the government agency responsible for rural electrification. The actual owners of the power plants end up being the rural electric cooperatives (RECs) in the areas where the plants are located. The RECs administer tree farm associations which grow the trees. The farmers are paid when the wood is delivered to the power plant but are given low interest loans to establish the plantations. This appears to be a workable model, resulting in new jobs for the remote areas. There are problems that have surfaced in 1983-1984 in the ability of the Philippines government to keep the plantation money flowing due to the current economic crisis the country is experiencing. It is hoped that this will be a temporary situation.

Economics

Six of the power plants under construction were analyzed extensively from an economic and financial perspective. The analysis considered the original plant as built and a replication of the plant incorporating changes recommended by the evaluation team. The analysis also considered the cost of fuel that must be purchased on the local market due to plantation failure.

The analysis indicated that most of the plants will be able to produce power at costs below those of the aggregate costs experienced by the National Power Corporation (NPC). One of the primary reasons for this low price production cost is that NPC depends almost exclusively on imported oil for power generation. Table 2 gives details on the economic analysis. As can be seen from the table, the cost of power produced by the wood-fired power plants should be in the range of 5.5 to 10.5 cents per kWh. The small gasifier systems exhibit the highest cost of power production. The cost of NPC power to the RECs varies widely, but in most cases it is in the range of 6 to 9 cents per kWh.

The analysis also concluded that a financial internal rate of return (FIRR) acceptable to the Asian Development Bank could be realized. This

Table 2. Summary of economic and financial analyses for six power plants

| | FIRR %* | | EIRR %** | | kWh Cost*** | |
	B****	R*****	B	R	B	R
CASURECO IV-A	10.6	11.9	14.7	15.6	1.09	1.09
ZAMECO	12.5	13.8	13.2	14.6	1.14	1.09
OMECO	15.8	17.1	6.4	9.3	1.97	2.10
ANTECO	11.9	12.6	13.2	13.7	1.21	1.24
PANELCO	14.0	14.9	21.0	21.0	0.83	0.81
ISECO	17.4	18.3	23.1	24.0	0.76	0.74

*FIRR = Financial Internal Rate of Return.
**EIRR = Economic Internal Rate of Return.
***kWh cost shown in Philippine pesos. Exchange rate at time of study approximately 20/1.
****"B" denotes baseline case.
*****"R" denotes replication case.

funding is of extreme importance to the RECs as they will be responsible for the power plant construction loans.

This FIRR could be significantly improved if an acceptable price for sale of power to NPC can be negotiated by NEA. The size of most of the power plants is larger than the average REC load, and the REC load factor is usually less than 50 percent. This is primarily due to the absence of any industry in the REC to establish a base load. An enabling agreement similar to PURPA is needed to encourage NPC to purchase the excess power plant output at an acceptable price. Such an agreement will benefit not only NEA but the nation as a whole.

Conclusion

Under certain conditions electric power can be generated from wood fuel economically and effectively. In the United States, more small power plants will be built if locations can be found which have high avoided costs from the power companies and low wood fuel costs. The Philippines are now establishing a model for other tropical developing countries in the world, and this situation can probably be duplicated in other Pacific islands and in certain nations in Africa, Central America, and Asia.

REFERENCES

1. P. N. Benson. GRI Programs on Methane from Biomass and Wastes. Presented at National Meeting on Biomass R & D. Arlington, VA. October. 1984.
2. J. L. Trimble. The Production of Wood by Short Rotation Forestry, presented at National Meeting on Biomass R & D. Arlington, VA, October. 1984.
3. B. W. Riall, et al. An Assessment at Increased Biomass Denied Energy Use in the Southeastern U.S., Final Report on Project B-551. Georgia Tech. Atlanta, GA. 1983.
4. Public Power Innovation. Public Power, November/December 1983. published by American Public Power Association. Washington, D.C. 1984.
5. J. L. Easterly and Elizabeth C. Sams. Electric Power from Biofuels: Planned and Existing Projects in the U.S., U.S. Department of Energy Contract No. DE-ACO1-83-CE-30784, August. 1984.
6. W. S. Bulpitt, et al. Biomass Fuel Resource Study. Final Report on Project A-3397, prepared for the City of Tallahassee, FL, by Georgia Tech. March 1983.
7. W. S. Bulpitt, et al. Dendro Thermal Power Generation Program Evaluation Study. Final Report on Georgia Tech Research Institute Project A-3717 for the Government of the Philippines. October. 1984.
8. F. K. Denton. Wood for Energy and Rural Development: The Philippine Experience. Washington, D.C. 1983.

A SYSTEM FOR PRODUCING BIOMASS FUEL FOR A

MULTIUSE INDUSTRIAL PARK

R. Edward Burton

Kleensmoke, Incorporated
222 Franklin Avenue
Willits, California 95490

ABSTRACT

This paper describes a system for cutting and bundling slash or brush into 0.9 Mg (1 t) bundles using a low-cost brush bundler. The bundles, about 1.2 m (4 ft) by 4 m (13 ft) long, are strapped with steel strapping and allowed to dry for 1-3 months. The bundles and logs are hauled over a four-foot wide "road" by a small but very powerful tractor connected to a powered trailer. The combination can carry two-ton loads up a 40% grade. The bundles and logs are loaded onto a self-loading trailer with a unique cross-haul system.

At the semi-portable Biomass Center, the bundles are segregated into:

1. Brush to 15 cm (4 in) diameter is made into charcoal

2. Wood 15-35 cm (4-9 in) into chips or firewood

3. 35 cm (9 in) and larger into logs or firewood

The heat from the charcoal production is used to dry the chips or firewood. The dry chips can then be shipped to permanent Multiuse Industrial Parks where a permanent system can supply energy for cogeneration and other uses.

Keywords. Brush bundles, charcoal chips, multiuse industrial park.

WOOD AND FUEL AND CHARCOAL

The burning of wood is actually a sequence of events. As wood is heated, free water and burnable gases are driven off along with particulates (smoke). This leaves a burning char which burns clean and hot. In the Kleensmoke Inverse Pile Burner, this sequence takes place in a long horizontal refractory tube (see Figure 1). The new material is pushed under the burning char so the particulates and gases are completely burned as they pass upward through the char. By overfeeding the burner, the burning char falls out with the ashes. The ashes are separated and the char is dry quenched in a water cooled auger.

This sequence produces clean heat in the 1140 C (2000°F) range while also producing charcoal that is around 90% carbon. This charcoal producing equipment (Figure 2) is semi-portable with its own power supply. This system is called the Satellite Biomass Conversion System or Biovertor. It produces cut dry firewood, charcoal and dry chips from the bundles of brush. Any sawlogs are sorted out here and sent to sawmills. The charcoal may be used for making barbeque briquets and chunks or for air and water filtering, as well as other industrial processes.

GETTING WOOD TO THE BIOVERTOR

The trouble with wood is that it grows on hills and other areas that make it difficult to harvest. About 40% of the tree is brush and limbs. When freshly cut it contains about 50% water, which not only reduces the useful caloric value, but causes smoke. If wood is chipped green, it is difficult to dry the chips because wood is such a good insulator and catches fire if not properly controlled. It is difficult and expensive to haul, particularly from the stump to the roadside. A high percentage of the biomass in the forest is in the form of brush which must be harvested and disposed. Wood harvesting equipment is expensive to buy and maintain and can be used only during daylight and the dry season. Wood harvesting is hazardous which greatly increases the cost of workman's compensation. Heavy logging equipment cannot be used in many areas because of silvicultural needs and state regulations.

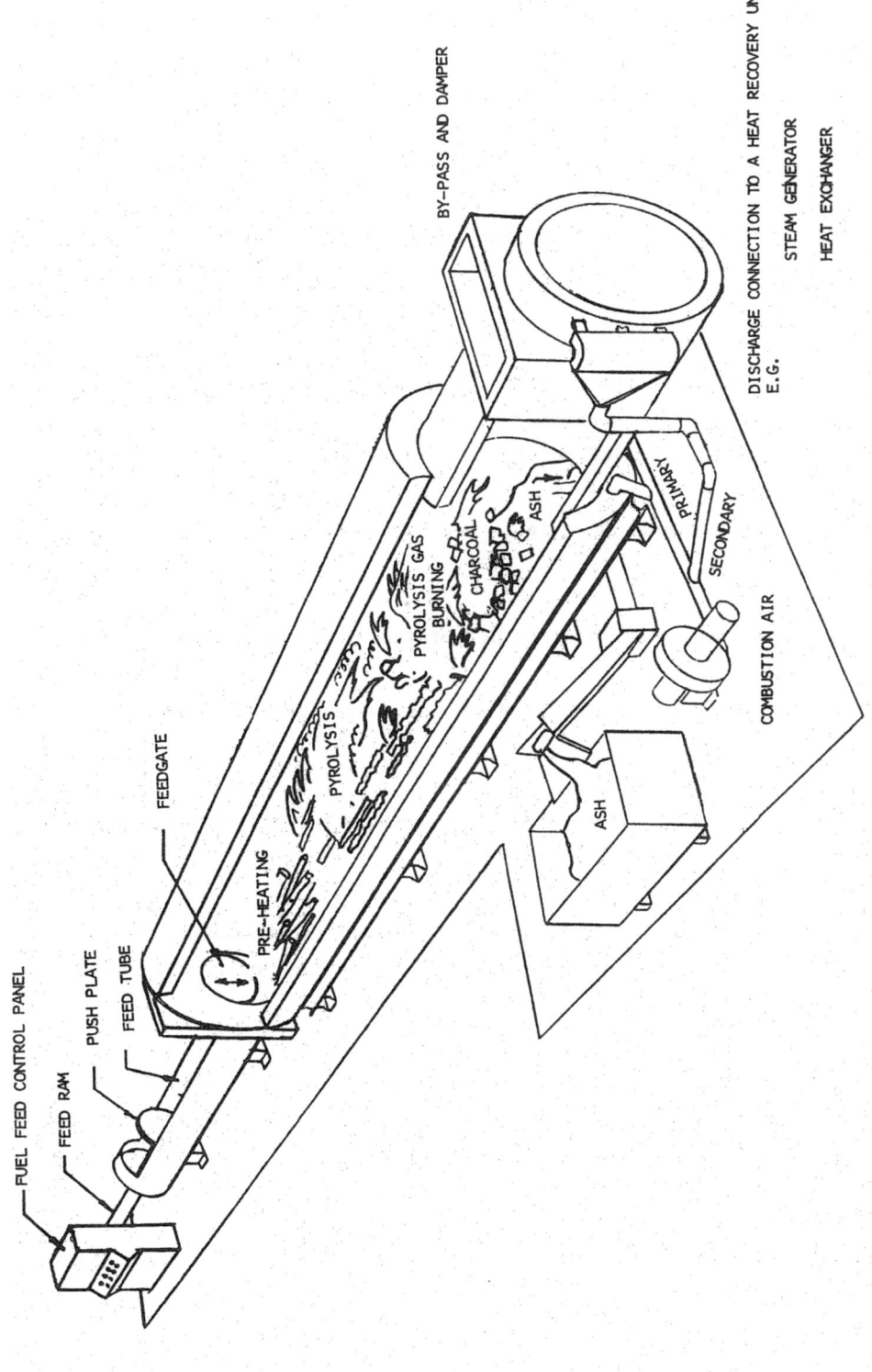

Figure 1. Schematic for the Inverse Pile Burner model 6032: 60" ID by 32' long—1.52 m ID x 9.76 m long. 30" Feed Port; 48" Discharge Port—0.76 m F.P.; 1.22 m DP. Refractory concrete rated at 2500°F—1371°C. Heat Release Range: 5 mm to 30 mm BTU/HR—5.27 to 31.63 million kiloJoules per hour

THE WOOD HARVESTING SYSTEM

This paper describes a system (Bioverter) that overcomes many of the difficulties by producing a variety of products using small, efficient units of production. To do this we set up a number of guidelines.

1. The entire tree or bush has to be removed and marketed.

2. Higher value species would be left for accelerated growth. In many situations this means that we would be paid to remove competing species.

3. The equipment had to be within the fiscal resources and skill levels of an owner-operator as much as possible. This means more jobs for semi-skilled workers working as independent operators.

4. The harvester must be able to harvest materials on slopes up to 35%.

5. The brush and limbs have to be unitized for hauling and storage.

6. The material should dry in the woods to reduce moisture and transportation costs.

7. The system should provide the opportunity to sort the components for the highest profit markets.

8. The installations must be semi-portable since the unprocessed material does not load or handle easily and is not readily hauled on public roads.

9. The material should be sorted and converted into highest value products that can be easily stored, handled, or transported to a central Biomass Industrial Park or to other markets.

THE REDWOOD REGION

The Redwood Region is small but the silvicultural problems get national attention. The Biovert system developed by Kleensmoke, Inc. meets very stringent environmental requirements.

Figure 2. Semi portable biovertor installation. Brush is fed into the
modified Kleensmoke Burner at the right. The char drops out and
is quenched and conveyed into the box. The Firewood Kiln is in
the background.

The Redwood (<u>Sequoia</u> <u>sempervirens</u>) grows principally with Douglas Fir
(<u>Pseudotsuga</u> <u>menziesie</u>) and Tanoak (<u>Lithocarpus</u> <u>densiflora</u>) on rugged coastal mountain ranges in northern California. Slopes range up to 60%, soils
erode easily and the average rainfall of about 55 inches comes mainly from
November to April. Tanoak sprouts from the stump and, along with other
hardwoods, aggressively take over the site. This has lead to herbicide
spraying as a method of control.

CUTTING THE MATERIAL

Gas-powered circular brush saws which cut fast but have considerable
maintenance have been used. The hydraulic chain saw and shears powered by
the Brush Bundler engine are slower-cutting but much more reliable. There
are more companies making hydraulic and air tools for the orchard and tree
service industry. Many can be made to work in cutting brush and slash.

Pieces over about 10 cm (4 in) are cut with a gas chain saw with a 75
cm (30 in) bar. It is important to note that the actual cutting takes only
10-20% of the time for the total operation of cutting and stacking brush.

367

Needed is something like compressed air or DC electricity that can create and store energy and then deliver it in short bursts.

BUNDLING THE SMALL MATERIAL

After cutting the brush and stems, they must be either processed on the spot or transported to a place where they can be converted into a useable product. Our yarding system can haul a bundle weighing about 1 megagram (2240 lb). These bundles are usually around 1.2 m (4 ft) diameter x 4 m (13 ft) long. It is difficult to find a landing area big enough for a chipper and a chip van. Standard chip vans cannot negotiate logging roads in the coastal mountain ranges. If a portable chipper is to be used, it must be set up near the main highway.

About four years ago, we began work on finding a method for compressing brush into a bundle that could be loaded, transported and stored easily. We worked with a number of hydraulic clamping and tieing systems that showed promise but were either too slow or clumsy or both.

The current Burton Brush Bundler, Model BB 900 (Figure 3) meets all of the parameters that we set out to meet. The "Bundler" is mounted on the Pac Trac 900 all-terrain vehicle which enables it to operate on slopes up to about 40%. On steeper ground, operation is off the four-foot wide road built by the Acutrac Tractor.

The Model BBT 4 fits on a standard three-point hitch on almost all farm tractors. The hydraulic power comes from the tractor system. This works well on level ground and enables the farmer or rancher to generate income in the off-season while clearing land of brush. Unfortunately, this system only works on slopes up to about 10%.

The operation is described as follows. The Burton Brush Bundler is positioned adjacent to the brush or trees being cut. The lower nylon straps are pulled out about 8 m (25 ft) and laid on the ground. Two steel straps are also pulled out and laid down. As the brush is cut into pieces that can be handled, it is piled on top of the straps making a pile about 8 m (25 ft) long, 2 m (6 ft) high and about 4 m (13 ft) wide. The upper nylon straps are now pulled out and connected to the lower straps. Hydraulic power now tightens the nylon straps with a pull of about 1600 kg (3500 lb) making a

Figure 3. The Burton Brush Bundler mounted on a Pac Trac walk behind
vehicle. Brush is piled on the straps on the ground then pulled
up to make a 1.2m (4 ft) diameter bundle.

bundle about 1.2 m (4 ft) in diameter by about 4 m (13 ft) long. The steel
straps are then connected, tightened and sealed. The nylon straps are then
released and the cycle repeated.

The bundles weigh from 700-1800 kg (1500 to 4000 lb) depending on the
type of material. Two men can produce 3/4 to 2 bundles per hour depending
on terrain and type of cover.

Notably, the cutters can always be working in a cleared area since the
bundles take about 1/4 the space of piled brush. In some cases the firewood
is cut and split from the boles at this point. Split wood dries much faster
than round wood. In other cases, everything that can be handled by hand is
put in the bundles. At this point it is best to let the bundles dry for one
to two months. In this time the small [under 18 cm (4 in)] branches with
the leaves dry to 17-25% moisture in the dry summer months and to 35-40% in
the winter due to transpiration drying (Figure 4). Larger material dries
much more slowly unless cut into short lengths and split. This on-the-spot
drying increases fuel value and cuts transportation costs. It is, of
course, possible to yard the bundles immediately.

Figure 4. Douglas Fir stand with hardwoods removed and bundled. The unclear area is in the background

YARDING AND ACCESS

The bundles cannot be skidded like logs because the steel straps catch on stumps. This difficulty was overcome by devising a self-loading trailer that was connected to the powerful little Acutrac Tractor.

The Acutrac Tractor, manufactured in Canada, is the invention of R. A. Triplett. A hydrostatic drive on this tractor delivers power to two drive sprockets in each track. The unique five-way tilt dozer blade enables the tractor to build a 1.2 m (4 ft) road through steep terrain at the average rate of about 91 m (300 ft) per hour.

This four-foot road has many advantages.

1. wide enough to haul out the bundles and logs.
2. wide enough to hike or ride horses on but too narrow for four wheelers.
3. cut bank and fill are low enough so there is access to the road almost anywhere in steep country.
4. little soil disturbance so healing is rapid.

370

5. later provides access for fire control or animal husbandry at a
    fraction of the cost of 3 m (10 ft) roads.
6. in the opinion of some authorities, the road could be used in some
    areas now closed to regular logging.

At first, we used the brush bundler to cross-haul the bundles up onto
the powered trailer behind the Acutrac. This led to installing the bundler
winches permanently on the Acutrac trailer. This combination resulted in a
system that could load a bundle weighing up to about 1800 kg (4000 lb) in
three minutes. The tracks under the trailer are powered by hydraulic motors
supplied by the tractor's system. Hence, the tracks are steered by the
operator and produce additional traction. It is impressive to see a tractor
weighing about 2400 kg (5300 lb) with the powered trailer behind taking an
1800 kg (4000 lb) load up a 32% slope. This tractor made the whole system
practical (Figure 5). The Bureau of Land Management has used the Acutrac
with powered trailer to haul hay into remote areas for erosion control.

At the roadside, the bundles are dumped on the ground or cross-hauled
onto a fifth-wheel trailer mounted on a heavy duty four-wheel drive pickup.
If the bundles and the logs are on the ground, they can be loaded on the
fifth-wheel trailer with a permanently mounted cross-haul system. Logs up
to 1 m (39 in) diameter and 4 m (13 ft) long have been loaded and hauled in
on the powered trailer. If there is room, a portable chip producing system
can be set up at the roadside and everything that is suitable can be
chipped.

THE SEMI-PORTABLE CHARCOAL PRODUCER

The bundles of brush and wood should be processed close to the woods
where they are developed since it is difficult to make bundles into a full
load for a large truck. This allows the operator to convert the low value
brush into high value charcoal, dry chips or firewood for transport to the
larger Biomass Multiuse Industrial Park.

This Biovertor System, see Figure 2, consists of a Kleensmoke Inverse
Pile Burner specially modified for producing charcoal from material up to 10
cm (4 in) diameter. To do this, the brush is fed in at one end which pushes
the burning material along a specially constructed refractory burner. There
are three holes in the bottom, allowing the burning char to drop into a

Figure 5. The Acutrac Tractor with powered self loading trailer carrying a bundle out over a 1.2 m (4 ft) road. The Redwood Forest is about 55 years old. The removal of the hardwoods releases the redwoods

conveyor which leads to a water cooled auger. Cooled exhaust from the diesel generator is piped into the water cooled auger. This combination completely quenches the char. The charcoal is then ready for sale to a briquetting company for filtration, toxic gas filtering and other uses.

This charcoal making process also produces heat at about 1140 C (2000°F). This heat can be used for a number of purposes such as drying firewood, drying chips or generating power for on-site use or for sale. In the installation shown in Figure 3, the heat is used for drying firewood. This is done in a 6.8 m (20 ft) long "oven" with a conveyor bottom. This firewood can be bundled or boxed for the supermarket trade.

It would be possible to chip the larger pieces and dry them with the heat from the charcoal process. Chips made from green material have 800 cal/kg (3000 BTU's/ lb) while chips at 15% moisture have 1800-2000 cal/kg (7-8000 BTU's/lb). These chips could be sold to large cogenerators or be used in the automatic chip fired burners. Regardless of end use, dried wood can more easily bear the cost of transportation to markets.

Most biomass-to-energy systems have been designed to generate electricity for sale to the utilities. Various laws and tax incentives have been set up to accomplish this. Many studies have shown that the minimum practical size for cogeneration alone is about 5 MW with most being built in the 8-15 MW range. In mountainous country with few good roads, the large demands for fuel in one location mean long hauls.

Willits, California is a town of 3,500 population, 150 miles north of San Francisco. In 1950, there were 22 sawmills in the valley and no other industry. Now, there are only 3 sawmills, but there are at least eight companies making computer parts, large hydraulic cylinders, toilets, automatic paint machines, and several other products. In addition, studies have shown that cold storage for the fishing and wine industries and other agricultural crops is needed. The Biomass Multiuse Industrial Park System can meet many of the energy requirements for these needs. Almost all small towns near the woods could do the same thing.

In every city the most expensive single installation and almost the biggest power user is the sewage treatment plant. There is also a desperate need to dispose of solid waste, preferably through recycling.

Almost all industries and service organizations require heating and cooling and waste disposal. Many, such as laundries, lumber dryers, tire recaps and greenhouses, require heat at different times and in different amounts. Figure 6 shows one possible configuration that could be set up around almost any sewage treatment plant or other industrial site with a variety of energy demands.

It is beyond the scope of this paper to go into detail on how this industrial park could be organized and funded. It could probably best be done as a joint public-private enterprise.

At present, Kleensmoke has all of the components described here operating, but not full time. This makes it difficult to produce reliable

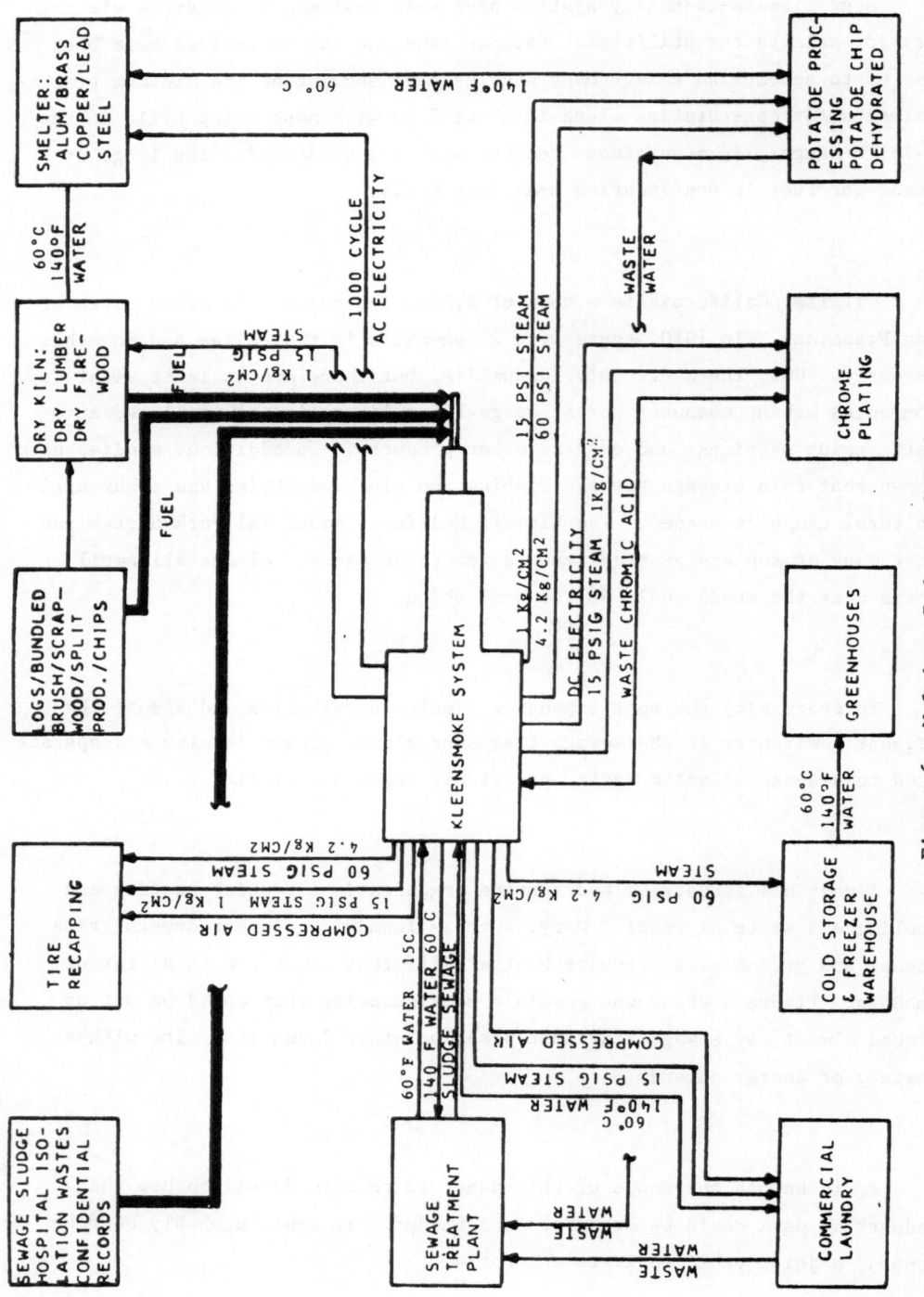

Figure 6. Multiuse Industrial Park

economic data. There are several reasons for this. We are attempting to develop higher markets such as the bundled firewood, charcoal and other products. We don't have firm prices for these. Some equipment, such as the system we use for cutting the bundles into 63 cm (16 in) lengths is in the prototype stage which means high production costs. These costs must be absorbed in the venture capital phase.

The following are the approximate current list prices for the various pieces of equipment.

Item	List Price
Burton Brush Bundler Mounted on the Pac Trac Walk Behind All-Terrain Vehicle	$ 5,000
Acutrac Tractor with Blade	30,000
Acutrac Powered Trailer with Self Loader	10,000
Kleensmoke Biovertor System Complete with Charcoal Quencher, Power Generator and Drying Oven	75,000
Bundle Cut-Off System for cutting bundles into short lengths	40,000 Est.
Bird Wood Splitter	4,000

The current costs in our area for clearing land of slash as for precommercial thinning range from $100 to $700 per hectare ($200 to $1500 per acre) with a good average for hand clearing at around $250 per hectare ($600 per acre). The actual green weight figures for material removed are around 20 Mg ha^{-1} (40 t A^{-1}).

Such variation in cost factors makes any economic analysis site specific and rather meaningless.

It is hoped that this paper will stimulate others to find a better way to create products and jobs for forest communities from locally abundant renewable resources.

WOOD GASIFICATION SYSTEMS FOR POWER GENERATION

Antonio A. Padilla

Florida Power Corporation
3201 - 34 Street South
St. Petersburg, Florida   33733

ABSTRACT

Florida Power Corporation (FPC) is an electric utility with a production capacity of 5993 MW_e, and a peak customer electrical demand of 5000 MW_e. Of the total capacity, 48 percent are produced by the use of coal, 28 percent by residual oil, 2.1 percent by coal-oil, and 18.3 percent are nuclear. In 1976, FPC started researching the potential applications of biomass to its system. Since then, FPC has studied the potential uses of wood waste in some of our smaller (100 MW and less) power plants. More concretely, FPC has undertaken demonstration projects on the use of methane derived from cow manure in a gas engine with the final goal of power production. FPC has also installed and operated a 26.4 million kilojoules per hour (25 million Btu per hour) wood waste gasifier at its Suwannee River power station located in Ellaville, Florida. The wood gasifier will be the main topic of this report and all aspects will be detailed and discussed, including gasifier design, wood handling, gas quantities and qualities, steam effects on gas quality, wood by-products, their positive and negative aspects, and the gasifier operation and maintenance requirements as well as the cost effectiveness of the project.

The wood gasifier is a fixed bed, updraft, air blown gasifier, operating at conditions of 0.21 kg cm^{-2} (3 psig) and 76.6°C (170°F). The boiler turbine combination is rated at 32 MW_e, with the gasifier being credited with 1.9 MW_e.

Keywords. Biomass, gasification, wood waste, electricity.

INTRODUCTION

FPC has a very active research and development and new technology philosophy. The corporation has progressed from a company which utilized 100 percent oil for the production of power, to a company which today, only eight years later, has substituted over half of that oil with coal and continues to investigate potential alternate fuels such as biomass. This effort led Florida Power Corporation to investigate the use of biomass gasification technology in existing older, smaller boilers. The gasification plant for woodwaste which was built for this effort, is the largest biomass project that the company has undertaken in the past 10 years. The gasification project has been a successful project for Florida Power Corporation, and the gasifier plant itself is now on a standby mode, operating when the host plant operates for long periods of time.

Gasifier Type

A large portion of time during the design phase of the gasification system was spent by FPC engineers in the selection of a gasifier from among existing commercially available gasifiers. Different types of gasifiers were evaluated, and contacts were made with several manufacturers as to the availability, operation, and cost of their units, as well as their capabilities to gasify our intended feedstock. The decision to purchase an updraft fixed bed gasifier was made from both an economical standpoint and from what we expected to be the ease of operation of a unit of that type. The closest competitor, a fluidized bed gasifier, exceeded the project budget.

Gasification System Components

The gasification plant has the typical components of a gasifier system. It consists of:

Fuel weighing station
Fuel unloading station
Fuel storage area
Conveyors
Live storage for weighing and batch feeding into the gasifier
Fuel weighing system
The gasifier
Ash collection system

Gas delivery system to the boiler

Flare system

Out of all the components and systems, most of which are quite typical
of systems utilized in other plants for conveyance of wood, receiving, and
storage, the only one unique is the gasifier itself. The gasifier, manufactured by Applied Engineering Co. (APCO) of Orangeburg, South Carolina, has
the capability of producing 26.4 million kJ hr^{-1} (25 million BTU hr^{-1}). It
is approximately 2.4 meters (8 feet) in diameter and 6.4 meters (21 feet)
high.

The ash collection system, located at the bottom of the gasifier vessel, under the air distribution grate, consists of an inverted vibrating
cone, capable of maintaining a uniform level of ash in the gasifier. The
ash level is usually kept at 5.1 to 7.6 cm (2 to 3 inches) above the grate,
to keep the grate metal cool. Thermocouples located on the grate serve as
an indication of ash level, by indicating what degree of insulation (thickness of ash) there is between the grate and the combustion zone.

A cooling jacket, located towards the lower portion of the gasifier,
surrounds the combustion area. This cooling jacket (water) prevents degradation of the brick lining, as well as of the metal shell of the gasifier.
The gasifier shell is brick lined from the top of the cooling jacket, to the
top of the gasifier.

Located inside the gasifier is a rake which serves to both maintain a
uniform fuel level in the gasifier, and to indicate the fuel level. The
hydraulic system pressure which turns the rake, also serves to indicate the
level in the gasifier. As the fuel level increases, so does the pressure
required to rotate the grate. We have found that the gasifier is at optimum
fill level when the hydraulic pressure reaches 77.33 kg cm^{-2} (1100 psig).
We also have determined that fuel should be introduced into the vessel when
hydraulic pressure falls to 28.12 kg cm^{-2} (400 psig).

The fuel is fed into the gasifier by a rotary valve. The rotary valve
is supplied from a live bottom bin located above the gasifier. This live
bottom bin has a set of two screws in series.

A very important component of the gasifier is the grate. In our case,
the grate consists of a set of four horizontal pipes, eight inch in diameter, with clinker grinders. This grate can be rotated a few degrees to

break up any potential clinkers contained in the ash bed. The grate also serves as the gasifier air distributor.

The Gasification Process

The gasification process in an updraft gasifier takes place in four different steps (see Figures 1 and 2); drying, devolatilization, reduction, and combustion. A general, simple explanation of the gasification steps can be accomplished if we follow 448 g (one pound) of wood waste as it travels down the gasifier vessel. The first zone the one pound of wood will encounter is the drying zone, and since wood is 50 percent water and the temperature in this zone ranges from 93.2 - 259.7°C (200-500°F), the 224 g (1/2 pound) of water in the wood becomes water vapor. The original 448 g (one pound) of wet wood now weighs only 224 g (1/2 pound) dry, and is ready to enter the devolatilization zone where the volatile part of the wood that usually accounts for 20 to 30 percent of its weight is released. The temperature in this zone is 259.7 - 481.7°C (500-900°F). The original pound of wet wood now weighs approximately 112 g (1/4 pound) and is mainly carbon when it enters the reduction zone, this zone reaches temperatures of over 1093°C (2000°F), where the majority of the remaining carbon is converted into a gas. After the reduction zone, a small percent of the original 448 g (one pound) of wood, now only carbon and ash, is left to the enter the combustion or burning zone located near the bottom of the gasifier, where air is admitted to achieve its combustion. This combustion provides the heat to carry out the gasification steps in the drying, devolatilization, and reduction zones. After the combustion zone, the only remainder of the original 448 g (one pound) of wood is ash which is removed from the bottom of the gasifier.

Gas Quality

The gas produced is a low energy gas containing approximately 126.6 kilojoules (120 Btu's per standard cubic foot). It leaves the gasifier typically at 76.6°C (170°F) and at a pressure of 0.14 Kg cm^{-2} (2 psig). Its wet volume composition is of 4.56 percent carbon dioxide, 7.33 percent hydrogen, 31.56 percent nitrogen, 1.36 percent methane, 20.61 percent carbon monoxide, and 34.47 percent water vapor (water and hydrocarbons).

Liquid By-Products

The wood gas is centrifuged as it is conveyed from the gasifier to the boiler. This removes moisture and heavy hydrocarbons which otherwise would drip off the end of the gas pipe/burner. Through this procedure, we achieve

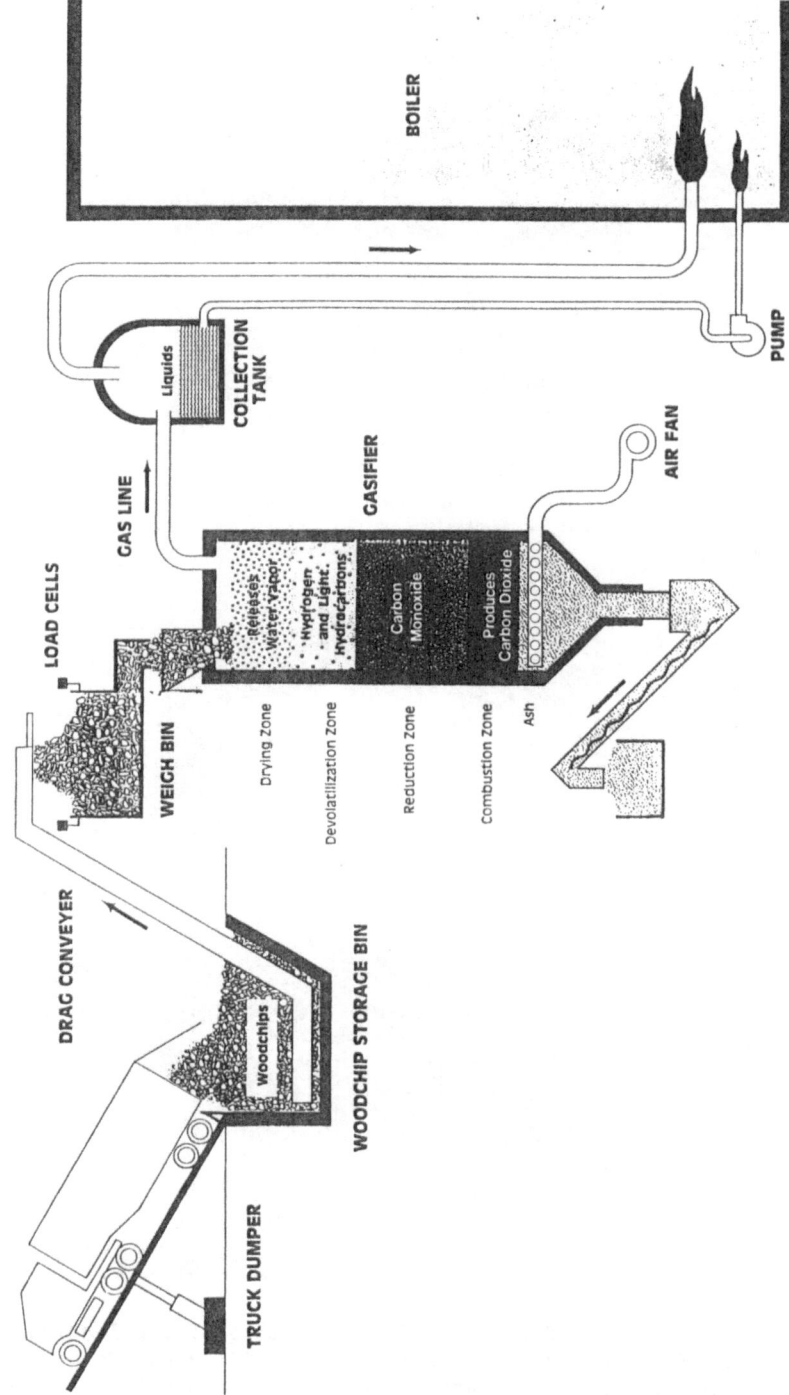

Figure 1. Schematic array of the units comprising the Florida Power Corporation wood gasifier system

Figure 2. An overall view of the updraft fixed bed gasifier leading to the boiler (inside plant), installed at Florida Power Corporation, Suwannee River power plant. Photos by Bob South (Audio-visual services, St. Petersburg)

a significant reduction of moisture in the gas. We recover liquids amount-
ing to approximately 0.68 cubic meters hr^{-1} (3 gal min^{-1}), at gasifier full
rating.

The liquid by-products of gasification are very hard to handle. Their
composition is 70 percent water and 30 percent hydrocarbons, acids, etc.
The liquids will not sustain a flame, and therefore need to be burned inside
the wood-gas flame. The liquids, when mixed with the solids entrained in
the gas, cost the operation the largest amount of downtime. The mixture of
heavy oils and solids would continually plug the burner. Of the many potential solutions we tried, only one had positive results. We purchased an
Arde-Barinco homogenizer, which made the liquid-solid mix quite easy to
handle. The homogeneous quality also improved the flame quality. By pre-
venting stratification of the liquids, we no longer encounter times when
stratified water would be the only liquid going through the burner.

Operation

The gasifier system requires full time attention, even in a fully
automated system such as ours. We used personnel from the host plant to
operate the gasifier. The lead operators were experienced power plant
operators, and their training was relatively easy.

The biggest operational problems we encountered were the liquid handling, and the problem was caused by oversized wood waste. The liquids
problem was resolved (solution outlined elsewhere in the liquid by-products
section of this report), and the oversized wood waste is a problem that may
only be resolved by either obtaining the cooperation of the fuel supplier,
or by installing screens to classify the fuel. We opted for obtaining a
good, qualified fuel supplier, and resolved our problem that way.

A problem with our gasifier, which persists until this day, has been
one of gas leakage through the clearances of the rotary valve and the rake
shaft. A resolution to this problem was attempted by the gasifier manufacturer with no success. Short of changing the feed method to the gasifier,
we believe that this is a problem that we will have to endure.

Fuel Supply

The basic specifications of our fuel were wood waste chips, with no more than 55 percent moisture, less than 6 percent ash, and with individual chip sizes less than 7.6 x 1.3 cm (3" x 1/2").

Florida has no real infrastructure to supply fuelwood for energy to users other than large paper mills. The broker we originally contracted as a fuel supplier proved to have very little control over the quality of the fuel. Consequently, we were receiving pieces of lumber mixed in with our wood chips. Large pieces of wood caused bridging at the underground storage, and many oversized pieces had to be manually removed from the conveyor (Figure 3) before they reached the rotary valve feeding the gasifier. The last fuel supplier had his own crews collecting our fuel, and a marked improvment was achieved in fuel quality.

We gasified fuels other than wood chips in the gasifier, such as pelletized peanut hulls, and were going to gasify pelletized wood. The pelletized peanut hulls cost was very competitive, and had we been successful at gasifying, we would have used them exclusively, but we were not. Shortly after the pellets entered the gasifier vessel, the compact structure of the fuel was lost. There might have also been some additives in the pellets which caused sintering in the combustion zone. Also, since the pellets have a very low moisture content, the temperatures of the gas leaving the gasifier were high, about 315.2°C (600°F). We had to use water to cool the exiting gases. We tried a mixture of wood and pellets to minimize the temperature problem, but we were never able to get around the fact that the pellets did disintegrate, and were sintering in the combustion zone.

Fuel Cost and Gasifier Economics

The gasifier economics depends mainly on the raw fuel cost. We were able to determine that in order to break even or to operate the gasifier with a profit, we would have to: 1) be competing against oil or gas, 2) be operating the gasifier a minimum of 3,000 hrs yr^{-1}, and 3) have a raw fuel cost of under $16 ton^{-1}, and preferably in the neighborhood of $12 to $14 ton^{-1}. Our fuel costs varied anywhere from $13-25 ton^{-1}.

The plant which uses the gas produced by the gasifier is not one of the continuously-run units in Florida Power Corporation. Power is being imported from out of state through that particular area of Florida, causing a high

384

Figure 3. Conveyor for delivering chips from chip storage to the weighing bin prior to entering the gasifier

voltage situation which more or less dictates to us that the plant should not run. That situation was not, of course, foreseen when we built the gasifier at this location.

The cost of FPC's wood gasification system, including the cost of a weight station, wood handling, controls, burner, and all ancillaries was approximately $1 million.

The cost of the gas, not including capital and 0 & M costs, is approximately 2.18 per GJ ($2.30 per million Btu). The gasifier requires one full time operator per shift.

Effects of Steam Enhancement Gas Composition

There was no measured change on gas heating value due to the addition of saturated steam into the air supplied to the gasifier. The amount of hydrogen increased proportionally to the steam added to the process. As the steam to air ratio was increased from 0 to 0.3, the hydrogen in the gas increased from 11 percent to 16.2 percent. The gasifier efficiency with no steam addition was measured at 91 percent. With a steam to air ratio of 0.1 the efficiency dropped to 90 percent, for a ratio of 0.2 the efficiency dropped further to 88 percent, and at a ratio of 0.3 the efficiency equaled the 91 percent achieved with no steam addition.

Effects of Wood Gas on the Overall Boiler Performance

The unit where the wood gas was tested has a 32 MW_e capacity, but in order to maximize the wood gas effect on the operating parameters, we did all our performance testing at 12 MW_e.

The efficiency of the boiler used to combust the wood gas was tested using the ASME PTC 4.1 heat loss method. The operation of the gasifier decreased boiler efficiency by an average of 0.63 percent points. A great amount of the decrease was due to the large amount of moisture in the wood gas and the wood liquids. Losses due to moisture were 6.88 percent where wood gas was being used versus 5.93 percent measured without wood gas. Boiler gas losses were approximately 0.3 percent lower with wood gas. The composition of the boiler flue gas changed slightly with the wood gas, compared to residual #6 oil: the oxygen decreased to 9.19 percent from 10.19 percent, and the carbon dioxide increased to 9.19 percent from an 8.14 percent level. Carbon monoxide levels in the flue gas were negligible.

The net heat rate of the unit increased by 384 kJ (364 Btu Kwh^{-1}), which is equivalent to a 2.3 percent increase in heat rate. This 2.3 percent increase in unit heat rate is directly attributable (at least 99 percent of it) to the 0.63 percent drop in boiler efficiency.

Summary

Florida Power Corporation has proven that a gasifier can be operated, with a certain degree of difficulty, in a steady, permanent mode. We have found that the burning of the gas is a relatively simple task once the problem of liquid entrainment and drop-out have been resolved.

The by-products of gasification (liquids, tars, alcohol, acids, entrained solids, etc.) should be homogenized before combustion. Adapting a gasifier to an existing facility to provide a portion of the total fuel requirement can be accomplished with only minor effort.

FPC has done what it set out to do, and any further steps will be dictated by the economy, the availability of reliable gasifiers, the availability of wood waste, and the cost escalation of competing fuels.

GAS CLEANING SYSTEMS FOR SMALL SCALE GASIFIERS

Robert H. Hargrave

Rocky Creek Farm Gasogens, Inc.
P. O. Box 326
LaCrosse, Florida 32658

Dale A. Lundgren

Department of Environmental Engineering Sciences
University of Florida
Gainesville, Florida 32611

Lawrance N. Shaw

Department of Agricultural Engineering
University of Florida/IFAS
Gainesville, Florida 32611

ABSTRACT

This paper describes five years of experience in building and operating small scale biomass gasifier/engine systems. At Rocky Creek we have operated gasifiers with two tractors, one pickup truck and other systems in cooperation with gasification specialists from the University of Florida who assisted in gasifier design, fuel preparation and handling, and other areas of gasifier research and development.

During 1984, Rocky Creek conducted a series of filter testing trials for small gasifiers including: 1) in-stack cascade impactor sampling of gas particulate and field testing of one primary cyclone separator, and 2) four secondary filter and particle collection systems. Fiberglass cloth filters gave the longest service time between filter servicing. Fiberglass batt filter systems gave satisfactory gas cleaning but required more service time. A new gas cleaning system was planned following this test series and constructed for testing during 1985. Results of earlier gas cleaning experience is described along with other gasification information.

Keywords. Gasifiers, thermo-chemical gasification, wood, producer gas.

INTRODUCTION

Rocky Creek Farm Gasogens, Inc. is a family owned small business formed to conduct research and development in biomass energy systems. Small scale gasifier/engine systems have been the primary area of work. The gas cleaning work done by Rocky Creek during the past five years has been directed mainly towards gas cleaning for mobile engine operation. The background for gas cleaning efforts is outlined in Chapter 5 of the book Generator Gas - The Swedish Experience from 1939-1945.[1] With the cooperation of specialists from the University of Florida and others, we have built and tested several kinds of gasifier filters and gas particulate collection systems. Our current project is testing a new mini-cyclone particle collection system for medium sized trucks, tractors and engines. This paper is a summary of gas cleaning experiences over the past five years.

EARLY EXPERIENCES

After studying the Swedish generator gas book, we built a modified cyclone particle collector, a gravity type cooler/cleaner and a fiberglass filter. During 1980 and 1981 we experienced considerable filter clogging with wet tar substances and soot and dust particles. The fiberglass batt material filter systems were tested on both a John Deere 1010, 22.4 kw (30 HP) gasoline two-row farm tractor and a 1968 Ford, 3936 cm^3 (240 in^3), 6 cylinder, pickup truck. The filters usually allowed from 10 to 15 hours service time before they appeared so clogged and matted that we changed them. On a number of occasions filters would be dripping with tarry water after a thirty-minute to two-hour run. Other gasifier/engine system problems experienced at Rocky Creek, the University of Florida and elsewhere were stuck engine valve push rods and stuck hydraulic valve lifters.

The number one cause of gasifier failure in many 1980 and 1983 American efforts was low quality fuel. Harry LaFontaine, Miami, Florida and Don Post of the University of Florida, were instrumental in establishing quality standards for gasifier fuel by monitoring gasifier performance when fed various fuels. The most important fuel standard is that gasifier fuel must be _dry_. Wet fuel, whether it is wood chips, wood blocks or any other biomass material, can cause severe filter and/or engine problems. The fuel must be below 20% moisture content; 14% to 16% moisture content is preferable. The fuel should also be as uniform and dense as practical and free from sand, trash, dust and mildew. Air dried, sifted wood chip fuel from water oak, laurel oak and sweetgum have given good operation. We have established Gas Cleaning Rule No. 1: "No gas cleaning apparatus, system or effort can make engine quality gas from wet or low quality fuel."

LATER EFFORTS

Once the fuel problem was realized and adjusted the next filter results were more satisfactory. All successful projects in Florida now include a primary cyclone particle separator through which the gas is passed when it leaves the generator.

In a filter testing project carried out by Rocky Creek Farm Gasogens, Inc. for the U. S. Department of Energy under the direction of Battelle Pacific Northwest Laboratory, four secondary filter systems were evaluated. Fiberglass batt material in two filter housings, fiberglass cloth in a baghouse and a gravity separator were tested. The fiberglass cloth baghouse was designed and built based upon recent Swedish practice in gasification for trucks and tractors. Style 1607 "Huyglas" filter fabric was used for the bags.

The fiberglass batt type filters were designed and built based upon gasification experience and by reading and visiting other gasification projects. The gravity type filter seemed to be a possibility because it had functioned well in earlier work. The batt material used was unfaced roll house attic insulation material.

TESTING RESULTS

Impactor Testing

A very useful set of results came from the in-stack cascade impactor tests. An in-stack cascade impactor is a piece of analytical equipment used to classify particles according to their aerodynamic diameter. This is accomplished by isokinetic sampling of the moving gas stream. The sampling is influenced by flow rate, temperature and viscosity of the gas stream. Preparation of the Impactor apparatus particulate collection plates and analysis can easily require one-half day of work by a skilled operator. The book, Use and Limitations of In-Stack Impartors,[2] contains further detailed information about impactors and their use.

The impactor test result summary includes the following information:

(1) Primary Cyclone: removes only the large pieces of charcoal, 1.6 mm (1/16") and removes practically none of the particles less than 5 microns.

(2) Barrell Filter: a) removes most of the particles greater than 3 micron diameter; b) removes most of the prticles less than 0.3 micron diameter; c) removes relatively little of the intermediate size particles (0.3 micron to 3 micron).

(3) Gravity Separator: a) removes very little of the greater than 5 micron diameter particles; b) removes practically none of the less than 5 micron diameter particles.

(4) Two Stage or Two Sleeve Fiberglass Filter: a) removes much of all size particles.

(5) Baghouse Filter, Fiberglass Cloth: a) removes much of the greater than 1 micron particles; b) removes much of the less than 0.3 micron particles; c) removes little of the 0.3 to 1.0 micron particles in this test.

(6) Engine: appears to burn off the small (less than 1 micron) carbon aerosol.

Carbon aerosol of less than 5 micron diameter can probably be burnt in the engine and therefore need not be removed by the filtration or gas cleaning system prior to combustion. Full data on the impactor testing is included in the final report on the Battelle administered project: (Field Testing of Small Gasification Filter Systems, Project B-F0463A-Q. Battelle Pacific NW Laboratories. Richland, WA.)

Important Questions for which answers are needed include:

(1) How much fuel value is gained or lost in gas being used to run an engine by the inclusion of less than about 5 micron diameter particles?

(2) How much particulate can safely be left in the gas stream for spark ignition engines?

(3) How much particulate can safely be left in the gas stream for diesel engines?

(4) How do the gas particulates from low value eastern U.S. hardwoods affect engine life as compared to engine life results from Swedish gasifiers tested since 1976?

(5) What is the effect of particulates on combustion by-products, oil quality, wear on rings, valves and bearings?

A project to be proposed will seek answers to questions 1, 2 and 3. Questions 4 and 5 will require a longer term of testing.

Field Testing

In the field, testing portion of the 1984 Rocky Creek filter testing project a two-row wheel type farm tractor retrofitted for producer gas operation with the Swedish/Florida design gasifier built and retrofitted by Rocky Creek was used to test the four secondary filters under actual field conditions. The filter media was changed when the pressure differential across the filter reached 12.5 cm (5 in) of water pressure. The fiberglass batt filters required changing about half way through the 25-hour field test cycle and the fiberglass baghouse reached only about 7.5 cm (3 in) of water pressure diffential after the 25 hour cycle. The gravity separator showed no pressure drop but in removing it from the tractor for other testing it was found that grape size masses of soot and dust had collected in the gas pipes going from the separator to the engine. It is very likely that engine problems would have occurred with continued use of the gravity separator without a filter of some type.

The service time for gas cleaning systems averaged approximately 15 minutes per each 7 hours of engine operation. Added to service time for ash and condensate removal the service time was about 35 minutes for 7 hours of engine time with 25 minutes required for wood chip fuel handling.

CONTINUING EFFORTS

As a result of observations and test results from this project, it is evident that improvements must be made if these small gasifiers are to reach their full potential.

The development of a milti-cyclone particle collection system is proposed for gasifier use. The multi-cyclone would be placed in the gas stream
immediately following the primary conventional cyclone. It is anticipated
that the multi-cyclone can remove dust particles greater than 3 microns in
size. The primary cyclone should remove dust particles greater than 10
microns. This project was submitted by Rocky Creek in the 1984 SBIR/DOE.
It was selected for an award and work is currently in progress.

Dr. Kay Eoff, Physics Department, University of Florida, has been
involved in development of American gasifiers since the beginning of the
work in Florida in 1978. During 1984, he directed the construction of three
small gasifier units in Honduras. These were stationary units which were
equipped with water bath scrubbers for gas cleaning. Scrubbers for gas
cleaning are shown and described in both the Swedish generator gas book[1] and
the newer book by A. Kaupp.[3] Other developments, and gasifier history is
discussed in a book published by the National Academy Press.[4]

CONCLUSIONS

Small scale (to 50KW) gasifier gas cleaning systems cannot make engine
quality gas out of low quality gas resulting from improper fuel or gasifier
design.

For small scale gasification work the Swedish/Florida design gasifier
is equal in performance to any design for which performance data are available for use in tropical and sub-tropical climates.

The in-stack cascade impactor can provide information about the size
and amount of gas particulates in determining the particle size efficiency
of gas cleaning systems.

Fiberglass cloth and batt filter materials have both given satisfactory
gas cleaning results in several systems used for mobile small-scale gasifier
engine operations. The batt type cannisters were less bulky to mount for
tractor field service but required more frequent service time than the bag
house container used for the fiberglass cloth material.

The development of low cost, convenient, safe gas cleaning systems must
be accompanied by improvements in air/gas mixing devices, automated fuel
feeding systems and other devices which make gasifier systems more "user

friendly" if small gasifier systems are to move out of the emergency technology category.

REFERENCES

1. Generator Gas: The Swedish Experience From 1939-1945, Translated From
        the 1950 Swedish Edition by the Solar Energy Research Institute,
        Golden, Colorado.
2. D. A. Lundgren and W. D. Balfour. Use and Limitations of In-Stack
        Impactors. EPA - 600/2/80-048. Research Triangle Park, N.C. 27711.
        1980.
3. A. Kaupp and J. R. Goss. State-of-the-Art Report for Small Scale (to
        50KW) Gas Producer Engine Systems, for the U. S. Agency for Inter-
        national Development. Washington, D.C. 1981.
4. Producer Gas: Another Fuel for Motor Transport, Report of an Ad Hoc
        Panel of the Advisory Committee on Technical Innovation National
        Research Council. National Academy Press. Washington, D.C. 1983.

REFERENCES

SIMPLE INEXPENSIVE GASIFIERS FOR EMERGENCY USE APPLICATIONS

Harry La Fontaine

Biomass Energy Foundation, Inc.
1995 Keystone Boulevard
Miami, Florida 33181

ABSTRACT

In the last 75 years, most producer gas-generator units were built and used when severe liquid fuel emergencies arose. Little was recorded, therefore, the next builder had to start mainly from memory. During World War II, it is estimated that 3 million wood, peat, coal and seaweed gas-generators were built and operated in Europe when liquid fuel became unavailable to the civilian population. The Biomass Energy Foundation is doing research to find the most simple, inexpensive design to make such biomass gasifiers available for farm machinery, fishing boats and trucks, to get food from the farm and sea to the table, in case of a severe fuel emergency caused by natural disaster, or, civil or other human mediated disruption. Thomas B. Reed, at the USA Solar Energy Research Institute, has after 14 years of gasification research, made a breakthrough from the long established, conventional biomass gasification by designing and testing a new concept called "Downdraft Stratified Gasification". In cooperation with Dr. Reed, the Biomass Energy Foundation has fabricated a 7½ kW, electrical generator that operates totally on biomass fuel. Our research indicates that this design could lead the way for mass-producing simple, inexpensive biomass gas-generators to be build by any mechanic, when and not IF a severe fuel emergency is forced upon our society.

Keywords. Gasifiers, small scale gasification, producer gas, wood, fuelwood, emergency power, downdraft gasifier.

INTRODUCTION

The world research community is currently working on nuclear energy, fuel cells, solar energy, magnetohydrodynamics, and geothermal power. Each of these technologies hold great promise for the future but no significant immediate relief from present potential shortages. Also, electric cars and efficient mass transit systems are, undoubtedly, part of the future. An entire generation probably will be born and grow to adulthood while we wait for these changes. Are alternatives available in the meantime? The reciprocating gasoline engine is here now, it works, and we have an enormous investment in it. The serious problem is the high price and short supply of gasoline and diesel fuels. If these engines can be run on alternative, renewable fuels, the transition into the future will be eased.

One of the available choices is the inexpensive portable gas-generator units designed to gasify virtually any solid fuel. Suitable fuels include the non-renewable fossil fuels from anthracite, through all of the bituminous coals, to lignite and peat, as well as the various cokes that can be derived from these. Further included are all of the wood species which are renewable and, therefore, can be considered inexhaustible, if properly managed. Wood derivatives, such as paper and other low-density biomass materials, are also useable with some preparatory processing. For example, the "fuel pellets" which are currently being produced experimentally from garbage could fuel gasifiers.

The potential for gas-generator units lies, not just in wheeled vehicles, but in virtually every combustion engine used. Other applications include space heating, commodity drying and various heat dependent operations although emphasis here is on mobile power systems.

EUROPEAN BEGINNINGS

Gas produced from the reduction of coal, charcoal and peat was used for heat production as early as 1840.[1] Aside from industrial heat production, the reports of using air-gas (also called producer gas) to fuel engines are found in England in about 1884. By 1890, the "suction gas" engine was developed from the English beginnings. In the suction gas engine, the air is drawn through the hearth for combustion, then through the subsequent filtering apparatus by the action of the engine's pistons. Because of simplicity, this is the basis process followed by all of the subsequent gas-generator units.

The generator is efficient enough to be constructed as self-contained, readily transported units. These suction gas engines were produced in numbers by German firms around the turn of the century. The first producer-gas engines were at least on a par with steam engines of the same relative power in terms of economy and efficiency. In addition, they could be run on small sized fuels which caused difficulties in the furnaces of steam boilers. They could also consume materials, such as plant fiber, cotton bolls, rice and wheat chaff, etc., along with their normal diet of coal and coke. However, the handier liquid fuels, quickly gained acceptance in the growing gas vehicle industry and air-gas fueled units realized little progress, despite their economy and efficiency.

In 1892, Rudolf Diesel, A German engineer born in Paris, took out an English patent for a diesel engine. The diesel engine's obvious advantages over steam doomed the heavy, stationary steam engine throughout the industrial world. The diesel similarly pushed out the gas-generator unit in its early industrial application, and it was lost from the mainstream of engineering interest in the race to improve the gasoline engine.

By the 1920's, some European manufacturers were fabricating parts from which gas-generator installations could be assembled, but nobody was mass producing the entire apparatus. Furthermore, each make of engine required its own peculiar adaption to non-liquid fuel. In 1930, gas-generator units were familiar to virtually every automotive manufacturer in Europe.[2] None considered them vital parts of their industry, but they were kept on file by many of the engineering staffs.

Europeans were always somewhat uneasy about their dependence upon foreign petroleum resources--a situation unfamiliar to oil-rich Americans. When war brought the anticipated shortages to reality, European manufacturers were able to come forth with mechanical alternatives which the populations were psychologically ready to accept.

Many different types of gas-generator units were designed.[3,4,5] Most were hand built to adapt different makes of cars or trucks to the curious business of running on coal or whatever could be found that would burn fairly well. Only two types were actually mass produced in Germany and distributed throughout the axis world.

Some American servicemen remember seeing these strange looking devices on taxicabs in Japan just after the capitulation. They were also made in

France and many American soldiers rode in Parisian taxis which were fueled by gas-generators, during the early days of the allied occupation.

Part of the data presented here was salvaged from the German occupation of Denmark.[6,7] It is amazing how little remained of this technology in view of the important role it had played in securing the Nazi's power. Such is the violent revulsion with which the world wanted to put World War II behind it.

GAS-GENERATOR OPERATION

A schematic view of a typical vehicle gas-generator is depicted in Figure 1. From a one-way inlet valve, air is drawn through the hearth zone by the start-up blower before the engine is started. After the engine is started, the air stream is maintained by suction from the manifold. Gasification occurs in the hearth zone, and the gas is sucked by the engine, through the exit pipe, from the stack into the gas cooler. Here the gas is cooled and course dust filtered off.

From the cooler, the gas is led through the filter apparatus and cleansed of remaining dust. From the filter, the gas goes to the carburetor where it is mixed with air in ratio of 1:1, arriving in the engine as a gas-air mixture. In the engine's combustion chambers, the gas-air mixture is compressed and brought to ignition by the spark plugs, exactly like the well-known gasoline-air mixture.

It is now useful to look in more detail at what transpires in the stack. First, the fuel is heated and subjected to degassing prior to gasification of the carbon. The high hearth temperature causes the fuel to be transformed into solid, liquid and gaseous products, depending on temperature and air control (the generator is closed air-tight). The proportions of these products depend on the pressure and on the temperature in the hearth zone. According to the nature of the fuel, combustion products are formed which are solids, such as charcoal and coke; liquids such as acids · and tars; and gases such as carbon dioxide, carbon monoxide, methane and other hydrocarbons.

The following reactions occur in the stack both sequentially and simultaneously. First, oxidation of the carbon:

$$C + O_2 \text{---------- } CO_2 \text{ (carbon dioxide)} \qquad (1)$$

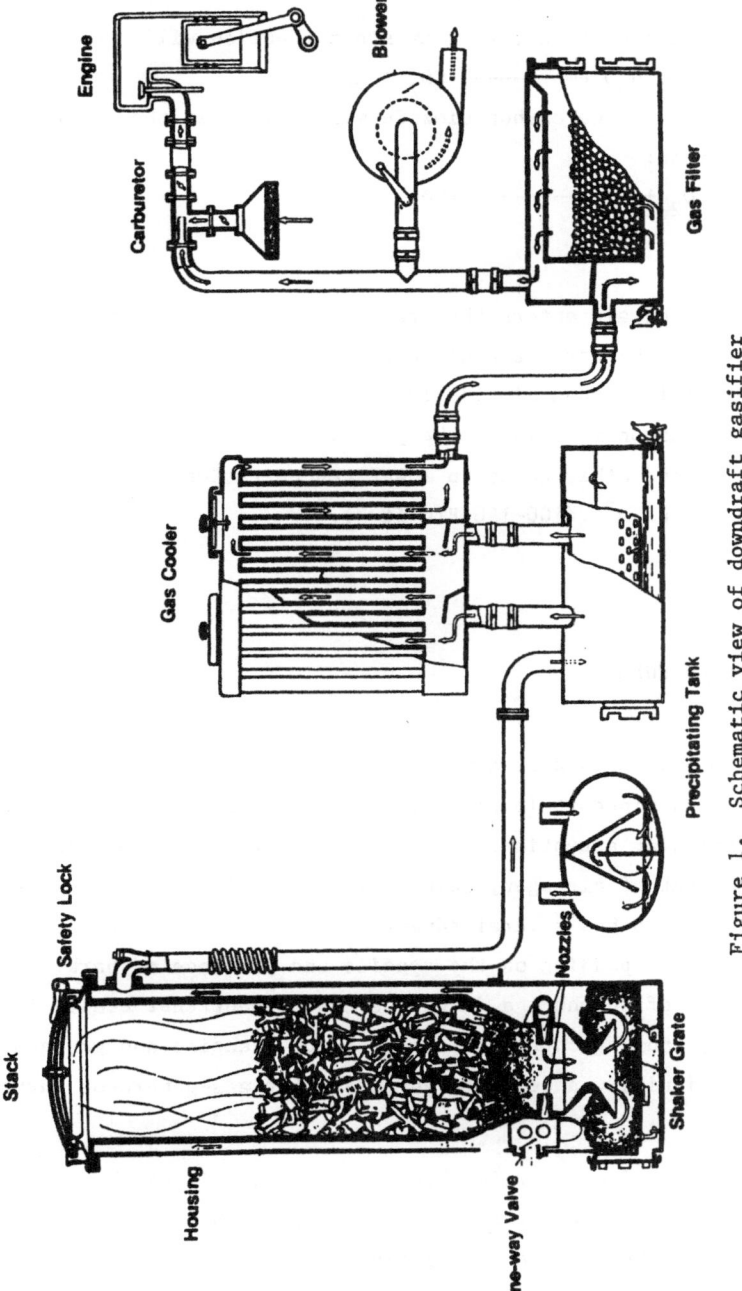

Figure 1. Schematic view of downdraft gasifier

Hereby, carbon dioxide is formed. There is also some partial combustion in the absence of sufficient oxygen:

$$C + O \ ---------- \ CO \ (carbon \ monoxide) \qquad (2)$$

At high temperatures, steam will dissociate in the presence of carbon:

$$C + H_2O \ ---------- \ CO + H_2 \qquad (3)$$

At intermediate temperatures, further steam reaction will occur:

$$CO + H_2O \ ---------- \ CO_2 + H_2 \qquad (4)$$

Further reduction of the carbon dioxide will follow in the presence of high temperature carbon:

$$CO_2 + C \ ---------- \ 2CO. \qquad (5)$$

In other words, the air enters the stack and is led to the hearth through the air nozzles. The hot charcoal near the nozzles burn to carbon dioxide. The non-combustible carbon dioxide is sucked through the hot charcoal below and reduced to carbon monoxide. The principal fuel gases are H_2 and CO diluted with N_2 and other gases in lower concentrations. The gas mix contains 100 to 150 kJ m^{-3} (100-150 BTU ft^{-3}).[8]

WOOD AS GASIFIER FUEL

Since wood was used extensively as gas-generator fuel in Europe during World War II, and since it is plentiful in many parts of the United States, it merits particular attention. Wood consists of carbon, oxygen, hydrogen, and a small amount of nitrogen, some 0.5% - 1.5%.[8,9] Viewed as a gas-generator fuel, wood has several advantages. The ash content is quite low, only 0.5% - 2.0%, depending on the species and presence of bark. Wood is essentially free of sulphur, a dangerous contaminant that easily forms sulphuric acid, causing corrosion damage to both engine and gas-generator. Wood is easily ignited, a definite advantage for gas-generator purposes.

Hardwood is decidedly preferable, but softwood is useable. The favored species in Europe was beech (Fagus), but several other species are as good, some even slightly better. A comparative overview of some common American species is displayed in Figure 2. The listed calorific values are averages and will vary a good deal within species, as a function of density which varies with age, growth rate and position in the tree. For example, a slow growing tree attains a higher density and caloric value than a fast grower.

402

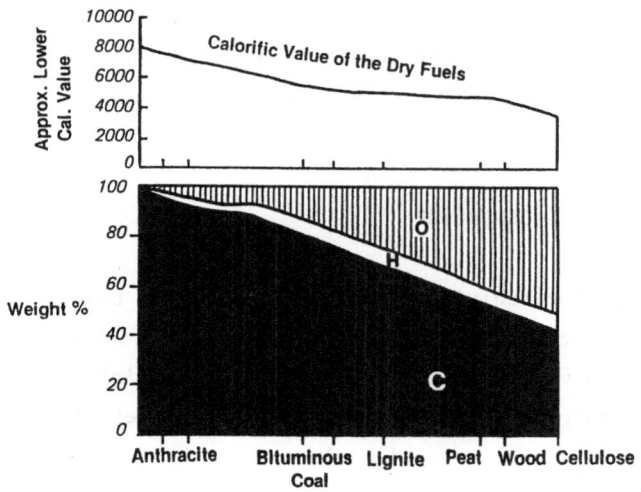

Figure 2. Fuel table – comparison of natural fuels

COAL AS GASIFIER FUEL

Bituminous coal is the most abundant fossil fuel in the United States, and gas production for cooking and heating from both lignite, soft coal, and hard coal is a known technology. Consideration of these fuels for gas-generator gas has never aroused much interest, as oil has been plentiful and seemingly inexhaustible. For the vehicle gas-generator, anthracites are the most desirable of the coals, as they have low sulphur and very high carbon contents. High density of coal makes it possible to carry three times as much anthracite as wood in the same space. One cubic meter accommodates about 800 kg anthracite. The ash content is about 4%. Comparisons of various fuels are shown in Figure 2.

Well over a million gas-generators were built and operated in Europe from 1939-1945, according to estimates.[10] Sweden alone built 75,000 units, and the people learned some significant facts. Trucks and buses fueled by gas-generator units turned out to be very efficient, especially when they were used in long distance hauling, requiring few stops. Sweden's agriculture was dependent on its tractors. In 1942, there were several thousand gasoline tractors standing idle for need of fuel. By 1945 there were 15,000 tractors working daily, thanks to the gas-generator units.

Boats have also proven to be suitable carriers for the installation of gas-generators, but the best and most efficient uses were fixed or semi-stationary installations. In sawmills, rock-crushers, pumping stations, etc., gas-generators proved invaluable. They were the key to Sweden's survival in many critical areas where power was needed.

CURRENT STATUS OF GAS-GENERATORS

America lags far behind Europe in experience or readiness to employ these devices in a crisis. Until other dramatic and radical technological devices provide us with more efficient fuels for vehicles, we need to know more about the gas-generators. Since 1979, researchers in the Biomass Energy Foundation have been looking closely at the feasibility of using organic plant materials (biomass) and designing gas-generators to operate existing diesel and gasoline engines, turbo or naturally aspirated. Producer gas technology has figured prominently in this work and, during the course of formulating and executing projects on this subject, considerable background information has been collected. It became apparent that much of this would have to be interpreted for adaptation to small-scale gasification. It was also necessary to augment this information with data obtained from experimental work.

Highlights of our findings are as follows:

There are basically three methods of generating motive power with producer gas using:

(1)   Spark ignition engines fueled with producer gas alone;

(2)   Compression diesel engines fueled with producer gas and a fossil fuel;

(3)   Gas turbines with or without recuperators.

The choice of one or the other of these methods will almost entirely depend upon the size of operation.

With a spark ignition engine, the engine must be turned over at a sufficient speed to draw enough gas through the system to start it. Alternatively, a foot-operated bellows, or a hand-driven or battery-powered fan can be used to achieve the same purpose, although the latter would entail the provision of a storage battery and charging system.

404

The high compression diesel engine will not start or run on producer gas alone, but can only operate in a pilot or dual fuel system, i.e. producer gas can only partially, not wholly, replace fossil fuels in the diesel engines.

Some diesel engines can be modified for operation on producer gas, through conversion to spark ignition, by removing the fuel injector system and replacing it with a spark plug, and reducing the compression ratio to about 10:1. On the other hand, the conversion of the selected spark ignition engine for operation on producer gas is relatively easy and can be carried out by altering the timing and replacing the carburetor, without a major change in the fundamental design of the engine. These units are generally cheaper than their diesel equivalents. It is important to note that this technology fits into the existing motor car technology already widely established in the main towns of most developing countries.

Furthermore, the spark ignition/producer gas fueled power units are easier to maintain and more suitable for operation in developing countries than dual units, where electricity is not to be supplied to the grid and running is intermittent for several hours a day.

Modern, small, high-speed engines, with small inlet valves are not suitable for operation with producer gas. Slow speed engines with large inlet valves are, in general, suitable, as the entry of the gas is less restricted and the rate of combustion of the gas mixture (air/producer gas) is very much lower than that of petrol vapour/air.

CURRENT ACTIVITIES IN BIOMASS GASIFICATION

Apparent from our literature search and personal site-visits was that most of the important research work on small-scale gas generators, after 1945, was done in Europe, Russia, and the Phillipines.[1,11,9] Highlights of some recent developments and the firms involved follow:

England

Tropical Products Institute (Berkshire), is just beginning it's research work, with interest in developing an extremely low cost method of converting crop residues into electrical power. A. J. Brockwell, Consulting Engineer, Darlington, has designed, constructed and tested a number of large-scale up-draft gas-generators, in Brazil, Kenya, Portugal, Egypt, Sudan, and India. Fuels used include cottonseed husks, sawdust, wood logs, and olive residues.

France

DuVant Motors manufactures a dual fuel engine which operates on 10% diesel fuel and 90% generator gas. They indicated that they have several installations in Africa, one on the Ivory Coast, running on gas being generated from coconut shells. Unit sizes range from 400 to 1000 kW.

The firm at Distibois, Orchamps, has an extremely interesting history. During World War I, this plant gasified wood, using an oxygen blown gas generator which supplied gas to a 1,000 kw dual fuel diesel. The gas producer was 10 meters high (30 ft) and 2.5 to 3 m (8 to 10 ft) in outside diameter. The shell was constructed of mild steel and lined with light duty firebrick. This particular gas generator operated over 30 years with no repairs.

Russia

Since World War II, the Soviet Union has researched and used biomass gasification. The latest development indicates a breakthrough in using green wood with up to 100% moisture (dry basis) as fuel for tractors and portable electrical power stations.

The ZNIIME-17 producer (Figure 3) is intended for operation with split wood, freshly cut, about 50 cm (20 in) long, from 7.0 cm to 9.1 cm (2.75 in to 3.6 in), or on round wood from 3.0 cm to 9.1 cm (1.2 in to 3.6 in) in diameter. The upper bunker (Figure 3) is of rectangular cross section, with sides of 0.55 m x 0.38 m (21.5 in x 15.0 in). The total capacity of the upper and lower bunkers is 0.24 m^3 (8.8 cubic feet); this suffices for the engine's continuous operation at its maximum capacity for an hour, without additional fueling of the producer.

In the gasification zone, the air is used for fuel gasification and for partial combustion of wood fuel. The heat produced in wood combustion facilitates wood drying in the bunker and water evaporation, while the steam goes out through the escape pipe (k), equipped with a damper (L), in the upper part of the bunker. In the upper part of the fuel chamber, is a removable cast iron diaphram (M) with an opening of 15.7 cm (6.2 in). The diaphram facilitates decomposition of tar and other products of dry distillation which, along with producer gas, get into the gasification zone. The gas passes over the grate between the wall of the inner bunker and the producer shell and is sucked in by the engine through the pipe (N).

At the same time the producer gas cools, the walls of the bunker are heated, thereby eliminating the condensation of the products of dry distillation. Because of this, the inner walls of the bunker stay free of tar and the fuelwood does not stick to them, so that the wood settles down uniformly into the gasification zone.

Such a design is far ahead of any wood-gasifier known in the "Free World", and the technology of operating on green wood with over 100% moisture (dry basis) could increase the production of the hydrogen component of the producer gas, which alone merits further research and verification.

Sweden

The National Swedish Testing Institute has been active in generator gas design and testing since approximately 1957. This extensive and long-term research program was initiated after the Suez crisis, when Sweden, again, realized that they were totally dependent on foreign oil. In fact, during World War II, Sweden had sustained its commercial, as well as agricultural, vehicles with wood gas generators. This research work is supported by the Swedish Department of Economic Defense, and as a result, very little of their research results have been published.

Philippines

In the Philippines, the use of producer gas was pioneered by Professor Anastacio Teodoro of the Philippine Bureau of Standards. Their efforts led to the shift to producer gas during the Japanese occupation when gasoline for motor vehicles became scarce. However, when oil became cheap and plentiful in the post-war period, the technology was shelved.

In 1967, gasifier technology was revived with experiments on various agricultural waste at the University of the Philippines, College of Engineering. The National Science Development Board and Ministry of Energy piloted a gas producer project in Barrio Wawa, Siniloan, Laguna, to deter-

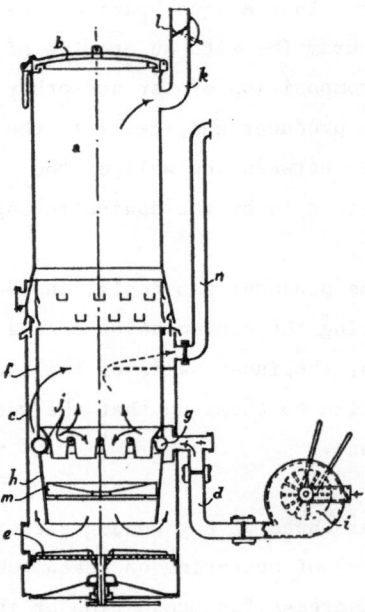

Gas producer ZNIIME-17, operating on green fuelwood, for powering skidding
tractor KT-12. a—Upper bunker; b— cover; c—case; d—air pipe; e—sectional grate; f—lower
bunker; g—air main; h—fuel chamber; i—fan; j—short pipes from air main to gasification
zone; k—steam escape pipe; l—damper; m—diaphragm; n—gas pipe to engine.

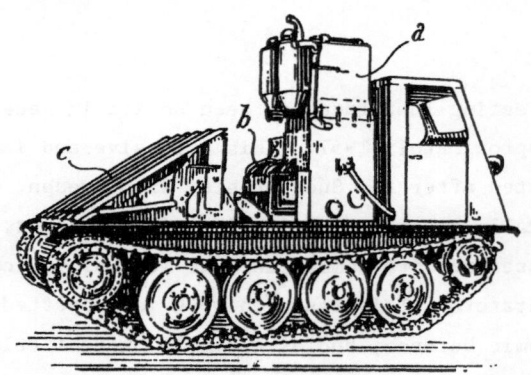

KT-12 tractor for skidding semi-suspended timber, operated on wood (producer gas),
a—Gas producer; b—winch; c—steel plate for guiding and supporting load, in
raised position (34). (Approx.)

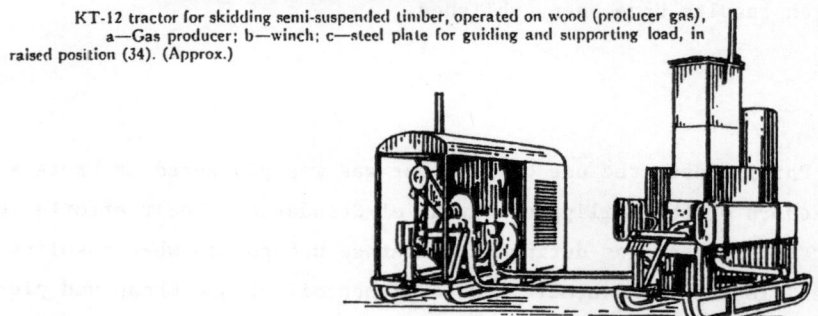

Portable electric power station PES-12 for operating motor saws and other small
logging machinery, with ZNIIME-18 gas producer, using green fuelwood of 1.2-in. – 3.5-in. ×
20-in. (34). (Approx.)

Figure 3.   ZNIME gas producer: KT 12 tractor; PES-12 portable electric
power station. Source: Logging mechanization in the USSR,
Chapter IV: Gas Producers and Electric Power Stations. Pulp and
Paper Research Institute of Canada. 1952.

mine the applicability of producer gas as diesel supplement for an irrigation system.

In 1979, to extend further the use of producer gas to other applications, particularly for rural electrification, a two-year project of the Board of Energy Development, National Science Development Board, and the Energy Development Foundation was inaugurated in Sitia Kalayaan, Laguna. The project provides 30 kW of electric power to the community, using producer gas from coconut husk waste. The installation provides a backstop to the regular lighting power source of the upland village.

In late 1980, the President and First Lady directed the Farm Systems Development Corporation, in cooperation with the University of the Philippines Engineering Research and Development Foundation, Inc., and the Project Station Barbara, to look into the applicability of gasifiers for diesel and gasoline driven vehicles.

The recently concluded "Alay sa Tao Development Caravan" and the Sampaquita Gasifier Rally marked the culmination of this endeavor. In the caravan, gasifier vehicles traversed the roads of Luxon, Visayas and Mindanao for one month, delivering goods and services to people in the rural areas. The gruelling rally, a sporting event which covered approximately 2,500 kilometers of road stretching from Tacloban, Leyte, to Tuguegarao, Cagayan, within six days, also served to test the endurance of gasifier- equipped vehicles. Using gasifiers, speed of 80 kilometers per hour was easily achieved on flat roads and the vehicles were able to climb steep grade roads without difficulty. Findings showed that two kg of dried wood chips (at half the cost of one kg of charcoal) can be used as a substitute for one liter of petroleum fuels. Driven by the most experienced rally drivers in the Philippines, over varying road conditions involving speed and endurance, the gasifier fitted vehicles reaffirmed their practicality as one answer to the country's ever worsening energy problem.

United States of America

The research and development of portable or stationary low energy producer gas units in the United States are far behind the rest of the World. R and D on large-scale, stationary gasification units are well financed in the United States and proceeding ahead of other countries.[12] US government agencies may still be suffering from the "gasohol intoxification

and the costly synfuel syndrome," thereby showing very little interest in
how to transport and produce food to sustain our people in case of a severe
emergency, caused by civil upheaval, war or natural disaster.

The same situation existed in Europe, 1935-1939, when the scientific
communities repeatedly informed their government of the need for emergency
power systems. They pointed out that when the Nazi leader in Germany got
going a starvation problem for their people was unavoidable with farm machinery, tractors, fishing boats, and the total transportation unable to
operate, with all diesel fuel and gasoline confiscated for the war effort.

Many thousands of civilians died of starvation, but could have survived
if research and development had made a small, inexpensive gasifier technology readily available before the war started. After loss of fuels, it took
nearly 12-14 months before power systems were available to move food from
the farms and sea to the consumers.

Many individuals and USA firms have, since the fuel crisis in 1978,
given up their project on low energy gas generators because of the lack of
interest and funds from responsible USA agencies, and the bureaucratic
delays in any dealing with them. In spite of such disinterest by government
agencies, some private or semi-private organizations and individuals and a
few academic institutions have tried to bring the United States up to par
with the rest of the world on this technology.[13,14]

One of the recent breakthroughs in low energy gas generation has been
formulated by Dr. Thomas Reed, a Senior Scientist with the Solar Energy
Research Institute (SERI) in Golden, Colorado. Dr. Reed has presented a new
concept in simple, inexpensive gasification, published as "Stratified Gasification."[15] The Biomass Energy Foundation has researched this approach for
advancing small-scale gasification technology.

STRATIFIED DOWNDRAFT GASIFICATION

After working with many low energy gas generator units all over the
world in the last 45 years, we are convinced that Dr. Reed's stratified
concept is a major breakthrough. The technical and chemical data were
revealed in detail in 1984 at several seminars around the world. Therefore,
we will only discuss our mechanical utilization of the development in this
project.

The stratified downdraft gasifier in which air enters the top of the gasifier bed and passes down through successive reaction strata is shown in Figures 4 and 5. The stratified downdraft gasifier has the advantage of simplicity and the open top permits a continuous flow of feedstock, and the measurement of temperature and gas composition at all points in the bed during operation. The gasifier is conceptually simple and easy to model mathmatically. A simple 7½ kW gasifier, built in a 19 L (5 gal) can, using these principles is shown in Figure 5.

In the last 12 months, we have built several units based on Dr. Reed's concepts, and have demonstrated a unit with a standard, mass-produced 7½ kW ONAN Generator, (110/220 ac). This unit consists of our simple, inexpensive, stratified downdraft gasificater a spark-ignition engine and the electricity generator. These have functioned well in several test-runs.

CONCLUSION

The only known solution for operating engines, when liquid fuel is not available, has in the last 100 years been simple, inexpensive, portable gas

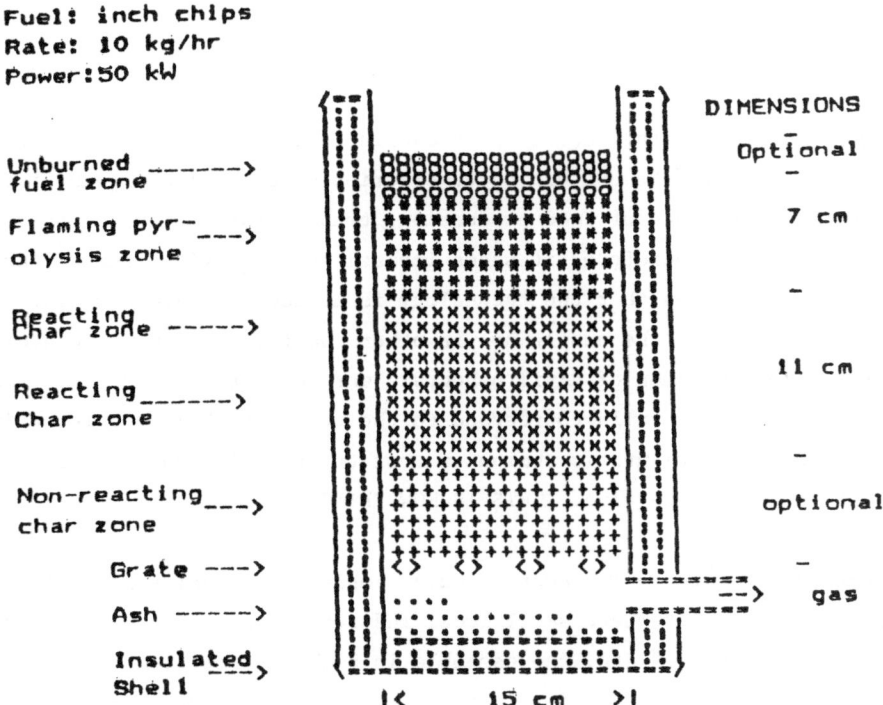

Figure 4. Dimensions of a 50 kW stratified downdraft gasifier produced from model. Source: T. B. Reed. Solar Energy Research Institute. Golden, CO.[15]

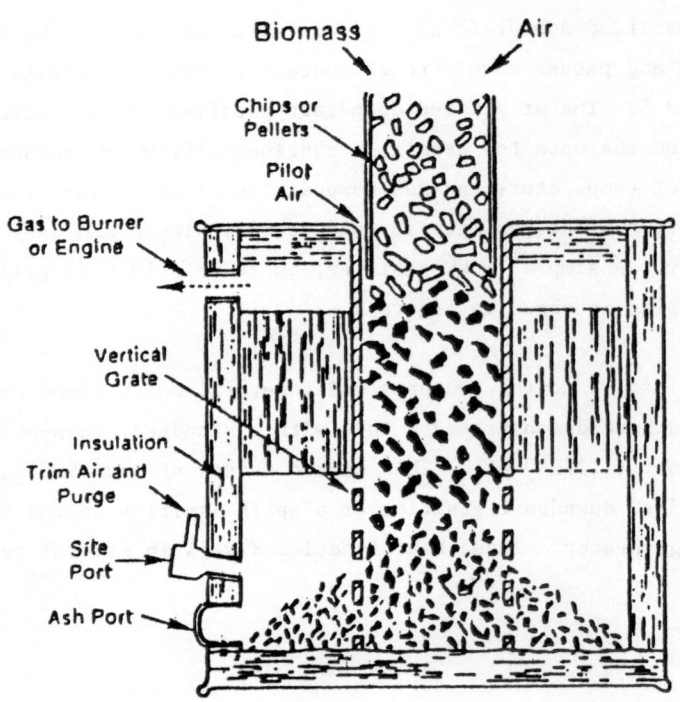

Figure 5.   Ten hp stratified downdraft air gasifier built in a 19-liter (5-gallon) solvent can.

generator units. No other solution is available or being considered any-where in the world. The USA has never experienced a total emergency fuel crisis with a breakdown of food production, transportation, and electric power production. Therefore, it seems impossible to convince the different United States agencies that the technology is needed for informing our civilian population, like people are in other countries, of ways to avoid mass starvation in times of civil upheaval, war, or natural disaster. As always, if government fails to respond to the needs of their people, private enterprise and the scientific community must try to solve the problem with their own resources and time. THIS IS NOW BEING DONE.

When the private/public sectors have simple, inexpensive gas units developed to the point where small local industry, gas stations, farm shops, "shade-tree mechanics," etc. can assemble in a short time from ordinary hardware, the gasifiers needed to get the food from the farm and sea to the consumer, the question of how to get the information to the people remains. The preparation and distribution of this information probably can best be undertaken by the respective governmental agencies.

412

REFERENCES

1. D. C. Field. Internal Combustion Engines. History of Technology. Edited by Ch. Singer, et al. Vol. V. Oxford University Press. 1958.
2. World Power Conference. Transactions. Stockholm. 1934.
3. H. Fiebelhorn. Oberinspektor an der Panzertruppenschule in Weunsdorf. Fahreug-Dieselmotoren Wid Fahrzeug Gas-Generatoren. Union Deutsche Verlagsgesellschaft Berlin. Roth & Company. 1938.
4. F. Jantsch. Zentralstelle fur Generatoren beim General-bevollmachtigten fur Rustungsaufgaben. Reichmaster Speer. Fajrzeigsgeneratoren. Joh. Kasper & Co. Berlin. 1943.
5. H. Finkbeiner. Hochleistungs Gaserzeuher Fur Fahrzeugs-Betrieb Und Kleinangetriebe. Springer. Berlin. 1937.
6. Fem Aars Kamp for Friheden. Edited by Niels Jørgensen. Alfred B'Hassings Forlag A/S. Copenhagen. 1966.
7. Motor. Copenhagen. April, 1974.
8. Chemical Engineers Handbook. McGraw-Hill. New York. 1963.
9. T. B. Reed, D. E. Jantzen. Energy from Wood. Solar Energy Research Institute. Golden, Colorado. 1975.
10. Academy of Sciences of Swedish Engineering. Gengas. Stockholm. 1950.
11. Producer Gas: Another Fuel for Motor Transport. National Academy Press, Noel Vietmeyer. Study Director.
12. G. G. Schiefelbein, D. J. Steven, M. A. Gerber. Biomass Thermochemical Converison Program. 1984 Annual Report. PNL-5442:UC-61D. Pacific Northwest Laboratory Betelle, Memorial Institute. Richland WA. 1984.
13. R. H. Hargrave, D. A. Lundgren and L. N. Shaw. Gas Cleaning Systems for Small Scale Gasifiers. In: Proceedings of the Third Southern Biomass Energy Research Conference. Ed. W. H. Smith. 1985.
14. L. N. Shaw. Gasification: Yesterday's Solution to Tomorrow's Problem. USDA Federal Extension Service Bulletin. Washington, D.C. January, 1984.
15. T. B. Reed. Stratified Gasification Solar Energy Research Institute, Golden CO. 1984.

MICROBIOLOGY AND BIOCHEMISTRY OF

METHANE FERMENTATIONS

Eugene L. Iannotti, Richard E. Mueller and James R. Fischer

Department of Agricultural Engineering
University of Missouri
Columbia, Missouri 65211

ABSTRACT

Anaerobic digestion is a biological process in which organic matter is
converted to methane and carbon dioxide. The process involves a bacterial
community with complex nutritional requirements and specialized ecological
roles. The primary substrates which supply carbon for methane production
are complex polysaccharides (hemicellulose and cellulose), lipids, protein,
short chain acids and starch. The organic matter is for the most part
insoluble and a significant portion is found in the form of three dimensional plant cell structures. The hydrolysis of these insoluble polymers is
considered rate limiting for the overall process. Bacterial entrance into
these cellular structures is gained through broken ends, natural openings
such as stomata, and disruption of the cell wall. The bacteria form microcolonies that are in close contact with the substrate. Highly refractile
tissues have very few adherent bacteria and are either sloughed off or left
intact. Intrinsic factors and organic and inorganic toxic compounds limit
the rate of digestion. Lignin concentration is most often correlated with
extent of digestion. Digestion is initiated with exoenzymes; soluble sugars
are metabolized through pyruvate to acetate, propionate, butyrate, carbon
dioxide and hydrogen. Propionate and butyrate are subsequently converted to
acetate by obligate hydrogen producing bacteria. Thus, acetate is the most
important microbial intermediate in terms of carbon flow. Methanogens
convert acetate, carbon dioxide and hydrogen to methane. Interspecies
hydrogen transfer modifies the overall fermentations such that there are
lesser quantities of reduced intermediates and a more efficient flow to
methane.

INTRODUCTION

Anaerobic digestion is a biological process in which organic matter is converted to methane and carbon dioxide. The process is carried out by a relatively large (10^9 to 10^{10} bacteria ml^{-1}) and complex bacterial community with specialized ecological roles.[1,2,3,4] Protozoa and fungi, though present in the rumen and in aquatic sediments, are only rarely found in anaerobic digesters.[5,6,7,8] Aerotolerant bacteria are normally only a small part of the total population; the actual percentage varies with the pre-treatment of the waste.[9,10,11] For the sake of simplicity, the process is usually depicted as occurring in four stages, including metabolism of: 1) complex molecules (cellulose, hemicellulose, lipids, protein, etc.) to volatile fatty acids, carbon dioxide, hydrogen, ammonia and sulfide; 2) acids with three or more carbons to acetate, carbon dioxide and hydrogen; 3) hydrogen and carbon dioxide to acetate; and 4) carbon dioxide and hydrogen or acetate to methane.[12,13,14] However, the anaerobic process is much more complex than a simple "food-chain" as implied by the listing of these four steps and involves co-metabolism, fermentation interactions and cross-feeding of nutrients. The stages are closely coupled resulting in a more efficient process.

In this review, the groups of organisms that carry out the four stages will be referred to as acidogens, acetogens, homoacetogens, and methanogens, respectively. The anaerobic process will be discussed in terms of the bacteria involved and their nutritional requirements, the role of bacterial attachment in and the biochemistry of degradation, carbon flow to methane and the interactions between bacterial groups and, finally, in terms of factors limiting the extent of digestion.

Bacterial Species Involved

Acidogens. While many bacterial species are found in a digester, only a few species are found in high concentration. The predominant organisms in a digester change with the type of substrate. The bacterial strains isolated from a swine manure digester were predominantly gram positive anaerobes and included Peptostreptococcus, Eubacterium, Bacteroides, Lactobacillus, Peptococcus, Clostridium and Streptococcus.[15] These bacteria produced acetate, hydrogen, ethanol, propionate, lactate, succinate and butyrate; acetate was the sole end product for some. Streptococci become predominate in lightly loaded digesters.[11,16] A high concentration of gram positive organisms appears to be characteristic of swine manure, swine manure handling systems and the swine manure digester.[15,17,18] Gram negative, non-spore forming anaerobic rods predominate in domestic sludge and thermophilic cattle waste digesters.[19,2,20,21,4] The predominate organism in a domestic digester is a Bacteroide that produces acetate from carbohydrates.[2] An organism similar to Bacteroides ruminicola has been found in large numbers in several anaerobic ecosystems.[22,7,15] Toerien[20] found a higher percentage of gram positive organisms in a variety of synthetic wastes; also, propionate and butyrate were produced by many of the strains. In thermophilic digesters, there is less species diversity with only rod-shaped organisms present; pleomorphic, gram-negative, anaerobic rods predominate.[4] Thirteen percent of isolates were spore formers. Ethanol, lactate, formate, acetate and hydrogen were common fermentation products.

Cellulolytic organisms isolated from digesters include Bacteroides cellulosolvens, Acetivibrio cellulolyticus, Clostridium thermocellum, C. papyrosolvens and C. stercorarium.[23,24,25,26,27,28,29,30,31] These organisms vary in the amount and type of cellulose degraded, end products utilized, formation of spores, etc. Since these organisms were isolated from low dilutions, there is still a need to confirm that the most important cellulolytic bacteria have been identified by performing isolations from high dilutions.

Acetogens. There have been very few studies of the acetogens. Syntrophobacter wolinii has the ability to oxidize propionate to acetate, hydrogen and carbon dioxide.[32] There were at least two groups of propionate-utilizing organisms in an enrichment culture as indicated by differences in Ks and Umax.[33] Syntrophomonas wolfei and other similar organisms oxidize butyrate and longer chain acids.[34,35] Due to the energetics of the reactions they carry out, these organisms have a stringent requirement that hydrogen be maintained at a low concentration by other organisms. Syntrophus buswellii is a benzoate catabolizer that has a similar requirement for low hydrogen concentration.[36] There are probably many compounds which would be undegradable if their degradation were not coupled to hydrogen utilizers.

Methanogens. In a mixed digester, methanogens comprise approximately 5% of the total population.[9,10,11] These organisms have unique characteristics, such as their cell wall, ribosomal structures, etc., that distinguish them from other bacteria in the digester.[37] The majority of methanogens utilize carbon dioxide and hydrogen for the formation of methane.[37,38] Methanobacterium formicicum, M. thermoautotrophicum, M. arboriphilus, M.

smithii, Methanobrevibacter ruminantium, and Methanospirillum hungatti have been isolated from anaerobic digesters.[23,39,40,11,41,42,43]

Some methanogens such as Methanosarcina barkeri, are also able to utilize acetate. This organism is considered the most important bacterium converting acetate to methane in digesters and is very versatile, having the ability to also metabolize hydrogen and carbon dioxide, methanol, acetate, carbon monoxide, and several methylamines.[44,45] Their distinctive cellular arrangement, composed of large masses of coccoid subunits, makes it possible to identify them visually.[46,47] These masses are reportedly bouyed by trapped gases.[48]

Recently several strains of acetate utilizing methanogens that do not utilize hydrogen and carbon dioxide have been isolated. Methanococcus mazei, Methanothrix soehngenii and Methanothrix concilii are examples of such organisms isolated from a digester.[49,50,51,52,53,54]

Methanothrix, a sheathed, filamentous organism, may be as important as Methanosarcina in the digester.[55,47] We have isolated rod shaped methanogens from our digesters using acetate enrichments (unpublished data). Methanothrix utilizes fewer substrates than Methanosarcina.

Homoacetogens. Homoacetogens are found in lower numbers than methanogens, but they still compete with methanogens for a small amount of hydrogen.[56,57] Eubacterium limosum, which also utilizes carbohydrates, amino acids, and methanol, produces acetate and butanol from carbon dioxide and methanol.[58] Butyribacterium methylotrophicum which was isolated from a sewage digester is either homoacetic, homobutyric or heteroacidic and can grow on hydrogen, monosaccharides, and lactate depending on growth conditions.[59] Clostridium thermoaceticum, C. aceticum, C. formicoaceticum, C. thermoautotrophicum, Acetobacterium woodii, Acetoanaerobium noterae, and Acetogenium kivui have been isolated from a variety of anaerobic ecosystems.[60,61,62,63,64,65,57,66,67] We have found that organisms with a homoacetogenic fermentation comprise 15% of the population;[15] however, we have not determined if acetate is produced from hydrogen and carbon dioxide.[15]

Nutritional Requirements

Most acidogens have relatively complex growth requirements.[68,7,69] While many methanogens have simple nutritional requirements, many times they are stimulated by certain organic compounds and some species have complex

418

requirements.[37,70,41,71,43] In order for these digester organisms to carry out their physiological role, these required growth factors must be present. These growth factors are either derived from the degradation of the influent solids or they are produced by other bacteria. The latter is an example of one type of interaction which occurs between the various bacterial populations in the digester. The longer retention time of an anaerobic digester facilitates more complex interactions. This fact is reflected in the more complex requirements of the microbial population.[72,7,1,67]

The most important known growth factors of anaerobes include sources of nitrogen, volatile acids, vitamins and minerals. In the rumen, ammonia is the most important source of nitrogen.[73,74] The ammonia comes from the catabolism of protein and is used to produce cellular protein. A few rumen anaerobes require pre-formed nitrogen sources. B. ruminicola requires amino acids in the form of peptides with four or more amino acids units.[75]

Ammonia is also the most important source of nitrogen for methanogens.[37,70,38,41,43] Several methanogens either require or are stimulated by other forms of nitrogen. For example, M. thermoautotrophicum and mesophilic strains of Methanobacterium can utilize glutamine, although less efficiently than ammonium.[76] Methanococcus voltae requires isoleucine and M. ruminantium is stimulated by isoleucine and leucine.[77,78] M. barkeri can use methionine as a nitrogen source.[79]

Volatile fatty acids are important nutrients for many anaerobic bacteria.[74,7,69] For example, the cellulolytic ruminococci require branched-chain fatty acids which are used to synthesize protein and long branched-chain fatty acids and aldehydes.[80,81,82] These acids are synthesized primarily from branched-chain amino acids in the feed. Hungate and Stack[82] reported that selected strains of R. albus also require phenylpropionic acid as a growth factor. Acetate is the single most important growth factor for methanogens, since up to 60% of their cell carbon is derived from acetate.[70,84,76,85] M. ruminantium has a requirement for methylbutyrate.[70]

Bacteroides are known for their requirement of hemin, vitamin K and menadione.[38,86,87] These compounds result primarily from the degradation of plant material and are important as a component of cytochromes. Succinivibrio has a requirement for naphtoquinone.[88] Bacteria from a swine manure digester were stimulated by the addition of hemin and vitamin K.[69]

Most anaerobic bacteria require vitamins, especially B-vitamins.[37,22,74,38,58,82,89,90] The vitamin requirement changes with the substrate and with the cell carbon source added to the medium.[91] Some methanogens require Coenzyme M, a unique cofactor associated only with these organisms.[37,70,92]

Very little is known about the mineral, and in particular, the trace metal requirements of anaerobes. Hydrogen sulfide is the most important inorganic source of sulfur.[68,70,82,93,94,95,96,90] Cysteine is the most important organic sulfur source.[76,68,7] A mesophyllic methanogen can also utilize methionine as a sulfur source.[76]

Inorganic forms of phosphorus can be used over a wide range of pH. B. ruminicola, B. succinogenes and B. amylophilus and several other genera have an absolute requirement for sodium.[97,98] Some bacteroides have an absolute requirement for potassium. Sodium, potassium, magnesium, nickel, cobalt, molybdenum, iron, selenium and tungsten are required by methanogens and in some cases homoacetogens.[99,100,101,102,103,79,104,105,54] The addition of nickel to a digester has stimulated the production of methane.[106]

Many rumen bacteria require low levels of carbon dioxide which are used as a cellular constituent; succinate producers use large amounts of carbon dioxide.[107] A cellulolytic sludge bacterium was inhibited by 100% hydrogen or 20% carbon dioxide in hydrogen.[89] Of course, many methane bacteria require carbon dioxide and hydrogen as substrates and may also need exogenous carbon dioxide for cell synthesis.[52]

Many digester isolates require unknown growth factors from crude or semidefined extracts for maximal growth.[39,69,108,52,4,59] We believe that at least some of these unknown growth factors arise from the activity of the microbial population and reflect complex interactions. This is partially based on the fact that solidifying agents were found to be inhibitory during initial isolation, a finding that can be best explained by proposing that the agar blocked diffusion of growth factors.[1] In addition, we have found that the spent media from the culture of a dominant strain isolated from a swine manure digester promotes the growth of many other strains, including those associated with fiber degradation. This indicates that certain growth factors and certain types of bacterial interactions are still unknown. It further substantiates that the metabolic activity which occurs in the digester is the result of specialized teams with many interactions.

Role of Bacterial Attachment in Degradation

Less than 10% of swine manure is soluble; the remainder is undigested plant structures, animal cells or insoluble substances from the animal.[109] Thus, an important aspect of the anaerobic process is conversion of these insoluble structures to soluble intermediates. The sequence of events in the degradation of three dimensional plant matter has been determined in the rumen. Physical disruption during mastication by the animal exposes tissues covered by resistant barriers, produces openings for entrance of bacteria, and increases surface area.[110] The bacteria can also enter at the stomata and spread through intercellular spaces. For example, Lachnospira colonizes spaces between parenchyma cells in legumes and metabolizes the thin pectin layer between the cells.[111] Entrance into a cell is gained by disruption of the cell wall. The bacteria proliferate on the intracellular contents forming microcolonies of single or mixed morphological types. Easily degraded structures, such as mesophyl and phloem, are digested without attachment. Moderately resistant outer bundle sheaths and epidermal tissues are broken down more slowly.[112] Highly lignified vascular and scherenchyma tissues and cuticle are either sloughed or left intact and have very few adherent bacteria.

A majority of the bacteria in the rumen are associated with particles.[113] The bacteria that initiate the degradation adhere to the plant cell wall and specialize in what they degrade.[111,7] B. succinogenes closely adheres and conforms to the shape of the substrate. Ruminococcus albus and R. flavefaciens are separated by a thick polysaccharide layer that fills crevices; the glycocalyx of R. flavefaciens is not as thick as that of R. albus.[111] These three organisms are usually associated with pits in the cell. Butrivibrio fibrisolvens adheres but is not associated with pits. The dominant organisms change with the type of plant matter degraded. Lachnospira and Ruminococcus have been found to predominately colonize legumes; R. flavefaciens, grass cells; and B. succinogenes, straw cell walls.[111]

The microscopic examination of both particles and fluids indicates a close physical association of microorganisms with different morphologies. A large number of noncellulolytic bacteria are associated with the cellulolytic organisms on particles. Microcolonies are also found in rumen fluid.[114] These are held together by an exopolysaccharide or glycocalyx with a large variation in morphology. Functionally, the glycocalyx probably protects the cell from toxins, traps nutrients, and possibly enhances interactions.

Biochemistry of Degradation

Hemicellulose and cellulose are the primary substrates in a swine manure digester.[115] The probable existence of polysaccharides other than hemicellulose and cellulose (mucopolysaccharides, etc.) was indicated by the quantities of individual sugars released by acid hydrolysis. Glucose, xylose and arabinose were the major metabolized sugar components of the polysaccharides; free sugars were found only in low concentrations. Lipids were next in importance. The changes in fatty acid patterns indicated that the bacterial lipids made up a greater percentage of the lipids in the effluent that in the influent; thus, the actual metabolism of lipids may be more important than is indicated by a simple mass balance.

Sixty-five percent or more of the components in swine manure are converted to biogas with the exception of protein, which exhibits about a 47% reduction during digestion.[115] Lignin is not degraded in the digester. Equations derived from the mass balance data in this study predicted biogas production to within 1%.

Dairy cattle wastes contain more cellulose and hemicellulose than swine manure; these components are degraded approximately 44%.[116] Domestic sludge contains high levels of lipid.[117] Plants contain high levels of cellulose and hemicellulose; tubors contain up to 35% soluble sugar. Because of the animal metabolism, animal manures are enriched for the more resistant fractions of plants.

The breakdown of complex substances is initiated with exoenzymes. The cellulases in the rumen are produced whether cellulose is present or not; however, the control mechanisms for cellulase production are not entirely clear.[118,119] There are cellulases in the rumen which are both bound to the cell and released into the medium.[120] The cellulase of R. albus is cell bound when grown in the presence of phenylpropionate and is probably attached to the capsule.[121]

After the breakdown is initiated, other acidogens with weak or no hydrolytic capabilities proliferate on the polymer fragments produced.[68,120,122] While the latter organisms are able to metabolize a

422

variety of substances, they generally have a higher affinity for one specific soluble substrate such as glucose or xylose.[123,124]

Thus, complex polysaccharides are degraded to oligosaccharides and simple sugars which are transported into the bacterial cells. For the most part, the soluble sugars are metabolized via the Embden-Meyerhof-Parnas pathway to pyruvate and nicotinamide adenine dinucleotide (NADH). Pure cultures then use one or more pathways to convert the pyruvate to acetate, propionate, butyrate, valerate, succinate, lactate, ethanol, etc., with the subsequent regeneration of NAD.[125]

As stated above, the process is not limited to carbohydrates. Lipids are degraded to long chain fatty acids, glycerol, choline, sugars, and other components. The unsaturated fatty acids are probably used as an electron sink.[115] Proteins are degraded to normal and branch chain volatile acids, phenylacetate, ammonia, etc. In addition, a wide variety of aromatic, industrial and halogenated compounds have been found to be degraded under anaerobic conditions.[126,127,128,129,130,131,132,133,134,35,135,136,137] Even chloroform, which has been used to inhibit methane production, can be degraded if added in low concentrations.[138,139,140]

Carbon Flow to Methane

For the most part, volatile acids are not found in high concentrations in the digester, with the acids being converted to methane as rapidly as they are formed.[117,141,142,143] There is a peak in the concentration of volatile acids shortly after loading in a digester loaded once per day.

Acetate is the most important intermediate in the flow of carbon to methane. Depending on the time after loading, 60 to 80% of the methane comes from acetate.[144,145,146,143,142,147,148] Propionate accounts for approximately 15% of the methane produced. Approximately 8% of the methane is derived from butyrate. A small quantity of acetate is converted to carbon dioxide because of exchange reactions, indirect metabolism to methane or a terminal electron acceptor other than methane.[149] A similar amount of acetate is incorporated into bacterial cells.[141]

Bacterial Interactions

Hydrogen is involved in all of the principal reactions and has a dominant influence on the overall process. Interactions involving interspecies hydrogen transfer result in increased substrate utilization, shifts in the fermentation, and a higher yield of microbial cells.[150,92,151,152]

Hydrogen is produced by acidogens in the cleavage of pyruvate to acetate. Hydrogen can also be formed from NADH by acidogens. This normally thermodynamically unfavorable reaction is made possible because the partial pressure of hydrogen is maintained at a low level by methanogens and homoacetogens. As a result, NADH is not used in the synthesis of such end products as ethanol, lactate, propionate and succinate. Thus, the interaction in the total population results in lesser quantities or the absence of these reduced products and increases the proportion of acetate.[122,151] Furthermore, since more adenosine triphosphate often results from production of acetate, more microbial biomass results.

Propionate and butyrate are subsequently metabolized to acetate and hydrogen.[32,12,34] Like the above conversion of hydrogen from NADH, these reactions are thermodynamically unfavorable unless hydrogen concentration is very low; moreover, the necessity for a low level of hydrogen is absolute. We know little of the importance of intermediates other than acetate, propionate and butyrate. Glucose is converted to lactate and acetate in a cattle manure digester; the lactate is then converted to propionate via the acrylate pathway.[152] However, the importance of glucose in the overall steps to methane is not known. Succinate is rapidly metabolized to propionate as in the coculture of B. succinogenes and Selenomonas ruminantium.[153,147]

Increases in hydrogen can inhibit every stage of the process, from cellulolysis to the formation of methane from acetate.[154,144,155,156,89,45] Based on the differences in the thermodynamics of each reaction, Bryant[12] has predicted the consecutive increase in propionate, butyrate, ethanol and lactate as the concentration of hydrogen is increased. We have for the most part, found this to be the case in swine manure digesters. The tenacity of the propionate to acetate, hydrogen and carbon dioxide reaction is evident by the difficulty of its maintenance in a digester with a synthetic feed.[145]

The degradation of lipids, proteins, etc., also results in low molecular weight substances that are converted to methane and involve interactions between organisms including coupling to methanogenesis. For example, choline, a lipid component, was metabolized to methane by a coculture from a digester enrichment. The choline was metabolized by a desulfovibrio to trimethylamine which was then converted in a simple food chain to methane by Methanosarcina.[157] E. limosum converts choline to dimethylethanolamine.[158]

Presumably M. barkeri converts methylamines to methane in the digester. Methanogens that grow only on methylamines have recently been isolated from other environments and could possibly be present in the digester.[159,160] Interspecies hydrogen transfer can also cause shifts in the fermentation of proteins analogous to those described for carbohydrates. In the Strickland reaction, one amino acid is a hydrogen donor while the other is an acceptor. In the digester the hydrogen acceptor can be the methanogens.[161,162]

Each of the stages in the metabolism of complex components to methane has been demonstrated with accompanying interactions in cocultures of 2 or 3 known organisms or in enrichments; the interactions involve not only coupling of fermentation steps but also nutritional exchanges.[163,154,164,165, 150,166,167,156,44,168,34,169,171,153,171,172,122] For example, a homoacetogen can bring about a shift in the fermentation of an acidogen similar to that effected by a methanogen.[169,31] Also, a combination of two organisms can grow on a mixture of cellulose and protein which neither can utilize alone. B. ruminicola can supply nitrogen sources to R. albus while receiving partially hydrolized cellulose fragments in return.[122] A succession of amino and volatile acids transfers were demonstrated with a triculture of B. amylophilus, Megasphaera elsdenii, and R. albus.[173] For the most part, the mixed cultures degraded substrates faster and more efficiently than single organisms carrying out the individual steps.

Limits to Digestion

Methanogenesis from short chain acids is normally considered the rate-limiting step in the degradation of dissolved organics, while the hydrolysis of insoluble polymers is rate-limiting for the overall process.[12,174,175, 176,156,177] For approximately 90% of the day, the rate-limiting step in a stable swine manure digester loaded once daily is the degradation of particles to acids.[109] The enzyme system which converts acetate to methane was found to be about half saturated in well-digested sludge.[176] However, we have found that for the first 10% of the day following loading, this enzyme system is saturated, and the rate-limiting step becomes the metabolism of volatile acids to methane.[109]

The digestion of the insoluble particles is limited by a variety of factors, including inherent characteristics of the molecule, structural blockage by components that are not readily degraded, and toxic components. Lignin is most often correlated with the ultimate extent of digestion; however, the best correlations result if the same plant strain is followed through a growth season.[178] The correlations become weaker with increased differences such as types or strain of plant, growth conditions, etc. Lignin, if degraded at all, is degraded very slowly under anaerobic conditions and apparently blocks degradation of other components.[179,180,181,182]

Silica, essential oils, cutin, and polyphenols are also related to the extent to which complex structures are degraded under anaerobic conditions.[178] The importance of multiple factors with respect to the ultimate extent of digestion can be made evident by examining the role of lignin. For example, plants with the same lignin content are degraded to different extents. Furthermore, treatments which do not change lignin concentration may increase the extent of digestion.[183] Many delignified celluloses exhibit long lag times and slow rates of fermentation; for example, solka floc has a lag time of 17-20 hours.[184] These studies indicate that we must increase our knowledge of the complex structure of lignin and develop assays that more accurately predict extent of breakdown. For instance, Reeves[185] recently found that lignin composition (using multiple lignin components as determined by gas chromatography following nitrobenzene oxidation) was a better predictor of cell wall and fiber digestibility than combinations of fiber variables such as acid and neutral detergent fiber. Hydration capacity and cation exchange capacity are other factors that need to be studied further, as they may affect attachment of cellulolytic organisms.[186]

The factors that affect the rate of cellulose and hemicellulose degradation include arrangement in the plant, association with other molecules, physical environment, and presence of toxic components.[178,187] A gradation from easily degraded to more resistant compounds is evident in swine manure as reflected by the rates of biogas production. There is a peak of digestive activity in the digester immediately after loading followed by a period of slower digestion and gas production.[109,143,94] The highest rates of biogas production occur within the first four hours of a 24-hour loading cycle.

Toxic compounds such as ammonia and plant phenolics may also affect the rates of degradation. Ammonia is released during catabolism of nitrogen containing components such as proteins. High ammonia concentrations reduce the growth and activity of both pure cultures (acidogens, acetogens, and methanogens) and the total population in digesters.[188,189,190,191,192] Plant phenolic acids have been shown to inhibit the cellulolytic activity of bacteria to varying degrees.[193,194,195] These phenolics, which are components of lignin or intermediates in the metabolism of protein and other compounds, can be detoxified and degraded to methane.[129,196,197,198,199]

Chemical, physical and biological assays have been developed in an effort to predict the rate and extent of digestion _in vivo_. The biological assays have included direct batch determinations, measurements using enzymes, etc. No one method is available that will consistently and accurately predict _in vivo_ digestibility.[200] In depth studies including a combination of anatomy, histology, chemistry, microbial attachment and analysis of anti-quality components are necessary to adequately predict the rate and extent of digestion.

The extent of digestion can in some cases be increased by physical and chemical treatment. Dehority and Johnson[201] found that fine grinding of forages increased digestibility, and the benefits of grinding increased with plant maturity. The extent of degradation of material which had previously undergone prolonged digestion was also increased by grinding. The benefits of grinding are dependent upon the type of organic material; e.g., particle size did not affect the rate of digestion of orchardgrass; however, it almost doubled the rate of digestion for alfalfa.[202] The alfalfa had nearly five times as much lignin in this study. As would be expected, reduction of particle size is most beneficial when the initial particle size is large.[203]

Delignification with or without heat treatment increased the digestion of substrates containing lignocellulose in anaerobic digesters.[204,205] However, the treatment necessary for a noticeable change in a digester must be more extensive than that in the rumen. We found that sodium hydroxide pretreatment of wheat straw had little effect on the amount of biogas produced from the straw when digested with swine manure.[206]

REFERENCES

1. E. L. Iannotti, J. R. Fischer and D. M. Sievers. Media for the enumeration and isolation of bacteria from a swine waste digester. Appl. Environ. Microbiol. 36:555-556. 1982.
2. E. J. Kirsch. Studies on the enumeration and isolation of obligate anaerobic bacteria from digesting sewage sludge. Dev. Ind. Microbiol. 10:170-176. 1969.
3. R. A. Mah and C. Sussman. Microbiology of anaerobic sludge fermentation: I. Enumeration of the nonmethanogenic anaerobic bacteria. Appl. Microbiol. 16:358-361. 1968.
4. V. H. Varel. Characteristics of some fermentative bacteria from a thermophilic methane-producing fermenter. Microbiol. Ecol. 10:15-24. 1984.

5. T. Bauchop. Rumen anaerobic fungi of cattle and sheep. Appl. Environ. Microbiol. 38:148-158. 1979.

6. T. M. Fenchel and B. B. Jorgensen. Dextritus food chains of aquatic ecosystems: the role of bacteria. In: M. Alexander (ed.), Advanc. Microbiol. Ecol. Vol. 1. Plenum Press, NY. 1977.

7. R. E. Hungate. The Rumen and Its Microbes. Academic Press, Inc., New York, NY. 1966.

8. P. L. McCarty. Anaerobic waste treatment fundamentals: I. Chemistry and microbiology. Public Works 95:107-112. 1964.

9. J. R. Fischer, E. L. Iannotti, J. H. Porter and A. Garcia. Producing methane gas from swine manure in a pilot-size digester. Trans. of ASAE 22:370-374. 1979.

10. J. R. Fischer, N. F. Meador, C. D. Fulhage, D. M. Sievers and E. L. Iannotti. Design and operation of a full-size anaerobic digester for swine waste. Trans. of ASAE 22:1129-1136, 1144. 1979.

11. P. N. Hobson and B. G. Shaw. The bacterial population of piggery waste anaerobic digesters. Water Research 8:507-516. 1974.

12. M. P. Bryant. Microbial methane production - theoretic aspects. J. Anim. Sci. 48:193-201. 1979.

13. E. J. Kirsch and R. M. Sykes. Anaerobic digestion in biological waste treatment. Progr. Ind. Microbiol. 9:155-176. 1971.

14. R. S. Wolfe and I. J. Higgins. Microbial biochemistry of methane - a study in contrast. Part I. Methanogenesis. p. 267-300. In: I. J. R. Quayle (ed.), Microbial Biochemistry, Vol. 21. University Park Press. Baltimore, MD. 1979.

15. E. L. Iannotti, J. R. Fischer and D. M. Sievers. Characterization of bacteria from a swine manure digester. Appl. Environ. Microbiol. 43:136-143. 1982.

16. A. Ueki, E. Miyagawa, H. Minato, R. Azuma and T. Suto. Enumeration and isolation of anaerobic bacteria in sewage digester fluids. J. Gen. Appl. Microbiol. 24:317-332. 1978.

17. J. P. Salanitro, I. G. Blake and P. A. Muirhead. Isolation and identification of fecal bacteria from adult swine. Appl. Environ. Microbiol. 33:79-84. 1977.

18. S. F. Spoelstra. Enumeration and isolation of anaerobic microbiota of piggery wastes. Appl. Environ. Microbiol. 35:841-846. 1978.

19. D. P. Chynoweth and R. A. Mah. Bacterial populations and end products during anaerobic sludge fermentation of glucose. J. Water Pollut. Control Fed. 49:405-412. 1977.

20. D. F. Toerien. Population description of the non-methanogenic phase of anaerobic digestion: I. Isolation, characterization and identification of numerically important bacteria. Water Res. 4:129-148. 1970.

21. D. F. Toerien and W. H. J. Hattingh. Anaerobic digestion: I. The microbiology of anaerobic digestion. Water Res. 3:385-416. 1969.

22. M. P. Bryant. The microbiology of anaerobic degradation and methanogenesis with special reference to sewage. p. 107-117. In: H. G. Schlegel and J. Barnea (ed.), Microbial Energy Conversion. Pergamon Press, New York, NY. 1977.

23. J. C. Converse, R. E. Graves and G. W. Evans. Anaerobic degradation of dairy manure under mesophilic and thermophilic temperatures. Trans. of ASAE 20:336-340. 1977.

24. A. W. Khan, E. Meek, L. C. Sowden and J. R. Colvin. Emendation of the genus Acetivibrio and description of Acetivibrio cellulosolvens sp. nov., a nonmotile cellulolytic mesophile. Int. J. Syst. Bacteriol. 34:419-422. 1984.

25. R. H. Madden. Isolation and characterization of Clostridium stercorarium sp. nov., a cellulolytic thermophile. Int. J. Syst. Bacteriol. 33:837-840. 1983.

26. R. H. Madden, M. J. Bryder and N. J. Poole. Isolation and characterization of an anaerobic, cellulolytic bacterium, Clostridium papyrosolvens sp. nov. Int. J. Syst. Bacteriol. 32:8791. 1982.

27. W. D. Murray, L. C. Sowden and J. R. Colvin. Bacteroides cellulosolvens sp. nov., a cellulolytic species from sewage sludge. Int. J. Syst. Bacteriol. 34:185-187. 1984.

28. T. K. Ng, P. J. Weimer and J. G. Zeikus. Cellulolytic and physiological properties of Clostridium thermocellum. Arch. Microbiol. 114:1-7. 1977.

29. G. B. Patel, A. W. Khan, B. J. Agnew and J. R. Colvin. Isolation and characterization of an anaerobic, cellulolytic microorganism, Acetivibrio cellulolyticus gen. nov., sp. nov. Int. J. Syst. Bacteriol. 30:179-185. 1980.

30. R. Sleat, R. A. Mah and R. Robinson. Isolation and characterization of an anaerobic, cellulolytic bacterium, Clostridium cellulovorans sp. nov. Appl. Environ. Microbiol. 48:88-93. 1984.

31. P. J. Weimer and J. G. Zeikus. Fermentation of cellulose and cellobiose by Clostridium thermocellum in the absence and presence of Methanobacterium thermoautotrophicum. Appl. Environ. Microbiol. 33:289-297. 1977.

32. D. R. Boone and M. P. Bryant. Propionate-degrading bacterium Syntrophobacter wolinii sp. nov. gen. nov., from methanogenic ecosystems. Appl. Environ. Microbiol. 40:626-632. 1980.

33. R. H. Hayes and R. J. Hall. Kinetics of two subgroups of propionate-using organisms in anaerobic digestion. Appl. Environ. Microbiol. 46:710-715. 1983.

34. M. J. McInerney, M. P. Bryant, R. B. Hespell and J. W. Costerton. Syntrophomonas wolfei gen. nov. sp. nov., an anaerobic syntrophic fatty acid-oxidizing bacterium. Appl. Environ. Microbiol. 41:1029-1039. 1981.

35. D. R. Shelton and J. M. Tiedje. Isolation and partial characterization of bacteria in an anaerobic consortium that mineralizes 3-chlorobenzoic acid. Appl. Environ. Microbiol. 47:840-848. 1984.

36. D. O. Mountfort, W. J. Brulla, L. R. Krumholz and M. P. Bryant. Syntrophus buswellii gen. nov., sp. nov.: a benzoate catabolizer from methanogenic ecosystems. Int. J. Syst. Bacteriol. 34:216-217. 1984.

37. W. E. Balch, G. E. Fox, L. J. Magun, C. R. Woese and R. S. Wolfe. Methanogens: reevaluation of a unique biological group. Microbiol. Rev. 43:260-296. 1979.

38. R. E. Buchanan and N. E. Gibbons. Bergey's Manual of Determinative Bacteriology. 8th ed. The William and Wilkins Co., Baltimore, MD. 1974.

39. T. J. Ferguson and R. A. Mah. Isolation and characterization of an H_2 oxidizing thermophilic methanogen. Appl. Environ. Microbiol. 45:264-265. 1983.

40. J. G. Ferry, P. H. Smith and R. S. Wolfe. Methanospirillum, a new genus of methanogenic bacteria, and characterization of Methanospirillum hungatii sp. nov. Int. J. Syst. Bacteriol. 24:465-469. 1974.

41. R. L. Mylroie and R. E. Hungate. Experiments on the methane in sludge. Can. J. Microbiol. 1:55-64. 1954.

42. P. H. Smith. The microbial ecology of sludge methanogenesis. Develop. Ind. Microbiol. 8:156-161.

43. J. G. Zeikus and R. S. Wolfe. Methanobacterium thermoautotrophicus sp. n., an anaerobic autotrophic, extreme thermophile. J. Bacteriol. 109:707-713. 1972.

44. R. A. Mah, M. R. Smith and L. Baresi. Studies on an acetate-fermenting strain of Methanosarcina. Appl. Environ. Mirobiol. 35:1174-1184. 1978.

45. S. H. Zinder and R. A. Mah. Isolation and characterization of a thermophilic strain of Methanosarcina unable to use H_2-CO_2 for methanogenesis. Appl. Environ. Microbiol. 38:996–1008. 1979.

46. M. Harvey, C. W. Forsberg, T. J. Beveridge, J. Pos and J. R. Ogilvie. Methanogenic activity and structural characteristics of the microbial biofilm on a needle-punched polyester support. Appl. Environ. Microbiol. 48:633–638. 1984.

47. R. W. Robinson, D. E. Akin, R. A. Nordstedt, M. V. Thomas and H. C. Aldrich. Light and electron microscopic examinations of methane-producing biofilms from anaerobic fixed-bed reactors. Appl. Environ. Microbiol. 48:127–136. 1984.

48. T. N. Zhilina. Biotypes of Methanosarcina. Mikrobiologiya 45:481–489. 1976.

49. B. A. Huser, K. Wuhrmann and A. J. B. Zehnder. Methanothrix soehngenii gen. nov. sp. nov., a new acetotrophic non-hydrogen-oxidizing methane bacterium. Arch. Microbiol. 132:1–9. 1982.

50. R. A. Mah. Isolation and characterization of Methanococcus mazei. Current Microbiol. 3:321–326. 1980.

51. R. A. Mah and D. A. Kuhn. Transfer of the type species of the genus Methanococcus to the genus Methanosarcina, naming it Methanosarcina mazei (Barker, 1936) comb. nov. et emend. and conservation of the genus Methanococcus (Approved lists, 1980) with Methanococcus vannielii (Approved lists, 1980) as the type species. Int. J. Syst. Bacteriol. 34:263–265. 1984.

52. G. B. Patel. Characterization and nutritional properties of Methanothrix concilii sp. nov., a mesophilic, aceticlastic methanogen. Can. J. Microbiol. 30:1383–1396. 1984.

53. J. P. Touzel and G. Albagnac. Isolation and characterization of Methanococcus mazei strain MC3. FEMS Microbiol. Letters 16:241–245. 1983.

54. A. J. B. Zehner, B. A. Huser, T. D. Brock and K. Wuhrman. Characterization of an acetate-decarboxylating, non-hydrogen-oxidizing methane bacterium. Arch. Microbiol. 124:1–11. 1980.

55. B. Z. Fathepure. Isolation and characterization of an aceticlastic methanogen from a biogas digester. FEMS Microbiol. Letters 19:151. 1983.

56. M. Braun, S. Schoberth and G. Gottschalk. Enumeration of bacteria forming acetate from H_2 and CO_2 in anaerobic habitats. Arch. Microbiol. 120:201–204. 1979.

57. K. Ohwaki and R. E. Hungate. Hydrogen utilization by clostridia in sewage sludge. Appl Environ. Microbiol. 33:1270–1274. 1977.

58. B. R. S. Genthner, C. L. Davis and M. P. Bryant. Features of rumen and sewage sludge strains of Eubacterium limosum: a methanol and H_2-CO_2 utilizing species. Appl. Environ. Microbiol. 42:12–19. 1981.

59. J. G. Zeikus, L. H. Lynd, T. E. Thompson, J. A. Krzycki, P. J. Weimer and P. W. Hegge. Isolation and characterization of a new, methylotrophic, acidogenic anaerobe, the Marburg Strain. Current Microbiol. 3:381–386. 1980.

60. A. D. Adames and C. T. M. Velzeboer. Features of a Clostridium, strain CV-AA1, an obligatory anaerobic bacterium producing acetic acid from methanol. Antonie Leeuwenhoek J. Microbiol. Serol. 48:305–313. 1982.

61. J. R. Andreesen, G. Gottschalk and H. G. Schlegel. Clostridium formicoaceticum nov. spec. Isolation, description and distinction from Clostridium aceticum. Arch. Microbiol. 72:154–174. 1970.

62. W. E. Balch, S. Schoberth, R. S. Tanner and R. S. Wolfe. Acetobacterium, a new genus of hydrogen-oxidizing carbon dioxide-reducing anaerobic bacterium. Int. J. Syst. Bacteriol. 27:355–361. 1977.

63. M. Braun, F. Mayer and G. Gottschalk. Clostridium aceticum (Wieringa), a microorganism producing acetic acid from molecular

hydrogen and carbon dioxide. Arch. Microbiol 128:288–293. 1981.

64. F. E. Fontaine, W. H. Peterson, E. McCoy, M. J. Johnson and G. J. Ritter. A new type of glucose fermentation by Clostridium thermoaceticum n. sp. J. Bacteriol. 43:701–715. 1942.

65. J. A. Leigh, F. Mayer and R. S. Wolfe. Acetogenium kivui, a new thermophilic hydrogen-oxidizing, acetogenic bacterium. Arch. Microbiol. 129:275–280. 1981.

66. R. Sleat, R. A. Mah and R. Robinson. Acetoanaerobium noterae gen. nov., sp. nov.: an anaerobic bacterium that forms acetate from H_2 and CO_2. Int. J. Syst. Bacteriol. 35:10–15. 1985.

67. J. Wiegel, M. Braun and G. Gottschalk. Clostridium thermoautotrophicum species novum, a thermophile producing acetate from molecular hydrogen and carbon dioxide. Current Microbiol. 5:255–260. 1981.

68. M. P. Bryant. Microbiology of the rumen, p. 287. In: M. J. Sevenson (ed.), Duke's Physiology of Domestic Animals. 9th ed., Cornell University Press. Ithaca, NY. 1977.

69. E. L. Iannotti, J. R. Fischer and D. M. Sievers. Medium for enhanced growth of bacteria from a swine manure digester. Appl. Environ. Microbiol. 43:247–249. 1982.

70. M. P. Bryant, S. F. Tzeng, I. M. Robinson and A. E. Joyner. Nutrient requirements of methanogenic bacteria. p. 33–40. In: A. G. Gould (ed.), Advances in Chemistry. Series 105. Amer. Chem. Soc. Washington, D.C. 1971.

71. G. T. Tayler and S. J. Pirt. Nutrition and factors limiting the growth of a methanogenic bacterium (Methanobacterium thermoautrophicum). Arch. Microbiol. 113:17–22. 1977.

72. C. Eller, M. R. Crabill and M. P. Bryant. Anaerobic roll tube media for nonselective enumeration and isolation of bacteria in human feces. Appl. Microbiol. 22:522–529. 1971.

73. M. J. Allison. Biosynthesis of amino acids by ruminal microorganisms. J. Anim. Sci. 29:797–807. 1969.

74. M. P. Bryant and I. M. Robinson. Some nutritional characteristics of predominate cultural ruminal bacteria. J. Bacteriol. 84:605–614. 1962.

75. K. A. Pittman and M. P. Bryant. Peptides and other nitrogen sources for growth of Bacteroides ruminicola. J. Bacteriol. 88:401–410. 1964.

76. L. Bhatnagar, M. K. Jain, J. P. Aubert and J. G. Zeikus. Comparison of assimilatory organic nitrogen, sulfur, and carbon sources for growth of Methanobacterium species. Appl. Environ. Microbiol. 48:785–790. 1984.

77. D. R. Lovely, R. C. Greening and J. G. Ferry. Rapidly growing rumen methanogenic organism that synthesizes Coenzyme M and has a high affinity for formate. Appl. Environ. Microbiol. 48:81–87. 1984.

78. W. B. Whitman, E. Ankwanda and R. S. Wolfe. Nutrition and carbon metabolism of Methanococcus voltae. J. Bacteriol. 149:852–863. 1982.

79. P. Scherer and H. Sahm. Influence of sulfur-containing compounds on the growth of Methanosarcina barkeri in a defined medium. Europ. J. Appl. Microbiol. Biotechnol. 12:28–35. 1981.

80. M. J. Allison, M. P. Bryant and R. N. Doetsch. Conversion of isovalerate to leucine and Ruminococcus flavefaciens. Arch. Biochem. Biophys. 84:246–247. 1959.

81. M. J. Allison, M. P. Bryant, I. Katz and M. Keeney. Studies on the metabolic function of branched-chain volatile fatty acids. Biosynthesis of higher branched-chained fatty acids and aldehydes. J. Bacteriol. 83:1084–1093. 1962.

82. J. Herbeck and M. Bryant. Nutritional features of the intestinal anaerobe Ruminococcus bromii. Appl. Microbiol. 28:1018–1022. 1974.

83. R. E. Hungate and R. J. Stack. Phenylpropanoic acid: growth factor for Ruminococcus albus. Appl. Environ. Microbiol. 44:79–83. 1982.

84. L. Daniels and J. G. Zeikus. One-carbon metabolism in methanogenic bacteria: analysis of short-term fixation products of $^{14}CO_2$ and $^{14}CH_3OH$ incorporated into whole cells. J. Bacteriol. 136:75-84. 1978.

85. G. Fuchs, E. Stupperich and R. K. Thauer. Acetate assimilation and synthesis of alanine, aspartate and glutamate by Methanobacterium thermoautotrophicum. Arch. Microbiol. 117:61-66. 1978.

86. D. R. Caldwell, D. C. White, M. P. Bryant and R. N. Doetsch. Specificity of the heme requirement for the growth of Bacteroides ruminicola. J. Bacteriol. 90:1645-1654. 1965.

87. M. Lev. The growth promoting activity of compounds of the vitamin K group and analogues for a rumen strain of Fusiformis nigrescens. J. Gen. Microbiol. 20:697-703. 1959.

88. R. Gomez-Alarcon, C. O'Dowd, J. A. Z. Leedle and M. P. Bryant. 1,4-Naphthoquinone and other nutrient requirements of Succinivibrio dextrinosolvens. Appl. Environ. Microbiol. 44:346-350. 1982.

89. J. N. Saddler and A. W. Khan. Cellulose degradation by a new isolate from sewage sludge, a member of Bacteroidaceae family. Can. J. Microbiol. 25:1427-1432. 1979.

90. V. H. Varel and M. P. Bryant. Nutritional features of Bacteroides fragilis subspecies fragilis. Appl. Microbiol. 28:251-257. 1974.

91. B. Linehan, C. C. Scheifinger and M. J. Wolin. Nutritional requirements of Selenomonas ruminantium for growth on lactate, glycerol, or glucose. Appl. Environ. Microbiol. 35:317-322. 1978.

92. R. A. Mah, D. M. Ward, L. Baresi and T. L. Glass. Biogenesis of methane. Ann. Rev. Microbiol. 31:309-341. 1977.

93. A. W. Khan and T. M. Trottier. Effect of sulfur-containing compounds on anaerobic degradation of cellulose to methane by mixed cultures obtained from sewage sludge. Appl. Environ. Microbiol. 35:1027-1034. 1978.

94. D. W. Mountfort and R. A. Asher. Effect of inorganic sulfide on the growth and metabolism of Methanosarcina barkeri strain DM. Appl. Environ. Microbiol. 37:670-675. 1979.

95. G. B. Patel, C. Breuil and B. J. Agnew. Sulfur requirements for growth of Acetivibrio cellulolyticus. Can. J. Microbiol. 38:772-777. 1982.

96. P. H. Ronnow and L. A. H. Gunnarsson. Sulfide dependent methane production and growth of a thermophilic methanogenic bacterium. Appl. Environ. Microbiol. 42:580-584. 1981.

97. D. R. Caldwell and R. F. Hudson. Sodium, an obligate requirement for predominant rumen bacteria. Appl. Microbiol. 27:549-558. 1974.

98. C. J. Smith and E. L. Iannotti. Enumeration and characterization of heterotrophic bacteria from anaerobic marine sediments. Dev. Ind. Microbiol. 25:727-739. 1984.

99. G. Diekert and M. Ritter. Nickel requirement of Acetobacterium woodii. J. Bacteriol. 151:1043-1045. 1982.

100. J. B. Jones and T. C. Stadtman. Methanococcus vannielii: culture and effects of selenium and tungsten on growth. J. Bacteriol. 130:1404-1406. 1977.

101. J. B. Jones and T. C. Stadtman. Selenium-dependent formate dehydrogenase and selenium-independent formate dehydrogenase of Methanococcus vannielii: separation of the two forms and characterization of the purified selenium-independent form. J. Biol. Chem. 256:656-663. 1981.

102. G. B. Patel, A. W. Khan and L. A. Roth. Optimum levels of sulfate and iron for the cultivation of pure cultures of methanogens in synthetic media. J. Appl. Bacteriol. 45:347-356. 1978.

103. H. J. Perski, J. Moll and R. K. Thauer. Sodium dependence of growth and methane formation in Methanobacterium thermoautotrophicum. Arch. Microbiol. 130:319-321. 1981.

104. P. Schonheit, J. Moll and R. K. Thauer. Nickel, cobalt, and molybdenum requirements for growth of Methanobacterium thermoauto-trophicum. Arch. Microbiol. 123:105-107. 1979.

105. G. D. Sprott and K. F. Jarrell. Potassium ion, sodium ion and magnesium ion content and permeability of Methanospirillum hungatii and Methanobacterium thermoautotrophicum. Can. J. Microbiol. 27:444451. 1981.

106. W. D. Murray and L. van den Berg. Effects of nickel, cobalt, and molybdenum on performance of methanogenic fixed-film reactors. Appl. Environ. Microbiol. 42:502-505. 1981.

107. B. A. Dehority. Carbon dioxide requirement of various species of rumen bacteria. J. Bacteriol. 105:70-76. 1971.

108. E. L. Iannotti, M. K. Wulfers, J. R. Fischer and D. M. Sievers. The effect of digester fluid, swine manure extract, rumen fluid and modified digester fluid on the growth of bacteria from an anaerobic digester. Dev. Ind. Microbiol. 22:565-576. 1981.

109. E. L. Iannotti, R. E. Mueller, J. R. Fischer and D. M. Sievers. Changes in a swine manure anaerobic digester with time after loading. p. 79-81. Third Annual Solar and Biomass Workshop. Atlanta, GA. 1983.

110. K. R. Pond, W. C. Ellis and D. E. Akin. Ingestive mastication and fragmentation of forages. J. Anim. Sci. 58:1567-1574. 1984.

111. K. J. Cheng, C. S. Stewart, D. Dinsdale and J. W. Costerton. Electron microscopy of bacteria involved in the digestion of plant cell walls. Anim. Feed Sci. & Tech. 10:93-120. 1984.

112. D. E. Akin. Microscopic evaluation of forage digestion by rumen microorganisms -- a review. J. Anim. Sci. 48:701-710. 1979.

113. C. W. Forsberg and K. Lam. Use of adenosine 5'triphosphate as an indicator of microbiota biomass in rumen contents. Appl. Environ. Microbiol. 33:528-537. 1977.

114. K. J. Cheng and J. W. Costerton. Adherent rumen bacteria - their role in the digestion of plant material, urea and epithelial cells, p. 227-250. In: Y. Ruckebusch and P. Thivend (ed.), Digestive Physiology and Metabolism in Ruminants. AVI Publishing Co., Inc. Westport, CT. 1980.

115. E. L. Iannotti, J. H. Porter, J. R. Fischer and D. M. Sievers. Changes in swine manure during anaerobic digestion. Dev. Ind. Microbiol. 20:519-529. 1979.

116. J. E. Wohlt, R. A. Frobish, R. E. Charron, C. L. Davis and M. P. Bryant. Chemical components and methane production from dairy wastes. Amer. Soc. Anim. Sci. Abstr. 1978.

117. D. P. Chynoweth and R. A. Mah. Volatile acid formation in sludge digestion. Advanc. Chem. Series 105:41-54. 1971.

118. P. Hiltner and B. A. Dehority. Effect of soluble carbohydrates on digestion of cellulose by pure cultures of rumen bacteria. Appl. Environ. Microbiol. 46:642-648. 1983.

119. W. R. Smith, I. Yu and R. E. Hungate. Factors affecting cellulolysis by Ruminococcus albus. J. Bacteriol. 114:729-737. 1983.

120. R. E. Hungate. The rumen microbial ecosystem. Ann. Rev. Ecol. Syst. 6:39-66. 1975.

121. R. J. Stack and R. E. Hungate. Effect of 3-phenylpropanoic acid on capsule and cellulases of Ruminococcus albus. Appl. Environ. Microbiol. 48:218-223. 1984.

122. M. J. Wolin. Interactions between the bacterial species of the rumen, p. 134-148. In: I. W. McDonald and A. C. I. Warner (ed.), Digestion and Metabolism in the Ruminant. University of New England Publishing Unit, Armidale, Australia. 1975.

123. J. B. Russell and R. L. Baldwin. Substrate preferences in rumen bacteria: evidence of catabolite regulatory mechanisms. Appl. Environ. Microbiol. 36:319-329. 1978.

124. J. B. Russell and R. L. Baldwin. Comparison of substrate affinities among several rumen bacteria: a possible determinant of rumen bacterial competition. Appl. Environ. Microbiol. 37:531-536. 1979.

125. M. J. Wolin and T. L. Miller. Carbohydrate fermentation, p. 147-165. In: D. J. Hentges (ed.), Human Intestinal Microflora in Health and Disease, Academic Press, New York, NY. 1983.

126. M. T. Balba and W. C. Evans. The methanogenic fermentation of aromatic substrates. Biochem. Soc. Trans. 5:302-304. 1977.

127. S. A. Boyd and D. R. Shelton. Anaerobic biodegradation of chlorophenols in fresh and acclimated sludge. Appl. Environ. Microbiol. 47:272-277. 1984.

128. S. A. Boyd, D. R. Shelton, D. Berry and J. M. Tiedje. Anaerobic biodegradation of phenolic compounds in digested sludge. Appl. Environ. Microbiol. 46:50-54. 1983.

129. W. Chen, K. Ohmiya, S. Shimizu and H. Kawakami. Degradation of dehydrodivanillin by anaerobic bacteria from cow rumen fluid. Appl. Environ. Microbiol. 49:211-216. 1985.

130. W. C. Evans. Biochemistry of bacterial catabolism of aromatic compounds in anaerobic environments. Nature (London) 270:17-22. 1977.

131. J. G. Ferry and R. S. Wolfe. Anaerobic degradation of benzoate to methane by a microbial consortium. Arch. Microbiol. 107:33-40. 1976.

132. D. Grbic-Galic. Anaerobic degradation of coniferyl alcohol by methanogenic consortia. Appl. Environ. Microbiol. 46:1442-1446. 1983.

133. D. O. Mountfort and M. P. Bryant. Isolation and characterization of an anaerobic syntrophic benzoate degrading bacterium from sewage sludge. Arch. Microbiol. 133:249-256. 1982.

134. P. M. Nottingham and R. E. Hungate. Methanogenic fermentation of benzoate. J. Bacteriol. 98:1170-1172. 1969.

135. E. R. Shlomi, A. Lankhorst and R. A. Prins. Methanogenic fermentation of benzoate in an enrichment culture. Microbiol. Ecol. 4:249-261. 1978.

136. J. M. Suflita, A. Horowitz, D. R. Shelton and J. M. Tiedje. Dehalogenation: a novel pathway for the anaerobic biodegradation of haloaromatic compounds. Science 218:1115-1117. 1982.

137. Y. Wang, M. T. Suidan and J. T. Pfeffer. Anaerobic biodegradation of indole to methane. Appl. Environ. Microbiol. 48:1058-1060. 1984.

138. E. J. Bouwer and P. L. McCarty. Transformations of 1- and 2-carbon halogenated aliphatic organic compounds under methanogenic conditions. Appl. Environ. Microbiol 45:1286-1294. 1983.

139. R. M. Sykes and E. J. Kirsch. Accumulation of methanogenic substances in CCi_4 inhibited anaerobic sewage sludge digester cultures. Water Res. 6:41-55. 1972.

140. P. G. Thiel. The effect of methane analogues on methanogenesis in anaerobic digestion. Water Res. 38:215-223. 1969.

141. E. L. Iannotti, R. E. Mueller, J. R. Fischer and D. M. Sievers. Biochemistry of the conversion of swine manure to methane, p. 143-145. Fourth Annual Solar and Biomass Workshop. Atlanta, GA. 1984.

142. D. O. Mountfort and R. A. Asher. Changes in proportions of acetate and carbon dioxide used as methane precursors during the anaerobic digestion of bovine waste. Appl. Environ. Microbiol. 35:648-654. 1978.

143. R. E. Mackie and M. P. Bryant. Metabolic activity of fatty acid-oxidizing bacteria and the contribution of acetate, propionate, butyrate and CO_2 to methanogenesis in cattle waste at 40° and 60°C. Appl. Environ. Microbiol. 41:1363-1373. 1981.

144. D. R. Boone. Terminal reactions in the anaerobic digestion of animal waste to methane. Appl. Environ. Microbiol. 43:57-64. 1982.

145. D. R. Boone. Mixed-culture fermentor for simulating methanogenic digesters. Appl. Environ. Microbiol. 48:122-126. 1984.

146. J. S. Jeris and P. L. McCarty. The biochemistry of methane fermentation using ^{14}C tracers. J. Water Pollut. Control Fed. 37:178192. 1965.

147. B. Schink. Mechanisms of succinate and propionate degradation in anoxic freshwater sediments and sewage sludge. J. Gen. Microbiol. 131:643-650. 1985.

148. P. H. Smith and R. A. Mah. Kinetics of acetate metabolism during sludge digestion. Appl. Microbiol. 14:368-371. 1966.

149. S. H. Zinder and M. Koch. Non-aceticlastic methanogenesis from acetate: acetate oxidation by a thermophilic syntrophic coculture. Arch. Microbiol. 138:263-272. 1984.

150. E. L. Iannotti, D. Kafkewitz, M. J. Wolin and M. P. Bryant. Glucose fermentation products of Ruminococcus albus grown in continuous culture with Vibrio succinogenes: Changes caused by interspecies transfer of H$_2$. J. Bacteriol. 113:1231-1240. 1973.

151. M. J. Wolin. Hydrogen transfer in microbial communities, p. 323-356. In: A. T. Bull and J. H. Slater (ed.), Microbial Interactions and Communities, Vol. 1. Academic Press, London, England. 1982.

152. E. A. Runquist, E. H. Abbot, M. T. Armold and J. E. Robbins. Application of ^{13}C-nuclear magnetic resonance to the observation of metabolic interactions in anaerobic digesters. Appl. Environ. Microbiol. 42:556-559. 1981.

153. C. Scheifinger and M. J. Wolin. Propionate formation from cellulose and soluble sugars by combined cultures of Bacteroides succinogenes and Selenomonas ruminantium. Appl. Microbiol. 26:789-795. 1973.

154. L. Baresi, R. A. Mah, D. M. Ward and I. R. Kaplan. Methanogenesis from acetate: enrichment studies. Appl. Environ. Microbiol. 36:186-197. 1978.

155. T. J. Ferguson and R. A. Mah. Effect of H$_2$-CO$_2$ on methanogenesis from acetate or methanol in Methanosarcina spp. Appl. Environ. Microbiol. 46:348-355. 1983.

156. V. M. Laube and S. M. Martin. Conversion of cellulose to methane and carbon dioxide by triculture of Acetovibrio cellulolyticus, Desulfovibrio sp. and Methanosarcina barkeri. Appl. Environ. Microbiol. 42:413-420. 1981.

157. K. Fiebig and G. Gottschalk. Methanogenesis from choline by a coculture of Desulfovibrio sp. and Methanosarcina barkeri. Appl. Environ. Microbiol. 45:161-168. 1983.

158. E. Muller, K. Fahlbusch, R. Walther and G. Gottschalk. Formation of N, N-dimethylglycine, acetic acid and butyric acid from betaine by Eubacterium limosum. Appl. Environ. Microbiol. 42:439-445. 1981.

159. H. Konig and K. O. Stetter. Isolation and characterization of Methanolobus tindarius sp. nov., a coccoid methanogen growing only on methanol and methylamines. Zentralbl. Bakteriol. Parasitenkd. Infektionskr. Hyg. Abt. 1 Orig. Reihe C 3:478-490. 1982.

160. K. R. Sowers and J. G. Ferry. Isolation and characterization of a methylotrophic marine methanogen, Methanococcoides methylutens gen. nov., sp. nov. Appl. Environ. Microbiol. 45:684-690. 1983.

161. M. Nagase and T. Matsuo. Interactions between amino acid degrading bacteria and methanogenic bacteria in anaerobic digestion. Biotechnol. Bioeng. 24:2227-2239. 1982.

162. E. Naumann, H. Hippe and G. Gottschalk. Betaube: new oxidant in the Stickland reaction and methanogenesis from betaine and L-alanine by a Clostridium sporogenes-Methanosarcina barkeri coculture. Appl. Environ. Microbiol. 45:474-483. 1983.

163. D. B. Archer. Hydrogen-using bacteria in a methanogenic acetate enrichment culture. J. Appl. Bacteriol. 56:125-129. 1984.

164. M. P. Bryant, E. A. Wolin, M. J. Wolin and R. S. Wolfe. Methanobacillus omelianskii, a symbiotic association of two bacterial species. Arch. Microbiol. 59:20-31. 1967.

165. M. Chen and M. J. Wolin. Influence of CH_4 production by Methanobacterium ruminantium on the fermentation of glucose and lactate by Selenomonas ruminantium. Appl. Environ. Microbiol. 34:756759. 1977.

166. A. W. Khan. Degradation of cellulose to methane by a coculture of Acetivibrio cellulolyticus and Methanosarcina barkeri. FEMS Microbiol. Letters 9:233–235. 1980.

167. M. J. Latham and M. J. Wolin. Fermentation of cellulose by Ruminococcus flavefacieus in the presence and absence of Methanobacterium ruminantium. Appl. Environ. Microbiol. 34:297–301. 1977.

168. M. J. McInerney and M. P. Bryant. Anaerobic degradation of lactate by syntrophic association of Methanosarcina barkeri and Desulfovibrio species and effect of H_2 on acetate degradation. Appl. Environ. Microbiol. 41:346–354. 1981.

169. P. L. Ruyet, H. C. Dubourguier and G. Albagnac. Homoacetogenic fermentation of cellulose by a coculture of Clostridium thermocellum and Acetogenium kivui. Appl. Environ. Microbiol. 48:893–894. 1984.

170. C. Scheifinger, B. Linehan and M. J. Wolin. H_2 production by Selenomonas ruminantium in the absence and presence of methanogenic bacteria. Appl. Microbiol. 29:480–483. 1975.

171. D. M. Ward, R. A. Mah and I. R. Kaplan. Methanogenesis from acetate: a nonmethanogenic bacterium from an anaerobic acetate enrichment. Appl. Environ. Microbiol. 35:1185–1192. 1978.

172. J. V. Winter and R. S. Wolfe. Complete degradation of carbohydrate to carbon dioxide and methane by syntrophic cultures of Acetobacterium woodii and Methanosarcina barkeri. Arch. Microbiol. 121:97–102. 1979.

173. H. Miur, M. Horiguchi and R. Matsumoto. Nutritional interdependence among rumen bacteria Bacteroides amylophilus, Megasphaera elsdenii and Ruminococcus albus. Appl. Environ. Microbiol. 40:294–300. 1980.

174. S. Ghosh, J. R. Conrad and D. L. Klass. Anaerobic acidogenesis of wastewater sludge. J. Water Poll. Control Fed. 47:30–45. 1975.

175. S. Ghosh and F. G. Pohland. Kinetics of substrate assimilation and product formation in anaerobic digestion. J. Water Poll. Control Fed. 46:749–759. 1974.

176. H. F. Kaspar and K. Wuhrmann. Kinetic parameters and relative turnovers of some important catabolic reactions in digesting sludge. Appl. Environ. Microbiol. 36:1–7. 1978.

177. J. T. Novak and D. A. Carlson. The kinetics of anaerobic long-chain fatty acid degradation. J. Water Pollut. Control Fed. 42:1932–1943. 1970.

178. P. J. Van Soest. Physico-chemical aspects of fibre digestion, p. 351–365. In: I. W. McDonald and A. C. I. Warner (ed.), Digestion and Metabolism in the Ruminant. University of New England Publishing Unit, Armidale, Australia. 1975.

179. R. Benner, A. E. Maccubbin and R. E. Hodson. Anaerobic biodegradation of the lignin and polysaccharide components of lignocellulose and synthetic lignin by sediment microflora. Appl. Environ. Microbiol. 47:998–1004. 1984.

180. W. F. Hackett, W. J. Connors, T. K. Kirk and J. G. Zeikus. Microbial decomposition of synthetic ^{14}C-labeled lignins in nature: lignin biodegradation in a variety of natural materials. Appl. Environ. Microbiol. 33:43–51. 1977.

181. E. Odier and B. Monties. Absence of microbial mineralization of lignin in anaerobic enrichment cultures. Appl. Environ. Microbiol. 46:661–665. 1983.

182. P. Porter and A. G. Singleton. The degradation of lignin and quantitative aspects of ruminant digestion. Brit. J. Nutri. 25:3–14. 1971.

183. P. J. Van Soest. The uniformity and nutritive availability of cellulose. Fed. Proc. 32:1804–1808. 1973.

184. P. J. Van Soest, A. M. Ferreira and R. D. Hartley. Chemical properties of fibre in relation to nutritive quality of ammonia-treated forages. Anim. Feed Sci. Technol. 10:155-164. 1984.

185. J. B. Reeves, III. Lignin composition and in vitro digestibility of feeds. J. Anim. Sci. 60:316-322. 1985.

186. M. I. McBurney, P. J. Van Soest and L. E. Chase. Cation exchange capacity of various feedstuffs in ruminant rations, p. 16-23. In: Proc. Cornell Nutr. Conf. for Feed Manufacturers, October, 1981. Syracuse, NY. 1983.

187. P. J. Van Soest. Some physical characteristics of dietary fibres and their influence on the microbial ecology of the human colon. Proc. Nutr. Soc. 43:25-33. 1984.

188. D. Georgacakis, D. M. Sievers and E. L. Iannotti. Buffer stability in manure digesters. Agri. Wastes. 4:427-441. 1982.

189. P. N. Hobson and B. G. Shaw. Inhibition of methane production by Methanobacterium formicicum. Water Research 10:849-852. 1976.

190. E. L. Iannotti and J. R. Fischer. Effects of ammonia, volatile acids, pH and sodium on growth of bacteria isolated from a swine manure digester. Dev. Ind. Microbiol. 25:741-747. 1984.

191. E. J. Kroeker, D. D. Schulte, A. B. Sparling and H. M. Lapp. Anaerobic treatment process stability. J. Water Pollut. Control Fed. 51:718-727. 1979.

192. L. Van den Berg, G. B. Patel, D. S. Clark and C. P. Lentz. Factors affecting rate of methane formation from acetate by enrichment cultures. Can. J. Microbiol. 22:1312-1319. 1976.

193. A. Chesson, C. S. Stewart and R. J. Wallace. Influence of plant phenolic acids on growth and cellulolytic activity of rumen bacteria. Appl. Environ. Microbiol. 44:597-603. 1982.

194. H. G. Jung, G. C. Fahey, Jr. and J. E. Garst. Simple phenolic monomers of forages and effects of in vitro fermentation on cell wall phenolics. J. Anim. Sci. 57:1294-1305. 1983.

195. V. H. Varel and H. G. Jung. Influence of forage phenolics on cellulolytic bacteria and in vitro cellulose degradation. Can. J. Anim. Sci. 64(Suppl):39-40. 1984.

196. P. J. Colberg and L. Y. Young. Biodegradation of lignin-derived molecules under anaerobic conditions. Can. J. Microbiol. 28:886-889. 1982.

197. J. B. Healy, Jr. and L. Y. Young. Catechol and phenol degradation by a methanogenic population of bacteria. Appl. Environ. Microbiol. 38:216-218. 1978.

198. J. B. Healy, Jr. and L. Y. Young. Anaerobic biodegradation of eleven aromatic compounds to methane. Appl. Environ. Microbiol. 38:84-89. 1979.

199. J. B. Healy, Jr., L. Y. Young and M. Reinhard. Methanogenic decomposition of ferulic acid, a model lignin derivative. Appl. Environ. Microbiol. 39:436-444. 1980.

200. R. F. Barnes and G. C. Marten. Recent developments in predicting forage quality. J. Anim. Sci. 48:1554-1561. 1979.

201. B. A. Dehority and R. R. Johnson. Effect of particle size on the in vitro cellulose digestibility of forages by rumen bacteria. J. Dairy Sci. 44:2242-2249. 1961.

202. A. Y. Robles, R. L. Belyea, F. A. Martz and M. F. Weiss. Effect of particle size upon digestible cell wall and rate of in vitro digestion of alfalfa and orchardgrass forages. J. Anim. Sci. 51:783-790. 1980.

203. R. F. Ehle, M. R. Murphy and J. M. Clark. In situ particle size reduction and the effect of particle size on degradation of crude protein and dry matter in the rumen of dairy steers. J. Dairy Sci. 65:963-971. 1982.

204. P. L. McCarty, L. Y. Young, D. C. Stuckey and J. B. Healy. Heat treatment for increasing methane yields from organic materials, p.

179–199. In: H. G. Schlegel and J. Barnea (ed.), Microbial Energy Conversion, Pergamon Press, New York, NY. 1977.

205. J. E. Robbins, M. T. Armold and S. L. Lacher. Methane production from cattle waste and delignified straw. Appl. Environ. Microbiol. 38:175–177. 1979.

206. J. R. Fischer, E. L. Iannotti and C. D. Fulhage. Production of methane gas from combinations of wheat straw and swine manure. Trans. of ASAE 26:546–548. 1983.

PREDICTION OF METHANE YIELDS FROM BIOMASS

Aziz Shiralipour and Paul H. Smith

Department of Microbiology & Cell Science
University of Florida/IFAS
Gainesville, Florida 32611

Kenneth M. Portier

Department of Statistics
University of Florida/IFAS
Gainesville, Florida 32611

ABSTRACT

Our standard bioassay procedures for determining methane yields from non-woody biomass species require up to 18 weeks of incubation. However, only small quantities of gas are produced during the final weeks. Results obtained from numerous assays of specific species demonstrate a high degree of correlation between methane yield during the early weeks and the ultimate yield. An equation determined for individual species can be used to calculate a predicted final yield from the yields of early weeks. The predicted value is sufficiently accurate to use in making management decisions requiring a short turn-around time.

Standard regression techniques were used to quantify the correlations between early yields and ultimate yields, and to generate prediction equations. Data for elephantgrass or Napiergrass (Pennisetum purpureum L.), water hyacinth [Echhornia crassipes (Mart.) Solms], and morado sweet potato (Ipomoea batatas), indicates the predicted ultimate yields were within 0.013, 0.014, and 0.004 std m^3 kg^{-1} volatile solid (VS) of the observed ultimate yield, with 95% confidence. Mean observed yields for these biomass species were 0.30, 0.28, and 0.40 m^3 kg^{-1} volatile solid.

Keywords. Bioassay, predicted yield, biomass, early yield, standard regression.

INTRODUCTION

Our standard bioassay procedure[1] for determining methane yield from
biomass is an accurate and useful method. This procedure is a modification
of the procedure first described by Owens et al.[2] However, non-woody
species require up to 18 weeks of incubation while woody species require
even longer period of incubations. Since the biomass samples are assayed
in sealed serum bottles[1] several gas measurements and gas bleedings are
performed during the incubation period to prevent rupture from the gas
pressure. If the ultimate yields of methane could be predicted from the
early stages of incubation a great deal of time, manpower, and money could
be saved. The objective of this study was to examine the correlation
between methane yields during the early weeks and ultimate yields in an
attempt to calculate a predicted final yield from the yields of early weeks.

MATERIALS AND METHODS

Biomass species used in this study included PI 300086 Napiergrass
(Pennisetum purpureum L.), water hyacinth [Eichhornia crassipes (Mart)
Solms], and morado sweet potato (Ipomoea batatas). The plant samples were
dried at 60°C, ground (1 mm mesh) then sent to our laboratory by the
investigators involved in the production projects of "Methane from Biomass
and Waste" program.[1]

A mixture of vitamins, minerals, buffer, reducing agents, and liquor
from a digester containing methanogenic bacteria was added to an equivalent
of 0.2 g volatile solid (VS) plant material in a 250 ml serum bottle. All
assay procedures were performed under anaerobic conditions. The serum
bottles were transferred to a 35°C incubator.[1]

Methane production was measured at weeks 1, 3, 5, and 18. The biogas
pressure produced inside the serum bottles was measured and methane
concentration in the biogas was determined by gas chromatography. These
measurements were then used to calculate methane production, as standard
(Methane production was calculated at 15.5°C and 1 atm according to the
standard of gas industries.) m^3 kg^{-1} volatile solids (VS) added.

Standard regression model building techniques[3] were used to explore the
relationship between readings at weeks one, three and five with the ultimate
methane yields. Linear quadratic and cross-product terms were included in a

stepwise regression analysis to determine those models which explained the largest portion of variability in the ultimate yields among samples tested. The final model for each species or cultivar under study was determined based upon the R-square value and characteristics of the residuals.

The 95% confidence level of the error for prediction was computed using the variance of the residual terms and assuming a normal distribution.

RESULTS AND DISCUSSION

Methane production of Napiergrass, water hyacinth and morado sweet potato at 1st, 3rd, 5th, and 18th weeks of incubation periods are shown in Tables 1-3. Approximately 36, 66, and 84% of the ultimate methane yields of Napiergrass samples collected from different locations were produced at the 1st, 3rd, and 5th weeks of incubations, respectively. Methane production for the same incubation periods was 37, 75, and 85% of

Table 1. Methane Production of Napiergrass During Early Weeks of Incubations Compared to Ultimate Yields (at 18 weeks).

| Location | Methane Yields (std m^3 kg^{-1} VS added)[a] | | | |
	Week 1	Week 3	Week 5	Week 18
Gainesville, Florida:				
	0.11	0.21	0.29	0.30
	0.12	0.21	0.26	0.31
	0.11	0.18	0.25	0.28
	0.11	0.20	0.27	0.31
	0.09	0.16	0.21	0.25
	0.09	0.18	0.22	0.26
Ona, Florida:				
	0.12	0.22	0.27	0.32
	0.12	0.22	0.28	0.34
	0.13	0.23	0.28	0.34
	0.13	0.25	0.30	0.37
	0.12	0.23	0.28	0.35
	0.12	0.23	0.28	0.33
Jay, Florida:				
	0.10	0.19	0.24	0.29
	0.10	0.18	0.22	0.26
	0.09	0.16	0.20	0.25
	0.10	0.20	0.25	0.30
	0.10	0.19	0.23	0.28
	0.11	0.20	0.25	0.29

[a]Value for each sample is the mean of three replicates. For conversion of std m^3 kg^{-1} VS to std ft^3 lb^{-1} VS divide by 0.0623.

Table 2. Methane Production of Water Hyacinth During Early Weeks of
          Incubations Compared to Ultimate Yield (at 18 weeks).

Location	Methane Yields (std m^3 kg^{-1} VS added)[a]			
	Week 1	Week 2	Week 3	Week 18
Sanford, Florida:				
	0.12	0.30	0.33	0.36
	0.11	0.27	0.30	0.33
	0.10	0.23	0.25	0.29
	0.10	0.28	0.30	0.33
	0.11	0.24	0.27	0.31
Gainesville, Florida:				
	0.12	0.19	0.22	0.26
	0.11	0.19	0.21	0.28
	0.12	0.20	0.23	0.29
	0.12	0.19	0.23	0.29
	0.12	0.19	0.22	0.28
	0.12	0.21	0.23	0.30
	0.12	0.21	0.24	0.30
	0.12	0.20	0.23	0.29
Gainesville, Florida:				
	0.07	0.17	0.19	0.21
	0.06	0.18	0.20	0.22
	0.06	0.17	0.19	0.22
	0.08	0.19	0.22	0.24
	0.09	0.20	0.23	0.26

[a]Value for each sample is the mean of three replicates. For conversion
of std m^3 kg^{-1} VS to std ft^3 lb^{-1} VS divide by 0.0623.

Table 3. Methane Production of Morado Sweet Potato During Early Weeks of
          Incubations Compared to Ultimate Yields (at 18 weeks).

Age	Methane Yields (std m^3 kg^{-1} VS added)[a]			
	Week 1	Week 3	Week 5	Week 18
95 day:				
	0.13	0.35	0.37	0.40
	0.12	0.32	0.35	0.39
	0.11	0.31	0.38	0.39
	0.10	0.29	0.35	0.36
	0.12	0.30	0.34	0.36
75 days:				
	0.12	0.34	0.36	0.40
	0.12	0.34	0.36	0.39
	0.13	0.33	0.35	0.39
	0.13	0.34	0.37	0.40
	0.13	0.33	0.37	0.40

[a]Value for each sample is the mean of three replicates. For conversion
of std m^3 kg^{-1} VS to std ft^3 lb^{-1} VS divide by 0.0623.

the ultimate yields for water hyacinth and 31, 84, and 93% for morado
sweet potato. Since over 80% of the ultimate yields of all three species
were produced during the first 5 weeks of incubation and methane
production for each individual species during this period was fairly
constant, an attempt was made to obtain a predicted final yield from the
yields of early weeks.

The final models for Napiergrass (PI 300086), water hyacinth, and
morado sweet potato are given in Table 4. Statistics on the observed
ultimate yields, predicted yields, and residuals are given in Table 5.
Each species will be discussed separately.

Based upon these models and the associated unexplained variability,
the level of error in prediction can be determined. These values are
given in Table 6. These numbers may be interpreted as the absolute
difference between the predicted value for yield and the value observed.
This is only valid if the species or cultivar is the same as those used
to build the model.

Linear regression techniques are usually used to explore the
relationships between measurements when no mechanistic or intuitive model
is available. Such empirical models provide a basis upon which
additional study into the functional form of the relationship is possible.
In this study, equations have been found which explain almost perfectly
the correlation between ultimate yield and measurements taken in the
earlier weeks of the assay. As shown in Table 5, averages, standard
errors, and ranges of the observed ultimate yields and the predicted
yields are virtually identical. An examination of the residual
variability (observed-predicted yields) indicated little heterogeniety of
variance or non-normality, suggesting that, from the statistical view at
least, the equations are acceptable. There is some evidence to suggest
that the perfect equation has not been established for water hyacinth,
since the residuals did not satisfy the normality assumptions. If the
errors in prediction are compared to the mean levels of observed ultimate
yield, the relative error in prediction is less than 5% in all cases.

While the regression models presented here provide a good fit to the
observed ultimate yields, the equations are somewhat nonintuitive. This
can happen in a strictly empirical study as has been done here. No
attempt has been made to make the equations conform to some theory of
methane production. The equations are specific to the species or

Table 4. Final Regression Models for Napiergrass, Water Hyacinth and Morado Sweet Potato

---

Napiergrass PI 300086:

$$\text{YIELD} = 0.157 + 2.778 \times \text{WEEK 3} \times \text{WEEK 5}$$
$$(0.0071) \quad (0.1321)$$

$$\text{R-square} = .965 \qquad \text{MSE} = 0.0198$$

Water Hyacinth:

$$\text{YIELD} = 0.150 - 1.402 \times \text{WEEK 1} \times \text{WEEK 1} + 6.70 \times \text{WEEK 1} \times \text{WEEK 3}$$
$$(0.007) \quad (0.769) \qquad\qquad (0.46)$$

$$\text{R-square} = .966 \qquad \text{MSE} = 0.0130$$

Morado Sweet Potato:

$$\text{YIELD} = 0.165 + 0.353 \times \text{WEEK 1} + 1.55 \times \text{WEEK 3} \times \text{WEEK 5}$$
$$(0.012) \quad (0.120) \qquad (0.102)$$

$$\text{R-square} = .986 \qquad \text{MSE} = 0.0012$$

Note: WEEK 1 represents the reading for week 1, YIELD is the predicted yield, the values in parenthesis are the standard errors for the associated model coefficients.

Table 5. Basic Statistics on Ultimate Yield, Predicted Yield and Residuals

Statistic	Yield	Predicted	Residual
(std m^3 kg^{-1} VS added)			
Napiergrass PI 300086:			
mean	0.302	0.302	0.00
standard error	0.0081	0.0080	0.0015
range	0.123	0.120	0.0263
normal test	n.s	n.s	n.s
Water Hyacinth:			
mean	0.281	0.281	0.00
standard error	0.0094	0.0092	0.0017
range	0.1477	0.1465	0.0293
normal test	n.s	n.s	>.05
Morado Sweet Potato:			
mean	0.390	0.390	0.00
standard error	0.0052	0.0052	0.0006
range	0.0492	0.0503	0.0057
normal test	n.s	n.s	n.s

n.s. implies not statistically significant at the 0.05 level.

Table 6. Error in Prediction, by Species, 95% Confidence

Species	Error in Prediction	Percent of Mean
Napiergrass	.0127	4.21%
Water Hyacinth	.0140	4.98%
Morado Sweet Potato	.0038	0.97%

Table 7. Effect of Substrate Concentrations on Methane Productions of Napiergrass and Water Hyacinth During Early Weeks of Incubations Compared to Ultimate Yields

Plant Species	g VS/ bottle	Methane Yields (std m^3 kg^{-1} VS added)[a]			
		Incubation Periods (Weeks)			
		1	3	5	18
Napiergrass:					
	0.10	0.10	0.21	0.24	0.28
	0.20	0.10	0.21	0.24	0.28
	1.00	0.01	0.12	0.17	0.23
Water Hyacinth:					
	0.10	0.10	0.18	0.23	0.28
	0.20	0.10	0.18	0.23	0.28
	1.00	0.02	0.11	0.17	0.26

[a]Value for each sample is the mean of three replicates. For conversion of std m^3 kg^{-1} VS to std ft^3 lb^{-1} VS divide to 0.0623.

cultivars for which they were developed. Unfortunately, no one equation form was found to hold for all three species. Within a species or cultivar combination, location of feedstock did not have an affect on the equation form. More work needs to be done to determine if some other equation form, specifically some nonlinear form, might be appropriate for all species.

If some error in prediction of ultimate yield from early week data is acceptable, the equations presented here should be useful, especially in shortening the assay time for samples from different locations or fertilization trails of the species or cultivars discussed.

For new cultivars, enough data will need to be collected to identify the equation structure and residual variance. It is hoped that some consistant equation form for each species will emerge as additional cultivars are examined.

The equations are only as good as the data upon which they are based. If the standard assay is not performed correctly, then the model predictions will not be correct. This might happen, for example, if the substrate concentrations in the assay bottles is not controlled (Table 7), leading to a different correlation structure between early weeks of incubation and ultimate yield. In this case the parameters and possibly the form of the prediction equation would be invalid.

Napiergrass and water hyacinth have been selected as potential energy crops in Florida.[4,5] Since morado sweet potato also has the potential for such selection, considerable research is being conducted on the convertabilities of these species in our laboratories. Shortening the time of assay and predicting the ultimate yield from early measurements will speed up our goals with less expense and manpower.

ACKNOWLEDGMENTS

The technical assistance of Ms. Phyllis Hansen and Mr. F.M. Bordeaux throughout the course of this study is gratefully acknowledged. This research is part of the joint program between the Gas Research Institute and the Institute of Food and Agricultural Sciences, University of Florida.

REFERENCES

1. A. Shiralipour and P. H. Smith. Conversion of biomass into methane gas. Biomass 6:85-92 1984.
2. W. F. Owens, D. C. Stuckey, J. B. Healey, L. Y. Young and P. L. MacCarty. Bioassay for monitoring biochemical methane potential and toxicity. Water Res. 13:405-92 1979.
3. N. R. Droper and H. Smith. Applied regression analysis. P. 709. In: Second Edition, John Wiley and Sons, Inc., New York 1981.
4. W. H. Smith. Methane from biomass and waste. Annual Report for Gas Research Institute Contract No. 5080-323-0423, January - December 1983 1984.
5. J. W. Mishoe, M. N. Lorber, R. M. Peart, R. C. Fluck and J. W. Jones. Modeling and analysis of biomass production system. Biomass 6:119-130 1984.

PREDICTION OF FERMENTABILITY OF BIOMASS

FEEDSTOCKS FROM CHEMICAL CHARACTERISTICS

Karen A. Bjorndal and John E. Moore

Department of Animal Science
University of Florida/IFAS
Gainesville, Florida 32611

ABSTRACT

Fermentability of biomass resources has been estimated using an in
vitro rumen fermentation technique for 1235 samples representing a survey of
over 100 species from terrestrial and freshwater habitats throughout the
state of Florida. Samples have also been analyzed for cell wall constituents, cell contents, lignin, nitrogen, phosphorus, and tannin in order to
identify the characteristics that determine fermentability.

Summaries of data on fermentability and chemical constituents are
presented from the survey of biomass feedstocks. The prediction of fermentability from chemical constituents using multiple regression analysis is
discussed. The ultimate aim of this study is to provide a rapid method for
predicting the potential methane production of a wide variety of biomass
resources.

Keywords. Rumen fermentation, methanogenesis.

INTRODUCTION

The rumen fermentation has many similarities to the fermentation in
methane digesters, particularly to the fermentation in the first stage of
two-stage digesters in which the end products are largely volatile fatty
acids. Our approach has been to study the relationship between fermentability of biomass feedstocks and chemical constituents that are known to have a
major effect on the fermentation of feedstocks in the rumen. Results of

447

these analyses would improve our understanding of the factors that control fermentability of biomass and production of methane, and provide the basis for predicting potential fermentability of biomass.

Samples were analyzed for cell wall constituents, also called neutral detergent fiber ash-free (NDFA). This analysis separates the total volatile solids into two fractions: cell contents, which are rapidly and completely fermentable, and cell walls, which vary in rate and extent of fermentability. Because many of the potential biomass feedstocks are high in cell walls, it is necessary to predict the extent of fermentability of the cell wall fraction. For this reason, the cell wall fraction was analyzed for lignin. It is well established that lignin is the major determinant of cell wall fermentability in ruminants.[1]

Tannins are polyphenolic compounds that form insoluble complexes with protein and are found in a wide variety of plants.[2] Because tannin is known as a common anti-quality component in sorghum,[3] sorghum samples were analyzed for tannin.

METHODS

Samples of biomass feedstocks were collected from freshwater and terrestrial habitats throughout the state of Florida by cooperating investigators as part of the GRI/IFAS project "Methane from Biomass and Waste." Feedstocks represent natural vegetation, crops grown specifically as energy crops, and crop residues. Samples were dried at 60°C and ground in a Wiley mill to pass through a 1 mm screen.

The assay used to assess the fermentability of feedstocks is in vitro total volatile solids digestibility (IVTVSD) (modified from Tilley and Terry).[4] For this analysis, rumen fluid is removed from a fistulated steer that is maintained on a constant diet of bermudagrass hay and protein supplement. The fluid is filtered through glass wool and diluted with a phosphate-bicarbonate buffer (4:1 buffer:rumen fluid). After CO_2 is bubbled through the inoculant, 50 ml is added to 90 ml tubes containing pre-weighed samples (approximately 0.5 g). The tubes are flushed with CO_2, stoppered with pressure-release valves and placed in a 39°C incubator. The tubes are swirled three times a day for two days to ensure thorough mixing of sample and inoculant. After 48 hours, HCl is added to the tubes to lower the pH and to kill the microbes, and pepsin is added to partially digest microbial and substrate protein. The tubes are returned to the 39°C incubator and swirled three times a day for the next two days. After an additional 48 hours (96 hours total), residues are recovered on glass wool in Gooch cruci- bles.[5] The crucibles are dried overnight at 105°C, weighed, ashed at 500°C for three hours and weighed. IVTVSD equals the percent of total volatile solids that disappears during the fermentation and acid-pepsin treatment. Two standards of known digestibility, one high and one low, are included in every run to test for between-run variation in activity. Four blanks are included in every run to correct for any rumen solids and microbial matter in the inoculant not digested by the acid-pepsin.

Analyses for cell wall constituents (neutral detergent fiber ashfree, NDFA) and permanganate lignin followed Goering and Van Soest[6] with sodium sulfite and decalin omitted from the NDFA procedure. Two analyses for tannins were performed. The modified vanillin-HCl method[7] measures the concentration of condensed tannins (in catechin equivalents), and the pro- tein precipitation method,[8] by measuring the amount of bovine serum albumin precipitated, gives the concentration of hydrolyzable and condensed tannins as well as phenolic compounds with similar protein-binding capacity.

RESULTS AND DISCUSSION

Values of fermentability (IVTVSD) and chemical composition for all samples analyzed and for some sub-groups are shown in Table 1. There is considerable variation in fermentability and chemical composition. A few patterns emerge, however. In general, groups with higher percentages of cell wall constituents and lignin have lower fermentability values. Al- though sample numbers for plant parts of both sorghum and Napiergrass are small, in both species leaves have higher fermentabilities, lower percent- ages of cell wall constituents and lignin, and higher percentages of nitro- gen than stems. In sorghum, seed heads are more fermentable than stems and leaves, have lower or equal percentages of cell wall constituents and lignin, and have higher or equal percentages of nitrogen and phosphorus.

Tannin values in the sorghum samples (Table 1) are quite low. These low values may be a result of oxidation of the tannin after the samples were ground.[7] The purpose of the tannin analyses was not to determine the tannin content of fresh plant material, but of the feedstock when fermentability was measured. Therefore, because samples were tested for fermentability

Table 1. Number of samples (n), means and standard deviations (SD) for fermentability (IVTVSD), cell wall constituents (or neutral detergent fiber ash-free, NDFA), lignin, nitrogen, phosphorus, vanillin-HCl tannin (VTannin) and protein-precipitation tannin (PTannin). All are expressed as a percent of total volatile solids, except phosphorus is percent of dry matter.

	IVTVSD			NDFA			Lignin			Nitrogen			Phosphorus			VTannin			PTannin		
	n	Mean	SD	n	Mean	SD	n	Mean	SD	n	Mean	SD	n	Mean	SD	n	Mean	SD	n	Mean	SD
All Samples	1235	55.4	19.2	1193	59.9	19.0	1102	10.6	7.5	886	1.75	1.19	902	.28	.16	---	---	---	---	---	---
Terrestrial	966	57.1	18.7	949	60.9	19.3	914	9.9	6.4	693	1.48	.92	694	.25	.12	---	---	---	---	---	---
Woody	19	9.5	7.7	18	86.1	7.9	15	22.6	5.5	15	.43	.27	15	.05	.04	---	---	---	---	---	---
Herbaceous	947	58.0	17.6	931	60.4	19.2	899	9.7	6.2	678	1.51	.91	679	.25	.12	---	---	---	---	---	---
Legumes	47	49.7	20.0	47	63.0	15.6	44	14.8	4.8	40	2.44	1.14	40	.24	.12	---	---	---	---	---	---
Brassicas	51	80.7	14.3	49	33.0	6.8	46	11.6	6.1	37	3.12	1.42	37	.48	.12	---	---	---	---	---	---
Grasses	537	54.5	11.8	532	69.2	11.6	523	7.2	3.1	317	1.26	.51	318	.22	.09	---	---	---	---	---	---
Napiergrass	203	52.0	10.9	200	77.7	4.6	195	7.3	2.7	73	1.30	.61	73	.25	.09	---	---	---	---	---	---
Stems	8	54.4	10.5	8	80.0	3.7	8	8.7	2.6	8	1.47	.79	8	.29	.10	---	---	---	---	---	---
Leaves	9	56.1	6.8	9	76.9	2.8	9	6.1	1.1	9	1.83	.57	9	.28	.06	---	---	---	---	---	---
Sorghum	208	59.2	9.1	208	60.1	9.8	208	6.3	2.0	179	1.16	.42	179	.20	.07	132	.11	.21	132	.05	.08
Stems	10	55.9	8.7	10	64.2	12.7	10	6.9	1.7	10	.52	.09	10	.11	.02	10	.02	.02	10	.02	.01
Leaves.	3	68.4	1.8	3	60.9	1.3	3	4.2	.3	3	1.71	.07	3	.24	.03	3	.08	.01	3	.01	.01
Seed Heads	10	70.1	10.8	10	40.5	12.0	10	4.2	1.6	10	1.71	.14	10	.31	.03	10	.24	.51	10	.09	.24
Aquatics	201	49.7	19.4	177	59.0	16.5	152	13.3	9.6	148	2.73	1.46	160	.41	.25	---	---	---	---	---	---
Emergents	56	59.0	17.6	55	48.2	22.2	54	9.0	5.7	42	2.01	1.23	54	.24	.13	---	---	---	---	---	---
Floating	145	46.1	18.9	122	63.9	9.9	98	15.6	10.5	106	3.02	1.45	106	.49	.25	---	---	---	---	---	---
Water Hyacinth	80	43.7	16.7	69	67.0	7.4	55	10.2	3.6	64	3.02	1.55	64	.52	.26	---	---	---	---	---	---

weeks after thev had been ground, tannin analyses were conducted on similar samples.

As discussed above, fermentability of a feed in ruminants is largely determined by the percent of cell wall constituents (NDFA) in the feed and the extent to which the cell wall is lignified (lignin/NDFA). Therefore, an initial model for predicting fermentability of biomass feedstocks would include NDFA and lignin/NDFA; the results of regression analyses of fermentability against NDFA and against NDFA and lignin/NDFA are given in Table 2. The predictive value of NDFA alone varies widely among the groups, and the effect of adding lignin/NDFA to the prediction equation is great in some groups (e.g., brassicas and water hyacinths) and is small in others (e.g., legumes and emergent aquatics). In addition, the ability of NDFA and lignin/NDFA to account for the variation in fermentability within each group

Table 2. Number of samples (n), R^2 and $S_{y \cdot x}$ values for predictions of fermentability (IVTVSD) from cell wall constituents (or neutral detergent fiber ash-free, NDFA) and lignin as a percent of NDFA

		IVTVSD vs. NDFA		IVTVSD vs. NDFA & Lignin	
	n	R^2	$S_{y \cdot x}$	R^2	$S_{y \cdot x}$
All samples	1102	.574	12.6	.752	9.5
Terrestrial	914	.663	10.9	.789	8.5
Woody	15	.465	5.9	.483	6.5
Herbaceous	899	.665	10.2	.788	8.1
Legumes	44	.891	6.7	.926	5.7
Brassicas	46	.420	11.2	.902	4.8
Grasses	523	.456	8.8	.735	6.1
Napiergrass	195	.731	5.7	.833	4.6
Stems	8	.837	4.6	.979	1.8
Leaves	9	.828	3.1	.829	3.2
Sorghum	208	.342	7.4	.595	5.8
Stems	10	.911	2.8	.951	2.2
Leaves	3	.180	2.3	--	--
Seed Heads	10	.322	9.4	.423	9.3
Aquatics	152	.572	12.8	.661	11.5
Emergents	54	.513	12.6	.519	12.5
Floating	98	.741	9.6	.881	6.5
Water Hyacinth	55	.699	9.1	.843	6.5

Table 3. Number of samples (n), R^2 and $S_{y \cdot x}$ values for predictions of fermentability (IVTVSD) from cell wall constituents (or neutral detergent fiber ash-free, NDFA), lignin as a percent of NDFA, nitrogen (N) and phosphorus (P)

		IVTVSD vs. NDFA & Lignin		IVTVSD vs. NDFA, Lignin & N		IVTVSD vs. NDFA,Lignin N & P	
	n	R^2	$S_{y \cdot x}$	R^2	$S_{y \cdot x}$	R^2	$S_{y \cdot x}$
All samples	857	.775	9.7	.783	9.5	.784	9.5
Terrestrial	690	.803	8.9	.837	8.1	.856	7.6
Woody	15	.483	6.5	.666	5.4	.688	5.5
Herbaceous	675	.802	8.4	.830	7.8	.847	7.4
Legumes	40	.933	5.3	.934	5.4	.947	4.9
Brassicas	37	.935	4.3	.964	3.3	.964	3.3
Grasses	317	.672	6.6	.739	5.9	.739	5.9
Napiergrass	73	.786	5.5	.848	4.7	.896	3.9
Stems	8	.979	1.8	.984	1.4	.997	0.9
Leaves	9	.829	3.2	.858	3.2	.942	2.3
Sorghum	179	.585	5.9	.606	5.8	.671	5.3
Stems	10	.908	2.8	.924	2.7	.999	0.3
Leaves	3	--	--	--	--	--	--
Seed Heads	10	.423	9.3	.461	9.7	.512	10.1
Aquatics	136	.674	11.1	.709	10.5	.710	10.5
Emergents	42	.444	13.9	.616	11.7	.677	10.9
Floating	94	.881	6.5	.884	6.5	.886	6.5
Water Hyacinth	54	.849	6.4	.850	6.4	.874	6.0

differs greatly among the groups. For example, NDFA and lignin/NDFA account for 93% of the variation in fermentability in legumes, but only 48% of the variation in woody samples.

As can be seen in Table 3, adding nitrogen and phosphorus to the multiple regression analysis has little effect in most cases. However, nitrogen increases the R^2 value for woody biomass, and phosphorus increases the R^2 values for the sorghum groups.

Adding tannin to the prediction equations for sorghums (Table 4) increases the R^2 values only slightly in all groups except for the sorghum seed heads, in which the R^2 value increases significantly. It would be expected that tannin would have the greatest effect on the prediction of fermentability of seed heads, because the highest concentrations of tannins were measured in sorghum seed heads (Table 1).

Table 4. Number of samples (n), R^2 and $S_{y \cdot x}$ values for predictions of fermentability (IVTVSD) from cell wall constituents (or neutral detergent fiber ash-free, NDFA), lignin as percent of NDFA, vanillin-HCl tannin (VTan) and protein-precipitation tannin (PTan).

	n	IVTVSD vs. NDFA & Lignin		IVTVSD vs. NDFA, Lignin & VTan		IVTVSD vs. NDFA, Lignin & PTan	
		R^2	$S_{y \cdot x}$	R^2	$S_{y \cdot x}$	R^2	$S_{y \cdot x}$
Sorghum	132	.788	4.6	.819	4.3	.856	3.8
Stems	10	.951	2.2	.960	2.1	.951	2.4
Leaves	3	--	--	--	--	--	--
Seed Heads	10	.423	9.3	.818	5.6	.924	3.6

The rumen fermentation model discussed here probably has greater application to the first stage of two-stage methane digesters--substrate disappearance and volatile fatty acid formation--than to the second stage--conversion of volatile fatty acids to methane. Additional work is needed to determine the application of this model to the prediction of potential methane production.

ACKNOWLEDGEMENTS

The authors appreciate the contributions of Mr. Garry Foster and Ms. Phyllis Hansen in conducting the laboratory analyses. This paper reports results from a project that contributes to a cooperative program between the Institute of Food and Agricultural Sciences of the University of Florida and the Gas Research Institute titled, "Methane from Biomass and Waste."

REFERENCES

1. P. J. Van Soest. Nutritional Ecology of the Ruminant. O & B Books, Corvallis. Oregon. 1982.
2. T. Swain. Phenolics in the environment. Rec. Adv. Phytochem. 12:617-640. 1979.
3. H. B. Harris, D. G. Cummins and R. E. Burns. Tannin content and digestibility of sorghum grain as influenced by bagging. Agron. J. 62:633-635. 1970.
4. J. M. A. Tilley and R. A. Terry. A two-stage technique for the in vitro digestion of forage crops. J. Brit. Grassl. Soc. 18:104-111. 1963.

5. J. E. Moore and G. O. Mott. Recovery of residual organic matter from in vitro digestion of forages. J. Dairy Sci. 57:1258-1259. 1974.

6. H. K. Goering and P. J. Van Soest. Forage fiber analyses (apparatus, reagents, procedures and some applications). Agriculture Handbook No. 379. U.S. Dept. of Agriculture. Washington, D.C. 1970.

7. M. L. Price, S. Van Scoyoc and L. G. Butler. A critical evaluation of the vanillin reaction as an assay for tannin in sorghum grain. J. Agric. Food Chem. 26:1214-1218. 1978.

8. A. E. Hagerman and L. G. Butler. Protein precipitation method for the quantitative determination of tannins. J. Agric. Food Chem. 26:809-812. 1978.

ECONOMICS OF METHANE GENERATION FROM LIVESTOCK

AND POULTRY WASTES IN THE SOUTH

Harold B. Jones, Jr. and E. A. Ogden

Economic Research Service, USDA
University of Georgia
Athens, Georgia 30602

ABSTRACT

Economic prospects for methane gas production from anaerobic digestion of livestock and poultry wastes have changed substantially in the last several years. This study presents updated estimates of capital investment costs and economies of scale for anaerobic digestion systems for various size livestock and poultry farms in the South. Capital investment costs vary from $28,000 for small digesters to $1.3 million or more for large systems. Annual capital and operating costs decline from 12.5 to 3.0¢ L^{-1} (47.2 to 11.2¢ per gallon) of digester capacity as digester sizes increase. Methane gas output and value will vary with size and type of enterprise. With methane gas valued at its highest use, as a substitute for LP gas, fully utilized on-farm, the break-even point for economic usage would be about 40,000 laying hens, 260 dairy cows, production of 3,600 hogs annually, and 800 beef cattle fed per year. Availability of lower priced fuels, lack of off-site markets, and higher capital costs for scrubbers, compressors, or electrical generation equipment (which may be necessary for larger farms) will reduce the potential for producing methane gas from livestock and poultry wastes. Additional investment tax credits or other subsidies will be necessary to encourage expansion of methane generation unless conventional energy prices increase substantially.

Keywords. Biomass energy, animal wastes, anaerobic digestion, energy costs.

INTRODUCTION

The production of methane gas from anaerobic digestion of livestock and poultry wastes is one alternative energy source that has been explored in some depth since the energy crisis developed in the early 1970's. Earlier reports showed that this process was not economically feasible;[1,2,3] other studies showed that it was possible to develop viable systems that could be feasible in some situations.[4,5,6,7,8,9,10,11] A number of digesters have been built on commercial farms,[12,13,9,10,14,15] but not all of these have continued to operate over time. The economic prospects for methane gas production on farms has changed substantially in the last several years. Digester designs have been improved and long term conventional energy prices have continued to increase which have made the potential for systems more feasible. However, capital investment costs and interest rates have also increased substantially which are negative factors. Also, petroleum based energy prices have been dropping somewhat in the last year or two, reducing the economic incentive for developing anaerobic digestion systems.

This study assesses the economic feasibility of methane gas production from livestock and poultry wastes based on economic conditions as of 1983. The objectives of the study are to update capital investment and operating costs, determine the possible economies of scale in anaerobic digestion systems, and estimate the minimum size farms where such systems might become profitable.

PROCEDURE

A review of previous studies and operational practices of anaerobic digestion systems for livestock and poultry wastes was undertaken. Capital investment requirements and operating costs were compiled for different sizes and types of digester systems. Since these operations were established at different time periods, costs were updated to 1983 using various cost indexes. Annual fixed and operating costs for different size digesters were then estimated. Using digester yield factors from recent studies, biogas production rates were estimated for various types and sizes of farms. Methane gas derived from biogas was then valued at its highest on-farm use to determine net farm value for this type of energy for different size farm enterprises. This provided a basis for estimating break-even points where methane gas production may be feasible.

THE ANAEROBIC DIGESTION PROCESS

Anaerobic digestion is believed to be the most feasible process for converting manure into energy.[16] It is a relatively efficient conversion process producing medium Btu gas from biochemical processes. The digestive process produces biogas from the natural bacterial decomposition of organic matter in a controlled environment without access to air. Large airtight digesters are fed a slurry of wastes of certain consistency at prescribed time intervals, and these wastes are then held within certain temperature ranges for specific time periods. The nature and composition of the wastes determine the loading rate, temperature, and retention time of the system. Most systems are site-specific with these variables being determined individually for each operation.

In the anaerobic digestion process, the wastes are digested in three steps: (1) enzymatic hydrolysis of the animal matter, (2) conversion of the decomposed matter to fatty acids, and (3) conversion of the acids to methane and carbon dioxide by anaerobic bacteria.[17] Optimum gas yields require a proper chemical and physical environment with nutritionally balanced feedstocks. Digester temperatures and retention time must also be closely controlled since they are important factors determining biogas yields. Very low ambient temperatures will reduce the net gas available for other uses due to the higher supplemental heat required. Digester designs can vary from the relatively simple plug flow type to more complex multi-tank batch systems, two and three stage digesters, or variable feed types. Many small farm digesters are the plug flow type due to their low cost and ease of operation.[17]

The anaerobic digestion process produces a collectible giogas with an average methane content of 60% plus a digested sludge which can be used as a soil fertilizer or livestock feed ingredient. Because of pretreatment procedures which may be required for the manure, a comprehensive waste management system is necessary to feed the digester and handle its by-products. A complete manure disposal and biogas production system would include: (1) a manure collection method such as pit-scraping with front-end loaders as in many poultry houses and beef feedlots, or slotted floors and flushing as in many dairy and swine operations; (2) storage for wastes prior to digestion in the form of pits, bunkers, or stacks for dry manure, or in holding tanks, ponds, or lagoons for liquid systems; (3) a feedstock delivery system such as gravity flow, liquid pumping, or screw type conveyors for solids for pre-mixing and possibly pre-heating; (4) an anaerobic digester;

(5) a gas collection system which may involve scrubbing equipment; and (6) a system to handle the digester effluent or sludge which could be gravity flow, liquid pumping, liquid separation by centrifugation or lagooning (Figure 1).

The methane gas produced by the system could be used as a fuel for boilers or to replace natural gas, fuel oil, or LP gas for other uses on the farm. The various energy options and pathways are illustrated in Figure 1. Some of the energy (about 15%) will be required to heat the digester and run the system. Energy use on the farm will depend on current needs. In large-scale operations, the gas can be scrubbed to remove impurities and compressed for sale to pipeline companies. The gas can also be used to power engine-generators to produce electricity for on-farm operations or possibly for sale to electric utilities in some states. Methane gas is not practical as a mobile fuel due to the high pressure needed to keep it in liquid form, and the high cost of compression makes it very expensive to store for future use.

CAPITAL INVESTMENT AND OPERATION COSTS

Capital investment costs for anaerobic digestion systems will vary widely depending upon size of the digester, design type, site location, and other factors. Digester sizes generally range from 3.785×10^4 to 1.135×10^6 L capacity (10,000 to 300,000 gallons) although larger multiple tank systems are possible. Digesters of a given size will often have the capacity to digest manure from various species at somewhat different rates depending upon the characteristics of the manure and other factors. The relationship between number of animal units and digester capacity is given in Figure 2, based on data from previous studies of operating systems. Animal units are expressed in terms of laying hens, but animal number equivalents based on manure output rates are: 1,000 laying hens = 20 fed beef cattle, 6.5 dairy cows, or 180 hogs.[25] Since digester volumes vary significantly by size and type of enterprise, the mid-range line of the relationship in Figure 2 was used to establish digester sizes for various enterprises (Table 1).

The capital investment required to construct anaerobic digesters of different sizes was based on studies and reported costs from operations established in the mid to late 1970's and early 1980's. There were wide variations in actual costs depending on type of digester and year of

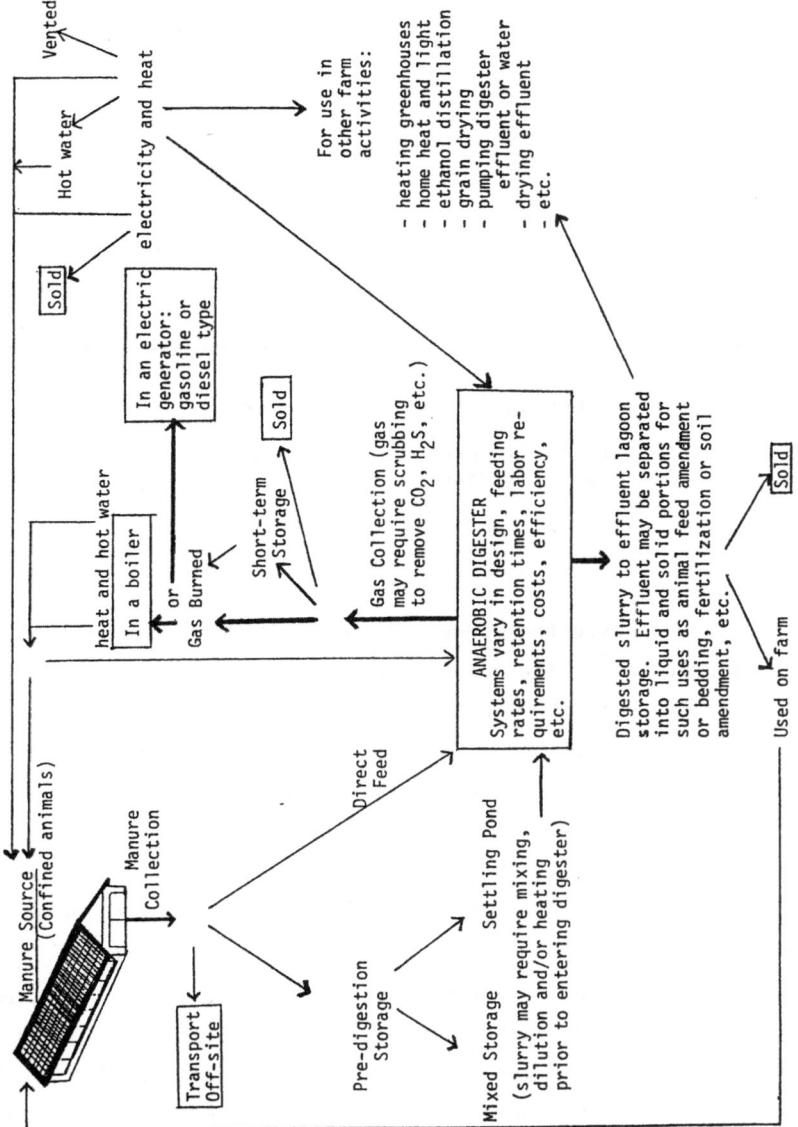

Figure 1. General diagram of anaerobic digestion system options and energy pathways

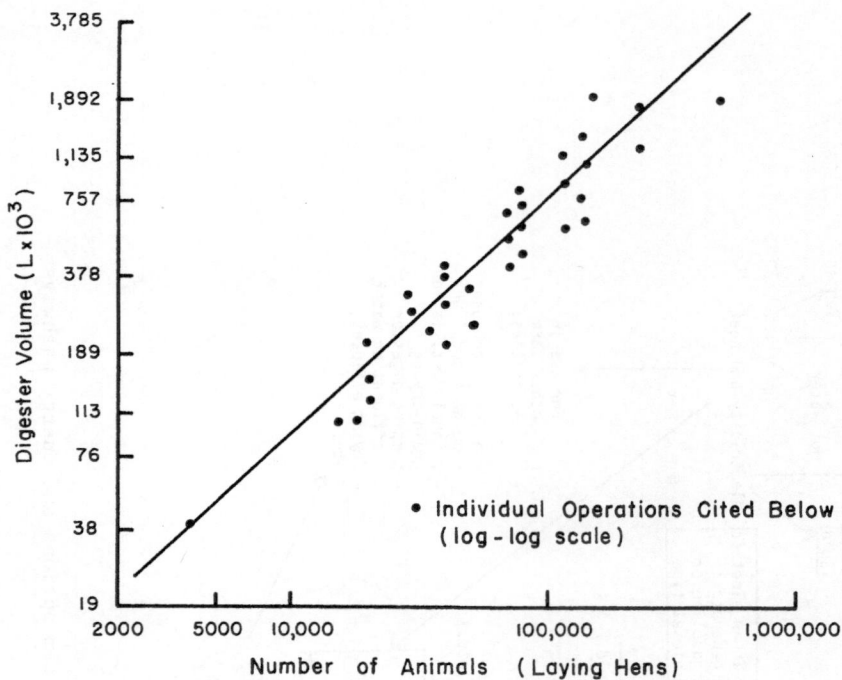

Figure 2. Relationship between digester volume and number of animal units, expressed in terms of laying hens, various studies, 1974-84.

construction. All costs were therefore updated to 1983 based on published construction cost indexes.[26] The adjusted costs are shown in Figure 3. They vary from $25,000 to $30,000 for very small digesters to over $1 million for digesters for large scale feedlot systems. Despite the obvious correlation between capital investment costs and size of digester, there is still considerable variation in costs due to differences in design factors, labor costs used in constructing the digesters, site locations, and the amount and types of other equipment included. The costs on or below the bottom line in Figure 3 were primarily for the basic digester itself, and not the related equipment such as engine-generators or scrubbers. The top line shows what these costs would be with most of the optimal equipment, not all of which would be useful or practical for every operation. The top line also assumes turn-key construction costs for digesters which are substantially higher than farmer constructed systems. Bottom line costs were used to calculate capital investment costs per liter and annual fixed costs (Table 1). Fixed costs were based on 12-year depreciation of digester facilities after an investment tax credit of 10%, an energy tax credit of 10%, plus tax savings from additional depreciation allowed based on the original investment minus one-half of the credits,[22] 14% interest rates, 1%

Table 1. Estimated capital investment costs and economies of scale in anaerobic digester systems, U.S. 1983.

Size of Digester[a]	Capital Investment Costs[b]		Annual Capital and Operating Costs				Annual Costs
	Total	Per Liter	Fixed[c]	Labor[d]	Utility[e]	Total	per Liter
(L x 10^3)	($ x 10^3)	($)	($ x 10^3)				($)
48.4	28	.578	4.0	1.2	0.8	6.0	.125
94.6	44	.465	6.3	1.8	1.0	9.1	.096
174.1	72	.413	10.4	2.8	1.2	14.4	.083
321.7	115	.357	16.6	4.3	1.5	22.4	.070
605.6	190	.314	27.4	7.0	1.8	36.2	.060
1,135.5	300	.264	43.2	11.0	2.2	56.4	.050
2,081.7	490	.235	70.6	16.0	2.6	89.2	.043
3,785.0	800	.211	115.2	24.0	3.0	142.2	.038
7,570.0	1,300	.172	187.2	34.0	3.4	224.6	.030

[a]Estimated from correlation of digester sizes from various studies (Figure 2).

[b]Based on cost relationships from previous studies updated to 1983 (Figure 3).

[c]Annual fixed costs based on 12-year depreciation of digester facilities and 14% interest rates, after tax credits of 20% on capital investment plus tax savings from additional depreciation allowed based on original investment, one percent of initial investment for taxes and insurance, and 2% of initial investment for repairs and maintenance.

[d]Labor costs estimated from previous studies. Reflects charge for loading and unloading the digester and monitoring the system. Wage rates at $4.75 per hour.

[e]Utility charges include electricity and water. Estimates based on previous studies. Assumes price of 6.5¢ kWh^{-1} for electricity.

of initial investment for taxes and insurance, and 2% of initial investment for repairs and maintenance. Annual labor costs and utility costs estimated from previous studies were also updated to 1983 (Table 1).

Results from Table 1 were used to derive the economies of scale relationship illustrated in Figure 4. This relationship shows that annual costs per liter of digester capacity decline from 12.5¢ in the smallest digester (48.4 x 10^{3}L or 12,800 gallons) to 3.0¢ in the largest digester (7.57 x 10^{6}L or 2 million gallons). Thus, there are substantial economies of scale in anaerobic digestion systems, but most of these economies are achieved when digester capacities reach the 321.7 x 10^3 to 378.5 x 10^{3}L level (85,000 to

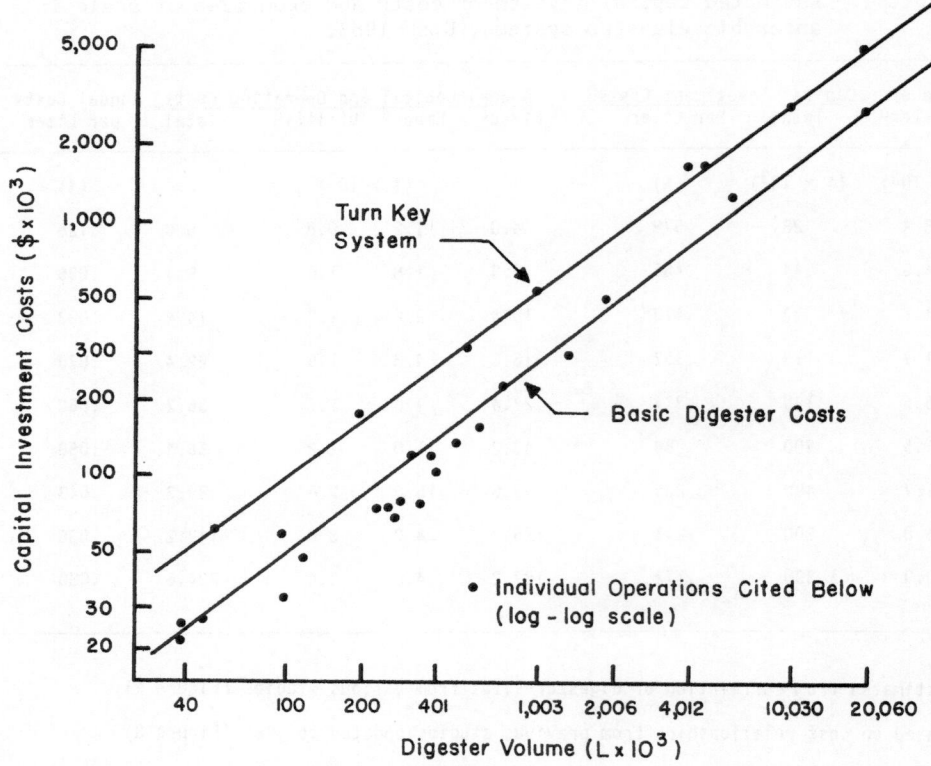

Figure 3. Relationship between digester volume and capital investment costs, various studies, 1974-84.

Sources: 12,27,19,4,20,13,21,28,29,22,5,6,30,7,8,9,23,10,14,24,3,31,11

100,000 gallons, or the capacity equivalent of 40,000 laying hens, 260 dairy cows, 800 fed beef cattle, and 7,200 hogs). Beyond that point, cost savings due to size become smaller. Costs for the digester operation itself continue to decline even for very large capacity operations, but the larger operations would probably need more auxiliary equipment such as scrubbers, compressors, or engine-generators, in order to make effective use of the methane gas produced. This will raise the total costs for larger operations unless they have very large on-farm energy requirements or well established market outlets.

ENERGY PRODUCTION AND NET FARM VALUE

Recoverable energy from the anaerobic digestion process will depend upon a number of factors: biogas yields, manure characteristics, storage

462

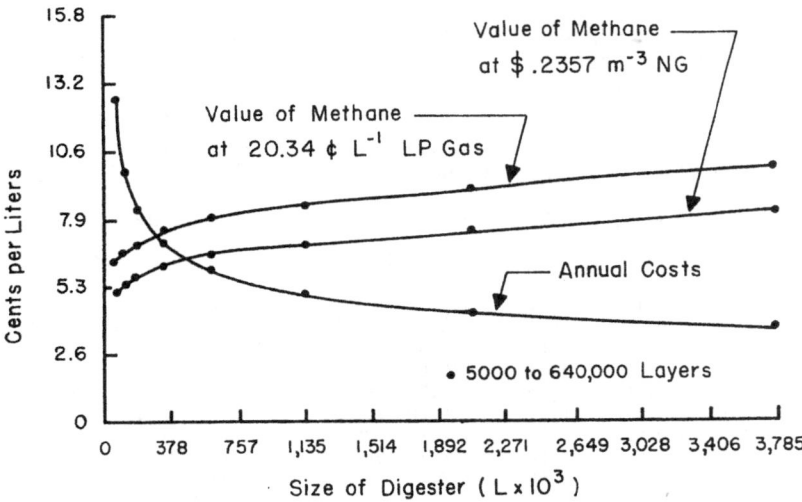

Figure 4. Relationship between annual costs and value of methane gas produced per liter of digester capacity for various size anaerobic digester systems, laying hen enterprises, 1983.

and handling losses, farm operating practices, and conversion losses in electrical or gas generation. Estimated manure yields on a dry weight basis were 11.43 Mg 1000 layers^{-1} (12.6 tons per 1,000 layers) annually, 1.77 Mg dairy cow^{-1} (1.95 tons per cow) annually, .063 Mg pig^{-1} (.07 tons per pig) raised, and .562 Mg beef cow^{-1} (.62 tons per cow) fattened in feed lots.[25] This manure, when converted to slurry form, becomes the input for the digester. The potential biogas yield will vary with different types of manure. Therefore, biogas production was calculated at three different yield levels for each type and size of enterprise (Table 2). These biogas yield levels were .240, .300, and .360 m^3 kg^{-1} (3.88, 4.85, and 5.82 ft^3 lb^{-1}) of total solids for laying hens; .233, .300, and .366 m^3 kg^{-1} (3.77, 4.85, and 5.92 ft^3 lb^{-1}) for dairy cows; and .263, .329, and .394 m^3 kg^{-1} (4.26, 5.32, and 6.38 ft^3 lb^{-1}) for swine and fed beef. These production rates were based on yields from operating digester systems.[19,29,7,25] Methane gas output was based on biogas with a 60% methane content with 22.628 x 10^6 J m^{-3} (600 Btu's per ft^3) of energy from biogas.

The methane gas produced can be used in a number of ways: as a substitute for other fuels on the farm, to power engine-generators for electricity, or sold to other users off-farm. This study considers only option one, use as a substitute for other fuels on-farm. The most common fuel used on farms in rural areas is LP gas, which is also the highest value fuel substi-

Table 2. Annual biogas production and estimated net farm values from anaerobic digestion of livestock and poultry wastes, various size farms, U.S. 1983.

Farm Type and Size	Biogas Production at Various Digester Yields[a]			Value of Methane at Various Digester Yields[b]			Net Farm Value at Various Digester Yields[c]		
	Low	Medium	High	Low	Medium	High	Low	Medium	High
	($m^3 \times 10^5$)			($ $ \times 10^3$)			($ $ \times 10^3$)		
Laying Hens									
5,000	0.14	0.17	0.20	2.38	2.97	3.57	-3.66	-3.07	-2.48
10,000	0.27	0.34	0.41	6.17	5.94	7.13	-4.38	-3.19	-2.00
20,000	0.55	0.68	0.82	9.51	11.89	14.26	-4.86	-2.48	-0.10
40,000	1.09	1.37	1.64	19.02	23.77	28.53	-3.34	1.42	6.17
80,000	2.19	2.74	3.28	38.04	47.55	57.06	1.88	11.30	20.90
160,000	4.38	5.47	6.57	76.08	95.10	114.12	19.68	38.70	57.72
320,000	8.76	10.95	13.14	152.16	190.20	228.24	63.00	101.04	139.08
640,000	17.52	21.90	26.28	304.32	380.40	456.48	162.12	238.20	314.28
1,280,000	35.04	43.80	57.56	613.40	766.74	920.09	388.79	542.14	695.49
Dairy Cows									
32	0.13	0.17	0.21	2.29	2.94	3.59	-3.75	-3.10	-2.45
65	0.27	0.34	0.42	4.65	5.98	7.30	-4.49	-3.16	-1.84
130	0.53	0.69	0.84	9.29	11.96	14.60	-5.07	-2.41	0.23
260	1.07	1.38	1.68	18.59	23.92	29.19	-3.77	1.56	6.83
520	2.14	2.75	3.36	37.18	47.83	58.38	1.02	11.67	22.23
1,040	4.28	5.51	6.72	74.36	95.67	116.77	17.96	39.27	60.37
2,080	8.56	11.02	13.45	148.73	191.33	233.54	59.57	102.17	144.38
4,160	17.12	22.03	26.89	297.45	382.66	467.09	155.25	240.46	324.89
8,320	34.25	44.06	53.76	594.90	765.33	934.17	370.30	540.73	709.57
Swine									
900	0.15	0.19	0.22	2.61	3.26	3.91	-3.43	-2.78	-2.13
1,800	0.30	0.38	0.45	5.22	6.52	7.82	-3.92	-2.62	-1.32
3,600	0.60	0.75	0.90	10.44	13.04	15.64	-3.93	-1.33	1.27
7,200	1.20	1.50	1.80	20.88	26.08	31.27	-1.48	3.72	8.91
14,400	2.40	3.00	3.60	41.77	52.16	62.55	5.60	16.00	31.39
28,800	4.81	6.00	7.20	83.53	104.32	125.10	27.13	47.92	68.70
57,600	9.62	12.01	14.41	167.06	208.63	250.20	77.90	119.47	161.04
115,200	19.24	24.02	28.81	334.12	417.26	500.40	231.92	275.06	358.20
230,400	38.47	48.05	57.62	668.25	834.53	1,000.80	443.65	609.93	776.21
Fed Beef									
100	0.15	0.18	0.22	2.57	3.21	3.85	-3.47	-2.83	-2.19
200	0.30	0.37	0.44	5.14	6.42	7.69	-4.00	-2.72	-1.44
400	0.59	0.74	0.88	10.28	12.83	15.39	-4.09	-1.54	1.02
800	1.18	1.48	1.77	20.55	25.66	30.78	-1.81	3.30	8.42
1,600	2.37	2.95	3.54	41.10	51.33	61.56	4.94	15.17	25.40
3,200	4.73	5.91	7.09	82.20	102.66	123.11	25.80	46.26	66.71
6,400	9.47	11.82	14.18	164.41	205.32	246.23	75.25	116.16	157.07
12,800	18.93	23.64	28.35	328.82	410.64	492.46	186.62	268.44	287.26
25,600	37.86	47.28	56.70	657.64	821.28	984.92	433.04	596.68	760.32

[a]Biogas yield levels for laying hens were .240, .300, and .360 m^3kg^{-1} of total solids; dairy cows were .233, .300, and .366 m^3kg^{-1}; swine and fed beef were .263, .329, and 3.94 m^3kg^{-1}. Manure yields were 11.43 Mg 1000 layer^{-1} annually dry weight, 1.77 Mg cow^{-1} annually, .063 Mg pig^{-1} raised, and .562 Mg beef cow^{-1} fattened in feed lots.

[b]Based on biogas at 60% methane content with energy value of 22.628 x 10^6Jm^{-3}. Methane gas value based on use as substitute for LP gas on farm with 2.66m^3 of methane equivalent to 3.785 L of LP gas. LP gas price was 20.34¢ L^{-1} (1983 price level). Assumes all biogas produced will be utilized as energy.

[c]Reflects net value from on-site utilization of all methane produced based on digester costs from Table 1. LP gas reflects highest value fuel substitute. Use of methane to generate electricity or selling methane off-farm would increase capital and operating costs, thereby reducing net value of gas production.

tute for methane. Based on a conversion rate of 2.66 m^3 (95 ft^3) of methane gas as equivalent to 3.785 L (one gallon) of LP gas and LP gas price of 20.34¢ L^{-1} (77¢ per gallon) (the 1983 level),[33] the value of methane gas at various digester yields was estimated (Table 2). These values range from $2,288 to $3,909 in the very smallest enterprises (5,000 laying hens or equivalent), and up to $1 million for large scale enterprises with high digester yields. Methane gas values would be lower if substituted for natural gas, fuel oil, or electricity. Also, as the size of enterprise increases, it often becomes harder to find effective ways to utilize the gas which will reduce the potential values below what is illustrated in Table 2.

The estimated net farm values (NFV) from anaerobic digestion of livestock and poultry wastes for various size enterprises are also given in Table 2. Net farm value reflects gross value of the methane gas produced minus annual capital and operating costs for the appropriate size digester (Table 1). Based on these values and assumptions, with methane gas valued at its highest use as a substitute for LP gas at 20.34¢ L^{-1} (77¢ per gallon), and assuming it can be fully utilized on-farm, the break-even point for economic usage would be about 40,000 laying hens, 260 dairy cows, production of 3,600 hogs annually, and 800 beef cattle fed per year (Figure 4). If methane were to replace natural gas (NG) at the rate of $.2357 m^{-3} (Figure 4) ($6.60 per 1,000 ft^3) (the 1983 retail price),[34] the break-even point would be closer to 80,000 hens due to the lower valued substitute. Availability of other lower priced fuels, lack of usage potential on farms, or lack of off-site markets, and higher capital costs for scrubbers, compressors, electrical generation equipment, or related transport and storage equipment which may be needed on larger farms will also reduce the potential returns from producing methane gas from livestock and poultry wastes.

The economic feasibility of anaerobic digestion systems is influenced by a number of important variables in addition to biogas yields. Changes in alternative energy prices, construction costs, interest rates, and utilization potential for the gas can have substantial impacts on NFV. If LP gas prices were to increase 10% to 22.46¢ L^{-1} (85¢ per gallon) compared to 20.34¢ L^{-1} (77¢ per gallon) in Table 2, the NFV for a 40,000 layer enterprise (at the highest digester yield) would increase by $2,964 annually or about 48%. If capital investment costs increased 10% for a digester for 40,000 layers, the NFV would decline $1,656 (or 27%) due to increased fixed costs (after allowing for 10% investment and 10% energy tax credits). If interest rates were to increase 2 percentage points, from 14 to 16%, annual fixed costs for a 40,000 layer digester would increase by \$920 which would reduce NFV by that amount (or 15%). If only 90% of the gas could be utilized on the farm (or sold) and the remaining 10% was lost, the NFV for a 40,000 layer operation would drop by \$2,853 annually for a 46% loss. A 10% decline in digester yield from .360 to 323 m^3 kg^{-1} (5.82 to 5.29 ft^3 lb^{-1}) of total solids would reduce the NFV for 40,000 layers by \$2,598 annually for a loss of 42%. Thus, biogas yields, gas utilization potential, and energy prices are highly significant factors influencing NFV, with changes in construction costs or interest rates having relatively smaller effects on NFV.

CONCLUSIONS

The economic feasibility of anaerobic digestion systems for livestock and poultry farms under present conditions thus appears to be marginal at best, particularly for smaller farms. These systems would be feasible for medium size farms where the methane could be effectively utilized and/or where energy production and fuel needs can be balanced on a daily basis. Energy balances could be most effectively achieved if the total output of gas could be utilized on farm by direct burning for water heating, space heating, air conditioning, refrigeration, or possibly electrical cogeneration with heat recovery. Another alternative would be to design a certain size system to meet the specific energy needs of a farm rather than digesting all of the manure produced. Larger farms would probably require off-site markets in order to generate maximum returns from anaerobic digestion which would require substantially higher capital investment for scrubbers, compressors, or other transport equipment. Construction costs will also be much higher for larger commercial turn-key systems as compared to farmer constructed systems, thereby raising fixed costs for these types of systems.

Due to the technical complexity of the processes involved, any system considered by farmers should be shown to be operationally effective before adoption. Anaerobic digestion systems are more effective in warmer climates where the manure is at a temperature of 4.4° - 10°C (40° - 50°F) or higher. Otherwise, the feedstock will be too cold and a greater heat load will be required to heat the influent to operating temperatures. The economic feasibility of these systems will be greater for farms that utilize high cost fuel and electricity that can be replaced. Since the price of fuel or displaced energy is a key factor in the economics of anaerobic digestion, expectations as to future increases in energy prices should be an important

466

consideration before investing in these systems. Additional investment tax credits or other government subsidies would of course increase the economic feasibility of these operations from the farmers' standpoint and may lead to greater self-sufficiency in energy production.

REFERENCES

1. J. A. Moore. Methane from poultry manure appears unprofitable at OSU. The Poultry Times. Mar. 24, 1980.
2. T. C. Slane, R. L. Christensen, C. E. Willis and R. G. Light. An Economic Analysis of Methane Generation: Internal Costs and External Benefits. Bul. 618, Mass. Agric. Expt. Sta., Univ. of Mass. Amherst, MA. 1975.
3. Ted Thornton. An assessment of anaerobic digestion in U. S. agriculture. ESCS-067. Econ., Statistics, and Coop. Service, U. S. Dept. of Agriculture. Washington, D.C. 1978.
4. David Baylon, E. Coppinger and J. Lenart. Economics and operational experience of full-scale anaerobic dairy manure digester. In: Biogas and Alcohol Fuels Production. Proc. Seminar on Biomass Energy for City, Farm, and Industry. J. G. Press. Emmaus, PA. 1980.
5. Edward L. Fulton. Methane gas production: Is value worth the cost? Paper presented at Texas Commercial Egg Clinic. College Station, TX. 1981.
6. A. G. Hashimoto, Y. R. Chen and V. H. Varel. Anaerobic fermentation of beef cattle manure. SERI/TR/98372-1. Solar Energy Research Institute. Golden, CO. 1981.
7. T. D. Hayes, W. J. Jewell, J. A. Chandler, S. Dell'Orto, K. J. Fanfoni, A. P. Leuschner and D. F. Sherman. Methane generation from small scale farms. In: Biogas and Alcohol Fuels Production. Proc. Seminar on Biomass Energy for City, Farm, and Industry. J. G. Press. Emmaus, PA. 1980.
8. E. S. Kebanli, R. W. Pike, D. D. Culley, Jr., and J. B. Frye, Jr. Fuel gas from dairy farm waste. In: Agricultural Energy. Vol. 2, Biomass Energy Crop Production. Paper from 1980 National Energy Symp. Am. Soc. Ag. Eng., St. Joseph, MI. 1981.
9. Karl Kessler. Feedlot manure keeps home fires burning. The Furrow. Sept. – Oct., 1978.
10. John H. Martin. Farmsite installations of energy harvester anaerobic digesters. In: Biogas and Alcohol Fuels Production. Proc. Seminar on Biomass Energy for City, Farm, and Industry. J. G. Press. Emmaus, PA. 1980.
11. L. P. Walker, R. A. Pellerin, M. G. Heisler, G. S. Farmer and L. A. Hibbs. Anaerobic Digestion on a Dairy Farm: Overview, Draft Report. Dept. of Agric. Engineering. Cornell University. Ithaca, NY. 1984.
12. David Amey. Electricity from chicken manure. Poultry Tribune. Watt Publishing Co., Mt. Morris, IL. Dec., 1982.
13. Boris E. Bravo-Ureta. Economic evaluation of electricity generation from dairy manure based on the Sunny Valley Foundation digestion system. Staff Paper 82-20. Dept. of Agric. Economics, Univ. of Conn. Storrs, CT. 1982.
14. Methane, Successful Farming. Feb., 1980.
15. Methane digester, Broiler Industry, Watt Publishing Co., Mt. Morris, IL. Mar., 1982.
16. Office of Technology Assessment. Energy from Biological Processes. Vol. 1, U.S. Congress, Washington, D.C. 1980.

17. Paul J. Driscoll, R. N. Boisvert and R. J. Kalter. Biomass potential for energy production in the northeast. A. E. Res. 84-2, Dept. Agric. Economics, Cornell Univ., Ithaca, N.Y. 1984.

18. T. P. Abeles and D. A. Ellsworth. Farm integrated utility systems: methods for farm scale recovery of methane. In: Biogas and Alcohol Fuels Production. Proc. Seminar on Biomass Energy for City, Farm, and Industry. J. G. Press. Emmaus, PA. 1980.

19. H. D. Bartlett, S. P. Persson and R. W. Regan. Energy production potential for a 100m^3 biogas generator. In: Agricultural Energy. Vol. 2, Biomass Energy Crop Production. Paper from 1980 National Energy Symp. Am. Soc. Ag. Eng. St. Joseph, MI. 1981.

20. M. Bery, R. Cassanova, R. Combes, J. Giles, R. Mattison, L. Moriarity and D. O'Neil. Poultry waste utilization. Georgia Poultry Industry Research, Final Report. Engineering Experiment Station. Georgia Institute of Technology. Atlanta, GA. 1978.

21. Boris Bravo-Ureta and Glen V. McMahon. The economic feasibility of electricity generation on cage layer operations. Res. Report 79. Storrs Agric. Experiment Station. Univ. of Conn. Storrs, CT. 1984.

22. J. R. Fischer, D. D. Osburn, N. F. Meador and C. D. Fulhage. Economics of swine anaerobic digester. Paper 79-4580. Am. Soc. Ag. Eng. Winter Meeting. New Orleans, LA. 1979.

23. J. T. Kiker. Anaerobic Digestion. Coop. Ext. Serv. Univ. of Georgia. Athens, GA. 1974.

24. D. A. Rockey, W. Turnacliff and R. J. Smith. A 1900 m^3 digester for laying hen manure. Paper 78-4569. Am. Soc. Ag. Eng. Winter Meeting. Chicago, IL. Dec. 18-20, 1978.

25. Harold B. Jones, Jr. and E. A. Ogden. Energy potential from livestock and poultry wastes in the south. Agric. Econ. Report 522. Economic Research Service. U.S. Dept. of Agric. Washington, D.C. 1984.

26. U. S. Bureau of the Census. Statistical abstract of the United States: 1984. 104th edition. Washington, D.C. 1983.

27. E. Ashare, D. L. Wise and R. L. Wentworth. Fuel Gas Production from Animal Residue. Engineering Report 1628. Dynatech R/D Co., Cambridge, MA. 1977.

28. L. F. Diaz and J. C. Glaub. Overview of selected biogasification installations in the U.S. In: Biogas and Alcohol Fuels Production. Proc. Seminar on Biomass Energy for City, Farm, and Industry. J. G. Press. Emmaus, PA. 1980.

29. J. R. Fischer, E. L. Iannotti and C. D. Fulhage. The engineering, economics, and management of a swine manure digester. In: Agricultural Energy. Vol. 2, Biomass Energy Crop Production. Paper from 1980 National Energy Symp. Am. Soc. Ag. Eng. St. Joseph, MI. 1981.

30. A. G. Hashimoto, Y. R. Chen and R. L. Prior. Thermophilic, anaerobic fermentation of beef cattle residue. In: Energy from Biomass and Wastes. Institute of Gas Technology. Chicago, IL. 1978.

31. M. B. Timmons. Biogas systems for poultry operations. Poultry Digest. Feb., 1984.

32. U. S. Dept. of Treasury. Farmers Tax Guide. Publ. 225, Internal Revenue Service. Washington, D.C. 1983.

33. U. S. Dept. of Agriculture. Agricultural Prices 1983 Summary. Pr 1-3(84). Crop Reporting Board. Stat. Reporting Serv. Washington, D.C. 1984.

34. U. S. Dept. of Energy. Monthly Energy Review. Energy Information Administration. Washington, D.C. 1983.

BIOGASIFICATION OF WATER HYACINTH AND PRIMARY SLUDGE

D. P. Chynoweth, R. Biljetina and V. J. Srivastava

Institute of Gas Technology
3425 South State Street
Chicago, Illinois 60616

T. D. Hayes

Gas Research Institute
8600 West Bryn Mawr Avenue
Chicago, Illinois 60631

ABSTRACT

This paper describes the results of research in progress to evaluate
the use of water hyacinth for wastewater treatment and subsequent conversion
of hyacinth and sludge to methane by anaerobic digestion. Laboratory
studies have been directed toward evaluating advanced biogasification concepts and establishing a data base for the design and operation of an experimental test unit (ETU) located at the water hyacinth wastewater treatment
facility at Walt Disney World (WDW) in Lake Buena Vista, Florida. Kinetic
experiments have been conducted using continuously stirred tank reactors
(CSTR) and a novel non-mixed vertical flow reactor (NMVFR) receiving a
hyacinth/sludge blend at retention times of 15 down to 2.1 days. The data
suggest that the best performance is achieved in the NMVFR which has longer
solids and organism retention. A larger-scale experimental test unit (4.5
m^3) was designed and installed at WDW in 1983 and started up in 1984. The
purpose of this unit is to validate laboratory experiments and to evaluate
larger-scale equipment used for chopping, slurry preparation, mixing, and
effluent dewatering. The ETU is currently being operated on a 2:1 blend
(dry wt basis) of water hyacinth and primary sludge. Performance is good
without major operational problems. Results are presented.

Keywords. Water hyacinth, sludge, anaerobic digestion, biogasification, wastewater treatment, digester scale-up.

This paper describes the status of research in progress which is part of a multidisciplinary program evaluating the use of water hyacinth for reclamation of domestic wastewater and anaerobic digestion for conversion of the hyacinth crop and primary sludge to methane.

The system concept, illustrated in Figure 1, employs water hyacinth ponds for secondary and tertiary treatment of effluent from primary treatment (removes settleable solids). Effluent supernatant is passed through water hyacinth ponds which effect organic and nutrient reduction. Collected primary sludge and harvested water hyacinth are added as a blend to the anaerobic digestion process, where a portion of the organic matter is converted to methane and carbon dioxide. The methane is separated from carbon dioxide and used as an energy source. Digester residue solids may be post-treated to improve biodegradability and recycled to the digester or used for land fertilization along with the digester supernatant.

The use of water hyacinth for secondary and tertiary treatment of domestic sewage sludge has been demonstrated to be technically feasible in several studies.[1-5] These emergent plants effect direct uptake of inorganic nutrients and other elements and provide a surface for microbial attachment and subsequent oxidation of organic matter. The prolific growth of hyacinth and ease of harvest make it a suitable feedstock for biological conversion to methane.

This scheme not only requires less energy than conventional wastewater treatment, but it has the potential for net energy production as well as the production of a higher-quality effluent. The process solves a waste disposal problem, reduces gas requirements for wastewater treatment, and adds gas as a process product to the overall gas supply.

This biomass waste treatment energy conversion scheme has served as a basis for the following overall program objectives:

* To determine the feasibility of using water hyacinth for treatment of sewage to secondary and tertiary standards

* To maximize hyacinth growth yields while maintaining effective wastewater treatment

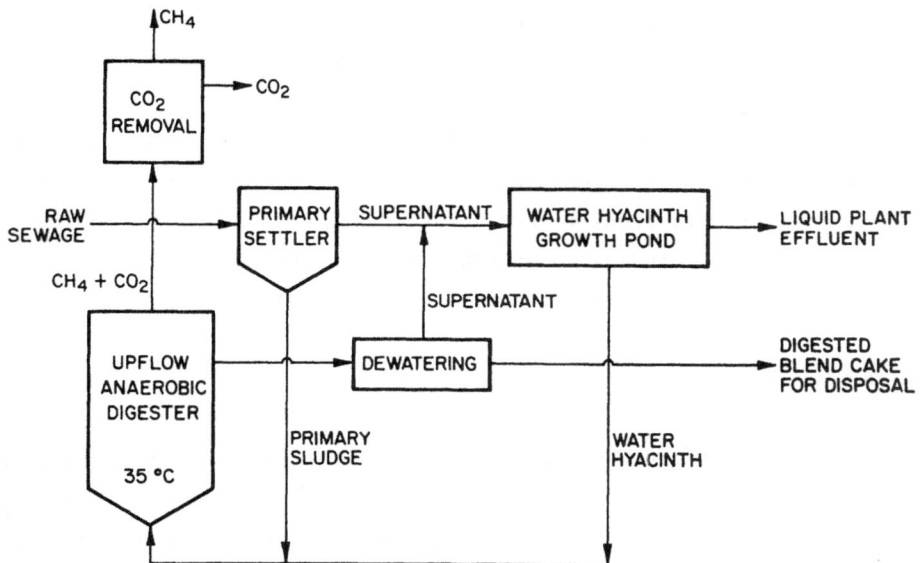

Figure 1. Schematic diagram of integrated hyacinth growth and digestion system

* To determine the feasibility of integrating a biogasification process with the hyacinth treatment scheme for conversion of the primary sludge and hyacinth to methane

The project is centered around the main growth and conversion facility located at WDW, and currently involves several participants including the University of Florida, Institute of Gas Technology (IGT), and several subsidiary companies of Walt Disney Productions. The current sponsor is the Gas Research Institute (GRI). Previous sponsors included the U.S. Environmental Protection Agency (EPA) and the U.S. Department of Energy (DOE). The facility includes five 1/10-ha hyacinth ponds, several smaller hyacinth test vaults, and a pilot-scale ETU that is used for anaerobic digestion studies.

LABORATORY-SCALE ANAEROBIC DIGESTION PROCESS DEVELOPMENT

Anaerobic digestion was selected as the process for energy conversion of the primary sludge and water hyacinth because it produces methane as the principal product, and the process is suitable for feedstocks with a high water content. The objective of this part of the study is to develop a data base for the design and operation of an optimized system for biogasification

471

of the hyacinth and sludge and to integrate this process with the hyacinth wastewater reclamation facility. It is expected that resulting process designs will be applicable to other aquatic species with high water content. The strategy for this work includes 1) baseline laboratory studies under conditions of conventional continuously stirred tank reactor (CSTR) mesophilic anaerobic digestion, 2) advanced digestion laboratory experiments, and 3) larger-scale studies in an ETU. The results and status of the laboratory studies are discussed elsewhere[5-11] and summarized below.

Feed Characteristics

Two types of feeds were utilized for these studies, either separately or as blends: water hyacinth and primary sludge. The physical and chemical characteristics of typical samples of these feeds are shown in Table 1. The hyacinth and sludge feeds contained low solids and ash. The heating values for hyacinth were in the range normally observed for herbaceous biomass: 19.2 to 19.7 MJ kg^{-1} VS (8250 to 8450 Btu lb^{-1} VS). The heating values for sludge ranged from 25.8 to 29.3 MJ kg^{-1} VS (11,100 to 12,600 Btu lb^{-1} VS). C/N and C/P ratios suggest that N and P were not limiting in the hyacinth and sludge feeds. Ultimate methane yields as determined by the anaerobic biogasification potential (ABP) assay (a 60-day batch anaerobic digestion test conducted with dilute feed (0.2%), buffer, and excess nutrients) were

Table 1. Characteristics of typical lots of water hyacinth and sludge used for experimental studies

	Hyacinth	Sludge
Total Solids (TS), %	5.09	4.54
Volatile Solids (VS), % TS	85.1	83.6
Carbon, % TS		
C/H	41.4	47.1
C/N	7.4	6.7
C/P	56	84
Heating Value, MJ kg^{-1} VS	19.2-19.7	25.8-29.3
Stoichiometric Methane Yield, S m^3 kg^{-1} VS*	0.52	0.69
Ultimate Anaerobic Biodegradability, S m^3 CH_4 kg^{-1} VS**	0.25-0.32	0.62

* Calculated from elemental analysis.
** Based on anaerobic biogasification potential assay.

in the range of 0.25 to 0.32 and 0.62 S m^3 kg^{-1} (S m^3 kg^{-1} - as used in this and similar notations throughout this paper, the S denotes "equivalent to volume measured under standard conditions of temperature and pressure") (4.0 to 5.1 and 10.0 SCF lb^{-1}) VS added for hyacinth and primary sludge, respectively. These yields represent an upper expected level anticipated from our bench-scale studies.

Conventional CSTR Studies

Several experiments were conducted to evaluate the performance of hyacinth and sludge and blends of these feeds under conditions of conventional anaerobic digestion; that is mesophilic, a loading of 2.4 kg VS m^3 day (0.15 lb VS ft^{-3}-day), and hydraulic retention time (HRT) of 15 days. Methane yields of 0.19, 0.55, and 0.28 S m^3 kg^{-1} VS (3.0, 8.8, and 4.5 SCF lb^{-1} VS) added were obtained from hyacinth, primary sludge, and a 3:1 blend (dry solids) of hyacinth and sludge, respectively. These yields correspond to 54% and 84% of the ultimate biodegradable yields for hyacinth and sludge as determined by the ABP assay. All three digesters exhibited stable performance with low concentrations of volatile acids, and no pH control was required. The above experiments, repeated in the presence of excess nitrogen and phosphorus, indicated that neither of these nutrients were limiting. Experiments with several blends of hyacinth and sludge showed that methane yields of blends can be predicted from the fractional content and methane yield of each component of the blend.

Several digester experiments were conducted to evaluate the digester performance on hyacinth, sludge, and a hyacinth/sludge blend at several different loading rates. The results indicated that digestion of these feeds adhered to the Monod kinetic model, which was modified to predict methane yield and production rate as a function of organic loading for a stirred tank reactor. The data indicate that maximum methane yields are achieved at lower loadings (longer retention times) and that the loadings giving maximum rate and yield do not correspond. These results may be attributed to the fact that at high loadings, active bacteria and unreacted particulate solids wash out due to the high water content of both feeds.

Unconventional NMVFR Studies

A non-mixed vertical flow reactor (NMVFR) was investigated to obtain improved performance through increased solids and microorganism retention. In this reactor, the feed is added to the bottom of a non-mixed vessel and solids and liquid are removed from the top on a semicontinuous basis. Sedimentation of particulate matter within this reactor results in a longer solids than liquid retention time. Performance data for this reactor type are compared with those from a CSTR for several different loadings in Table 2. Methane yields ranged from 13 to 31% higher in the NMVFR reactor.

Data from evaluation of the performance of the NMVFR at various loadings were used to develop kinetic parameters for the anaerobic digestion of a 3:1 blend of hyacinth/sludge in this reactor. The resulting theoretical plots of methane yield and production rate as a function of organic loading are compared with data from the NMVFR in Figures 2 and 3. The actual data points showed reasonable adherence to the theoretical plots and clearly indicated that performance in the NMVFR was better.

It is recognized that bench-scale units do not offer the opportunity to duplicate the various field operating conditions encountered in a full-scale hyacinth/sludge gasification system. Also, a laboratory system does not permit evaluation of the major unit operations and processes under actual conditions of feed availability, variability of feed characteristics, and other operating conditions (for example, continuous operation, digester solids concentration and recycling, intra-process, recycling, physical, chemical environments, etc.). Perhaps more importantly, a laboratory system does not allow concerted operation of all major processes and subsystems, which is vital to the development of an integrated and reliable biogasification system that will meet the ultimate program goal of methane production in an economically justifiable and energy-effective manner.

Table 2. Comparison of methane yield from 3:1 hyacinth/sludge blend (dry wt basis)

Loading, kg VS m^{-3} day	Methane Yield, S m^3 kg^{-1} VS added	
	CSTR	NMVFR
1.6	0.28	0.35
1.9	0.24	0.34
2.7	0.24	0.28
3.2	0.23*	0.27**
4.8	0.20	0.24**
6.4	0.16	0.21**

* Average of yield at loading rates of 2.77 and 3.7 kg VS m^{-3} day.
** Equipped with mechanical solids/gas separator.

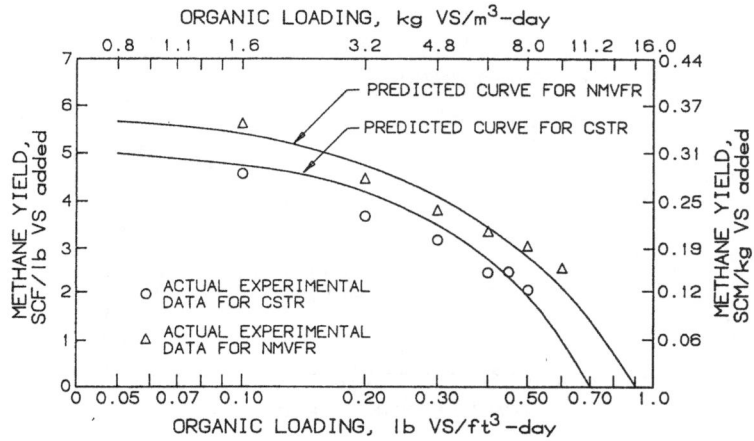

Figure 2. Comparison of actual and predicted methane yields for hyacinth/sludge blend (3:1 dry wt basis) in NMVFR and CSTR

In view of these limitations and the fact that a stable process was obtained on hyacinth/sludge at loadings up to 4.81 kg VS m^{-3} day (0.3 lb VS ft^{-3}-day) with a methane yield in the range of 0.25 to 0.35 S m^3 kg^{-1} (4.0 to 5.6 SCF lb^{-1}) VS added, it was decided to begin planning for design, construction, and operation of a larger-scale ETU. Our results supported selection of the NMVFR rather than the CSTR for design of this digester. Laboratory experiments are continuing to evaluate design and operating conditions to further improve performance.

BIOGASIFICATION EXPERIMENTAL TEST UNIT

Objective

The objective of this phase of the GRI project is to operate and collect data from an ETU of sufficient scale to demonstrate integrated operation of the advanced biogasification process with an existing wastewater treatment system using water hyacinth and other biomass feeds. A facility capable of processing 910 wet kg (1 ton) per day of water hyacinth/sludge was designed by engineers from the Institute of Gas Technology and Rexnord's EnviroEnergy Technology group for the WDW wastewater treatment ponds in Orlando, Florida. It is expected that a definitive bioconversion process design will evolve from the results of the ETU-scale studies for optimization, system evaluation, and development of prototype operation.

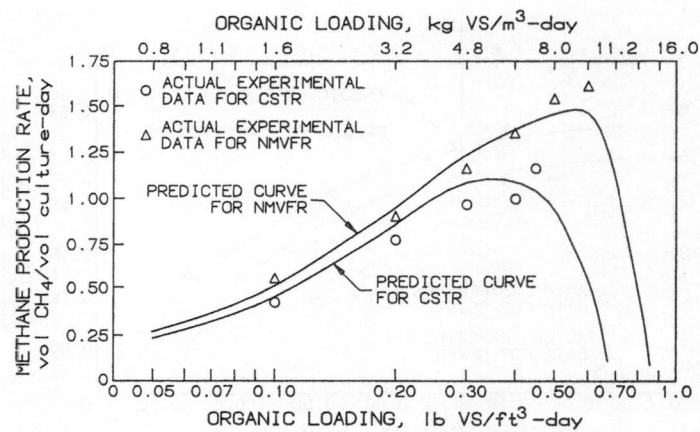

Figure 3. Comparison of actual and predicted methane production rates for hyacinth/sludge blend (3:1 dry wt basis) in NMVFR and CSTR

Design Basis

The scale of operation and the integration of the ETU with the waste-water treatment ponds provide data and operating experience for process development that cannot be obtained in the laboratory. The design of the ETU provides for –

*   Verification of laboratory observations of performance and kinetic data under field conditions using fresh and representative feedstocks.

*   Flexibility in testing selected feed-processing, pretreatment, and digestion configurations at a scale demand sufficient to delineate the best overall process design.

*   Field-scale evaluation of 1) key supporting physical and chemical unit operations for solids management; and 2) such process control problems as maintaining a selected hyacinth/sludge blend ratio, digester hydrau-lic retention time, and solids residence time, loading rate, culture temperature, and pH.

*   Integrated operation of the major physical, chemical, and biological unit operations and processes relevant to feed processing, pretreat-ment, biogasification, residue processing, and disposal.

*   Compilation of preliminary information on maintenance and operating problems, equipment safety, and scale-up effects.

The ETU was sized to guarantee sufficient feedstock from the existing 5.0 ha (1.25 acre) water hyacinth growth facility when operating at high loading rates. Variations in production had to be considered. Monthly yield variations from a high of about 6.7 dry metric tons ha^{-1} (3.0 dry tons acre^{-1}) to a low of 2.2 dry metric tons ha^{-1} (1.0 dry tons acre^{-1}) were included in the design basis.

Description

Figure 4 schematically represents the major processing units of the ETU. Water hyacinth plants are harvested from the ponds at the WDW waste-water treatment plant and transported to the loading platform at the ETU. The plants are first chopped to a nominal 50 mm (2 inch) size and then are fine-ground to a product size consistency variable between 50 and 2 mm (2 and 1/16 inch). The ground water hyacinth is stored in an insulated, cooled, enclosed tank to preserve the inventory during the intermediate storage period. A mixer and pump recirculation are provided to guarantee uniform product delivery to the blending operation. Equipment is sized to process a 3-day inventory in 1 to 2 hours.

Sludge for the ETU is made available from the primary clarifiers of the WDW wastewater treatment plant. A sludge grinding and pumping station is located at the clarifier to chop any foreign materials (plastic, paper, etc.) to a product size less than 3 mm (1/8 inch) in diameter and to pump the sludge through an underground main to the ETU. Collection is timed to interface with the cyclic operation of the plant clarifiers to optimize the sludge solids concentration.

Water hyacinth and sludge are blended to the proper weight ratios. The blend tank is mounted on a set of load cells to provide an accurate account of the solids loading rate to the digester. A variable speed, progressive-cavity pump and cycle timers allow feed to be added continuously or on an intermittent basis. A feed heat exchanger automatically preheats the blend to minimize temperature fluctuations in the digester. The nominal feed rate to the digester is 910 wet kg day^{-1} (2000 wet lb day^{-1}) of mixed feed containing 5% solids. Water hyacinth-to-sludge ratios can be varied from 4:1 to 1:1. Biogasification of the water hyacinth/sludge mixture occurs in a normally unmixed 4.5 m^3 (160 ft^3) digester having a height-to-diameter ratio of 2:1 to promote solids retention. The digester is initially being operated as un upflow solids reactor; however, the unit can be modified to alternative reaction systems. Internal rings have been provided to allow

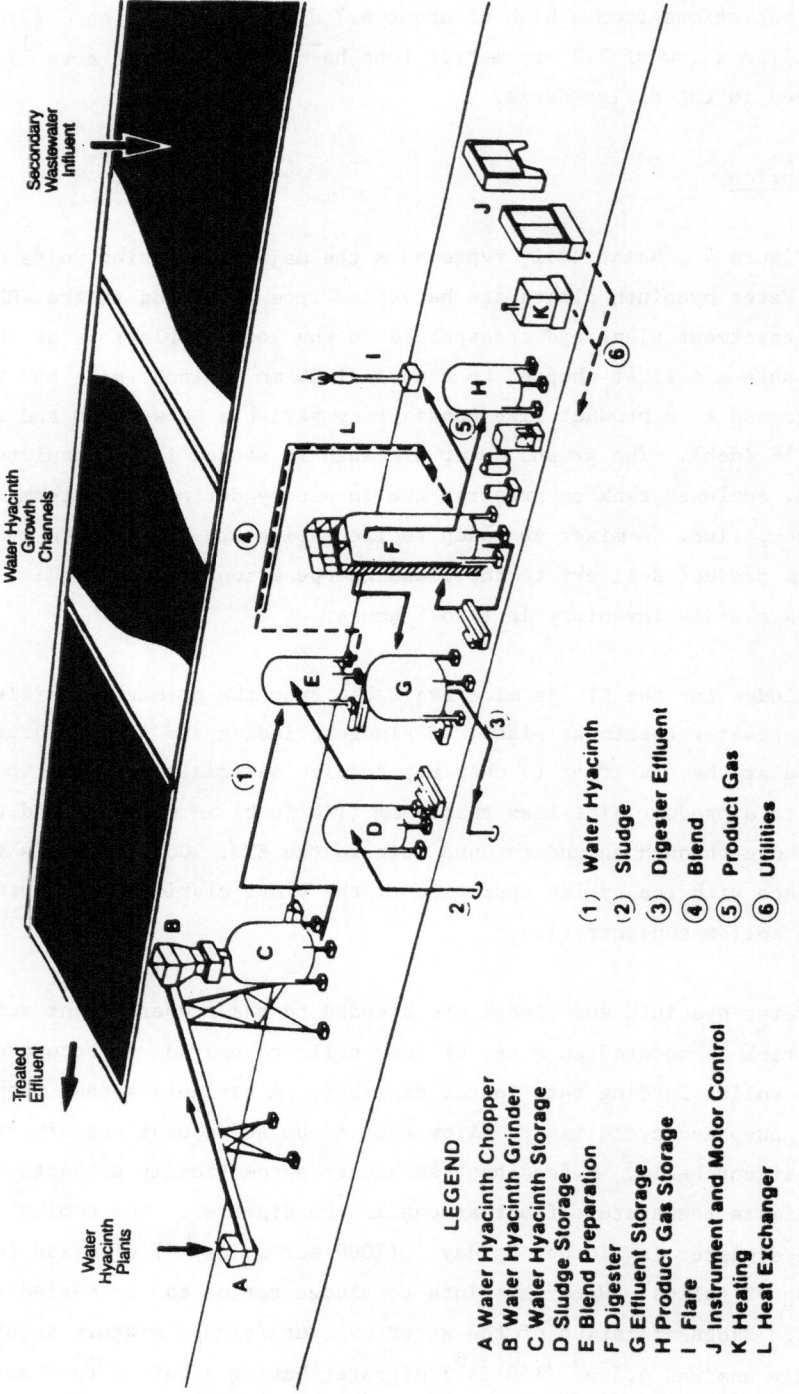

LEGEND

A Water Hyacinth Chopper
B Water Hyacinth Grinder
C Water Hyacinth Storage
D Sludge Storage
E Blend Preparation
F Digester
G Effluent Storage
H Product Gas Storage
I Flare
J Instrument and Motor Control
K Heating
L Heat Exchanger

(1) Water Hyacinth
(2) Sludge
(3) Digester Effluent
(4) Blend
(5) Product Gas
(6) Utilities

Figure 4. Biogasification ETU facility

attachment of baffles, and provisions for both hydraulic and gas mixing are included. The digester is fully jacketed and insulated to allow automatic temperature control of the contents. Twelve thermocouples monitor internal temperatures and sample ports are provided every 0.30 meter of vertical height. A complete gas conditioning storage and handling system is provided. Continuous analysis for methane and carbon dioxide and measurement of the gas flow are provided prior to gas flaring. All pertinent data are continuously recorded on strip charts located in a main control panel.

Effluent from the digester can be removed automatically through an overflow system or by pump withdrawal from a number of locations. Digester effluent can be stored in a 4.5 m^3 (160 ft^3) tank to allow further processing or can be returned to the plant clarifier.

The operating flexibility provided for the ETU is designed to maximize the experimental and data collection capability at this scale of operation within reasonable investment levels. It   expected that the ETU can accept other biomass feedstocks and be operated under different options such as separate processing of the water hyacinth juice and fiber fractions with a minimum of modification. Only minor changes are required to operate the digester as a completely mixed, packed-bed, fluidized-bed or expanded-bed reactor. Ample room for expansion would allow two-phase or multiple-reactor operations.

It is anticipated that these capabilities will lead to the orderly definition and scale-up to prototype operation and that such staged development will result in the construction of optimized commercial digesters with favorable process economics.

Research Plan

The primary objective for the ETU studies was to begin operating the unit in the upflow solids mode and to evaluate performance, scale-up, and materials handling parameters at several different loadings of hyacinth/ sludge blends. After an initial acclimation period on primary sludge, the digester is to receive a water hyacinth/sludge blend (2:1 dry weight basis) at a loading of 3.2 kg VS m^{-3} day (0.2 lb VS ft^{-3}-day). Loadings will be increased sequentially to 9.6 kg VS m^{-3} day (0.6 lb VS ft^{-3}-day) or until failure occurs. At each loading increment, the performance of the unit will be evaluated using the parameters listed in Table 3.

Table 3. ETU evaluation parameters

Performance Parameters	Engineering Parameters
Methane Yield	Materials Balance
Methane Production Rate	Solids
Gas Composition	Carbon
Volatile Fatty Acids	Nitrogen
Temperature	Phosphorus
pH	Energy Balance
Alkalinity	Materials Handling
Organic Matter Reduction	Scale-Up of Laboratory Performance

Long-term goals are to use the flexible design of the ETU to develop a low-cost process for the efficient conversion of water hyacinth/sludge blends and other community-derived wastes to methane.

Status and Results

Experimental operation of the ETU began in January 1984. Initial operation with water hyacinth and sludge blends were conducted in the upflow solids mode at loading rates of 3.2 kg VS m^{-3} day (0.2 lb VS ft^{-3}day) and blend ratios of 2:1 and 1:1 (water hyacinth/sludge solids weight basis). These blend ratios were based on expected commercial seasonal water hyacinth growth variations and secondary wastewater treatment efficiencies. The loading rate was nearly double the rate achieved in conventional stirred tank reactor systems. Recent steady-state data have shown that operation at these conditions can be achieved and have verified earlier laboratory obser-vations. Data validity has been confirmed by material balance calculations, which provided a closure of 105% for this steady-state data period.

Table 4 presents the data summary for a steady-state data period com-pleted between November 5 and December 2, 1984. Loading rates were con-trolled at 3.2 kg VS m^{-3} day and the water hyacinth-to-sludge ratio was maintained at 2:1. Daily gas production, methane yield, methane concentra-tion, and daily loading for the steady-state period are shown in Figure 5.

Performance as determined by methane yield, methane content of the gas, and volatile solid reduction was similar to that observed at the laboratory scale (Table 5) in an NMVFR in previous experiments. Methane yields and volatile solids reduction of the ETU were significantly higher in the ETU (NMVFR) than in a lab-scale CSTR receiving identical fresh feeds.

480

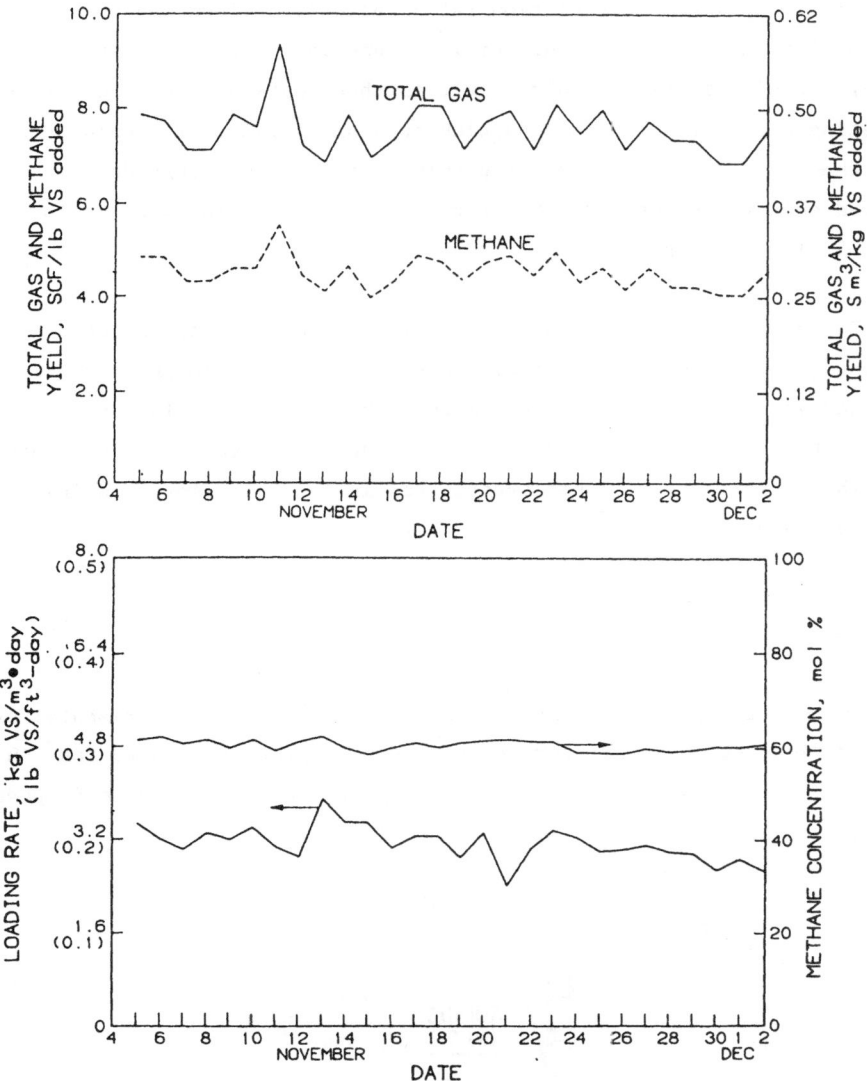

Figure 5. Steady-state performance data for experimental test unit operated
on a 2:1 blend (dry wt basis) of hyacinth sludge at a loading
rate of 3.2 kg VS m^{-3} day

It was also observed that a significant variation in pH, volatile acid,
and solids concentration existed in the unmixed digester for this water
hyacinth/sludge blend ratio. Solids concentration and pH increased with
vertical height in the digester, and volatile acids decreased as measured
from bottom to top in the digester. Figure 6 provides data results of a
sample profile taken on November 27. These data will be very useful in

determining optimum recirculation rates, as well as feed and withdrawal points, for the digester. Further improvements in digester gas yields and conversion efficiencies should result as these data are used to refine operating procedures for this particular biomass feed. It should be noted that these observations were not possible in the laboratory due to the limited culture volume available for such data determinations in the small 5- to 50-liter units.

The experimental plan for 1985 will focus on further increases in loading rate to the ETU digester. This should result in substantially lower digester volume requirements and improvement economics. In addition, more efficient processing techniques, such as separating the water hyacinth juice and fiber fractions, will be investigated in the laboratory and initiated at the ETU scale.

Future Plans

It is anticipated that favorable results from this experimental program will provide the basis for further scale-up to prototype operations in the late 1980's. GRI will actively seek the participation of waste disposers,

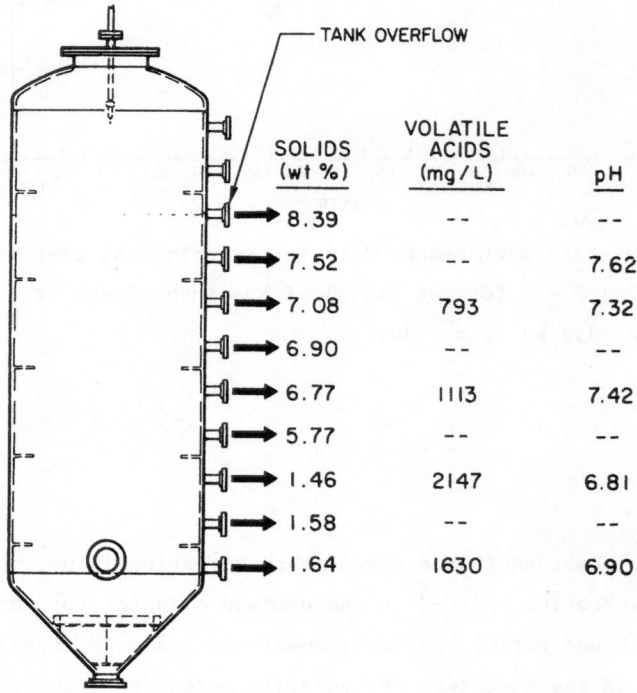

Figure 6. ETU digester solids, volatile acids, and pH profile on 11/27/84

Table 4. Biogasification experimental test unit steady-state operating and
performance data

Performance Period	11/5-12/2/84
	4 weeks

Water Hyacinth (WH) and Sludge (PS) Feed Characteristics

Average WH Solids Content, wt%[a]	5.74
Average WH Volatile Solids Concentration, wt % TS[a]	85.7
Average Sludge (PS) Solids Content, wt%[b]	4.22
Average PS Volatile Solids Concentration, wt % TS[b]	90.7
Average Blend Ratio WH/PS, TS wt ratio	2.0

Digester Operating Condition

Feed Injection	Bottom
Daily Loading Frequency	1
Average Daily Loading Rate, kg VS m^{-3} day	3.2
Culture Volume, m^3	3.88
Average Temperature, °C	35.3
pH Concentration	6.7-7.7
Volatile Acids Content, mg L^{-1} as acetic[c]	300-5000
HRT, days	12

Performance Data

Number of HRT's Completed[d]	6.0
Average Daily Gas Production, S m^3	5.89
Total Gas Yield, S m^3 kg^{-1} VS	0.47
Methane Content, vol %	60
Methane Yield, S m^3 kg^{-1} VS	0.29
Methane Production Rate, vol/vol-culture-day	0.92
Volatile Solids Reduction, wt %	43

Effluent Quality Data

Average Solids Content, wt %	3.02
Average Volatile Solids Concentration, wt % TS	74.5
pH	7.3
Average Volatile Acids Content, mg L^{-1} as acetic	2800
Average Alkalinity, mg L^{-1} CaCO$_3$	6000
Average NH$_3$-N Content, mg L^{-1}	890

Remarks: a-After grinding; b-From WDW Clarifier; c-Varies with reactor
height; d-Since 9/22.

Table 5. Comparison of steady-state performance of lab-scale CSTR and non-mixed vertical flow (NMVFR) and ETU-scale NMVFR

	Lab NMVFR*	Lab CSTR**	ETU NMVFR**
Culture Volume, m^3	0.007	0.05	3.88
Methane Yield, S m^3 kg^{-1} VS added	0.28	0.24	0.29
Methane Content, %	62	60	61
Volatile Solids Reduction, %	46	36	43

* Received a single lot of hyacinth and sludge frozen until use.
** Received freshly collected hyacinth. Both digesters received the same feeds.

sanitary districts, municipal and county governments, gas companies, and developers in the research and development project as its proceeds.

This research will ultimately benefit the gas consumer and the public at large by providing a lower cost alternative to wastewater treatment while simultaneously providing a local source of pipeline-quality gas. Engineering and architectural studies[12] of 10- and 50-million gallon day^{-1} (37,850 and 189,250-m^3 day^{-1}) facilities show that a water hyacinth/wastewater treatment facility could provide wastewater treatment at costs equal to conventional secondary wastewater treatment costs when the pipeline-quality gas produced by biogasification is sold for $2.64 to $5.06 GJ^{-1} ($2.50 to $4.80 10^{-6} Btu levelized constant 1982 dollars over a plant life of 30 years) depending on the size of the facility. If the gas is sold at higher prices, the wastewater treatment costs would be lower than those for conventional facilities.[12] With further improvements, costs can be reduced to extend applications of this technology to even small communities.

ACKNOWLEDGMENTS

The authors acknowledge the Gas Research Institute support of this work under GRI Contract No. 5082-223-0567. We further acknowledge our field scientists, Ahmad Razik and Susan Smallwood, for their diligence and skill in operating the ETU digester. In addition, we appreciate the interest and assistance of employees of WDW and Reedy Creek Utility Company in preparing feedstocks for the ETU system.

NOMENCLATURE

Technology

ABP	=	Anaerobic Biogasification Potential
C/N	=	Carbon:Nitrogen Ratio
C/P	=	Carbon:Phosphorus Ratio
CSTR	=	Continuously Stirred Tank Reactor
ETU	=	Experimental Test Unit
HRT	=	Hydraulic Retention Time
HV	=	Heating Value
NMVFR	=	Non-Mixed Vertical Flow Reactor
PS	=	Primary Sludge
TS	=	Total Solids
VS	=	Volatile Solids
VSR	=	Volatile Solids Reduction
WH	=	Water Hyacinth

Organizations

IGT	=	Institute of Gas Technology
DOE	=	U.S. Department of Energy
EPA	=	U.S. Environmental Protection Agency
GRI	=	Gas Research Institute
WDW	=	Walt Disney World

REFERENCES

1. E. S. Del Fosse. Water hyacinth biomass yield potentials, In: Symposium Papers, Clean Fuels from Biomass and Wastes. 73-99, Institute of Gas Technology. Orlando, FL. 1977.

2. B. C. Wolverton, R. C. McDonald and J. Gorden. NASA Technical Memorandum. TM-X2725. July, 1975.

3. C. Lee and T. McKim. Water hyacinth wastewater treatment system, In: Symposium Papers, Energy from Biomass and Wastes V. 26-30. Institute of Gas Technology. Orlando, FL. 1981.

4. D. A. Cornwall, J. Zobtek, Jr., C. D. Patrineby, C. O. Furman and J. I. Kim. Nutrient Removal by Water Hyacinths. J. Water Pollut. Control Fed., 49, 57. 1977.

5. D. P. Chynoweth, B. Schwegler, D. A. Dolenc and K. R. Reddy. Waste Reclamation and Methane Production Using Water Hyacinth and Anaerobic Digestion, in: Proceedings of Tenth Energy Technology Conf. 1293-1303. 1983.

6. D. P. Chynoweth, S. Ghosh, M. P. Henry and V. J. Srivastava. Kinetics and Advanced Digester Design for Anaerobic Digestion of Water Hyacinth and Primary Sludge, in: Biotech. Bioeng. Symp. No. 12. 381-98. 1982.

7. D. P. Chynoweth, D. E. Jerger, J. R. Conrad, A. Razik, V. J. Srivastava, S. Ghosh, M. P. Henry and S. P. Babu. Gasification of Land-based Biomass. Final Report for GRI Contract 5081-323-0733. July 1978-December. 1982.

8. D. P. Chynoweth, K. F. Fannin, D. E. Jerger, V. J. Srivastava, J. R. Conrad and J. D. Mensinger. Biological Gasification of Renewable Resources. Annual Report for GRI Contract 5083-323-0464. January 1983-February. 1984.

9. R. Biljetina, D. P. Chynoweth, J. Janulis and V. J. Srivastava. Biogasification of Walt Disney World Biomass Waste Blend. Annual Report for GRI Contract 5082-223-0567. January-December. 1983.

10. D. P. Chynoweth, R. Biljetina, V. J. Srivastava, J. E. Melanowski, T. D. Hayes and D. A. Dolenc. Biogasification of Water Hyacinth and Primary Sludge, in: Proc. 1984 Internat. Gas Res. Conf. Washington, D.C. In press.

11. D. P. Chynoweth, S. Ghosh and M. P. Henry. Biogasification of Blends of Water Hyacinth and Domestic Sludge, in: Proc. of 1981 Internat. Gas Res. Conf. 742-55. September, 1981.

12. Black and Veatch. Engineers-Architects. An Economic Systems Assessment of the Concept of a Water Hyacinth Wastewater Treatment/Methane Production. Final Report for GRI, Contract No. 5082-511-0629. 1983.

METHANE PRODUCTION AND UTILIZATION AT FUEL

ALCOHOL PRODUCTION FACILITIES

Enos L. Stover, Reinaldo Gonzales and Ganapathi
Gomathinayagam

School of Civil Engineering
Oklahoma State University
Stillwater, Oklahoma 74078

ABSTRACT

Research with both fixed-film and suspended growth anaerobic treatment
of stillage has shown very high treatment efficiencies in conjunction with
high quality methane gas production. The biological treatment and methane
production kinetics have been defined so that full-scale facilities can be
built successfully. The impact on energy balances and economics of fuel
alcohol production in a 4,000,000 L yr^{-1} plant are presented as a case
example. The actual impacts of day-to-day operations (plant shut-down and
start-up, non-operating periods and variable wastewater flows and loadings)
are also addressed relative to operations of the anaerobic treatment system
and production of fuel alcohol with the methane produced from the anaerobic
system as the primary energy source.

Keywords. Fuel alcohol, anaerobic treatment, methane, energy balance,
kinetics, operations.

INTRODUCTION

Two very serious problems associated with fuel alcohol production from
grain center around the energy requirements to produce the alcohol and
around the production of high-temperature, high-strength, acidic wastewaters
called thin stillage. Heat energy is required at every stage of the production process including grain drying, cooking, saccharification, fermentations, and distillation. Unless careful design and operational constraints
are imposed, the energy consumption per liter of alcohol produced can be

greater than the energy available from one liter of alcohol. Thus, the total energy balance around the production plant becomes a critical factor, and innovative improvements to reduce the total process energy demands are needed. Ethanol production by fermentation also requires considerable quantities of water with substantial amounts ending up in the thin stillage as wastewater. Although a portion of this water can be recycled, complete reuse is generally not possible due to the build-up of salts and toxic by-products of the fermentation reactions. These high strength wastewaters can be treated biologically to high treatment efficiencies in both aerobic and anaerobic systems.[1,2,3,4] The anaerobic systems offer the advantage of methane production which can be used as a fuel source in the alcohol production process.

Independently, these two problems can have a significant negative impact on the cost of fuel alcohol production. However, when considered jointly, a synergistic effect can be realized by application of anaerobic treatment of the stillage and use of the methane produced to supplement the fuel requirements of the alcohol plant. Research at Oklahoma State University with both fixed-film and suspended growth anaerobic treatment systems has demonstrated very high treatment efficiencies of thin stillage. The biological treatment and methane production kinetics have been defined such that full-scale facilities can be designed and operated for stable and reliable methane production. The impacts of employing anaerobic treatment at a four-million liter per year alcohol production plant and utilizing the methane from treatment of the thin stillage have been evaluated. The results of this investigation follow.

ANAEROBIC TREATMENT KINETICS

Extensive evaluation of anaerobic reactors (both suspended growth and fixed-film) by the authors over the past few years have shown that these systems comply with the same types of kinetic relationships developed by Stover and Kincannon for description of aerobic suspended growth and fixed-film reactors.[2,3,4,5,6,7,8,9] These kinetic models for substrate consumption or substrate removal are based on the observation that mass substrate removed is a function of the mass substrate applied or the applied substrate loading rate. Mathematical description of substrate utilization rate is expressed as a function of the loading rate by monomolecular kinetics.

A second very important observation has also been made relative to the methane production rate and quality of the gas produced by anaerobic reactors. Not only was substrate removal found to be a function of the mass substrate loading rate, but both the gas production rate and methane content of the gas were observed to be dependent on the applied substrate loading rate. Again, monomolecular kinetics were found to provide very accurate descriptions of these relationships. These relationships allow mathematical modeling and accurate prediction of substrate removal, effluent quality, total gas production, and methane content as the substrate loading rate to the reactor changes. The combined impacts of both the hydraulic flow rate and influent substrate concentration on the anaerobic reactor efficiency then become predictable by the kinetic analyses.

Determination of the kinetic constants, U_{max} and K_B, describing substrate utilization according to the model of Stover and Kincannon for anaerobic treatment of fuel alcohol wastewaters by suspended growth activated sludge, and submerged fixed film biological tower systems are shown graphically in Figures 1 and 2, respectively. In the top part of these figures, the specific substrate utilization rate is plotted as a function of the mass substrate loading rate in terms of BOD. In the bottom part of these figures, the reciprocal of the specific substrate utilization rate is plotted as a function of the reciprocal of the mass substrate loading ratio for determination of the kinetic constants. U_{max} is the reciprocal of the Y-axis intercept, and the slope of the line is equal to K_B/U_{max}. For more detailed explanation of this kinetic model and its use in biological wastewater treatment, the reader is referred to references 2-9, where the following terms are described in more detail:

U_{max} = maximum specific substrate utilization rate
K_B = proportionality constant or loading rate where the specific substrate utilization rate (U) is one-half of U_{max}
F = flow rate to reactor
V = suspended growth reactor volume
A = surface area of media in fixed film system
S_i = influent substrate concentration
S_e = effluent substrate concentration
X = mixed liquor volatile suspended solids concentration
F/M = food-to-microorganism ratio

Typical test results and methane production rates from the suspended growth and fixed film anaerobic systems treating fuel alcohol production

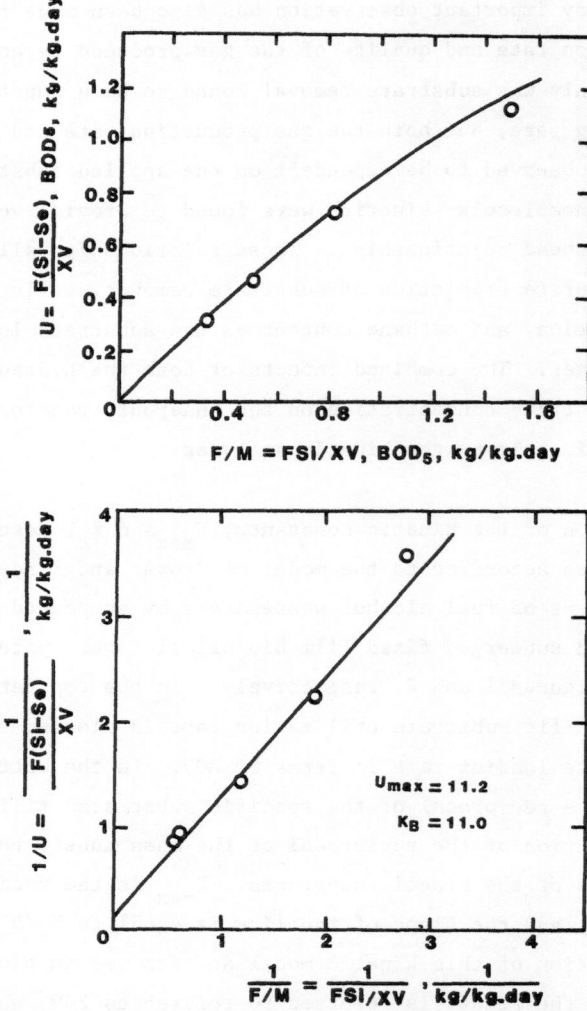

Figure 1.   Determination of kinetic constants for suspended growth system

wastewaters are presented in Table 1. As can be observed in Table 1, either type of anaerobic system is capable of achieving very high treatment efficiencies along with excellent methane production capabilities.

Both types of processes are capable of about the same levels of treatment and methane production when designed and operated properly. Therefore, the decision of which process to choose for a particular fuel alcohol production facility will probably depend on site specific constraints, operational considerations and flexibility, as well as engineering preferences.

490

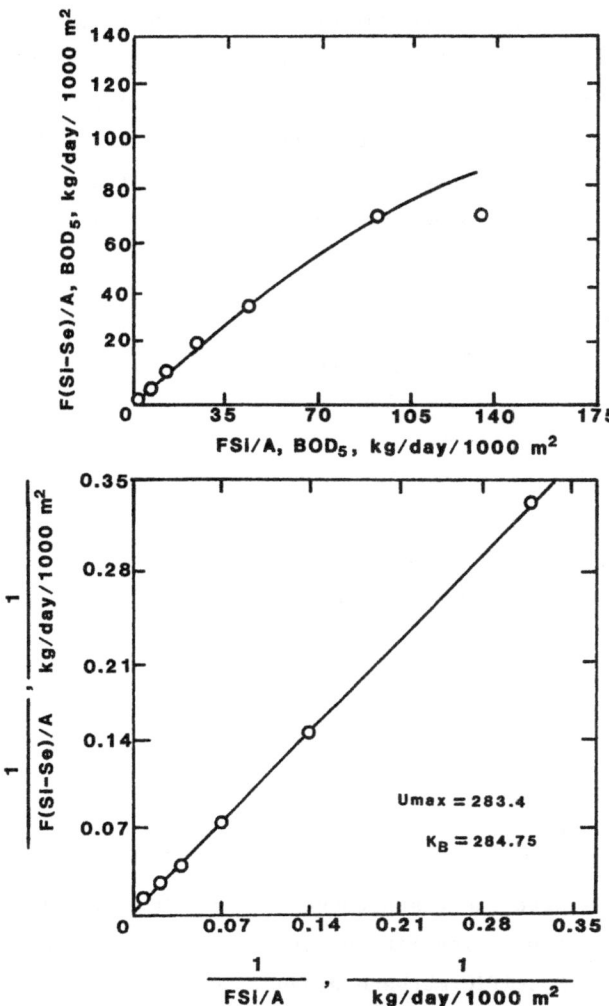

Figure 2. Determination of kinetic constants for fixed-film system

METHANE PRODUCTION KINETICS

As previously mentioned, the methane production kinetics of anaerobic
biological treatment systems can also be developed in terms of the mass
substrate loading rates to these systems. The total gas and methane production characteristics of the anaerobic fixed-film system treating fuel
alcohol wastewater are presented graphically in Figure 3.

The maximum specific gas production rate (G_{max}) and the maximum specific methane production rate (M_{max}) were determined as a function of the
applied BOD loading rate. The gas production kinetics summarized in Table 2

were determined in terms of both soluble BOD and COD removed. These kinetic constants were determined with correlation coefficients greater than 0.99. The maximum specific production rates were the same, irrespective of whether they were calculated in terms of BOD or COD. Mathematical description of the gas (methane) production rates can be modeled as the substrate loading rate changes by using monomolecular kinetics as follows:

$$G = \frac{G_{max} \ \frac{FSi}{A}}{G_B + \frac{FSi}{A}}$$

and

$$G = \frac{M_{max} \ \frac{FSi}{A}}{M_B + \frac{FSi}{A}}$$

where

G = Specific gas production rate, $M^3 \ day^{-1} \ 1000 \ m^{-2}$

G_B = Proportionality constant, kg substrate $day^{-1} \ 1000 \ m^{-2}$

M = Specific methane production rate, $m^3 \ day^{-1} \ 1000 \ m^{-2}$

M_B = Proportionality constant, kg substrate $day^{-1} \ 1000 \ m^{-2}$

A more detailed discussion of the gas production kinetics has been presented in Reference 9.

The gas production rates shown in Table 2 were higher than the expected stoichiometric values. Similar rates of higher gas production were observed in the suspended growth systems also. These high production rates were due to the fact that the rates were reported in terms of soluble BOD and COD removed instead of total BOD and COD. The wastewater was a highly complex wastewater containing long chain fatty acids, nucleic acids, compounds such as pyridine, and other compounds not readily amenable to the COD test. Additional testing of the raw wastewater under more stringent oxidation conditions toward the end of the study indicated that the COD was actually higher than that reported by the Hach test procedure employed during these

Table 1. Anaerobic treatment system performance in terms of BOD (COD)

A. Suspended growth system

Loading Rate Kg Kg^{-1}	Influent mgL^{-1}	Effluent mgL^{-1}	% Removal	MLVSS mgL^{-1}	Methane percent	Methane* Production m^3 kg^{-1}
0.22	2,300	15	99.3	3,380	78	1.32
(0.50)	(5,125)	(380)	(92.6)			(0.62)
0.23	4,100	28	99.3	5,380	71	1.30
(0.56)	(10,100)	(380)	(96.2)			(0.55)
0.31	8,880	35	99.6	8,500	70	0.98
(0.55)	(16,000)	(425)	(97.3)			(0.53)

B. Fixed-film reactor systems (Plastic media with 138 m^{2}m^{-3})

Loading Rate Kg(day 1000^2)$^{-1}$	Influent mgL^{-1}	Effluent mgL^{-1}	% Removal	Methane Percent	Methane* Production m^3 kg^{-1}
3.32	1,786	34	98	77	1.06
(4.64)	(2,512	(131)	(95)		(0.79)
7.12	3,968	57	98	75	0.80
(10.25)	(5,696)	(215)	(96)		(0.57)
13.80	7,485	140	98	70	0.80
(19.57)	(10,100)	(271)	(98)		(0.60)
25.08	12,167	290	97	60	0.81
(37.48)	(18,445)	(756)	(96)		(0.54)

*Based on soluble BOD (COD) removed.

All studies were conducted at around 35 C.

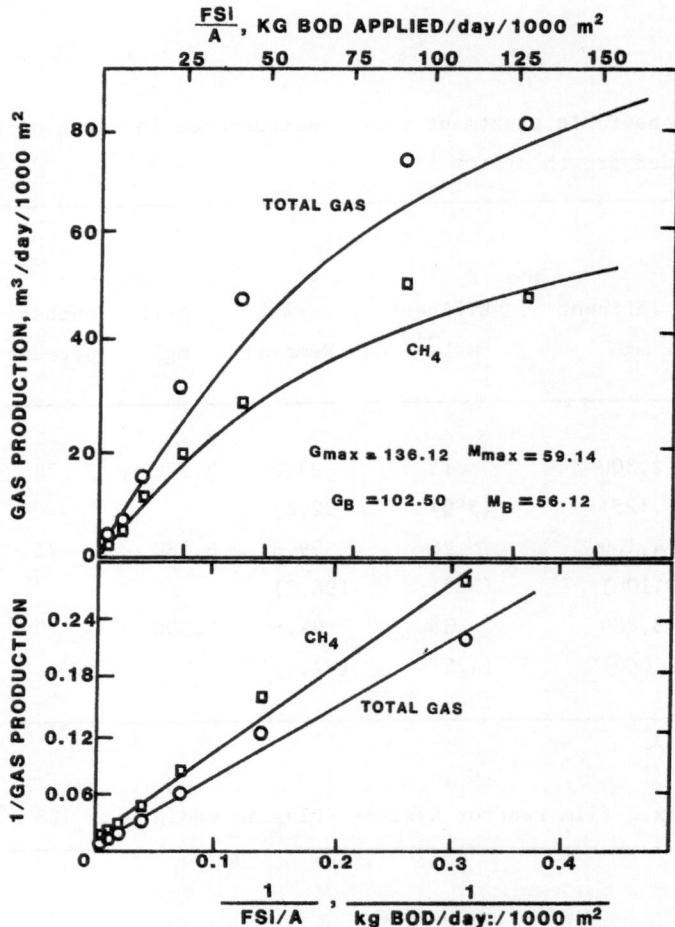

Figure 3. Total gas and methane production kinetics as a function of soluble BOD loading rate

studies. Based on these observations, an organic carbon (TOC) balance was conducted around the reactors, and the TOC balance confirmed the gas production rates to be correct.

Since this investigation centers on the energy balance around a typical four-million liters per year alcohol plant and its associated wastewater treatment plant, it is imperative to evaluate the impacts of shock loads and shut-down periods on methane generation from the wastewater treatment plant. Alcohol production has to be temporarily ceased whenever there are clean-up operations, mechanical break-down of equipment, or over-hauling of the plant. This may also occur if manufacturing is a seasonal operation. As a result of these time periods, no wastewater would be generated.

Table 2. Gas (methane) production kinetic constants

Kinetic Constants	BOD* Kinetics	COD* Kinetics
Maximum Gas Production Rate M^3 day^{-1} 1000 m^{-2}	136.12	137.65
Proportionality Constant kg day^{-1} 1000 m^{-2}	102.60	150.00
Correlation Coefficient	0.994	0.995
Maximum Methane Production Rate m^3 day^{-1} 1000 m^{-2}	59.14	59.41
Proportionality Constant kg day^{-1} 1000 m^{-2}	56.12	81.00
Correlation Coefficient	0.993	0.994

*Kinetics in terms of soluble BOD and COD removal.

Figure 4 is a graphical profile explaining the chronological impacts of feed-shutdown for a week on the performance of a pilot anaerobic activated sludge system, with emphasis on gas production. As seen from the graph, the gas production rate, which was in the range of 1300 to 1400 mL hr^{-1}, plunged down to 25 mL hr^{-1} within 24 hours and reached a low value of 10 mL hr^{-1} on the fifth day. Gas production continued at that rate until the seventh day when the feed was restarted. There was a spontaneous increase in the gas production rate which jumped back to 1200 mL hr^{-1} within 24 hours and reached the original range within 48 hours. The system did not suffer any serious set back due to the cessation of feed for a week. Similar studies and results were observed with the fixed-film reactors during feed-shut down periods. Thus, anaerobic treatment of alcohol wastes offers the advantage of spontaneous stop-and-start fexibility of the wastewater treatment facility while generating the necessary methane gas for use in the alcohol plant.

Various investigations on the effects of shock loads due to variations in feed concentration, flow rate, pH, temperature and nutrients also revealed that these anaerobic treatment systems could handle such shocks if the wastewater treatmen. system were properly monitored and controlled.

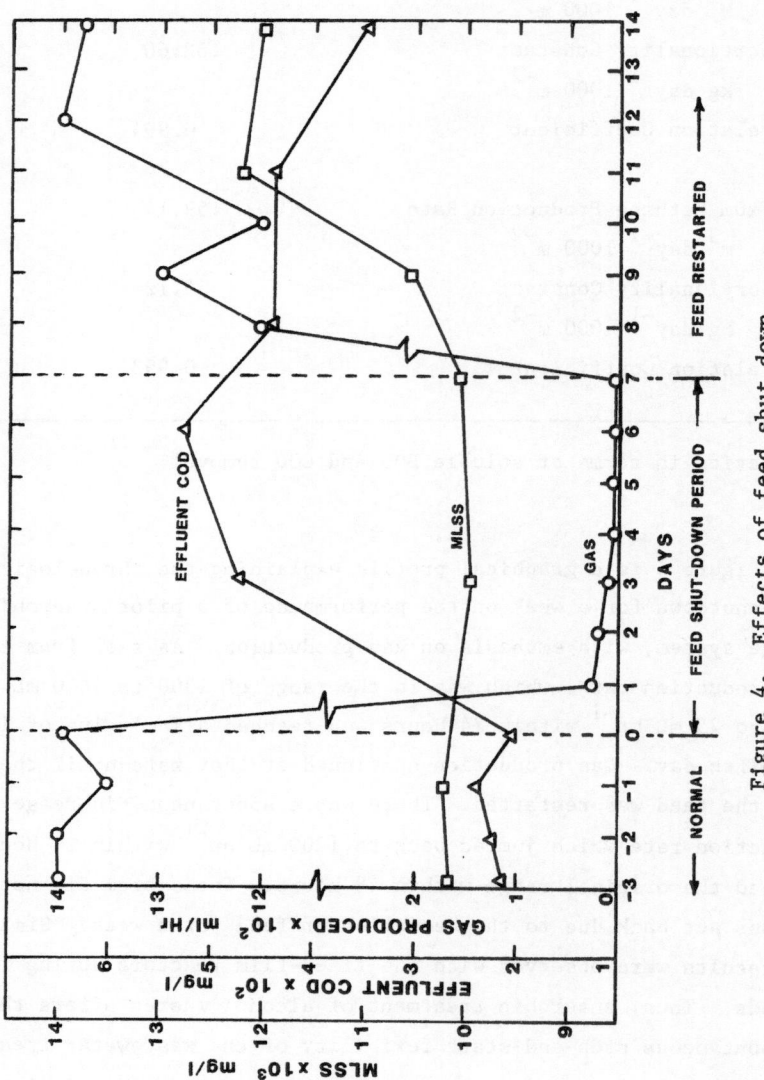

Figure 4. Effects of feed shut down

Material and energy balances were conducted around a small commercial or industrial ethanol (fuel alcohol) plant capable of producing four-million liters of anhydrous ethanol (200 proof) per year. When operating 24 hours per day for 300 days out of the year at the production rate of 568 liters hour^{-1} (13,626 liters day^{-1}), this plant can produce four-million liters of pure ethanol per year. The feedstock consists of corn and milo, which under good operating and management conditions yields around 9.5 liters of ethanol per bushel of grain. The 200 proof ethanol is sold to nearby petroleum refiners and distributors. The distiller's dried grain solids (DDGS) are recovered as by-products and sold to nearby feedlots for cattle feed.

Energy consumption at the various stages of the process along with the energy content of the product are shown in Table 3. Efficiencies of energy utilization are dependent on both plant design and operation and, thus, will vary at different facilities. The extent to which the energy input and

Table 3. Energy consumption for fuel alcohol production and fuel energy content.

Process Stage	kilojoule liter^{-1} alcohol
Precooker	1,530
Cooker	2,440
Flash Chambers	-1,980
Distillation	6,690
Grain Drying	2,030
Electrical	470
Losses & Miscellaneous	460
Subtotal	11,640
Dehydration	5,570
Subtotal	17,210
Stillage Evaporation	7,920
Stillage Drying	2,150
Total	27,280

Ethanol Produced	kilojoule liter^{-1}
Anhydrous	23,690
190 Proof	22,570
160 Proof	18,950
Gasohol	33,440

Input:Output = 27,280:23,690 = 1.15:1.00

output boundaries are limited will influence the input:output ratio. For example, when the energy to produce corn is considered, the total energy input can increase by as much as 12,544 to 16,725 kilojoule liter^{-1}; however, an energy credit for the distiller's dried grains may more than offset this energy input. Energy balances where input:output are at or near 1.0:1.0 can provide favorable overall economics, and every effort should be made to conserve energy at all points within the production system to keep the energy input as low as possible.

Since the energy credit for DDGS offsets the energy to produce the corn and milo, the energy input and output boundaries for the case history presented here have been drawn around the production plant and associated facilities, as shown in Table 3. Natural gas at 42.8 kilojoule liter^{-1} is used to fire the boilers to provide steam for alcohol production. The whole stillage containing the grain solids contains about 40% solids, and the thin stillage contains about 6.0 to 10.0% solids. Around 10% of this thin stillage or backset is recycled back to the mash mixing tank in the cooking and saccharification section, while the remaining thin stillage is presently treated by evaporation.

As can be readily observed in Table 3, thin stillage evaporation places a significant strain on the total facility energy balance. The total energy balance with thin stillage evaporation is input:output = 27,270 kilojoule liter^{-1}:23,690 kilojoule liter^{-1} = 1.15:1.00, whereas, if stillage evaporation was not required, the total energy balance would be input:output = 17,210 kilojoule liter^{-1}:23,690 kilojoule liter^{-1} = 0.72:1.00. When the previously developed kinetics of anaerobic treatment of fuel alcohol production wastewater and the wastewater characterization data are applied to the four-million liter per year production facility for 99 percent BOD removal, approximatley 3,682,000 liters of methane gas are produced.

At the heating value of 37.2 kilojoules liter^{-1} of methane, 1.37×10^{8} kilojoules are available for use in the alcohol production process. As can be observed in Figure 5, this is 60% of the total daily joule requirements (2.32×10^{8} kilojoules day^{-1}) in the production facilities. When treating the thin stillage anaerobically and utilizing the methane in the boilers, the total net energy consumption for fuel alcohol production is shown in Table 4. The total net energy balance now becomes input:output = 7,180 kilojoules liter^{-1}:23,690 kilojoules liter^{-1} = 0.30:1.00. The input:output ratio, when the stillage was treated by evaporation, was 1.15:1.00. The natural gas requirements will be reduced from 5.38×10^{6} liters day^{-1} to 2.21×10^{6} liters day^{-1}.

498

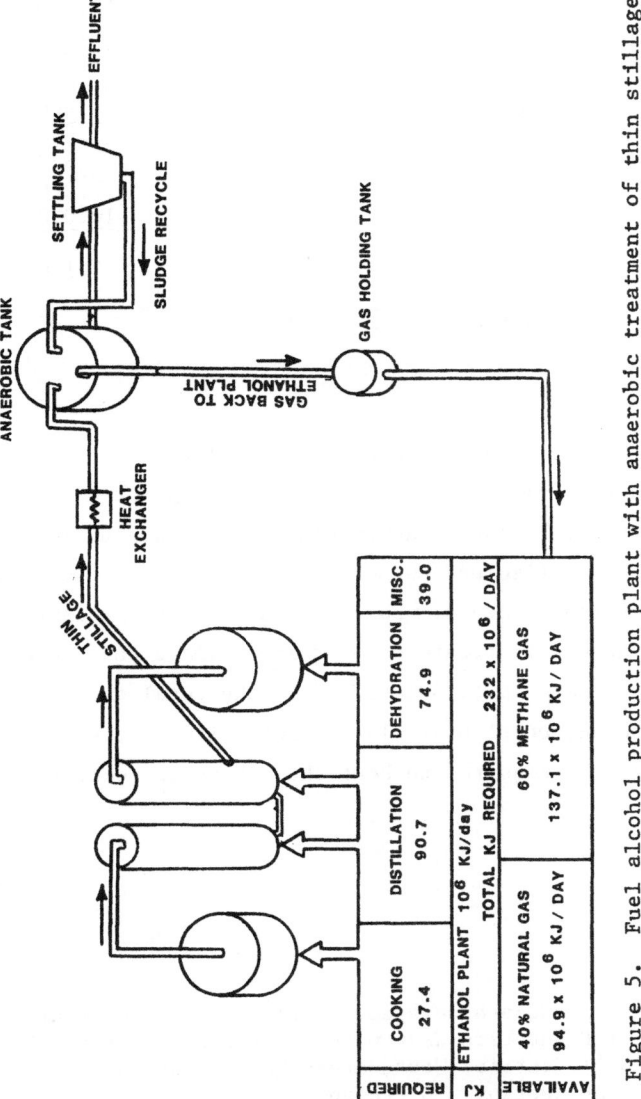

Figure 5. Fuel alcohol production plant with anaerobic treatment of thin stillage

Table 4. Energy consumption for fuel alcohol production using methane
produced during anaerobic treatment of stillage

Process Stage	kilojoule liter^{-1} alcohol
Precooker	1,530
Cooker	2,440
Flash Chambers	-1,980
Distillation	6,690
Grain Drying	2,030
Electrical	470
Losses & Miscellaneous	460
Subtotal	11,640
Dehydration	5,570
Subtotal	17,210
Anaerobic Treatment	-10,030
Total	7,180

Input:Output = 7,180:23,690 - 0.30:1.00

The actual amount of energy consumption per liter of ethanol produced
will vary from plant to plant due to the age of the plant and the design and
operation of a particular plant. Newer plants are designed more efficiently
than older plants, and efficient use of heat exchangers can save up to 25%
of the total energy requirements. Based on the studies presented here, it
is estimated that anaerobic treatment of thin stillage and use of the methane gas produced can save up to 50% to 75% of the total energy requirements
in fuel alcohol production plants. In northern climates, part of the
methane produced may be required to heat the anaerobic reactor in the
winter.

REFERENCES

1. E. L. Stover and G. Gomathinayagam. Activated Sludge Treatability of
    Fuel Alcohol Production Wastewaters. Presented at the Biological
    Treatment of Industrial Wastewaters Session of the 1982 Summer
    National AIChE Meeting. Cleveland, Ohio. August 29-Sept. 1, 1982.
2. E. L. Stover and G. Gomathinayagam. Biological Treatment of Synthetic
    Fuel (Alcohol Production) Wastewater. Presented at the Water Pollu-
    tion Control in Synfuels Production Session of the 55th Annual Water
    Pollution Control Federation Conference, St. Louis, Missouri.
    October. 1982.
3. E. L. Stover and G. Gomathinayagam. Biological Treatment Kinetics of
    Alcohol Production Wastewater. Presented at the 1982 Winter Meeting
    American Society of Agricultural Engineers, ASAE Paper #82-3601,
    Palmer House. Chicago, Illinois. December 14-17, 1982.

4. E. L. Stover, G. Gomathinayagam and R. Gonzalez. Anaerobic Treatment of Fuel Alcohol Wastewater by Suspended Growth Activated Sludge. Proceedings of the 38th Annual Purdue Industrial Waste Conference 95. West Lafayette, Indiana. May, 10-12, 1983.
5. E. L. Stover and D. F. Kincannon. Rotating Biological Contactor Scale-Up and Design. Proceedings of the First International Conference of Fixed-Film Biological Processes. Kings Island, Ohio (April 20-23, 1982), and Water & Engineering Management, Reference Handbook, R 48 (May 1982).
6. D. F. Kincannon and E. L. Stover. Design Methodology for Fixed-Film Reactors - RBC's and Biological Towers, Civil Engineering for Practicing and Design Engineers, 2,107. 1982.
7. E. L. Stover. EPA Priority Pollutant Treatability Studies. Presented at the Chemical Manufacturers Association Seminar on Biological Treatment, Priority Pollutants and BATEA. Washington, D.C. January, 1983.
8. D. F. Kincannon and E. L. Stover. Biological Treatability Data Analysis of Industrial Wastewaters. Presented at the 39th Annual Purdue Industrial Waste Conference. West Lafayette, Indiana. May, 1984.
9. E. L. Stover, R. Gonzalez and G. Gomathinayagam. Anaerobic Fixed-Film Biological Treatment Kinetics of Fuel Alcohol Production Wastewaters. Presented at the Second International Conference on Fixed-Film Biological Processes. Arlington, Virginia. July, 1984.

PRELIMINARY RESULTS OF A METHANE-PRODUCING,

ENERGY-INTEGRATED TROPICAL DAIRY FARM

Donald S. Sasscer and Thomas O. Morgan

Center for Energy and Environment Research
University of Puerto Rico
Mayaguez, Puerto Rico   00708

ABSTRACT

An energy-integrated, environmental-compliance farm system has been
designed and constructed on a dairy farm in Puerto Rico. Electrical power
is being produced by a 40 kw motor-generator fueled by biogas obtained from
the anaerobic fermentation of manure from 320 dairy cows. This well-managed
system complies with both the letter and the spirit of local environmental
laws.

Keywords. Manure, biogas, methane, dairy farm, tropical environment.

INTRODUCTION

Typical of tropical islands, Puerto Rico is wholly dependent upon
imported oil for the production of electrical power. When foreign oil was
less expensive than fossil fuel from the United States, electricity on the
island was relatively inexpensive. However, over the past 12 years, electrical costs have increased drastically, and electrical power in Puerto Rico
presently costs approximately $0.12 per kilowatt-hour. In order to be
economically competitive with imported food markets, Puerto Rican farmers
need to utilize electricity more efficiently and investigate less expensive
sources of electrical power.

Management of farm wastes has recently become a major environmental
issue in Puerto Rico. In its former mode of operation, the private farm
where this project is being conducted probably did not comply with the

environmental regulations stipulated under either Puerto Rican or federal laws.[1] Disposal of animal wastes was uncontrolled, and as a result, the area was unattractive and afflicted with nuisance levels of foul odor. In its new mode of operation, the unsanitary and nuisance conditions have been replaced with an attractive, well-managed system complying with both the spirit and the letter of applicable environmental laws. In Puerto Rico, there is an urgent need for a dairy farm which will demonstrate that it is both practical and cost effective to comply with environmental regulations.

The overall objectives of this project are to:

1. Increase the profitability of tropical dairy farms.
2. Bring tropical dairy farms into environmental compliance.

In this project, a method of increased farm profitability is demonstrated by the design, construction, and operation of a system which produces energy for a private dairy farm. Biogas is produced by the anaerobic fermentation of manure from approximately 320 dairy cows and used as an energy source to provide electricity for the farm. Environmental compliance is demonstrated by a system of waste management which utilizes farm wastes. Both the liquid and solid components of the digester effluent are utilized.

There are seven components of the system: manure collection, manure preparation, biogas production, electrical power production, liquid-solid waste separation, liquid waste storage and utilization, and green feed production.

Site of Demonstration

This research and demonstration project is being conducted on a private dairy farm owned and operated by Mr. Antonio Ubarri-Blanes and located in south-central Puerto Rico. Although this region is commonly described as "south coastal," it lies to the north of a flat agricultural plain immediately bordering the Caribbean Sea. It is an area of low hills and valleys with elevations from 45 to 210 meters (150 to 700 feet) above sea level. The hilly nature of the farm, while a disadvantage agriculturally, was used to advantage by utilizing gravity-flow throughout most of the system. The use of gravity-flow saves on both capital and operating costs while increasing reliability.

The system components are diagramed in Figure 1. Manure is collected from approximately 320 cows in four loafing barns. It is loaded into a cart, transported to a mixing sump, and unloaded there. In the mixing sump, it is diluted with water from the milking parlor, homogenized and then fed into two 334 m^3 (88,000 gallon) digesters where biogas is produced by anaerobic fermentation.[2] The biogas is used as fuel for a 40 kilowatt motor-generator unit which cogenerates electrical power with the Puerto Rico Electrical Power Authority.

Effluent from the digesters flows by gravity to a liquid-solid separator. The liquid fraction of the effluent then flows to an irrigation pond with a capacity of 1,500 m^3 (400,000 gallons) where it is used to irrigate 20 hectares (50 acres) of green feed being grown for the milking herd. The solid fraction is used for bedding material for the cows and marketed as a soil conditioner.

Manure Collection

To ensure efficient utilization of dairy manure, several conditions are imposed on manure collection and handling practices which may not normally

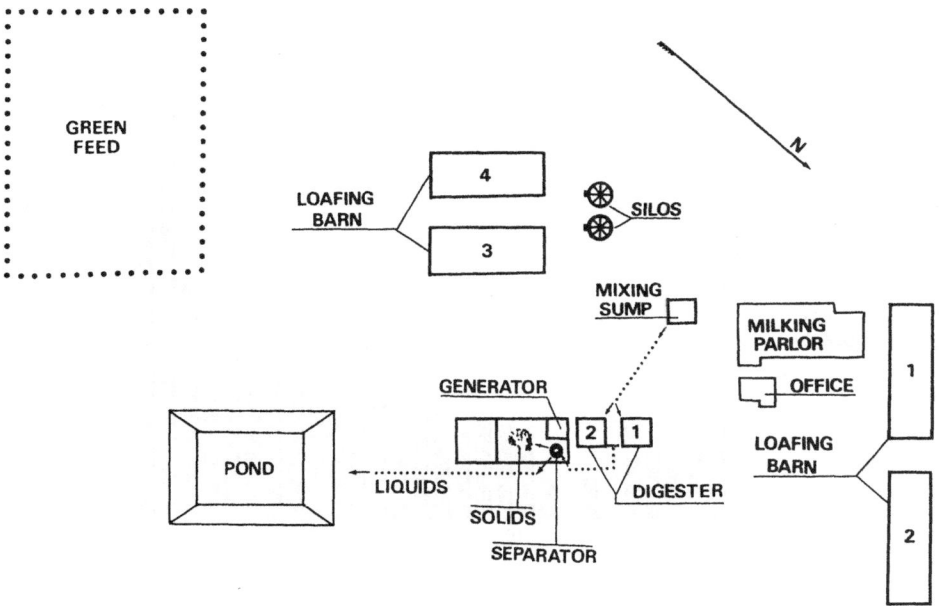

Figure 1. Diagram of the energy integrated farm system for the Ubarri-Blanes dairy farm – Center for Energy Environment Research, University of Puerto Rico.

exist. Manure recovery should be maximized since this is the fuel for the digesters, and the manure collected should be as free of grit and other inert material as possible since these materials will accumulate in the digester. Straw, hay and other low density materials which are not readily digestible can form a hard crust on top of the digester slurry and severely restrict biogas collection. For this reason, the quantity and size of these materials entering the digester should be minimized.

To facilitate manure collection and to minimize contamination of the manure, the four loafing areas were paved with concrete, allowing 5.6 m^2 (60 ft^2) per cow. Each paved area was shaded and surrounded with a curb and fence. Manure recovery began in the second week of August 1984, and Figure 2 shows the average manure recovery per cow-day for the first 28 weeks of operation. During the first four months, approximately 8 kg (18 lbs) of manure per cow-day was recovered. This represents only about 20 percent of the manure available. Principal reasons for this low recovery rate were:

1. The loafing areas were not scraped regularly.
2. The cows spent a considerable amount of time off the paved areas since food and/or water were available away from the pavement.
3. At times, the cows were fed from hay wagons parked on the pavement. This meant that manure from the vicinity of the wagons had considerable amounts of hay mixed in and was therefore unusable.

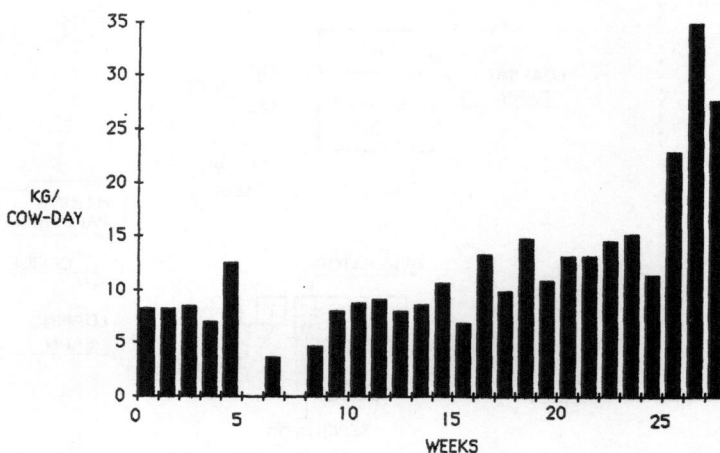

Figure 2. Manure collection during the first 29 weeks of operating showing a marked increase during the last three weeks due to modifications to the loafing areas and a more consciencious manure recovery effort.

During December and January, several improvements were made to the loafing areas in order to encourage the cows to spend more time on the pavement (the farm owner did not wish to confine his cows completely to the pavement unless bedded stalls were available). These improvements included constructing free stalls in one area so that the cows could be confined, constructing water troughs accessible only from the pavement, and constructing feed troughs around the perimeter of two loafing areas which, again, were accessible only from the pavement. In addition, farm workers were urged to scrape the loafing areas on a daily basis. As a result of these efforts, manure recovery increased to between 25 and 35 kg cow-day^{-1} (55 and 75 lbs cow-day^{-1}) during the month of February.

Manure Preparation and Handling

Manure collected from the four loafing barns is unloaded into a mixing sump which has a capacity of 45 m^3 (12,000 gallons). The manure arrives partially dehydrated, so it is diluted to 10 to 13 percent solid content utilizing water diverted from the milking parlor and homogenized with a 15 kw Flygt chopper pump. Homogenization of the slurry is important because when the manure is first unloaded it is generally lumpy and these lumps will not break up without the use of the pump.

Considerable care is taken to maintain solid content of the slurry in the mixing sump within the range of 10 to 13 percent. When solid content exceeds 13 percent, the slurry becomes excessively viscous and flows poorly. When solid content of the slurry falls below 9 percent, separation occurs and a hard scum layer forms on the surface. Since the digesters are not mixed, it is important to avoid conditions where a scum buildup might occur.

Biogas Production

Slurry from the mixing sump passes to two concrete digesters (Figures 3 and 4) through a 25-cm (10-inch) pipe. Normally, both digesters are in operation, but if it becomes necessary to drain one digester, a second digester is still available to process manure and biogas production will not be interrupted.

Total volume of each digester is 334 m^3 (88,000 gal). They are 9.1 m (30 ft) square with 3.5 m (11.5 ft) of vertical wall, and the bottom slopes from the walls at a 30 degree angle to a centrally located well which is 1.2

m (4 ft) square and 0.8 m (2.5 ft) deep. Grit accumulates in the wells and
can be cleaned out using pipes which access them from the outside.

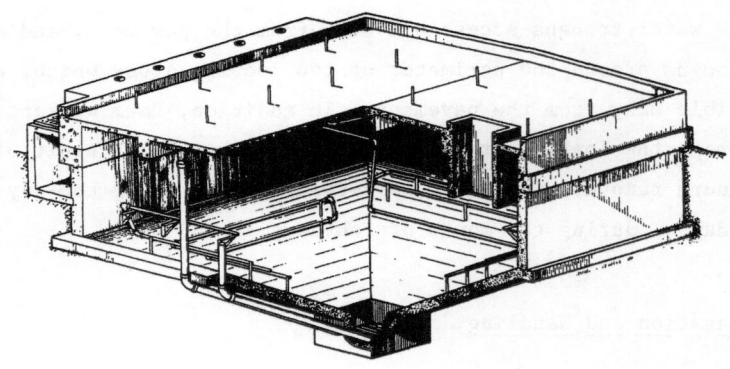

Figure 3. Biogas digester

    The digesters are covered by poured concrete roofs 13 cm (5 in) thick
with three support beams – each measuring 60 cm by 30 cm (2 ft by 1 ft).
The covers have a 1.2-m (4-ft) square access manhole and are penetrated by
13 sample tubes which are used to monitor the digesters and by four gas
collection tubes. Since the support beams for the covers divide the gas
space into four compartments, it is necessary to have a gas collection pipe
for each compartment. Covers are not monolithic with the walls of the
digesters, but a seal was provided by embedding PVC stripping at the
interface.

    Digester covers are designed to withstand an internal gas pressure of
60 cm (2 ft) of water column pressure when covered with 30 cm (1 ft) of
soil. To protect the concrete against the corrosive effects of biogas,
undersides of the covers and the upper 60 cm (2 ft) of the inner surface of
the digester walls were coated with Koppers Super Service Black. Design gas
pressure within the digesters is 38 cm (15 in) of water column pressure.
Originally, the covers were to be of fiberglass, but based on considerations
of durability and safety it was decided to use concrete covers.

    Slurry enters each digester by means of an inlet pipe 25 cm (10 inches)
in diameter which extends two-thirds of the width of the digester. The pipe
is open at the end and has a hole cut into it near the middle. Slurry
leaves the digesters through six vertical 25-cm (10-inch) effluent pipes

508

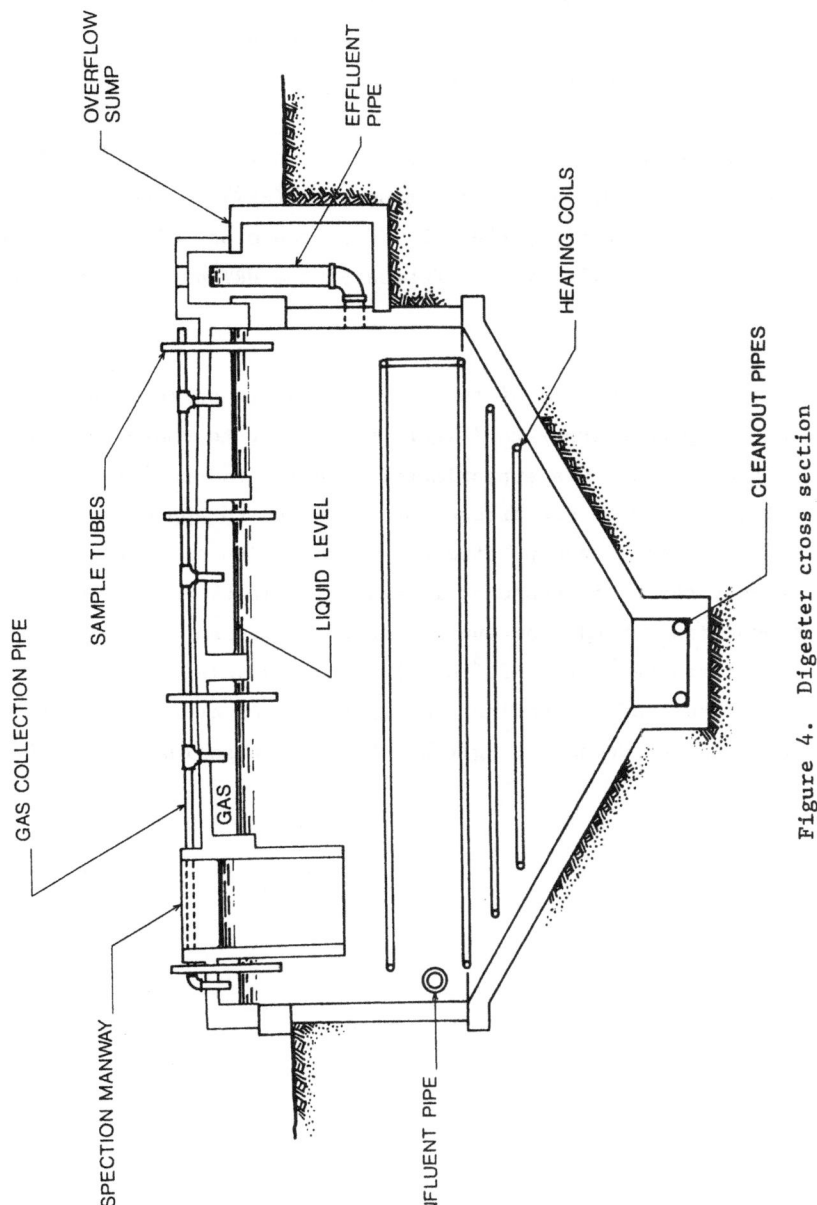

Figure 4. Digester cross section

509

located on the wall of the digesters opposite the inlet pipes and falls into an overflow sump which extends along the entire side of each digester.

To maintain a mesophilic temperature of approximately 38°C (100°F) within the digesters, 100 m (330 ft) of 5-cm (2-inch) galvanized heat exchanger pipe was installed around the circumference of the digesters at four levels. Water circulating within the heat exchanger pipes is heated by the cooling water from a biogas-fueled motor-generator.

As of the date of preparation of this manuscript, only one digester had been put into operation. Hydraulic retention time (HRT) for this digester is shown in Figure 5. Initially, the HRT was of the order of 180 days, but with the improved manure collection in February, the HRT dropped to near the design level of 30 days.

The most significant result of the improved manure collection can be observed in the gas production rate (Figure 6). Prior to February, the average gas production rate was approximately 1.6 liters sec^{-1} (200 ft^3 hr^{-1}), but during this month gas production increased to approximately 4.3 (550 ft^3 hr^{-1}). The relationship between gas production and digester loading over a five-day period in February is shown in Figure 7. During this period, gas production varied from approximately 2 liters sec^{-1} (250 ft^3 hr^{-1}) to over 7 liters sec^{-1} (900 ft^3 hr^{-1}) with peak values attained between four to six hours after loading the digester. An average of 0.08 m^3 (6 ft^3) of biogas was produced per kg (lb) of volatile solids fed to the

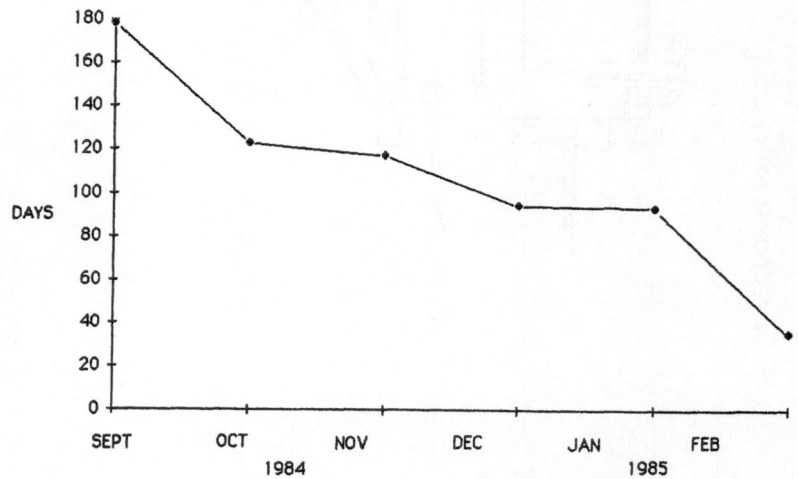

Figure 5. Hydraulic retention time

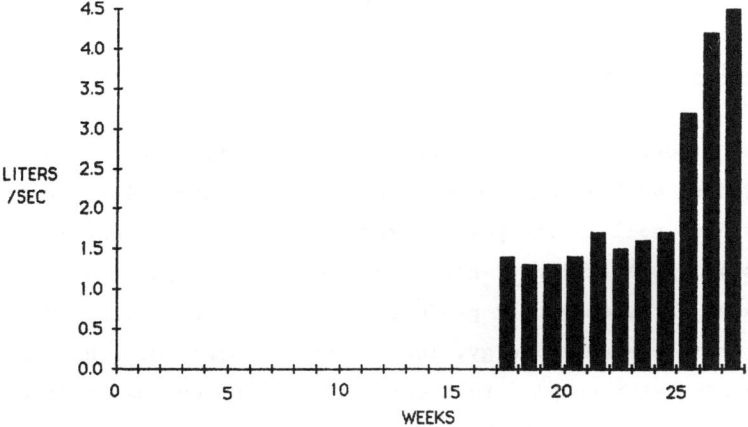

Figure 6. Gas production

GAS PRODUCTION WITH LOADING

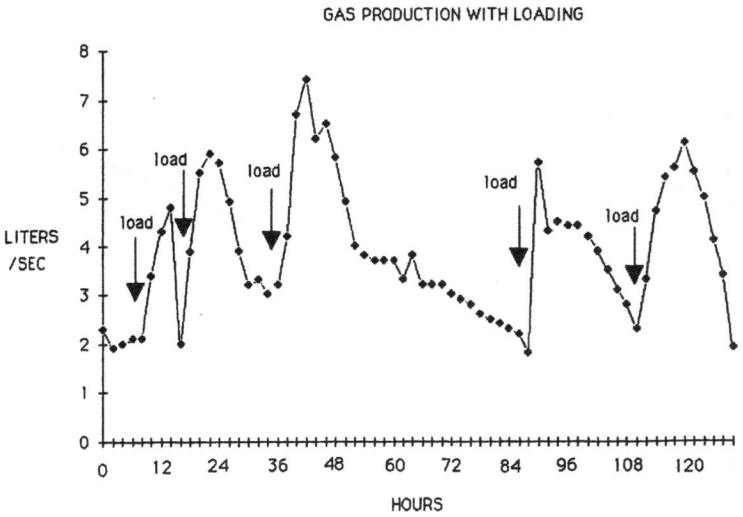

Figure 7. Gas production with loading

digester, which is an indication of good conversion.[3] Methane content of the biogas varied between 56 percent and 64 percent depending on loading frequency, but was generally about 60 percent.

Electrical Power Production

Biogas produced in the digesters is used as fuel for a synchronous, spark-ignition, 40-kilowatt motor-generator to produce electrical power for the farm and to sell to the power company. Currently, electricity in Puerto Rico costs $0.12 per kilowatt-hour, and the power company will pay $0.05 per kilowatt-hour for electricity produced through cogeneration. The unit was purchased from Perennial Energy, Inc. in West Plains, Missouri, and utilizes a Marathon generator coupled to a Caterpillar 3304 engine modified to run at 1200 rpm on either biogas or propane. Gas storage facilities, which are expensive and use a considerable amount of space, are not needed because the motor-generator is equipped with a system which automatically controls generator output in response to volumetric input of biogas from the digesters. Using this control system, the motor-generator can continue running until biogas fuel input drops to approximately 50 percent of design capacity. Below this level, the engine shuts down automatically until gas pressure builds again.

The electrical power production characteristics of the motor-generator provided by the manufacturer are shown in Figure 8. Using the maximum and

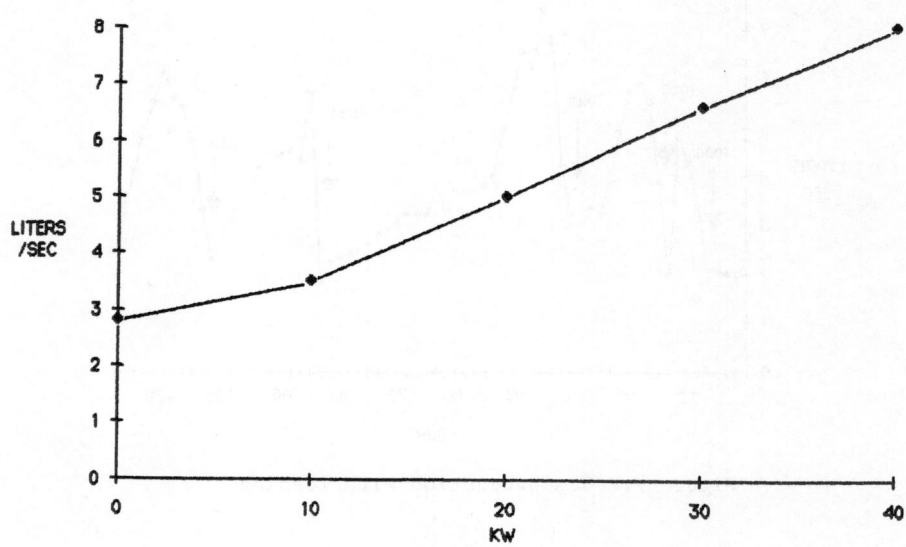

Figure 8. Power production

minimum values from Figure 7, electrical production during an average day would vary between 0 and 40 kw. Power usage on the farm is characterized by two daily peaks of 35 kw associated with milking periods, and between the peaks demand drops to 10 to 15 kw (Figure 9).

It is obvious that the present rate of electrical power production does not conform to variations in the power demand of the farm. Two schemes are presently being investigated to provide efficient utilization of electricity produced by the system. The quantity and timing of loading the digester could be controlled so that power produced by the motor-generator would correspond to the power demand of the farm. An alternative system is to reduce the peak in power usage by producing electrical power at a convenient schedule and store the energy in the form of ice, which by means of an appropriate heat exchanger system can then be used to chill the milk.

Liquid-Solid Waste Separation

Digester effluent having a solids concentration of approximately nine percent flows by gravity from the digester overflow sump into a well in the separator building. Mounted on top of the well is a centrifugal-type liquid-solid separator (a DeLaval Lisep) with a 30 millimeter screen. Twenty to thirty percent of the solid is removed from the effluent, producing a cake of 75 percent moisture content. Liquid fraction from the separa-

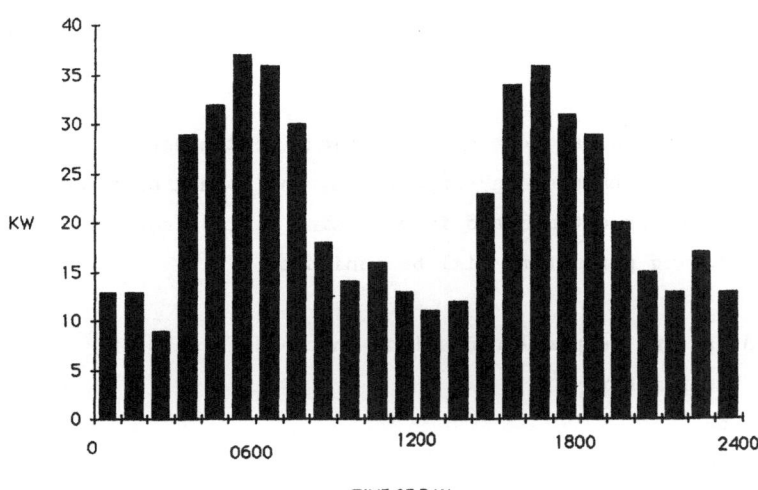

Figure 9. Power usage on the farm show two daily peaks corresponding to milking periods.

tor flows to the irrigation pond to be used to fertilize the green feed subsystem. The solid fraction is used as bedding material and marketed as a
soil conditioner.

Liquid Waste Storage and Utilization

The Soil Conservation Service of the U. S. Department of Agriculture
designed and supervised construction of the irrigation pond, which has a
capacity of approximately 1,500 m^3 (400,000 gallons). The bottom and sides
of the pond were covered with a blanket layer of compacted clay to prevent
seepage. Water from the pond is mixed with irrigation water to provide
fertilizer for the green feed subsystem.

Green Feed Production

The farm owner feeds his cows a mixture of forage, grain concentrate
and roughage extenders, and prior to the initiation of this project, no feed
was grown on the farm. By growing forage material on the farm year-round,
operating costs could be substantially reduced. For a herd size of 350
milking head, approximately $78,000 is spent each year on forage. It is
estimated that 20 hectares (50 acres) of green feed will provide all the
forage needed, and that the cost for these 20 hectares (50 acres), including
crop planting, maintenance and harvesting, would be $26,000 per year. Thus,
20 hectares (50 acres) of green feed would result in a savings of $52,000
per year.

PERFORMANCE MONITORING

It is vital to the success of the project to demonstrate that the
system developed for this energy-integrated, environmental-compliance farm
is operating near its optimum and is producing maximum cost savings. To do
this, the following parameters will be monitored:

1. Spatial and temporal variations in the digesters of:
    a. temperature
    b. pH
    c. total solids
    d. volatile solids
    e. volatile acids
    f. alkalinity

g. BOD

h. COD

2. Hydraulic retention time in digesters

3. Biogas production and quality

4. Electrical power production

5. Green feed production

6. Milk production

ACKNOWLEDGMENTS

    This work was performed under the auspices of the Center for Energy and Environment Research of the University of Puerto Rico. Funding for the project was received from the Office of Industrial Programs of the United States Department of Energy (DOE), and the Puerto Rico Office of Energy. Cost-sharing funding for the project was provided by the farm owner, Mr. Antonio Ubarri Blanes, and by the University of Puerto Rico. The authors wish to acknowledge the valued assistance of Messrs. T. Sadler and D. Corales and Ms. E. Cardona.

REFERENCES

1. Puerto Rican Public Law PL #9 and Federal Clean Water Act of 1977.

2. D. E. Hughes. What is Anaerobic Digestion? An Overview, in: Anaerobic Digestion, D. A. Stafford, B. I. Wheatley, D. E. Hughes, ed., Applied Science Publishers Ltd, London. 1980.

3. D. J. De Renzo. Energy from Bioconversion of Waste Materials. Noyes Data Corporation, Park Ridge, New Jersey. 1977.

METHANE PRODUCTION FROM ANAEROBIC

DIGESTION OF DISTILLERY RESIDUES

L. M. Szendrey and G. H. Dorion

Bacardi Corporation
P. O. Box G 3549
San Juan, Puerto Rico   00936

ABSTRACT

The Bacardi Corporation Anaerobic Treatment Plant has been in operation
since its startup in January, 1982. Since startup, the plant has met or
exceeded, all design performance criteria and the biogas generated is currently substituting a significant portion of the fuel consumed by the distillery steam boilers. Operation of the 13.24 million liters (3.5 million
gal) fixed film downflow anaerobic filter has been carefully monitored since
startup to determine the capabilities and characteristics of this novel
anaerobic treatment process. The data collected confirm the findings of the
pilot plant studies and demonstrate the stability and high efficiency of the
process on a very large scale. Periodic maintenance shutdowns of the distillation columns have interrupted the wastewater feed to the anaerobic
filter on at least six occasions for periods of 4 to 28 days. As predicted
by the pilot studies, the anaerobic filter suffered no adverse effects from
the interruption of feed and was readily restarted in each case within 24 to
48 hours. Data on the biodegradeability of distillery residues, methane
yields, loading rates, and other parameters are presented, and the relationship between parameters such as loading rate, efficiency, methane yield,
volatile acid levels, and pH are discussed. Also, the reasons for the high
degree of stability and efficiency are presented. Finally, a brief outline
of other wastewaters shown to be treatable by this process is given.

Keywords. Anaerobic digestion, biological filter, methanogenesis,
distillery waste, wastewater treatment, rum slops, energy recovery.

INTRODUCTION

Anaerobic digestion has been used as an integral part of wastewater treatment for over 80 years. Initially, and even to this day, anaerobic processes are mainly used to reduce the volume of the solid waste produced during aerobic treatment requiring land disposal, incineration, or other suitable means of disposal. The skyrocketing cost of energy over the last decade has prompted a careful reevaluation of the potential of anaerobic digestion not only to decrease the volume of solid waste requiring disposal, but also as a primary treatment process that is capable of producing significant quantities of usable energy. In fact, biomass grown solely for conversion to energy using anaerobic and other processes may soon become competitive with fossil fuels.

Bacardi Corporation, confronted with the challenge of finding an environmentally acceptable yet cost effective means of disposing of its strong distillery wastewater, carefully evaluated numerous conventional and advanced treatment technologies. After nearly a decade of studies, Bacardi Corporation concluded, with the concurrence of the U.S. Environmental Protection Agency (EPA), that anaerobic treatment was the preferred treatment process for protecting the environment and complying with its discharge requirements. The major factors considered in reaching the above conclusions were the high organic content in the wastewater, the degree of biochemical oxygen demand (BOD) removal required by EPA, the capital cost of the system and the value of the methane generated by the process. Studies conducted by Bacardi Corporation in 1979 by Shea, Ramos and others,[1] demonstrated that anaerobic digestion could remove the required 75% BOD from the rum slops generated by Bacardi Corporation. The 1972 studies, in conjunction with the work of McCarty, Young and later others,[2] convinced Bacardi Corporation that a more stable and efficient anaerobic treatment process could be developed. Research and development work conducted at Bacardi Corporation in early 1980 to 1981 led to the development of the Bacardi Corporation Anaerobic Treatment Process. The work leading to the development of the process, pilot scale studies and startup of the full scale treatment plant have been reported elsewhere.[3,4,5] This novel fixed film anaerobic filter has been in continuous operation for three full years without a single process upset. The operating data for the last 2 years of this highly efficient treatment process is presented.

WASTEWATER CHARACTERISTICS

The wastewater produced during the manufacture of rum is very high in BOD, salt and nutrients. The various options available for the environmentally acceptable treatment and disposal of this wastewater have been reviewed elsewhere.[4] The typical characteristics of rum slops are presented in Table 1.

The operation of the Bacardi Corporation distillery requires at least one, and sometimes two, extended shutdowns each year for routine maintenance; therefore, any effluent treatment process utilized must be capable of coping successfully with an extended interruption of the wastewater feed because of these distillery shutdowns.

PROCESS DESCRIPTION

The Bacardi Corporation Anaerobic Treatment Process (BCATP) utilizes acetogenic (facultative), and methanogenic bacteria, immobilized on a fixed rigid plastic media. These bacteria convert the soluble organic waste into a methane rich gas, thereby reducing the BOD of the wastewater treated.

Table 1. Ranges in composition for several parameters characterizing rum slops

Parameter	Unit	Range
BOD	$mg\ L^{-1}$	36,000 – 42,000
COD[5]	$mg\ L^{-1}$	80,000 – 105,000
TSS*	$mg\ L^{-1}$	3,000 – 8,000
TS	%	7.5 – 11.0
N (Kjehldahl)	$mg\ L^{-1}$	790 – 1,450
P (ortho)	$mg\ L^{-1}$	50 – 100
pH	SU	4.0 – 5.0
Alkalinity ($CaCO_3$)	$mg\ L^{-1}$	600 – 1,700
Vol. Acids (CH_3CO_2H)	$mg\ L^{-1}$	4,000 – 7,000
Specific Gravity	–	1.02 – 1.05
Color	APHA	75,000 – 100,000
Sulfates	$mg\ L^{-1}$	4,000 – 10,000

* With molasses clarification and yeast recycle

The biochemistry of BCATD, like other anaerobic processes, depends on the biochemical utilization of the organic waste material present in wastewater to achieve a BOD reduction in the wastewater treated. During anaerobic treatment, the soluble organic fraction of the wastewater is converted to mostly acetic and other volatile acids by acetogenic facultative bacteria. The volatile acids, in turn, are converted to methane and carbon dioxide by methanogenic bacteria in the same anaerobic reactor.[2] Unlike other anaerobic processes, both bacteria types are immobilized on a rigid, ordered plastic media that allows them to be retained in the reactor indefinitely. Wastewater enters through the top of the anaerobic filter, which is always full of liquid, and passes through the plastic media which is completely covered by a thin film of bacteria which utilize the organic matter in the wastewater. The contents of the anaerobic filter are recirculated to maintain a homogenous mixture and assure good mixing. The treated effluent

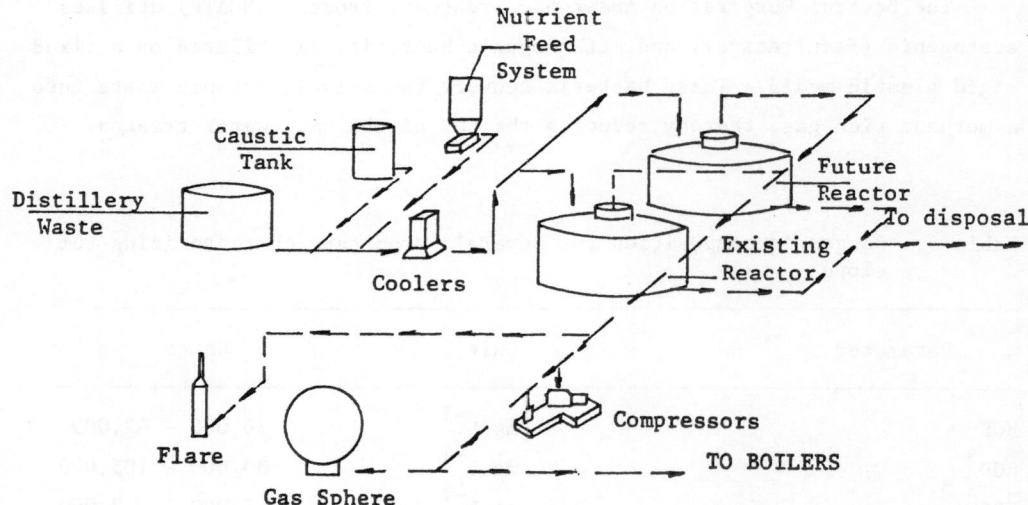

Figure 1. Major component of Bacardi Corporation - anaerobic treatment plant

from the anaerobic filter is discharged from the bottom of the reactor, allowing excess solids to be continuously removed and to maintain a constant inventory of active solids in the reactor. No sludge is required to be isolated for return to the anaerobic filter, therefore, the BCATP does not require a clarification step. The process is shown schematically in Figure 1.

Based on these studies of the operation of the anaerobic filter, it appears that the highest loading the anaerobic filter is currently capable of sustaining over an extended period of time is 14.45 kg of COD per m^{-3} of media (0.9 lb ft^{-3}). A loading 12.84 kg of COD per m^{-3} of media (0.8 lb ft^{-3}) was found to be the optimum. These values compare favorably with those reported by Dogue in 1982.[6] Under normal operating conditions, the pH in the anaerobic filter fell between 7.2 and 7.4. No caustic additions to the anaerobic filter or the rum slops fed were required under normal operating conditions. The rum slops were naturally rich in most nutrients, therefore, no nutrients were added on a regular basis. Additional nitrogen, phosphorous, iron, nickel, and cobalt were added to the system and no significant improvements in filter efficiency were observed. Studies of other micronutrients were conducted in laboratory scale units.[7,8,9] There was more than sufficient hydrogen sulfide present, as rum slops may contain up to 1% sulfate. It is interesting to note that the system acclimated to hydrogen sulfide levels that have been reported as inhibitory and even fatal to anaerobic systems.[10] The operating temperature of the anaerobic filter was 310.93°K (100±2°F), although excursions up to 353.54°K (106°F) have occurred without any noticeable effect on the efficiency. The volatile acids concentration was perhaps the most important operating parameter and averaged between 3,000 and 4,000 mg l^{-1} while concentrations above 5,000 mg $^{-1}$ become inhibitory.[11,12] The alkalinity was controlled by the characteristics of the rum slops fed and while there were considerable variations, no inhibitory effects have been noted.[13] Biogas production averaged 0.56 m^3 kg^{-1} (9 ft^3 lb^{-1}) of COD removed both the pilot and the full scale plant. This value was in the range predicted for this wastewater.[11,12]

The methane content of the gas was sensitive to the pH and the volatile acids concentrations. The optimum and maximum operating values determined thus far for the anaerobic filter are presented in Table 2.

The BOD removal rates achieved by the anaerobic digestion of rum slops approached 90% in the laboratory and led us to suspect that a small proportion of the organic constituents present were biodegradeable by aerobic but not by anaerobic organisms. To test our suspicions, we set up a 500 ml contact anaerobic digester in the laboratory and periodically analyzed samples for BOD and COD for a two month period. The results confirmed our suspicions as the BOD and COD values remained constant after 14 days of digestion, indicating that the remaining organic materials were not degradeable anaerobically.

Table 2. Minimum or maximum sustainable values for given parameters of the anaerobic filter

Parameter	Units	Optimum Value	Maximum or Minimum Sustainable Value
Loading	lbs COD/ft^3/day	0.8	0.9
pH	S.U.	7.3	6.9
Temp.	°F	98–102	106
Vol. Acids	mg l^{-1}	3,000	5,000
Sulfates	%	0.5	1
Alkalinity	mg l^{-1}	3000–4000	6,000
Methane	%	53	50–65

Other industrial effluents were tested using this procedure and were found to contain a lower proportion of non anaerobically digestable BOD. For example, over 98% of the BOD in the slops from a distillery using sugar beet molasses was found to be anaerobically digestable.

STEADY STATE OPERATION

As mentioned above, the BCATP has been in successful operation for three full years. During these first three years of operation, no process upsets occurred and all design performance parameters were met or exceeded. The prediction that this fixed film rigid media downflow reactor was not susceptible to plugging, as are the upflow and downflow reactors using random media, was unequivocally proven. The solids balance observed and reported in Figures 2 through 4 prove that once the anaerobic filter has reached a steady operating state, the solids inventory remains relatively constant in the reactor. During the first three years of operation, the high stability of the process was proven as shown by the monitoring of significant process and treatment parameters such as pH, influent and efflu- ent COD and BOD, gas production rates, and influent and effluents solids.

pH

As indicated earlier, the influent pH of the rum slops varied between 4.2 and 5.0 S.U. The pH of an anaerobic reactor should be above 7.0 S.U. and preferably between 7.2 and 7.8. The effluent pH for the Bacardi Corpora- tion Anaerobic treatment process for the past three years is shown in Figure 5. First, the pH in the anaerobic system never fell below 7.1 S.U. or

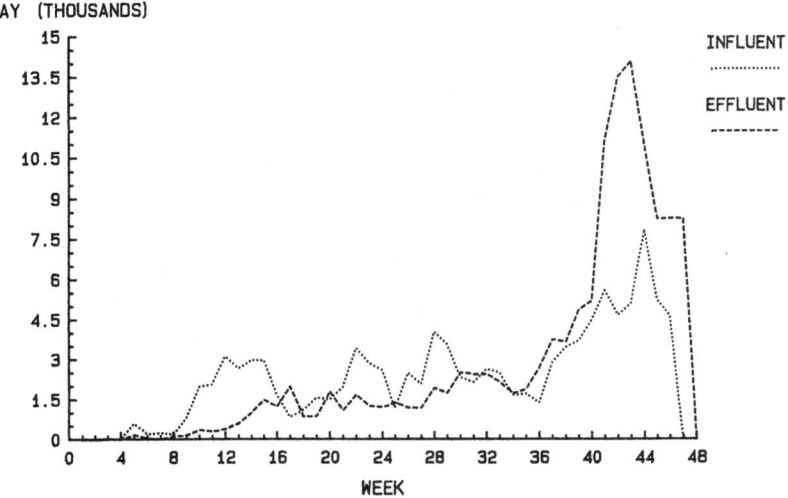

KG/DAY (THOUSANDS)

INFLUENT
....

EFFLUENT

Figure 2. Influent and effluent TSS history. 1982

exceeded 7.7, demonstrating that the reactor was highly buffered and very
stable. Second, the pH tended to rise at the end of each year of operation
because the annual December to January shutdown of the distillery required
the anaerobic filter to remain without feed for 6-8 weeks. When the anaerobic filter was "starved" during a distillery shutdown, the volatile acids
present in the reactor were consumed by the methanogens, causing a rise in
the pH. Finally, although not explicitly indicated in Figure 5, the anaer-

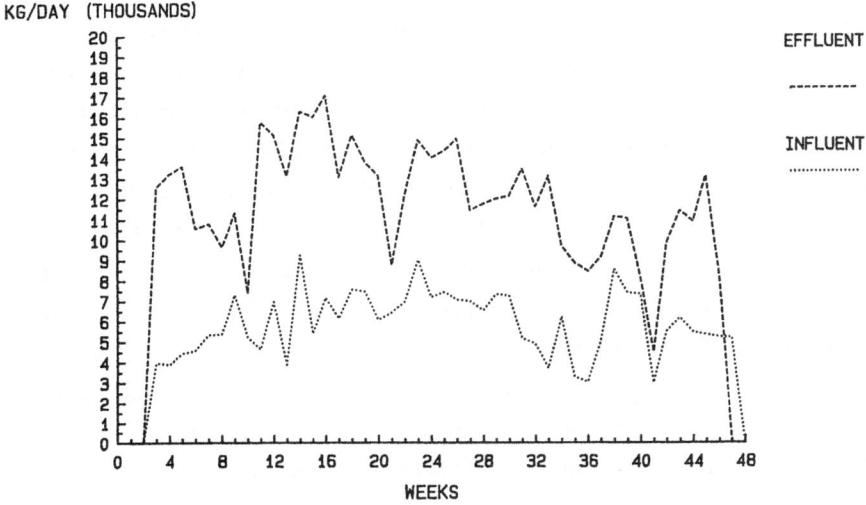

KG/DAY (THOUSANDS)

EFFLUENT

INFLUENT
....

Figure 3. Influent and effluent TSS history. 1983

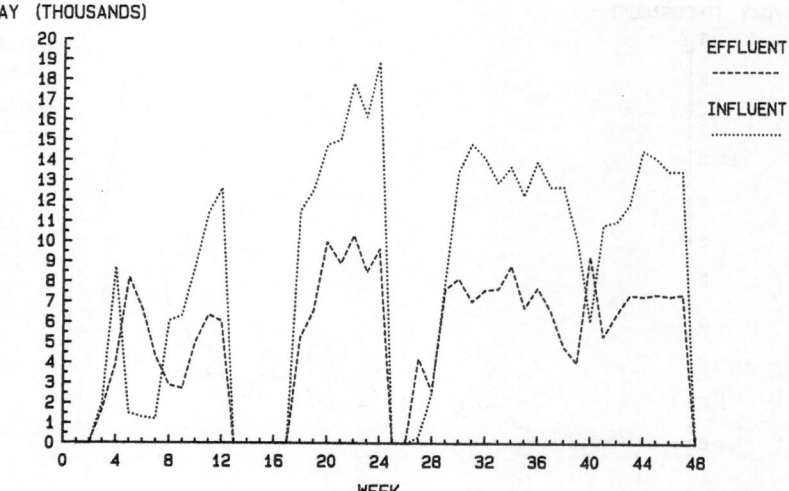

Figure 4.   Influent and effluent TSS history. 1984

obic filter was operated essentially without the need to add sodium hydroxide for pH control in spite of the fact that the pH of the influent was
between 4.2 and 5.0 S.U.

Total Suspended Solids

The influent and effluent suspended solids history of the anaerobic
filter is presented in Figures 2, 3, and 4 for the years 1982, 1983 and 1984
respectively. Figure 2 presents the data for 1982, the first year of operation. It is most interesting to observe that for the first 32 weeks of the
startup year, the suspended solids concentrations leaving the reactor were
less than those in the influent. We attribute this to two basic causes:
First, as shown during our pilot scale studies, the biomass (bacteria)
produced was immobilized on the fixed plastic media to build up an inventory
of active biomass. Careful monitoring of the pilot plant for two years has
shown that a biofilm of between 3.175 to 6.35 mm (1/8 in to 1/4 in) was
developed and maintained. Second, during the first months of the startup
year, the hydraulic retention time was greater than 10 days, allowing sufficient time for solids degradation to occur. The Figures 3 and 4 show a
clear, constant relationship between the influent and effluent suspended
solids, demonstrating that a steady state has been achieved. The constant
relationship between influent and effluent suspended solids shows that the
solids level in the reactor remains constant. We were very pleased by the
constant relationship observed as this confirms the findings of the pilot

524

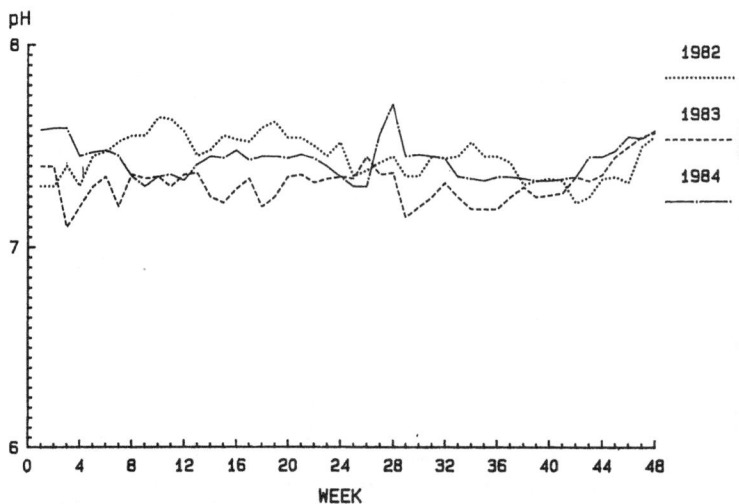

Figure 5. Bacardi Corporation anaerobic filter system--effluent pH history. 1982-1984.

studies; namely, that reactor plugging will not be a concern for the BCATP. Although the full scale anaerobic reactor cannot be opened for inspection, the pilot plant was opened and the media inspected periodically for a period of 2 years to study biofilm buildup.

Biochemical Oxygen Demand

The BOD removal efficiency of the anaerobic filter is, of course, significant because it is a measure of the organic material converted to methane, carbon dioxide and new cells as well as a measure of the perform- ance of the process. The influent and effluent BOD values for the first three years of operation for the Bacardi Corporation Anaerobic System are presented in Figures 6 and 7 respectively for the years 1982, 1983, and 1984. Figure 6a, which presents the data for the startup year of 1982 demonstrates that even during startup, the system was very stable to significant fluctuations in the BOD loading rate. Even though 1982 was a startup year, a constant feed rate to the treatment system could not be maintained due to feed interruptions dictated by the distillery. In spite of the fluctuations observed in 1982, the BOD removal rates, as shown in Figure 6a, exceeded expectations. The Figures 6b and 7, which present the BOD influent and effluent history for 1983 and 1984 respectively, only serve to underline the stability and efficiency of this process for BOD removal.

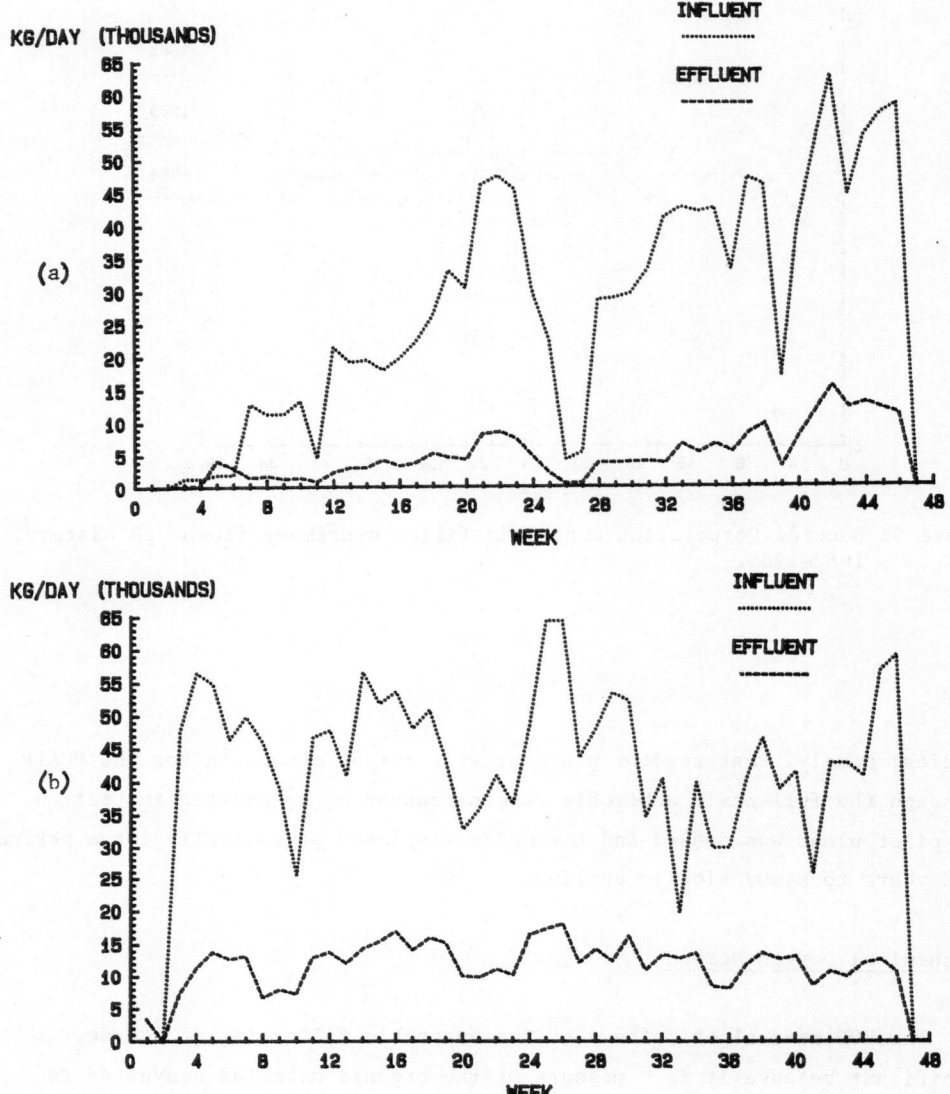

Figure 6. a. Influent and effluent BOD history. 1982
          b. Influent and effluent BOD history. 1983

During 1984, labor work-stops caused two distillery shutdowns interrupting
the feed to the anaerobic filter. In spite of the frequent shutdowns
experienced, the system was able to maintain a constant BOD removal
efficiency.

Chemical Oxygen Demand

The COD removal efficiency of the anaerobic filter is a very important
process control parameter as it is used to rapidly estimate and verify gas

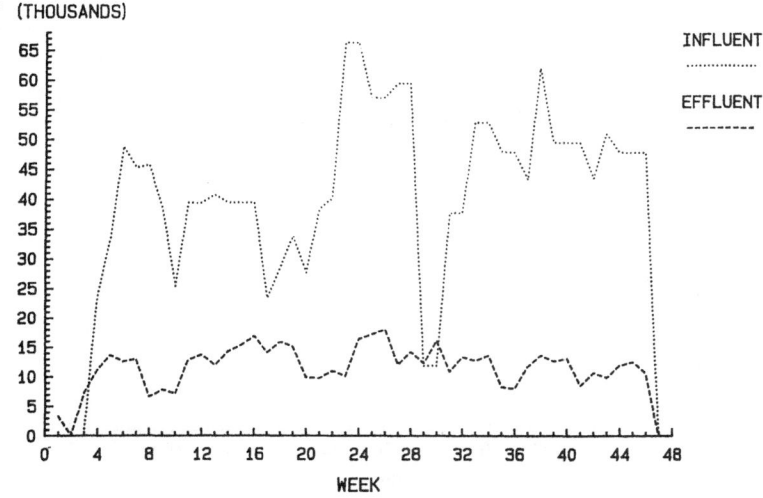

Figure 7.   Influent and effluent BOD history. 1984

production rates.   In fact, we prefer COD to BOD for process control as it does not require 5 days to yield analytical results.   Figures 8, 9 and 10 present the influent and effluent COD levels of the anaerobic filter. Again, the COD removal rates observed both during the startup and subsequent years of operation as presented in Figures 8, 9, and 10 clearly demonstrate the very high stability and efficiency of the downflow fixed film process.

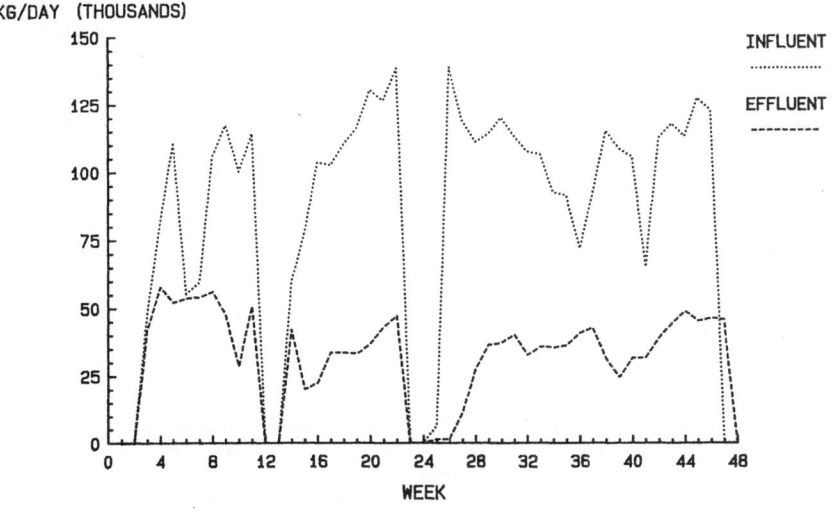

Figure 8.   Influent and effluent COD history. 1984

KG/DAY (THOUSANDS)

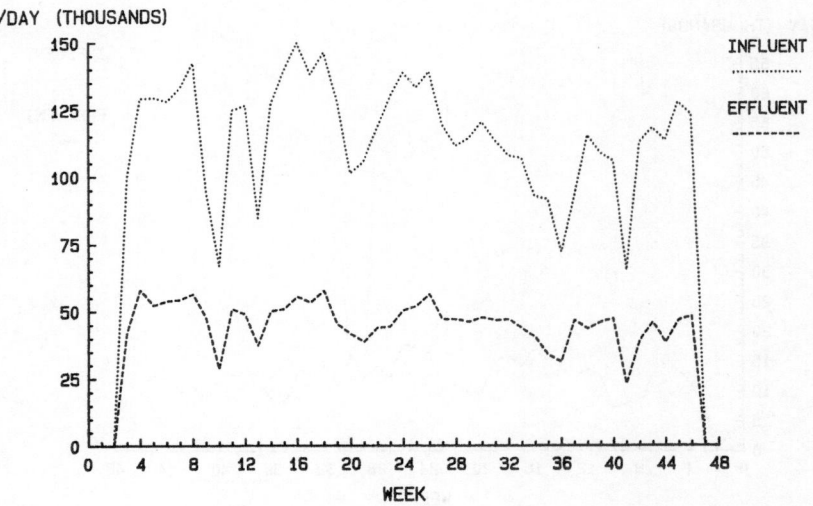

Figure 9.   Influent and effluent COD history. 1983

Biogas Production

Biogas production and quality are directly related to the pounds of BOD or COD removed, the hydraulic retention time and the level of volatile acids in the anaerobic reactor as well as the pH, temperature, alkalinity and nutrient levels present. Under normal operating conditions, we find that

KG/DAY (THOUSANDS)

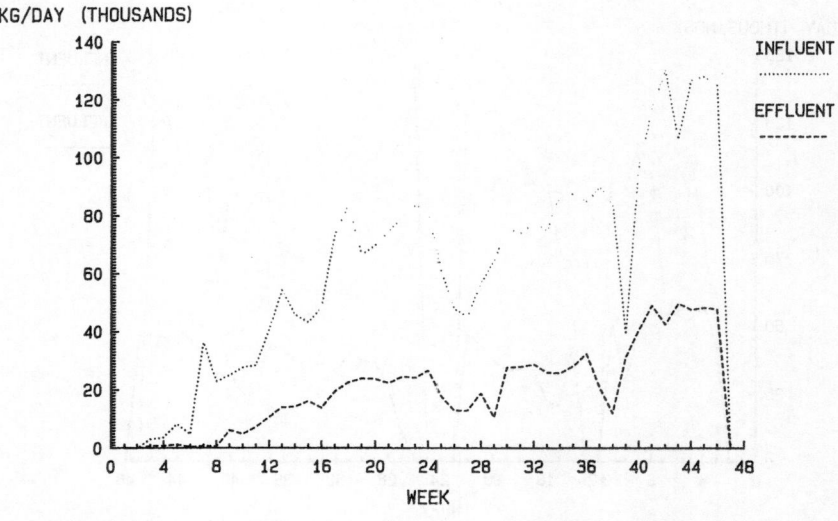

Figure 10.   Influent and effluent COD history. 1982

approximately 0.56 m^3 (9 ft^3 lb^{-1}) of biogas were produced for every kg of
COD removed. We chose to relate biogas production to COD rather than BOD
removal as COD was a more precise and rapid analytical tool.

Biogas production as a function of COD removal is shown in Figure 11
for 1984. As expected, there was a very good correlation between biogas
production and COD removal. It is important to point out that biogas
quality, besides being dependent on the parameters listed above, is also a
function of the type of material being digested. For example, the biogas
produced from a saturated hydrocarbon substrate will have a higher methane
content than that produced from a highly substituted unsaturated substrate.
This observation has been born out by the numerous types of substrate tested
thus far in our laboratory.

CONCLUSION

The Bacardi Corporation anaerobic treatment process has been demonstrated to be more efficient and stable than earlier projections. The most
significant finding reported in this paper is the well defined solids balance demonstrated over three full years of successful operation. Unlike
upflow and/or randomly packed anaerobic filters, this unique downflow design
reactor is a truly fixed film system that does not exhibit any tendencies to

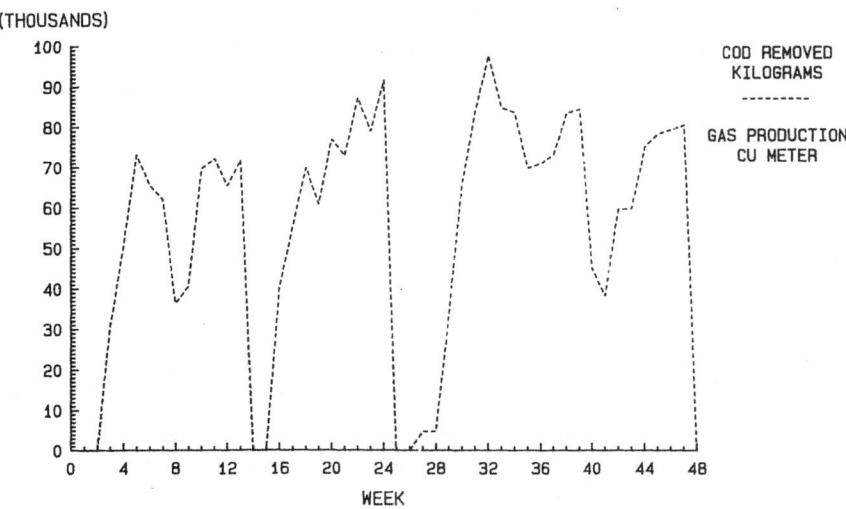

Figure 11. Bacardi Corporation anaerobic filter system--COD removed and gas
production

plug or channel as demonstrated by the data presented and previously reported pilot studies. Based on studies performed in our laboratory and literature reports, we believe that upflow anaerobic reactors are, to a large extent, sludge blanket type reactors where the suspended solids are retained in the reactor by the media. This fact readily explains the problems such as plugging, channeling, and variations in the efficiency common to upflow packed bed reactors.

The patented Bacardi Corporation anaerobic system is applicable to a wide variety of wastewaters and substrates. Pilot and/or laboratory scale studies have confirmed that the following streams are readily and cost effectively treatable using the Bacardi Corporation anaerobic system.

* Pharmaceutical Fermentation Wastewater
* Citric Acid Fermentation Wastewater
* Pulp Paper Mill Wastewater
* Cheese Whey
* Yeast Production Wastewater
* Beet Molasses Distillery Slops
* Brewery Wastewater
* Organic Chemical Manufacturing Wastewater
* Winery Wastewater
* Vegetable Processing Wastewater

All of the above were found to be amenable to anaerobic treatment using the BCATP. Of course, organic loading rates, removal rates and gas production varied with the biodegradeability of each wastewater.[14] In general, most wastewaters having a BOD greater than 500 mg 1^{-1} and a TSS content below 15,000 mg 1^{-1} can be successfully treated by the Bacardi Corporation anaerobic treatment process.

REFERENCES

1. T. C. Shea, G. H. Dorion, J. Rodriguez and Ramos. Rum Distillery Slops treatment by Anaerobic Process. U. S. Environmental Protection Agency. 1974.
2. J. C. Young and P. L. McCarty. The Anaerobic Filter in Waste Treatment. Proceedings of the 22nd Purdue Industrial Waste Conference, West Lafayette, Ind. 1966.
3. L. M. Szendrey. The Bacardi Corporation Anaerobic Process for Stabilizing Rum Distillery Wastes and Producing Methane. Proceedings 7th Symposium – Energy from Biomass and Wastes. Orlando FL. 1983.

4. S. M. Szendrey. Startup and Operation of the Bacardi Corp. Anaerobic Filter. Proceedings-Third International Symposium on Anaerobic Digestion, Boston, Mass. 1983.
5. L. M. Szendrey. Anaerobic Treatment of Fermentation Wastewaters. Environmental Progress. 1984.
6. R. R. Dogue. Principals of Anaerobic Filter Design. Proceedings, 26th Annual Great Plains Wastewater Design Conference, Omaha, Neb. 1982.
7. C. R. Kelly. Temperature and Nutrient Effects on the Anaerobic Extended Bed Treating of High Strength Wastes. Proceedings - 38th Annual Purdue Industrial Waste Conference, West Lafayette, Ind. 1983.
8. A. Diekert, U. Konheiser, K. Piechulla and R. K. Thaur. J Bacteriology. 148, 459-464. 1981.
9. W. P. Murray and L. Van den Berg. Appl. and Env. Microbiology. 42, 502-505. 1981.
10. A. W. Lawrence and P. L. McCarty. Air and Water Pollut. Int. J. 10, 207-221. 1966.
11. P. L. McCarty. Public Works. Sept. 107-112. 1964.
12. P. L. McCarty. Public Works. Oct. 123-126. 1964.
13. A. Li and P. M. Sutton. Determination of Alkalinity Requirements for the Anaerobic Treatment Process. Proceedings - 38th Annual Purdue Industrial Waste Conference, West Lafayette, Ind. 1983.
14. L. Van den Berg, K. J. Kennedy and M. K. Hamoda. Effects of Type of Waste on Performance of Anaerobic Fixed Films and Upflow Sludge Bed Reactors. Proceedings - 36th Annual Purdue Industrial Waste Conference, West Lafayette, Ind. 1981.

ETHANOL FERMENTATIONS

Mary Jim Beck

Tennessee Valley Authority
333 Chemical Engineering Building
Muscle Shoals, Alabama 35660

ABSTRACT

Ethanol is an example of a liquid fuel which can be produced by the
microbial conversion of plant biomass, a renewable resource. The biological
process is a well-known and commercially practiced art. The end-use product
has traditionally been in the spirits and other high-value chemical industries. Process efficiencies for these products, however, do not meet the
economic and energetic constraints necessary for use in the energy field.
Improvements in pretreatment of biomass to enhance bioconversion and in
microbial conversion rates and efficiencies can help overcome the economic
and energetic constraints. Cellulose is the most abundant organic compound
on earth and has great potential as a source of glucose. Physical/chemical
or enzymatic hydrolysis of plant biomass composed of cellulose, hemicellulose, and starch result in a sugar pool comprised of 60% glucose, 25%
xylose, 10% mannose, and smaller amounts of other monosaccharides. Unlike
the polymeric forms cellulose, hemicellulose, and starch, most monosaccharides are readily fermented by microorganisms. Saccharomyces cerevisiae and
Zymomonas mobilis ferment glucose, Pachysolen tannophilus and other yeasts
ferment xylose, while bacteria, including Clostridium thermocellum and Cl.
thermohydrosulfuricum as well as some thermophyllic non-spore forming
species such as Thermoanaerobacter ethanolicus, ferment monosaccharides and
polysaccharides to ethanol. Substrate limitations, rate of reaction,
ethanol tolerance, sensitivity to osmotic pressure, temperature, and other
factors influence bioconversion process efficiencies. Appropriate applications of biotechnology show promise in overcoming some of these constraints.

Keywords. Ethanol fermentation, monosaccharides, polysaccharides, biotechnology.

INTRODUCTION

There is growing interest in using renewable biomass resources for the production of liquid fuels and chemicals. Many of the biological conversions applied are analogous to those used in other processes such as the production of ethanol in the distilled beverage industry. However, the technologies required to produce liquid fuels and chemicals from renewable resources must be reviewed in a different light. Not only must potential for productivity be reexamined, but also the constraints which the use of natural resources has on the world political and economic situation must be borne in mind. Yet, the biological conversion steps lie at the heart of any process or scheme for the exploitation of biomass for liquid fuels and chemicals.

There are a number of liquid products derived through biological conversions which have potential as chemical feedstocks and/or fuels. Among these are butanol, acetone, acetic acid, butanediol, and ethanol. Ethanol serves as a model product for showing the areas of research which have been and are being pursued to improve the biological conversion aspect of the production of such biomass-derived liquids. Similar areas of research would apply for other liquids in addition to ethanol.

There are two roles which microorganisms can play in conversion of biomass to ethanol. Microorganisms or the enzymes they produce may be applied in the hydrolysis of cellulose, hemicellulose, or starch to monosaccharides. Various species of yeasts, fungi, and bacteria can be used to convert these monosaccharides to ethanol.

Hydrolysis

The carbohydrate content of biomass is about 60% glucose and 25% xylose, and the remainder is a mix of galactose, mannose, and arabinose. Sugars are present in biomass as complex mixtures of polymers. A hydrolysis step is required to make them available to most microbial conversions.

Starch is an important source of carbohydrate for ethanol production. However, cellulose is the more abundant glucose polymer. It is also a major waste from agricultural, forestry, and food processing activities. Considerable interest has been placed in the organisms and their enzymes capable of utilizing cellulose in the production of ethanol. The cellulolytic enzyme system is complex, and the crystalline structure of natural cellulose causes difficulty in obtaining rapid and efficient hydrolysis of cellulose.

Increased efficiency in the enzymatic conversion of cellulose as well as in the production of cellulases is needed for profitable application of this technology. A number of areas of research are making gains toward increased efficiency. Mutation and selection have brought about the identification of a number of strains of _Trichoderma_ which have high cellulolytic activities. Improvements in enzyme activity and production may be possible through genetic manipulations[1] and culture techniques[2,3] such as the addition of surfactants. Efficient means of pretreatment of cellulosic substrates such as size reduction, wet oxidation,[4] alkali treatment, irradiation, and steam extraction[5,6] are being examined. Costs may be reduced by enzyme recycle,[7] use of inexpensive substrates for production of cellulases, and solid state fermentation.[8]

The use of acid catalysts for the hydrolysis of cellulose has been given much attention because current costs are more favorable than for enzymatic hydrolysis. Some of the processes being investigated are a revival of technologies from earlier times, such as the dilute sulfuric acid processes.[9,10] The technologies, especially those utilizing the hydrogen halides, have shown improvements over older processes with the application of more current engineering and safety technology.[11,12,13] Acid hydrolysis remains the technology with the best near-term potential for commercial development.[14]

Fermentation

Conventional fermentations using either _Saccharomyces_ or _Zymomonas_ are limited because of their inability to utilize directly the polysaccharides cellulose, hemicellulose, and starch. Anaerobic bacteria including _Clostridium_ _thermocellum_ and _Cl_. _thermochydrosulfuricum_[15,16] as well as the thermophile _Thermoanaerobacter_ _ethanolicus_[17] not only ferment the simple sugars to ethanol with good yield, but also utilize polysaccharides. Their limitations lie in the facts that they cannot tolerate substrate concentrations over 1% and that they carry out mixed fermentations. The bacteria are especially sensitive to natural components of biomass, such as phenolics, which result from the pretreatment steps necessary to allow the bacteria access to polysaccharides from biomass sources.[18] The transfer of the enzyme systems of these bacteria to microorganisms better suited to ethanol production could produce the ultimate microbe for biomass processing.

The conversion of glucose via the Embden-Meyerhoff-Parnas pathway is well known in yeasts. A potential yield of 0.51 gram of ethanol for each gram of glucose metabolized is possible. Relatively high ethanol concentrations have been achieved. The bacterium Zymomonas mobilis makes use of the Entner-Doudoroff pathway for the production of ethanol. A number of characteristics possessed by this bacterium make it worthy of investigation for use in ethanol production from biomass sources, although it has a disadvantage in the lack of industrial experience with the organism. Among its favorable characteristics are higher specific rates of glucose uptake and ethanol production than yeasts and higher ethanol yield coefficients than yeasts.[19] The bacterium is relatively tolerant of heat, osmotic pressure, and ethanol when compared to yeasts. Although it has a limited substrate profile, improvements may be possible through genetic engineering. Genetic methodologies for cloning bacterial cellulase genes and improved pH, salt, and ethanol resistance are approaches being applied to Zymomonas mobilis.

Attention to the total utilization of biomass carbohydrates is an important consideration for the prudent use of such resources. The utlization of the pentoses, particularly xylose, has been the focus of much research. A significant contribution to the utilization of xylose from biomass has been the identification of the ability of certain yeasts to produce ethanol from xylose. This characteristic was only recently identified.[21,22,23,24,25] The metabolism of xylose proceeds through the pentose phosphate pathway. This pathway functions primarily for the production of NADPH and ribose-5-phosphate for nucleotide synthesis. It consists of an oxidative phase that converts hexose phosphates to pentose phosphates, and a non-oxidative phase that converts pentose phosphates back to hexose phosphates. Carbon exiting the non-oxidative phase can enter the Embden-Meyerhoff-Parnas pathway for ethanol production. The stoichiometry of the pathway shows a potential of five moles of ethanol from three moles of pentose. Yeasts must perform a reduction and reoxidation to form xylulose from xylose.[26] The initial steps have been the subject of much research since the intermediate, xylitol, accumulates during the bioconversion, taking its toll on the ethanol yield which has been reported to be about 0.34 gram ethanol per gram xylose consumed. This yield was found to result from fermentation of xylose using Pachysolen tannophilus. More recently it has been reported that strains of Pichia stipitis and Candida shehatae have been identified as having ethanol yield coefficients of up to 0.45 from xylose and accumulating no appreciable amounts of xylitol.[27] Inhibitors of the xylose fermentation are derived from degradation of xylose which can occur during its recovery through acid hydrolysis. This prevents the application of many of the xylose-fermenting yeasts in processes currently being pursued[28] due to the sensitivity of the yeasts.

Bacteria are known to possess an isomerase which converts xylose directly to xylulose. Xylulose can be converted to ethanol by a number of yeasts.[29] Schemes have been devised for a two-step bioconversion[30] with isomerization followed by fermentation. Such schemes have proved inefficient. A more recent approach has been the genetic engineering of yeasts capable of the xylulose to ethanol conversion.[31] The addition of the ability to isomerize xylose to xylulose could bring together many valuable attributes into one yeast.

Conclusion

Microbial conversions have potential in the production of liquid fuels and chemicals from biomass. Ethanol serves as an example of such a product of microbial conversion. The production of ethanol from simple sugars such as glucose or sucrose is well known. The recovery of simple sugars from the more complex polymers found in renewable biomass sources can present barriers to ethanol production. Microorganisms can be employed for the purpose of breaking down complex carbohydrate polymers to simpler sugars. Mineral acids can also serve as catalysts to bring about hydrolysis of the polymers.

Discoveries such as the production of ethanol from xylose have added potential worth to biomass sources and have initiated new research into the biochemistry of these microorganisms. The exploitation of biological conversions for ethanol has also introduced aspects of biotechnology previously applied in the production of economically favorable products, such as drugs, to the production of fuels and chemicals. These tools of biotechnology are in addition to genetic manipulations which are being attempted to improve bioconversion. Novel bioreactors for continuous production of ethanol with vacuum removal of product, means of enhancing ethanol yields through processing steps such as simultaneous saccharification and fermentation, immobilized cell technologies, and continuous fermentation processing have been examined for their potential to add greater value to biomass resources for ethanol production.

REFERENCES

1. H. Durand and M. Clanet. A genetic approach of the improvement of cellulase production by <u>Trichoderma</u> <u>reesei</u>, In: Bioenergy 84, Vol. III, H. Egneus and A. Ellegard, Eds., Elsevier Applied Science Publishers, London. 1985.

2. S. K. Tangu, H. W. Blanch and C. R. Wilke. Enhanced production of cellulase, hemicellulase, and beta-glucosidase by <u>Trichoderma</u> <u>reesei</u> (RUT C30), Biotechnol. Bioeng. 23:1837. 1981.

3. W. D. Murray and S. J. B. Duff. Studies on the production and use of cellulases for the conversion of cellulose to fermentable sugars and alcohol. In: Proceedings of the VI International Symposium on Alcohol Fuels Technology. Vol. III. May, 1984.

4. G. D. McGinnis, W. W. Wilson and C. J. Biermann. Biomass conversion into chemicals using wet oxidation. In: Bioconversion Systems. D. L. Wise, Ed., CRC Press, Boca Raton. 1984.

5. R. F. H. Dekker and A. F. E. Wallis. Enzymatic saccharification of sugarcane bagasse pretreated by autohydrolysis-steam explosion. Biotechnol. Bioeng. 25:3027. 1983.

6. J. N. Saddler and H. H. Brownell. Pretreatment of wood cellulosics to enhance enzymatic hydrolysis to glucose. In: Proceedings of the Royal Society of Canada International Symposium on Ethanol from Biomass. Winnipeg, Canada. Oct., 1982.

7. NYSERDA. Recycle of the enzyme complex after cellulase hydrolysis. Biomass Update. May, 1985.

8. D. S. Chahal. A new approach to produce cellulase complex with <u>Trichoderma</u> <u>reesei</u> on lignocelluloses for ethanol (fuel) production, In: Proceedings of the VI International Symposium on Alcohol Fuels Technology. Vol. II. May, 1984.

9. R. J. Burton. The New Zealand wood hydrolysis process. In: Proceedings of the Royal Society of Canada International Symposium on Ethanol from Biomass. Winnipeg, Canada. Oct., 1982.

10. J. E. Jordan. TVA's ethanol-from-hardwood project. In: Proceedings of the Fifth Annual Solar and Biomass Energy Workshop. Atlanta, Georgia. April, 1985.

11. C. M. Ostrovski and J. C. Aitken. Fuel ethanol from wood through hydrofluoric acid solvolysis of cellulose, Seventh Symposium on Biotechnology for Fuels and Chemicals, Gatlinburg, Tennessee. May, 1985.

12. F. leGrand. Conversion of agricultural fibers into fermentable sugars using sulfur trioxide as a catalyst. U.S. Patent No. 4, 427, 584. Jan., 1984.

13. M. Wayman, A. Tallevi and B. Winsborrow. Hydrolysis of biomass by sulfur dioxide. Biomass 6:183. 1984.

14. W. Hoagland. An overview of engineering research on acid hydrolysis processes. In: Biochemical Conversion Program Semi-Annual Review Meeting. SERI. Golden, Co. June, 1985.

15. T. K. Ng, A. Ben-Bassat and J. G. Zeikus. Ethanol production by thermophilic bacteria: fermentation of cellulosic substrates by cocultures of <u>Clostridium</u> <u>thermocellum</u> and <u>Clostridium</u> <u>thermo-hydrosulfuricum</u>. Appl. Environ. Mirobiol. 41:1337. 1981.

16. G. C. Avgerinos, H. Y Fang, I. Biocic and D. I. C. Wang. A novel. Single-step microbial conversion of cellulosic biomass to ethanol, In: Advances in Biotechnology. M. Moo-Young and C. W. Robinson, Eds., Pergamon, Toronto. 1981.

17. L. H. Carreira, J. Wiegel and L. G. Ljungdahl. Production of ethanol from biopolymers by anaerobic, thermophilic, and extreme thermo-philic bacteria: I. Regulation of carbohydrate utlization in mutants of <u>Thermoanaerobacter</u> <u>ethanolicus</u>, Biotechnol. Bioeng. Symp. 13:183. 1983.

538

18. G. C. Avgerinos and D. I. C. Wang. Selective solvent delignification for fermentation enhancement. Biotechnol. Bioeng. 25:67. 1983.

19. P. L. Rogers, K. J. Lee, M. L. Skotnicki and D. E. Tribe. Ethanol production by Zymomonas mobilis. Adv. Biochem. Engin. 23:27. 1982.

20. S. E. Buchholz, C. Coulson, N. M. Dooley, D. E. Eveleigh, A. Lejeune, Z. P. Shalita and M. D. Yablonsky. Genetic approaches in Zymomonas mobilis, In: Biochemical Conversion Program Semi-Annual Review Meeting. SERI. Golden, Colorado. June, 1985.

21. C. S. Gong, L. D. McCracken and G. T. Tsao. Direct fermentation of D-xylose to ethanol by a xylose-fermenting yeast mutant. Candida sp. XF217. Biotechnol. Lett. 3:245. 1981.

22. A. Margaritis and P. Bajpai. Direct fermentation of D-xylose to ethanol by a strain of Candida shehatae. Biotechnol. Lett. 5:357. 1983.

23. J. C. duPreez and J. P. van der Walt. Fermentation of D-xylose to ethanol by a strain of Candida shehatae, Biotechnol. Lett. 5:357. 1983.

24. H. Schneider, P. Y. Wang, Y. K. Chan and R. Maleszka. Conversion of D-xylose into ethanol by the yeast Pachysolen tannophilus, Biotechnol. Lett. 3:80. 1981.

25. P. J. Slininger, R. J. Bothast, J. E. VanCauwenberge and C. P. Kurtzman. Conversion of D-xylose to ethanol by the yeast Pachysolen tannophilus. Biotechnol. Bioeng. 24:371. 1982.

26. K. L. Smiley and P. L. Bolen. Demonstration of D-xylose reductase and D-xylitol dehydrogenase in Pachysolen tannophilus, Biotechnol. Lett. 4:607. 1982.

27. J. C. duPreez and B. A. Prior. A quantitative screening of some xylose-fermenting yeast isolates. Biotechnol. Lett. 7:241. 1985.

28. J. E. Fein. S. R. Tallim and G. R. Lawford. Evaluation of D-xylose fermenting yeasts for utilization of a wood-derived hemicellulose hydrolyzate. Can. J. Microbiol. 30:682. 1984.

29. R. Maleszka and H. Schneider. Fermentation of D-xylose, xylitol, and D-xylulose by yeasts, Can. J. Microbiol. 28:360. 1982.

30. L. C. Chiang, H. Y. Hsiao, P. O. Ueng, L. F. Chen and G. T. Tsao. Ethanol production from xylose by enzymic isomerization and yeast fermentation. Biotechnol. Bioeng. Symp. 11:263. 1981.

31. N. W. Y. Ho. Cloning the E. coli xylose isomerase gene for the improvement of xylose to xylulose conversion by microorganisms. 189th National Meeting of the American Chemical Society. Miami Beach, Florida. April, 1985.

ETHANOL PRODUCTION FROM BIRD-RESISTANT AND

NON-BIRD-RESISTANT GRAIN SORGHUM

J. T. Mullins and C. NeSmith

Department of Botany
University of Florida/IFAS
Gainesville, Florida 32611

ABSTRACT

Sorghum is recognized as a cultivated crop with high potential for biomass energy production. It can be grown over a wide range of climatic and soil conditions, and is a familiar crop to the farmer. In Florida, there is an interest in utilizing new varieties of grain sorghum for large scale ethanol production because yields are greater than or equal to corn. Bird-resistant (BR) varieties, containing higher levels of tannins, suffer less bird and disease damage than do non-bird-resistant (NBR) varieties. Depending upon planting time and location, average yields of the BR variety, Savanna 5, are 0.88 Mg ha^{-1} greater than the NBR variety DK-64, and 0.27 Mg ha^{-1} greater than corn. Thus Savanna 5 could produce 1020 L ha^{-1} more ethanol than DK-64, and 850 L ha^{-1} more than corn. Actual fermentations of both varieties gave about 11% wv^{-1} ethanol in the final beer and were equivalent to our values for corn. The time required for fermentation of the BR variety, Savanna 5, was much longer than for DK-64 or corn. In an effort to determine whether or not the high tannin content was responsible for the longer time of fermentation, viable yeast cell counts were made on the mash after 24 and 48 h. The mash prepared from DK-64 had twice as many viable cells as did that from Savanna 5. Washing the high tannin variety with water, prior to fermentation, had no effect on the rate of production of ethanol. The enzymes used for starch hydrolysis gave the same levels of glucose in mash minus yeast cells prepared from both varieties. These results suggest that the mash of Savanna 5 inhibits the cell cycle of yeast during fermentation, by a mechanism which may or may not involve tannin.

Keywords. Sorghum, fuel ethanol, tannin.

INTRODUCTION

Sorghum has been suggested as a major candidate for development as a multi-product crop.[1] The multi-product concept envisions sorghum as supplying: (i) fibers for paper, insulating and construction boards, furfural, charcoal, and alpha-cellulose; (ii) silage or dry forage for animal production; (iii) energy by direct combustion or by conversion into ethanol. As C-4 photosynthesis plants, they return a favorable energy balance, and additionally in terms of food energy calories produced per cultural energy invested, sorghum exceeds both sugarcane and corn.[1] This results partially from drought-resistance and a relatively low water requirement. The major species of sorghum is Sorghum bicolor (L.) Moench, but there are thousands of varieties distributed among grain, sweet, broomcorn and forage types. Sorghum ranks fifth in production among world grain crops.[1] Grain sorghum production peaked at about 26×10^6 Mg in 1973 in the U.S. and is currently about 21×10^6 Mg.[2] The use of hybrids that combine characteristics of sweet and grain types, termed high energy sorghums, has been proposed as one way to increase production potential for biomass energy.[3] The yields of grain sorghum are greater than those of corn in Florida.[4] Bird-resistant (BR) varieties, containing higher levels of tannins, suffer less bird and disease damage than do non-bird-resistant (NBR) varieties. Depending upon planting time, location and cultivation practices, average yields of the BR variety, Savanna 5, were 0.88 Mg ha^{-1} greater than the NBR variety DK-64.[4] Assuming normal production of ethanol from Savanna 5, 3570 L 100% ethanol ha^{-1} could be obtained. Such projections for ethanol production from grain sorghum are much higher than those given in the Biomass Energy Production and Use Plan developed in response to section 211 (b) of the Energy Security Act of 1980 (P. L. 96-294) and updated for 1982.[5]

The impact of high tannin in grain sorghum on the bioconversion to ethanol has not been fully explored. Objectives of the sorghum improvement program in Florida include increasing the biomass yields, improving animal feed efficiency of high tannin types and examining the influenece of high tannin levels on fermentation to ethanol. Specifically, this paper will address the ethanol fermentation of both BR and NBR grain sorghum and provide comparisons with corn.

MATERIALS AND METHODS

Cooking

Grain sorghum at 10% moisture was obtained from the Agricultural Research Center, Marianna, FL, and ground into particles ranging from 0.5-2.0 mm. Most studies were carried out on two varieties, Northrup-King, Savanna 5, a BR form with 2% tannin, and Dekalb, DK-64, a NBR form with a 0.1% tannin.[4] Each sample was cooked while being agitated with a Sorvall Omni-Mixer at a ratio of 7.5 L water and 3 kg sorghum (final volume of mash 10 L), to 45-50 C when 2 g (0.07%) Canalpha enzyme (Biocon U.S. Lexington, KY), was added. Canalpha is an alpha-amylase derived from *Bacillus subtilis* and is recommended for use at 0.05% of dry weight of feedstock in a pre- and 0.10% in a post-cook level. Cooking was continued until 93-95 C was obtained. The mash was cooled to 70-80 C and 4 g (0.13%) Canalpha was added to complete liquifaction. Saccharification was begun, after cooling to 30-40 C, by adding 4.5 g (0.15%) of Gasolase enzyme. Gasolase, Biocon U.S., has a wide spectrum of hydrolases derived from *Aspergillus niger* and *Rhizopus niveus* and is recommended for use at 0.05% of dry weight of feedstock. The starting pH of the mash was 4.5-5.5 and no adjustments were made. This mash represents a more concentrated form (63 L water added per 25.4 kg) than Biocon recommends (95 L water added per 25.4 kg) and the level of enzymes was increased as required to produce maximum ethanol, as in previous studies on corn.[6]

Inoculum Preparation

The yeast, *Saccharomyces cerevisiae* from Biocon U.S., distillers active dry form was used in all fermentations. A 250 ml inoculum was prepared by placing 5.4 g dry yeast in 0.67% Yeast Nitrogen Base (BBL Microbiology Systems Cockeysville, MD) with 2% D-glucose. The growth period was 12-18 h at 30 C with agitation on a New Brunswick Scientific Gyrotory water bath shaker, Model G76. A cell count of 1.7×10^8 cells ml^{-1} resulted and the cells were in late log phase when used.

Fermentation

Immediately after combining saccharification enzyme with 3.25 L cooked sorghum and 84 ml of yeast inoculum, triplicate samples (representing a final volume of 84.8 L total mash per 25.4 kg sorghum) were transferred to 9 L carboys fitted with fermentation locks. The total number of cells present at the beginning of fermentation was about 5.6×10^{11}. The containers were placed on a Model 6010 Eberbach Shaker and agitated at 140 excursions per minute at room temperature for 72 h or more.

Analytical Methods

At various intervals samples were removed for assays. Mash and beer samples were dried to 0% moisture before the starch was hydrolyzed with Biocon's Canalpha and exo-1,4-alpha-D-glucosidase (Sigma Chemical, St. Louis, MO). The glucose (wv^{-1}) produced was assayed by the glucose oxidase method in a YSI Model 27 Industrial Analyzer, (Yellow Springs Instrument Co. Yellow Springs, OH). Glucose in the beer was determined directly following clarification by centrifugation. Ethanol (wv^{-1}) was determined by temperature compensated hydrometer readings on 50 ml distillates from 100 ml samples of whole beer. The pH values were determined with an Orion Microprocessor, Model 811. Viable yeast cell counts were determined by serial dilution at various times during the fermentation. The standard method[7] was followed and colony counts were made with a Fisher, Model 133-8002.

RESULTS AND DISCUSSION

Figure 1 indicates the levels of glucose, ethanol and pH during the course of fermentation of the NBR variety DK-64 and the BR variety Savanna 5. The time course values for DK-64 are similar to those reported for corn.[6] After a lag of about 12 h, ethanol production appears to be linear until about 48 h and peaks at 11% (wv^{-1}) after 72 h of fermentation. Data beyond this point are not included because increases remained within one Standard Deviation of the 72 h value. The time course values for Savanna 5 are different from both DK-64 and corn. Glucose concentration peaks after about 12 h, but remains constant until 48 h, thereafter declining to about 0.05% (wv^{-1}) at 96 h. Ethanol production did not peak until 96 h (10.7% wv^{-1}), but the final value is the same as that of DK-64 and corn. Table 1 compares the fermentation time and final ethanol production, and projects yields for sorghum and corn. DK-64 has a somewhat slower fermentation time than corn, but final ethanol production and projected yields at various locations within the state of Florida are essentially the same.[4] In contrast, Savanna 5, has a much longer fermentation time, but produces the same ethanol in the final beer. The projected ethanol yields for this variety at various locations, however, are 9% higher than DK-64 and 7% higher than corn (Table 1). Using the best location, Ona, FL, and production for 1979-83,[4]

544

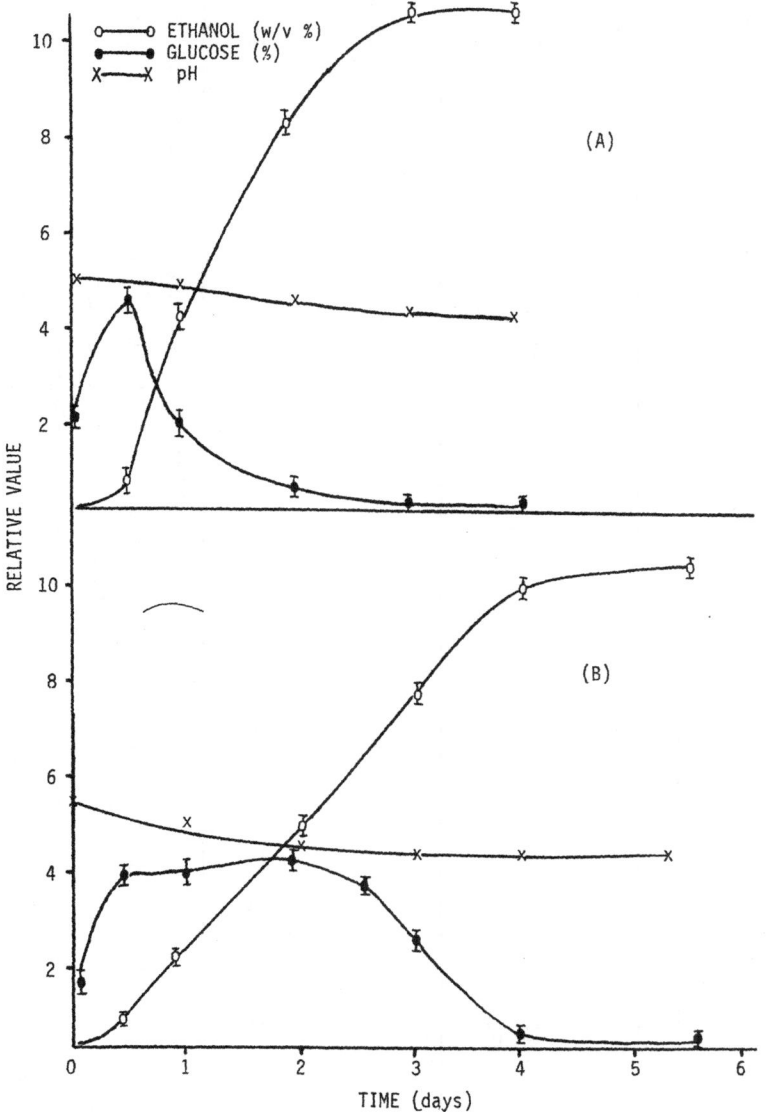

Figure 1. Glucose concentration, ethanol production and pH values during
fermentation of concentrated grain sorghum, non-bird-resistant
DK-64 (A) and bird-resistant Savanna 5 (B) mash (84.8 L total
mash 25.4 kg^{-1}), containing 15.25 g saccharification enzyme
(Gasolase) per 25.4 kg^{-1}. Verticle bars + SD. Symbols refer to
(o) ethanol, (•) glucose, and (x) pH.

Table 1. Fermentation time and ethanol yield of bird-resistant[a] and non-bird-resistant[b] sorghum varieties compared to corn

Feedstock	L 100% 45.4 kg feedstock[-1] fermentation time (h)					Ethanol final beer[c] (wv[-1])	Yield[d] (L 100% ha[-1])	Yield[e] (L 100% ha[-1])
	48	60	72	84	136			
Corn[f]	19.7	19.7	-	-	-	10.3	2720	2720
Sorghum								
DK-64	16.7	18.5	20.8	20.8	-	11.0	2660	2550
Savanna 5	9.5	12.9	15.1	17.0	20.1	10.8	2900	3570

[a]Northrup-King brand, variety Savanna 5. [b]Dekalb brand, variety DK-64. [c]Amount of ethanol obtained after actual distillation. [d]Ethanol yields based on average grain yields for Savanna 5 (6.54 Mg ha[-1] or 2.92 t acre[-1]) and DK-64 (5.80 Mg ha[-1] or 2.58 t acre[-1]) grown at Marianna, FL, during 1981-83 with April planting dates; Jay, FL, during 1983 with March 31 planting date; Ona, FL, during 1979-83, with March-April planting dates. Ethanol yield for corn based on 6.27 Mg ha[-1] or 2.8 t acre[-1]. [e]Ethanol yields based on average grain yields for Savanna 5 (8.06 Mg ha[-1] or 3.59 t acre[-1]) and DK-64 (5.56 Mg ha[-1] or 2.48 t acre[-1]) grown at Ona, FL, during 1979-83 with March planting dates. Ethanol yield for corn based on 6.27 Mg ha[-1] or 2.8 t acre[-1]. [f] Corn data from Ref. 6.

Savanna 5, would give ethanol yields 40% greater than DK-64 and 31% higher than corn (Table 1). These projections for ethanol yield from Savanna 5, are 150% higher than the averages given in the Biomass Energy Production and Use Plan by USDA and DOE.[5] Thus it is clear that Savanna 5 deserves close analysis in Florida. The Florida projections, however, are lower than some Texas projections for high energy sorghum hybrids when both grain and stalk components are included.[8] The Texas evaluation did not include cellulose or hemicellulose components.

The time required for complete fermentation of Savanna 5 is unacceptable when compared with DK-64 or corn (Table 1). An effort was made to determine whether or not the higher tannin content was responsible for the slow fermentation. A comparison of the numbers of viable yeast cells was made for DK-64 and Savanna 5 during the fermentation. The results in Table 2 show a marked reduction in the number of viable yeast cells in Savanna 5 mash as compared with DK-64 after 24 h into the fermentation. Doubling the initial yeast inoculum did not change the number of yeast cells capable of carrying out the cell cycle. It did, however, increase the ethanol value after 72 h into the fermentation by 15%, which was only 7% below that of DK-64 at 72 h. The lower glucose levels after 72 h were as expected when the ethanol values are considered.

Table 2. Comparison of yeast viability from bird-resistant (Savanna 5) and non-bird-resistant (DK-64) grain sorghum fermentations

Inoculum[b]	DK-64		Savanna 5		Washed Savanna 5[a]	
	1X	2X	1X	2X	1X	2X
VC 24h[c]	6.85	12.45	3.49	3.45	2.87	1.98
VC 48h	6.05	4.40	3.44	2.19	4.06	2.55
Ethanol 72h[d]	9.7	9.6	7.8	9.0	7.5	8.5
Glucose 72h[e]	53.0	43.0	490.0	43.0	563.0	50.0

[a]Ground grain, washed with three volumes water, soaked overnight, and re-washed twice. [b]Yeast inoculum: 1X contains 5.23×10^6 viable cells ml^{-1} or 1.36×10^5 ml^{-1} total mash; 2X contains 9.5×10^6 viable cells ml^{-1} or 2.46×10^5 ml^{-1} total mash. [c]Viable cells (VC) in millions ml^{-1} during fermentation. [d]Ethanol % wv^{-1} [e]Glucose % wv^{-1}.

Florida-grown grain sorghums are usually grouped into two classes: those with high digestability by animals and low tannins and those with low digestability and high tannins.[9] A number of methods have been used to reduce the extractable tannin content in sorghum grain, in order to improve their animal feed value. Imbibing the grain with water, acid, base, non-ionic detergents and polyvinylpyrrolidone have been suggested.[10,13] Among these methods we chose soaking the grain in five changes of water over 24 h as the most practical large scale method. As seen in Table 2, treating Savanna 5 in this way did not improve yeast viability or ethanol production during the time period used. In Bantu beer brewing, inhibition of ethanol production by Saccharomyces is said not to be associated with tannins but with some unknown compounds produced during the malting.[10] The addition of a range of phenolic compounds at 0.5%, consisting of tannic, galic, caffeic, ellegic acids or catechol did not inhibit yeast alcoholic fermentation in 10% suspensions of normal malt. It was suggested that starch degrading enzymes were rendered insoluble and inactive, although maltase retains its activity.[10]

We examined the activity of Canalpha and Gasolase in mash from DK-64 and Savanna 5, including the water washing of Savanna 5. The results in

Table 3. Glucose production in enzyme treated bird-resistant (Savanna 5) and non-bird-resistant (DK-64) mash[a].

Variety	Time (h)				
	0	18	42	72	86
DK-64	1.18[b]	6.86	13.10	16.00	17.66
Savanna 5	1.20	7.05	12.30	16.22	17.46
Washed Savanna 5[c]	1.26	6.83	12.22	16.94	17.54

[a]Standard amounts of Canalpha (0.07% ww^{-1} pre-cook and 0.13% ww^{-1} post-cook) and Gasolase (0.15% ww^{-1}), Biocon U.S. added without yeast. [b]Glucose % wv^{-1}. [c]Ground grain, washed with three volumes water, soaked overnight and re-washed twice.

Table 3 show no inhibition of glucose production from either variety in mash lacking yeast cells.

We conclude that the bird-resistant variety, Savanna 5, has great potential as a feedstock for ethanol production in Florida. This potential will not be realized, however, until the basis for the slow rate of yeast fermentation is determined and overcome. At this time, the inhibition seems to be centered in the cell cycle of yeast and not in the enzymes used for starch hydrolysis. Whether the inhibitory agent is tannin or an unidentified substance is yet to be determined.

CONVERSIONS

SI Units		Non-SI Units
Divide	By	To obtain
liter per hectare, L ha^{-1}	9.35	gallon per acre
megagram per hectare, Mg ha^{-1}	1.12×10^{-3}	pound per acre
kilogram, kg	0.454	pound
1 Mg = 1000 kg		

ACKNOWLEDGMENT

The authors wish to thank Drs. J. E. Marion, Department of Poultry Science, D. W. Gorbet and R. O. Myer, Agricultural Research Center, Marianna, Institute of Food and Agricultural Sciences, University of Florida, Gainesville, FL, for generously supplying the samples of grain sorghum used in these studies.

REFERENCES

1. J. W. Clark, F. R. Miller and R. A. Creelman. Sorghum - a versatile, multi-purpose biomass crop. Biomass Digest. January, 1981. p. 35-57. 1981.
2. F. R. Miller. Grain sorghum. In: CRC Handbook of Biosolar Resources, O. R. Zaborsky (Ed.). Vol. II, Resource Materials, T. A. McClure and E. S. Lipinsky (Eds.). CRC Press, Boca Raton, FL. p. 43-50. 1981.
3. R. A. Creelman, L. W. Rooney and F. R. Miller. Sorghum. In: Cereals: a renewable resource, Y. Pomeranz and L. Munch (Eds.). Am. Assoc. Cereal Chemists, St. Paul, MN. 1981.

4. Institute of Food and Agricultural Sciences, University of Florida. Florida Field and Forage Crop Variety Report. 1983. Florida Cooperative Extension Service. Agronomy Research Report, AY 84-9, February, 1984.

5. United States Departments of Agriculture and Energy. A biomass energy production and use plan for the United States, 1983-90. Agricultural Economic Report No. 505. U.S. Government Printing Office. Washington, D.C. November, 1981.

6. J. T. Mullins. Enzymatic hydrolysis and fermentation of corn for fuel alcohol. Biotechnol. Bioeng. 27:321-326. 1985.

7. F. Sherman, G. R. Fink and J. B. Hicks. Methods in yeast genetics. Cold Spring Harbor, NY. 1983.

8. R. L. Monk, F. R. Miller and G. G. McBee. Sorghum improvement for energy production. Biomass 6:145-153. 1983.

9. V. E. Green. Yield and digestibility of bird-resistant and non-bird-resistant grain sorghum. Soil and Crop Sci. Soc. Fla. Proc. 33:13-16. 1973.

10. T. G. Watson. Inhibition of microbial fermentations by sorghum grain and malt. J. Appl. Bact. 38:133-142. 1975.

11. M. L. Price, L. G. Butler, J. C. Rogler and W. R. Fetherston. Overcoming the nutritionally harmful effects of tannin in sorghum grain by treatment with inexpensive chemicals. J. Agric. Food Chem. 27:441-445. 1979.

12. R. D. Riechert, S. E. Fleming and D. J. Schwab. Tannin deactivation and nutritional improvement of sorghum by anaerobic storage of H_2O-, HCl-, or NaOH-treated grain. J. Agric. Food. Chem. 28:824-829. 1980.

13. R. Kumar and M. Singh. Tannins: their adverse role in ruminant nutrition. J. Agric. Food Chem. 32:447-453. 1984.

PRODUCTION OF ETHANOL FROM WOOD BY

ACID HYDROLYSIS AND FERMENTATION

E. C. Clausen and J. L. Gaddy

Department of Chemical Engineering
University of Arkansas
Fayetteville, Arkansas   72701

ABSTRACT

Fermentable sugars may be produced from woody biomass materials by
hydrolysis with mineral acids. After acid recovery, the sugars may be
converted by bacteria or yeast into a variety of chemicals, including ethanol. This paper presents the results of laboratory studies for the production of sugars from oak sawdust and develops an industrial process for
ethanol production. Preliminary economics for the process are projected and
discussed.

Keywords. Acid hydrolysis, biomass, wood, ethanol.

INTRODUCTION

Wood residue in the form of limbs, treetops, stumps, bark and sawdust
represent an abundant source of raw material for the biological production
of chemicals and energy. In addition to providing this raw material source,
the harvesting of wood residues allows for easier reforestation of timber-
lands. Many chemicals can be produced, including methane, alcohols, organic
acids, and solvents.

There are two major methods of hydrolyzing biomass to sugars in the
presence of a mineral acid catalyst: a high temperature dilute acid hydrolysis and a lower temperature concentrated acid hydrolysis. Early high
temperature dilute acid processes employed acid concentrations of less than
5 percent at temperatures exceeding 200°C.[1,2] Since low acid concentrations

were employed, there was no need for acid recovery, since neutralization with lime or other suitable base could be employed and process economics did not dictate the need for acid recovery. These processes suffered from low yields (often less than 50 percent of theoretical) due to sugar losses by degradation and reverse polymerization at the high temperatures and pressures employed. More recent research has concentrated on improving yields by modifying the reaction scheme to lessen the probability of toxic by-product formation.

Concentrated acid processes employ more concentrated acid catalysts (up to 13 N) at lower reaction temperatures for the hydrolysis.[3,4] Typically the hemicellulose and cellulose are degraded in separate steps because the hemicellulose is more easily hydrolyzed than cellulose, and the resulting xylose from hemicellulose is susceptible to decomposition to furans, toxins to microorganisms used in fermentations. Although yields from concentrated acid processes can be quite high, acid recovery in these processes is essential to make the process economically feasible.

Acid hydrolysis processes are receiving widespread attention as methods to produce sugars from biomass. However, little attention has been given to the fermentation of the hydrolyzates to chemical intermediates. When fermentation has been attempted, the toxic by-products formed during hydrolysis such as furfural and hydroxymethyl-furfural (HMF) have often inhibited the fermentation. Various techniques have been used to remove or "tie-up" the by-products with some success. However, it is still generally felt that the fermentation time for hydrolyzates will be much greater than when fermenting synthetic glucose.

A two-step hydrolysis process using concentrated mineral acid has been developed at the University of Arkansas which results in nearly 100 percent yields in the prehydrolysis of the hemicellulose and the hydrolysis of the cellulose, without significant degradation or reverse polymerization.[5] Several acid recovery processes have been developed and tested, yielding an energy efficient method of separating sugar and acid. The resulting sugar solution has been successfully fermented to ethanol and other chemicals without pretreatment. The purpose of this paper is to describe a process for producing chemical feedstocks from woody biomass. Data for the two-stage acid hydrolysis are presented along with preliminary fermentation results of the hydrolyzate to ethanol. A preliminary process design and economic analysis for a plant producing (75 million lyr^{-1}) of ethanol from hardwood residues is also presented.

PARAMETERS AFFECTING SUGAR YIELDS

The two major factors which control the hydrolysis reactions are temperature and acid concentration. Studies in the University of Arkansas laboratories have been made to determine the effect of these variables on the hydrolysis of biomass material.[6] Ten percent slurry concentrations are maintained. Complete xylan conversion can be achieved in the prehydrolysis at a low acid concentration (2N) at temperatures of about 100°C using reaction times of 2 hours or less. The use of higher acid concentrations (10N) results in complete conversion at about 60°C. Similarly, total conversion in the main hydrolysis can be achieved at room temperature with an acid concentration of 14N. High temperatures (i.e. 150-250°C) are required to give suitable conversions at low acid concentrations (3N or less). These high temperatures, however, promote sugar degradation and repolymerization.

In deciding upon which conditions to use for acid hydrolysis, several factors must be considered. The acid cost would be prohibitively high, even at low concentrations (2N), unless the acid was recovered for reuse. The degradation of xylose to furfural and glucose to HMF is more readily promoted by high temperature than by high acid concentration (in this range). Therefore, the preferred operating range is a moderate acid concentration and mild temperatures (i.e. 5-10N at 100°C or less).

The sugar concentrations and yields from a typical pre-hydrolysis and hydrolysis stagewise process for oak sawdust using a concentrated acid hydrolysis are given in Table 1. Sugar analysis was by UV-visible spectrophotometry and dinitrosalacylic acid (DNS) reducing sugar analysis. Specific acid conditions are not given due to proprietary considerations. Very dilute (2-8 percent) sugar concentrations result from these reactions. The prehydrolysis step yields 14 percent of the initial stover as xylose. The combined conversion of hemicellulose and cellulose to sugars.

Data for the reaction rate as a function of time have been compiled for both the prehydrolysis and hydrolysis of a number of substrates. Plots of these data show first order kinetics with respect to carbohydrate concentration at constant acid concentration. Using these expressions, a relationship can be developed for the dependency of reaction rate on acid concentration.

Table 1. Oak sawdust acid hydrolyzates

Prehydrolyzate
Xylose, g 1^{-1}		15.9
g $100g^{-1}$ sawdust		13.9
Glucose, g 1^{-1}		5.8
g $100g^{-1}$ sawdust		5.1

Hydrolyzate
Xylose, g 1^{-1}		0
g $100g^{-1}$ sawdust		0
Glucose, g 1^{-1}		80.3
g $100g^{-1}$ sawdust		57.0

Combined
Xylose, g $100g^{-1}$ sawdust	13.9
Glucose, g $100g^{-1}$ sawdust	62.1

Figure 1 is a plot of ln k, where k is the first order rate constant, (min^{-1}) as a function of ln C_A, acid normality. Straight lines result for both the prehydrolysis and hydrolysis, which dictate an equation of the form: $k = aC_A^b$, where a and b are found from the intercept and slope, respectively. As expected, the prehydrolysis is faster than the hydrolysis reaction. The following rate expressions result at 100°C for sulfuric acid:

Pre-hydrolysis
$$k = .020 \, C_A.67$$
Hydrolysis
$$k = .001 \, C_A.51$$

These rate coefficients can be used to design reactors to carry out these reactions. Reaction times of about thirty minutes are required to achieve a 90 percent conversion in a series of two reactors each for the prehydrolysis and hydrolysis at acid concentrations of 8-14N with 10 percent solids feed.

The proposed process for the stagewise acid hydrolysis of oak residue is shown schematically in Figure 2. Sawdust or other wood by-product is fed continuously to the prehydrolysis reactor. Residual solids are separated by filtration, washed with fresh acid and fed to the hydrolysis reactor. Solids from the hydrolysis reactor are filtered, washed and discarded. Acid

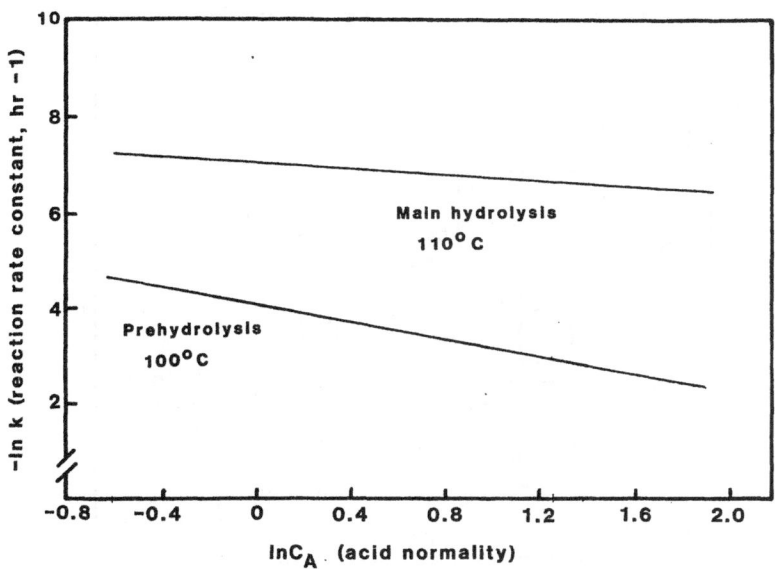

Figure 1. Kinetics of biomass hydrolysis

and sugars are separated and the acid is returned to the reactors. The acid recovery process is not disclosed due to pending patent rights. The dilute sugar streams result: the pre-hydrolyzate, containing primarily xylose, and the hydrolyzate, containing only glucose. These sugars can then be fermented to ethanol or other chemicals.

SUGAR DECOMPOSITION

The fermentability of the sugars is dependent upon the decomposition that occurs during hydrolysis. Xylose decomposes to furfural and glucose decomposes to HMF, which are both toxic to yeast. Tolerance can often be developed, and toxicity is difficult to define. However, the toxic limit of furfural on alcohol yeast is reported to be .03 - .046 percent.[7] HMF is reported to inhibit yeast growth at .5 percent, and alcohol production is inhibited at .2 percent.[8]

The rate of decomposition of xylose to furfural and hexoses to HMF were studied at varying sugar concentrations. Using the method of initial rates, these reactions were found to be first-order. The ratios of rate constants for decomposition to formation are given in Table 2. These ratios appear to

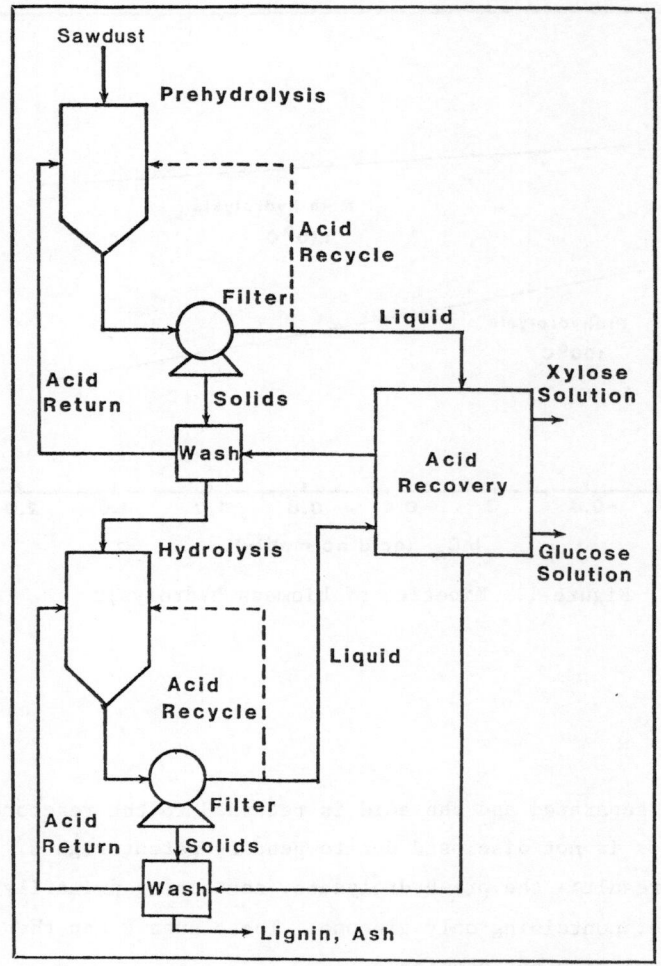

Figure 2. Schematic of acid hydrolysis process

be small, and subsequent calculations and experiments show that the rate of HMF appearance is insignificant. However, the rate of furfural appearance can reach toxic limits, especially if acid recycle is utilized.

ETHANOL FERMENTATION

The acid hydrolysis of cellulosic residues, if performed at high temperature (i.e. 100°C or greater) can produce a sugar/degradation product mixture that is difficult, if not impossible, to ferment. However, if the hydrolysis conditions are mild (i.e. less than 100°C), only small quantities

Table 2. Ratio of first-order rate constants for sugar decomposition to formation at 100°C

Sugar	Acid Concentration	Rate of Decomposition/ Formation
Glucose	2N	0.0053
	3N	0.0090
	4N	0.0074
Xylose	2N	0.0257
	3N	0.0402
	4N	0.0374

of toxic substances such as hydroxymethyl furfural and furfural are observed, and normal fermentation is expected.

Batch fermentation experiments were carried out at 30°C to compare the production rates ethanol from the main hydrolyzate and synthetic glucose. Saccharomyces cerevisiae ATCC 24860 was used in the study.

Inocula containing approximately 5×10^{-7} cells ml^{-1} from a 24-30 hour old seed culture were used in all experiments. The analysis of glucose and ethanol were made on a YSI 27 Industrial Analyzer.

Figure 3 shows the comparison of a batch fermentation with Saccharomyces cerevisiae of both synthetic glucose and corn stover hydrolyzate using 5 percent sugar concentrations. Identical results were found when fermenting synthetic glucose and hydrolyzate in the presence of yeast extract. Ethanol yields were also nearly identical. The concentrations of furfural and HMF in the hydrolyzates were found to be negligible. These very low levels of by-products are believed to be the major reason for the successful fermentation.

ECONOMIC PROJECTIONS

To illustrate the economics of this process, a design has been performed for a facility to convert oak residue into 75 million liters per year of ethanol, utilizing the acid hydrolysis procedures previously described. The capital and operating costs are summarized in Table 3.

Table 3. Economics of 75 million liter per year ethanol facility

A. Capital Cost	Million $
Feedstock Preparation	2.1
Hydrolysis	4.8
Acid Recovery	6.2
Fermentation and Purification	2.0
Utilities/Offsites	5.0
	20.1

B. Operating Cost	Million$ yr^{-1}	$ L^{-1}
Raw Material		
($22 Mg^{-1})	4.6	0.061
Utilities	2.1	0.029
Chemicals	5.0	0.066
Labor	3.0	0.040
Fixed Charges		
Maintenance (5%)	1.0	0.013
Depreciation (10%)	2.0	0.026
Taxes & Insurance (2%)	0.4	0.005
Profit (94%)	18.9	0.250
	$37.0	$0.49 L^{-1}

Oak residue such as sawdust and trimmings, would be delivered to the plant site as needed. Feedstock preparation consists of shredding, grinding and conveying to the reactors. The hydrolysis section, as shown in Figure 2, consists of continuous reactors. Acid resistant materials of construction such as FRP are necessary for this equipment. Acid recovery is accomplished by electrodialysis; although, other methods, such as evaporation, are equally applicable and cost about the same. Continuous fermentation and the typical ethanol distillation units are included. Laboratory studies indicate that feed concentrations of up to 30 percent solids can be utilized by the separate feeding of solids and liquids. Acid recycle is used to further maximize the sugar concentrations from the reactors. The total capital cost for this plant is $20.1 million, including all utilities, storage and offsites.

The annual operating costs are also shown in Table 3. These costs are also given on the basis of unit production of alcohol. Raw material is estimated to cost $22 Mg^{-1} for collection and transportation. A lignin boiler is used to reduce the energy requirements, and energy costs are 29¢ per gallon. Fixed charges are computed as a percentage of the capital investment, using the recent price of fuel grade ethanol of 49¢/liter, a 94

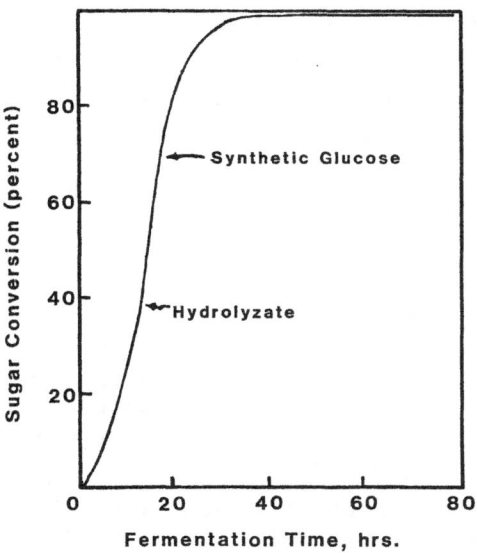

Figure 3. Fermentation of hydrolyzate and synthetic glucose

percent profit results. The break even cost of production is 24¢ per liter, leaving sufficient margin for commercialization.

It should be noted that this process does not include utilization of the pentose stream. Acid recovery is included, but fermentation of the pentoses is not provided. Xylose could be fermented to alcohol, acids or other valuable chemicals, which would improve the economics. However, since this technology is not perfected, such products have not been included.

The economics of this process are substantially improved by the use of acid recycle, high feed concentrations and continuous fermentation. These modifications in the standard process result in increased sugar concentrations, smaller equipment and reduced energy consumption. Sugar concentrations up to 25 percent are possible, which substantially reduces the hydrolysis and fermentations equipment and the energy required for purification. These high sugar concentrations are fermentable in immobilized cell column reactors.

CONCLUSIONS

The two-stage acid hydrolysis of residues, such as oak sawdust, requires mild temperatures, but large quantities and high concentrations of acid. The resulting hydrolyzates, containing primarily xylose and glucose, can be fermented to ethanol or other chemicals without pretreatment. Preliminary economic evaluations utilizing acid recycle, high solids feed, and continuous fermentation show that ethanol can be produced for 49¢ per liter with a profit of 94 percent.

REFERENCES

1. E. G. Locke. Working Party on Wood Hydrolysis. Presented at Annual Meeting of Forest Products Research Society. Louisville. June. 1961.
2. E. E. Harris and E. Boglinger. Ind. Engr. Chem., 38, p. 890. 1946.
3. F. Beguns. Ind. Engr. Chem., 29, p. 247. 1937.
4. I. S. Goldstein, et. al. The Hydrolysis of Cellulose with Superconcentrated Hydrochloric Acid. Presented at Fifth Symposium on Biotechnology for Fuels and Chemicals. Gatlinburg, TN. May. 1983.
5. R. B. Shah, E. C. Clausen and J. L. Gaddy. Production of Chemical Feedstockd by Acid Hydrolysis and Fermentation of Biomass. Chem. Eng. Prog., 80.1 1984.
6. E. C. Clausen and J. L. Gaddy. Economic Analysis of a Bioprocess to Produce Ethanol from Corn Stover. Presented at the Fifth Symposium on Biotechnology for Fuels and Chemicals. Galinburg, TN. May 1983.
7. N. Banerjee and L. Vishwanathan. Proc. Annual Sugar Conv. Tech. Conf., CA 61:1228f. 1972.
8. N. Banerjee and L. Vishwanathan. Proc. Annual Sugar Conv. Tech. Conf., CA 82:123277W. 1974.

CHEMICALS FROM BIOMASS: THE IDENTIFICATION AND SOLUTION OF INHIBITION
CAUSED BY WOOD EXTRACTIVES ON THE FERMENTATION OF A SOUTHERN PINE
PREHYDROLYZATE TO BUTANEDIOL

Ai V. Tran and Robert P. Chambers

Department of Chemical Engineering
Auburn University
Auburn, Alabama 36849

ABSTRACT

Dissolving pulp mills provide a substantial amount of by-product
hemicellulose in the form of water prehydrolysis liquor. Mannose, the
predominant sugar in southern pine water prehydrolysis liquor, has been
fermented to butanediol by Klebsiella pneumoniae AU-1-d3. Southern pine
extractives in the water prehydrolysis liquor, however, hindered the butanediol fermentation. Treatments with sequential lime-sulfuric acid or mixed
bed ion resin efficiently removed the inhibitors. Identified among the
inhibitors derived from southern pine extractives were: nonanoic acid,
palmitic acid, oleic acid, stearic acid, borneol, α-terpineol, d-verbenone,
and dehydroabietic acid.

Keywords. Butanediol, fermentation, mannose, Klebsiella pneumoniae,
inhibitor, southern pine extractives.

INTRODUCTION

The prehydrolysis liquors generated from conifer wood by the dissolving
pulping process are rich in mannose. Such carbohydrates, whether monomeric
or polymeric, are derived from the hemicellulose fraction of wood.[1] The
chemistry and mechanism of water prehydrolysis of southern pine wood have
been extensively investigated.[2,3] Processes to produce a number of mannose
chemicals, e.g., D-mannose, methyl α-D-mannopyranoside, sodium D-glycero-D-
galacto-heptonate, and D-mannitol, from water prehydrolysis liquors of
southern pine wood have also been developed.[4]

561

2,3-butanediol fermentation by <u>Klebsiella pneumoniae</u> (<u>Aerobacter</u> <u>aero-</u>
<u>genes</u>) has been solely studied so far with xylose or glucose.[5-10] Mannose,
on the other hand, has not been yet considered as the substrate for 2,3-
butanediol production although this hemicellulose is abundant in the effluents of dissolving pulp mills. Butanediol obtained by fermentation can be
used in the production of polyesters, butadiene, antifreeze agents and
methyl ethyl ketone.[5,11,12]

In the previous work, soluble lignin derivatives in southern pine water
prehydrolysis liquors, which inhibited the butanediol fermentation, were
identified.[13] The objectives of the present report are thus to examine the
inhibitory effects of wood extractives in southern pine water prehydrolysis
liquors on the butanediol fermentation, to identify these southern pine
extractives and to develop methods for effectively removing the inhibitory
effects of these agents.

EXPERIMENTAL

Water Prehydrolysis of Southern Pine

Air-dried southern pine wood meals (8-20 mesh, about 8% moisture) were
subjected to water prehydrolysis with conditions similar to those for ex-
tractive-free wood.[13] Namely, southern pine wood meals were cooked with
water at 170°C for 1 hour in a 7.6 liter (2 gallon) Parr reactor. Water to
wood ratio was 4:1 (by weight). After cooling and filtration, the filtrate
was treated with 1% (w/v) sulfuric acid at 120°C for 1 hour. Insoluble
materials were filtered and the water prehydrolysis liquors thus obtained.
The yield of the water prehydrolysis liquors, as the difference of wood
meals before and after water prehydrolysis, was 9.5% based on oven-dried
wood.

Fermentation System and Organism

Figure 1 depicts the fermentation system used throughout this work.
The bacterium <u>Klebsiella</u> <u>pneumoniae</u> AU-1-d3 was obtained from the Animal and
Dairy Science Department, Auburn University, and grown on brain heart infu-
sion agar.

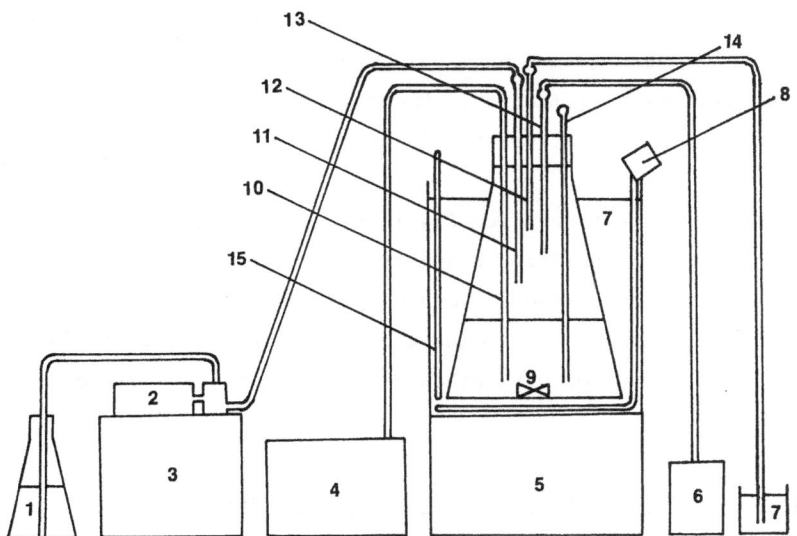

Figure 1. Butanediol fermentation system. 1. 5N NaOH; 2. pump; 3. speed controller; 4. pH controller; 5. magnetic stirrer; 6. air pump; 7. water; 8. heater; 9. stirrer bar; 10. pH probe; 11. inlet of NaOH; 12. air and CO_2 outlet; 13. air inlet; 14. sampling pipet; 15. thermometer

Inoculum Preparation

Solutions of mannose (1%, w/v) and phenol red broth (1%, w/v) were sterilized for 15 minutes at 120°C. A loopful of organism was aseptically added. The culture was incubated for 24 hours at 37°C then centrifuged.

Media Preparation

The nutrients were ammonium chloride (0.2%), sodium chloride (0.1%), magnesium sulfate heptahydrate (0.02%) and yeast extract (0.6%). All percentages were weight per volume basis. For the experiments of inhibitory effects of wood extractives model compounds, the nutrients were sterilized with mannose solutions as above. The nutrients, however, were added to the treated water prehydrolysis liquors without further sterilization. All fermentation volumes were 420 ml.

Effects of Southern Pine Extractives on the Butanediol Fermentation

Each of the following compounds: campesterol, α-pinene, D-limonene, (-)-borneol, pyrogallol, β-sitosterol, stigmastanol, linoleic acid, abietic acid and palmitic acid was added to solutions containing 100 gL^{-1} of mannose and nutrients prior to sterilization and fermentation. These compounds were commercially purchased (Sigma Chemical Company and Pfaltz & Bauer, Inc.) and used without further purification. Except α-pinene, they were solubilized in the mannose solutions after sterilization.

Ethanol-benzene extract from 114.1g southern pine wood meals was dried over anhydrous sodium sulfate and evaporated to dryness in vacuo (50°C). It was then mixed with a solution of mannose (100 gL^{-1}) and nutrients prior to sterilization and fermentation. The ethanol-benzene extraction was carried out in a Soxhlet extractor with ethanol-benzene (1:2,v/v) mixture for 8 hours.

Treatments of Southern Pine Water Prehydrolysis Liquors

From the previous results,[13] southern pine water prehydrolysis liquors were subjected to two different treatments.

Treatment 1: the water prehydrolysis liquors were adjusted to a pH of 11.5 with lime, filtered, then to a pH of 5.4 with concentrated sulfuric acid, and filtered.

Treatment 2: the water prehydrolysis liquors were stirred with mixed bed ion resin AG 501-X8(D) (Bio-Rad Laboratories) (100 gL^{-1}) for 30 minutes then filtered.

Analytical Methods

Ethanol, acetic acid, acetoin, and butanediol were analyzed on a Varian Gas Chromatograph Model 3700 equipped with a flame ionization detector using 1 m glass column packed with Chromosorb 101.

Carbohydrates in the water prehydrolysis liquors were gas-chromato-graphically analyzed.[14] Mannose in the broths of mannose fermentations was determined using a Waters Sugar Analyzer I Liquid Chromatograph.

Water prehydrolysis liquors (685 ml) from 186.1 g southern pine wood meal were diluted with acetone (200 ml) and methanol (50 ml). The mixture was then extracted with 150 ml of petroleum ether. The lower phase was transferred to a beaker and the upper phase washed with 25 ml of a mixture of water: acetone: methanol (1:2:1). The washing was combined with the lower phase, then extracted with petroleum ether in similar manner for two

more times. All the petroleum ether were combined, dried over anhydrous
sodium sulfate and evaporated to dryness in vacuo (35°C). The yield of the
petroleum ether extract was 0.03% based on oven-dried wood.

The residue was dissolved in ether-methanol (95:5, v/v) (1 ml) and
methylated with freshly prepared diazomethane (3 ml). The sample was then
analyzed at Florida State University by gas chromatography/mass spectrometry
for the wood extractives soluble in southern pine water prehydrolysis
liquors.

RESULTS AND DISCUSSION

Butanediol Fermentation from Mannose and Inhibitory Effects of Wood Extractives Model Compounds

Figure 2 indicates the fermentability of mannose by <u>Klebsiella pneumo-</u>
<u>niae</u> AU-1-d3. Yields of butanediol (29.7 gL^{-1}) and ethanol (10.7 gL^{-1})

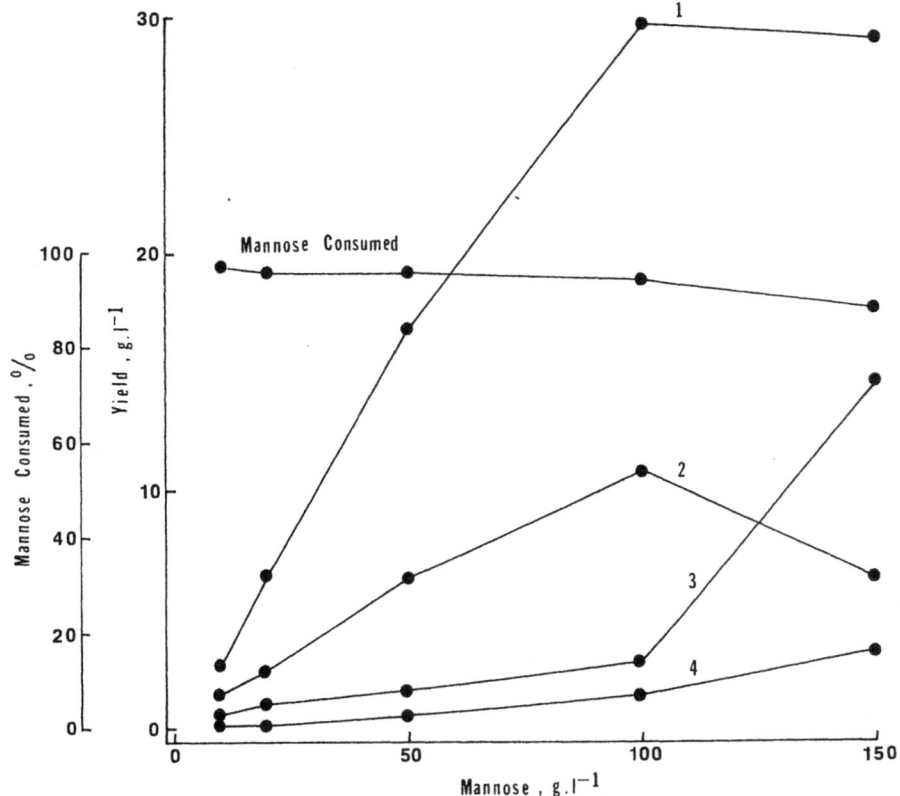

Figure 2. Yield of fermentation products and mannose consumed vs. mannose
concentration. 1. butanediol; 2. ethanol; 3. acetic acid; 4.
acetoin

reached maximum at a mannose concentration of 100 gL^{-1}. These results are similar to those reported for xylose.[10] Yields of other fermentation side-products, i.e. acetic acid and acetone, on the other hand, increased as mannose concentration increased.

Although the main extractives in softwoods and tall oils of kraft pulping black liquors were identified,[15,16] those in wood hydrolyzate of the dissolving pulping process have not been known. Therefore, model compounds of terpenes, sterols, fatty acids, and resin acids, and their quantities

Table 1. Inhibitory effects of wood extractives model compounds on the butanediol fermentation of mannose

Wood Extractives Model Compounds gL^{-1}	Maximum Butanediol			Maximum Ethanol	
	Yield gL^{-1}	theoretical yield* %	time hr	yield gL^{-1}	time hr
Blank (100 gL^{-1} mannose)	29.7	71.7	46.5	10.7	46.5
α-Pinene, 0.260	14.6	68.7	52.3	5.7	34.1
D-Limonene, 0.251	13.4	12.5	95.0	4.9	95.0
Campesterol, 0.006	14.9	18.0	95.3	4.9	70.4
(-) Borneol, 0.231	17.7	58.9	94.5	5.3	94.5
Pyrogallol, 0.235	28.9	57.8	121.3	10.3	121.3
β-Sitosterol, 0.233	10.4	22.1	151.7	5.0	122.7
Stigmastanol, 0.232	24.4	51.9	80.3	8.3	80.3
Linoleic acid, 0.245	10.0	23.5	75.0	5.6	75.0
Palmitic acid, 0.233	14.3	35.8	71.0	3.2	71.0
Abietic acid, 0.158	10.9	23.1	71.3	2.4	46.5
All above model compounds plus Oleic acid (0.177 gL^{-1})	11.3	34.2	95.5	3.6	95.5
Ethanol-Benzene Extract	9.8	19.6	98.2	2.3	98.2

*Theoretical yield of butanediol is 50% of sugar consumed

were selected on the basis of those investigations. Table 1 data show the
inhibitory effects of each model compound on the butanediol fermentation.
Generally, maximum yields of butanediol and ethanol decreased and fermentation times were longer. A combination of all individual model compounds
also exhibited considerable inhibitory effects. However, ethanol-benzene
extract was more inhibitory to the butanediol fermentation than the model
compounds tested. This suggests that the ethanol-benzene extract probably
contained other more inhibitory wood extractives than the model compounds
used.

Butanediol Fermentation of Water Prehydrolysis Liquors from Southern Pine

Southern pine water prehydrolysis liquors could not be fermented to
butanediol as indicated by negligible yields of fermentation products (Table
2). This is mainly due to the inhibition of soluble extractives and lignin
derivatives[13] in southern pine water prehydrolysis liquors. Treatment with
lime (pH 11.5) followed by sulfuric acid (pH 5.4) and with mixed bed ion
resin resulted in good yields of butanediol, 6.0 and 5.8 gL^{-1} or 78.1 and
77.6% of theoretical yields, respectively. The time course of butanediol
fermentation of treated southern pine water prehydrolysis liquors are shown
in Figure 3. Although the yields of the fermentation products from water
prehydrolysis liquors are comparable to those from mannose fermentations
(Tables 1 and 2, Figure 2), the prehydrolysis liquors fermentations needed
longer time than the mannose fermentations, 70.0 and 72.8 hours vs. 46.5
hours.

Carbohydrates present in southern pine water prehydrolysis liquors were
decreased, to some extent, by the treatments (Table 2). _Klebsiella_ pneumoniae had a tendency to metabolize glucose first and then mannose. Arabinose, xylose, and galactose were metabolized at much slower rates (Figure 4),
indicating that hexose sugars (except for galactose) are more favorable
substrates for _Klebsiella pneumoniae_ than pentose sugars although this
bacterium was originally screened to ferment xylose to butanediol.[6]

Identification of Extractives Soluble in Southern Pine Water Prehydrolysis
Liquors

The ethanol-benzene extract consisted of southern pine extractives
prior to the water prehydrolysis. This extract showed the overall inhibitory effect of southern pine extractives on the butanediol fermentation.
Due to the hydrolysis reaction in the water prehydolysis, the extractives in
the ethanol-benzene extract might differ from those in the petroleum ether

Table 2. Yields of fermentation products and carbohydrate compositions of various southern pine water prehydrolysis liquors

Water Prehydrolysis Liquors	Maximum Butanediol		Maximum Ethanol		Carbohydrate Composition, gL^{-1}				
	Yield gL^{-1}	Time hr	Yield gL^{-1}	Time hr	Arabinose	Xylose	Mannose	Galactose	Glucose
Neutralized with lime	0.3	117.5	-	117.5	1.34	5.05	10.12	1.85	2.08
Treated with lime and then with conc. sulfuric acid	6.0	70.0	1.3	101.0	1.21	4.65	9.37	1.52	1.82
Treated with mixed ion resin	5.8	72.8	1.4	72.8	1.14	4.46	8.79	1.30	1.58

extract of the water prehydrolysis liquors. This can be seen in the absence of diterpene and triterpene alcohols and sterols, i.e. campesterol, β-sitosterol, stigmastanol,[15] in the petroleum ether extract (Figure 5). The

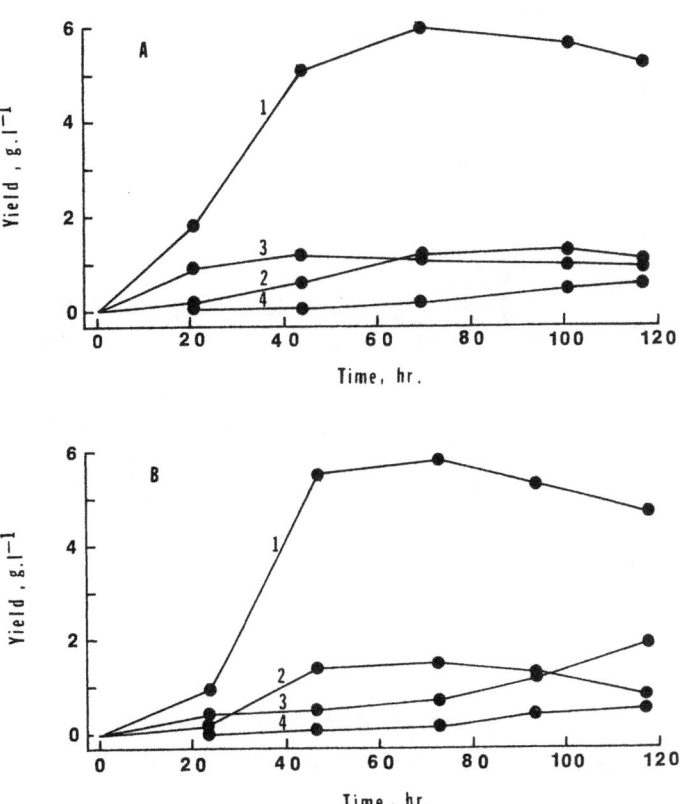

Figure 3. Time course of fermentation products of treated southern pine water prehydrolysis liquors. A: treated with lime-sulfuric acid. B: treated with mixed bed ion resin. 1,2,3,4: same as in figure 2

identification of southern pine extractives soluble in the water prehydrolysis liquors was hence performed with the petroleum ether extract. A study of inhibitory effect of the petroleum ether extract was deemed to be necessary. Since southern pine wood meals were not sufficient for this experiment, the inhibitory effect of the petroleum ether extract was unfortunately unknown.

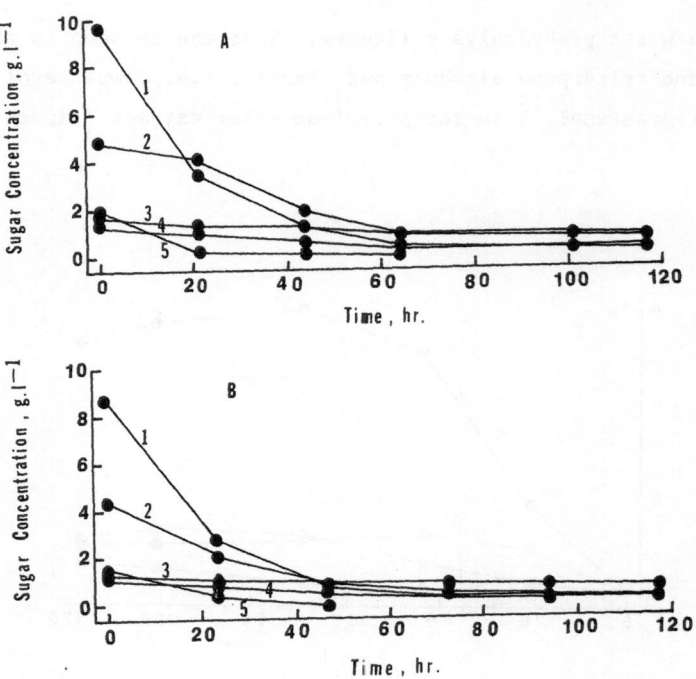

Figure 4. Fermentation time-course of sugars in treated southern pine water prehydrolysis liquors. A,B: same as in figure 3. 1: mannose, 2: xylose, 3: galactose, 4: arabinose, 5: glucose

The gas-chromatogram of the methylated petroleum ether extract is shown in Figure 5. Based on the mass spectra in the literature, the following monoterpene alcohols: borneol and α-terpineol,[17] monoterpene ketone: d-verbenone,[18] fatty acids: nonanoic acid, palmitic acid, oleic acid, stearic acid,[19] and resin acid: dehydroabietic acid[20] were identified.

From the above results, the toxic effects of southern pine extractives on the butanediol fermentation were demonstrated. The inhibitory effect of lignin derivatives were also examined.[13] However, since the water prehydrolysis liquors may contain other toxic compounds derived from carbohydrates such as furfural, 5-hydroxymethylfurfural, levulinic acid, formic acid, and acetic acid, it is not known whether all toxic materials derived from extractives, lignin, and carbohydrates will have synergistic enhancement of or cumulative toxicity. Further work on this question will be of considerable value, particularly with the view to the development of efficient treatments to detoxify the water prehydrolysis liquors.

570

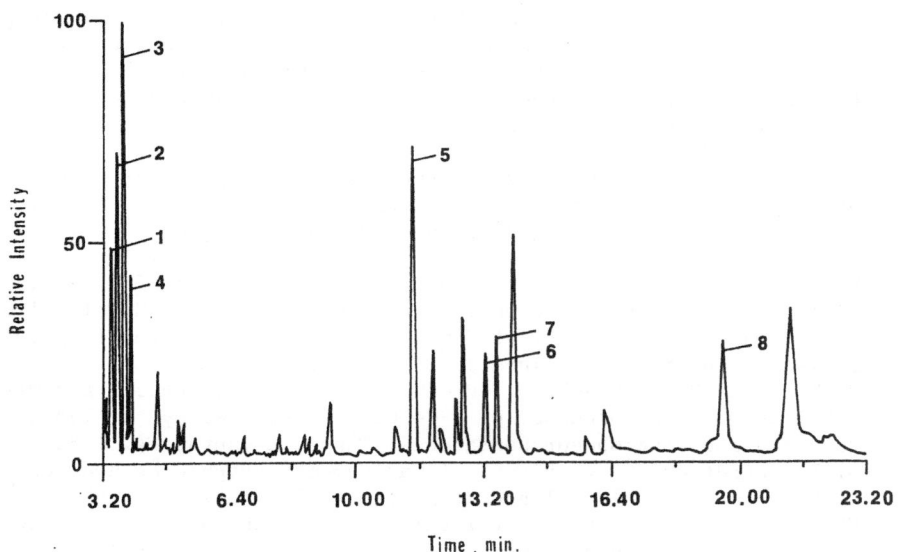

Figure 5. Gas chromatogram of the methylated petroleum ether extract from southern pine water prehydrolysis liquors. 1. borneol; 2. α-terpineol; 3. nonanoic acid; 4. d-verbenone; 5. palmitic acid; 6. oleic acid; 7. stearic acid; 8. dehydroabietic acid

REFERENCES

1. J. K. Hamilton. The Behavior of Wood Carbohydrates in Technical Pulping Process. Pure Appl. Chem. 5: 197-231. 1962.
2. R. L. Casebier, J. K. Hamilton and H. L. Hergert. Chemistry and Mechanism of Water Prehydrolysis on Southern Pine Wood. Tappi 52 (12):2369-2377. 1969.
3. K. D. Sears, A. Beelik, R. L. Casebier, R. J. Engen, J. K. Hamilton and H. L. Hergert. Southern Pine Prehydrolyzates: Characterization of Polysaccharides and Lignin Fragments. J. Polymer Sci. Part. C. 36:425-443. 1971.
4. F. W. Herrick, R. L. Casebier, J. K. Hamilton and J. D. Wilson. Mannose Chemicals. Appl. Polymer Symp. 28:93-108. 1975.
5. S. K. Long and R. Patrick. The Present Status of the 2,3-Butylene Glycol Fermentation. p. 135-155. In: Advances in Applied Microbiology. Vol. 5. W. W. Umbreit (Ed). Academic Press, New York. 1963.
6. S. Veeraraghavan, Y. Y. Lee, R. P Chambers and T. A. McCaskey. Ethanol-Butanediol Fermentation of Xylose. p. 171-173. In: Enzyme Engineering. Vol. 5. H. H. Weetall and G. P. Royer (Eds). Plenum Publishing Corporation, New York. 1980.
7. E. K. C. Yu and J. N. Saddler. Power Solvent Production by Klebsiella pneumoniae Grown on Sugars Present in Wood Hemicellulose. Biotechnol. Letters 4:121-126. 1982.
8. N. B. Jansen and G. T. Tsao. Bioconversion of Pentoses to 2,3-Butanediol by Klebsiella pneumoniae. In: Advances in Biochem. Eng. 27:85-99. 1983.

9. J. M. Sablayrolles and G. Goma. Butanediol Production by Aerobacter aerogenes NRRL B199: Effects of Initial Substrate Concentration and Aeration Agitation. Biotechnol. Bioeng. 26:148-155. 1984.

10. N. B. Jansen, M. C. Flickinger and G. T. Tsao. Production of 2,3-Butanediol from D-xylose by Klebiella oxytoca ATCC 8724. Biotechnol. Bioeng. 26:362-369. 1984.

11. G. A. Ledingham and A. C. Neish. Fermentative Production of 2,3-Butanediol. p. 27-93. In: Industrial Fermentations. Vol. 2. L. A. Underkofler and R. J. Hickey (Eds). Chemical Publishing Co., Inc. New York. 1954.

12. A. V. Tran and R. P. Chambers. Chemical Transformation of 2,3-Butanediol into Methyl Ethyl Ketone: Characterization of a Novel Solid Acid Catalyst. Poster 61 at Sixth Symposium on Biotechnology for Fuels and Chemical. Gatlinburgh, TN. 1984.

13. A. V. Tran and R. P. Chambers. Production of 2,3-Butanediol and Methyl Ethyl Ketone from Mannose. p. 199-209. In: Proceedings of Tappi Research and Development Conference. Tappi. Atlanta, GA. 1984.

14. L. G. Borchardt and C. V. Piper. A Gas Chromatographic Method for Carbohydrates as Alditol-Acetates. Tappi 53(2):257-260. 1970.

15. R. Ekman. Analysis of the Nonvolatile Extractives in Norway Spruce Sapwood and Heartwood. Acta Academiae Aboensis. Ser. B, 39(4):1-20. 1979.

16. B. Holmbom and E. Avela. Studies on Tall Oil from Pine and Birch. 1. Composition of Fatty and Resin Acids in Sulfate Soaps and in Crude Tall Oils. Acta Academiae Aboensis. Ser. B, 31(13):1-14. 1971.

17. E. Sydow. Mass Spectrometry of Terpenes II. Monoterpene Alcohols. Acta Chem. Scand. 17 (9):2504-2512. 1963.

18. G. Bunau, G. Schade and K. Gollnick. Massenspektrometrische Untersuchungen von Terpenen. I. Monoterpenaldehyde und-ketone. Z. Anal. Chem. 227:173-187. 1967.

19. G. Odham and E. Stenhagen. Fatty Acids. p. 211-249. In: Biochemical Applications of Mass Spectrometry, G. R. Waller (Ed). Wiley-Interscience, New York. 1972.

20. D. F. Zinkel, L. C. Zank and M. F. Wesolowski. Diterpene Resin Acids. p. C24. USDA, Forest Service, Forest Products Laboratory. Madison, WI. 1971.

IMPROVED FERMENTATION OF WOOD HYDROLYZATE TO ETHANOL BY REMOVAL

OF COMPOUNDS INHIBITORY TO PACHYSOLEN TANNOPHILUS

T. A. McCaskey, M. D. Rice and R. C. Smith

Department of Animal and Dairy Sciences
Alabama Agricultural Experiment Station
Auburn University, Alabama 36849

ABSTRACT

The yeast Pachysolen tannophilus metabolizes glucose and xylose, the principal sugars of hardwood, to ethanol. The sugars exist as polymers in hardwood which must be hydrolyzed to monomeric sugars before they can be metabolized by P. tannophilus and other yeasts. Hydrolysis of hardwoods with 0.5% sulfuric acid at 1.311 MPa gauge (190 psig) for 3 minutes yields a solution with 35 mg ml^{-1} D-xylose and 4 mg ml^{-1} D-glucose. In addition to minor levels of other sugars, the solution also contains 12 mg ml^{-1} acetic acid, 6 mg ml^{-1} phenolic compounds and 2 mg ml^{-1} furfural. These compounds are derived from the carbohydrate and lignin fractions of hardwood during the acid hydrolysis process. Pachysolen tannophilus NRRL Y-2460 at concentrations of about 5 mg ml^{-1} (dry cell mass) is killed by 12 mg ml^{-1} of acetic acid, and by 6 mg ml^{-1} of a 2:1 mixture (weight basis) of syringaldehyde and vanillin in a minimal medium at pH 4. Furfural at 2 mg ml^{-1} is not inhibitory when high yeast inoculation rates are used.

Calcium hydroxide and other chemical treatments were evaluated for the removal of acetic acid and phenolic compounds from a model system containing D-xylose, acetic acid and syringaldehyde at the concentrations representative of the levels in mild acid hydrolyzed hardwood. Calcium hydroxide treatment was effective in removing 80% or more of the syringaldehyde but had no effect on acetic acid. Loss of D-xylose due to the treatment was less than 2% when the model system was treated to pH 9 with calcium hydroxide. The treatment was effective, also, with acid hydrolyzed hardwood.

Keywords. Hardwood hydrolyzate, yeast fermentation, yeast inhibitors, inhibitor removal, ethanol.

INTRODUCTION

The conversion of carbohydrates in acid-hydrolyzed hardwood to ethanol, by the yeast Pachysolen tannophilus, is suppressed by microbial inhibitors in wood hydrolyzate liquor.[1] Four potential sources of microbial inhibitors in wood hydrolyzates were identified by Leonard and Hajny.[2] The metals or minerals from wood, soil, or hydrolysis equipment was one source. Another source was carbohydrate decomposition products such as furfural and acetic acid. Lignin degradation compounds and those derived from wood extractives were the third and fourth source, respectively. It was determined that the inhibitory effects of these compounds were dependent on pH of the medium, oxygen concentration, ratio of yeast population to inhibitory compounds, temperature, and the physiological condition of the yeast inoculum.[2] More recent studies of inhibitors in wood hydrolyzate have investigated specific effects of the inhibitors on microorganisms. It has been demonstrated that yeast grown in the presence of furfural showed lower levels of cytochromes[3] and lower concentrations of dehydrogenases.[4] The aromatic inhibitors derived from lignin and/or wood extractives are broadly grouped as the phenolics.[5] These compounds interact with proteins in the cell membrane of yeasts, thereby causing disruption of the membrane.[6] Polyphenolics such as tannins inhibit yeasts by binding to the cell wall,[7] and binding is more pronounced at pH 4 than at pH 7.[8] Acetic acid also is inhibitory to P. tannophilus. The acid is formed when O-acetyl groups are hydrolyzed from xylan in wood. As with phenolic compounds, acetic acid is more inhibitory at low pH values.[9]

The action of phenolic compounds and acetic acid on microorganisms has been studied primarily through their use as microbial inhibitors employed in foods.[10,11,12] Phenolic compounds in forages have received attention in recent years because they effect digestibility of forages; therefore, the performance of ruminant animals. The adverse role of tannins on the digestibility of forages and by-product feeds by ruminant animals was reviewed by Kumar and Singh.[13] Tannin-protein interactions, inhibition of digestive enzymes, and the action of tannin derivatives on rumen microorganisms were discussed as modes of action of how tannins adversely affect digestibility of feedstuffs. Various alkali treatments such as hydroxides of sodium, calcium, and ammonium have been evaluated for the removal of tannins from feedstuffs, especially high tannin grain. Treatment of cereal straws with sodium hydroxide has been demonstrated to enhance the degradation of structural polysaccharides by rumen microorganisms,[14] and sodium hydroxide treatment of high tannin sorghum grain reduces the content of tannin by as much as 86%.[15] Further research has been proposed to help elucidate the mechanisms by which treatment of feedstuffs remove phenolic monomers from plant cell walls and increase their digestibility.[16]

Although not much is known about the effect of lignin derivatives on rumen microorganisms, studies have been conducted on the effect of phenolic compounds on axenic cultures of yeast and bacteria cultivated in semidefined media. Zemek et al.[17] reported on the antibiotic properties of eleven lignin compounds on yeasts, bacteria, and a mold culture. Isoeugenol, syringaldehyde, and ferulic acid were inhibitory to Saccharomyces cerevisiae at concentrations of 0.4 mg ml^{-1} or more. The effect of phenolic compounds on the production of 2,3-butanediol (BDO) by Klebsiella pneumonia AU 1-D3 from mannose was demonstrated by Tran and Chambers.[18] The theoretical yield of BDO was decreased approximately 25% and the fermentation time was more than doubled with 0.19 mg ml^{-1} vanillin or 0.09 mg ml^{-1} syringaldehyde in the fermentation substrate.

Studies conducted by McCaskey and Martin[1] showed that hydrolysis of hardwoods with 0.5% sulfuric acid at 1.311 MPa gauge (190 psig) for 3 minutes yielded a solution with about 35 mg ml^{-1} D-xylose and 4 mg ml^{-1} D-glucose. In addition to minor levels of other sugars, the solution also contained 12 mg ml^{-1} acetic acid, 6 mg ml^{-1} phenolic compounds and 2 mg ml^{-1} furfural. They[1] demonstrated that acetic acid, furfural, and a 2:1 mixture of the phenolics syringaldehyde and vanillin (by weight), representing the total concentration of phenolics found in the above wood hydrolyzate were inhibitory to P. tannophilus in a minimal medium. The phenolic compound mixture killed the yeast at pH 4.0, 5.5 and 6.5 employing a high yeast inoculum. Acetic acid at 12 mg ml^{-1} killed the yeast at pH 4.0; however, at pH 5.5 with a high yeast inoculation rate (5 mg ml^{-1} dry cell mass), the inhibition was practically overcome. Furfural at 2.1 mg ml^{-1} was not toxic to P. tannophilus with the high inoculation rate at pH 4.0, 5.5, and 6.5. However, at low inoculation levels (about 10^4 viable cells ml^{-1}) the yeast was completely inhibited. Therefore, components of acid-hydrolyzed hardwood which are most inhibitory to P. tannophilus are phenolic compounds and acetic acid.

A study was conducted to evaluate chemical procedures for the removal of phenolics, acetic acid and furfural from a model acid wood hydrolyzate solution and to further evaluate the procedures for the removal of the inhibitors from actual wood hydrolyzate. Fermentation studies were conducted to determine whether the wood hydrolyzate treatment processes improved the production of ethanol from wood sugars by P. tannophilus.

MATERIALS AND METHODS

A model wood hydrolyzate solution was prepared in distilled water to represent the major carbohydrates and microbial inhibitors at the approximate concentrations found in first-stage, dilute sulfuric acid digest of oakwood chips. The solution was composed of D-xylose, 41 mg ml^{-1}; furfural, 2.1 mg ml^{-1}; acetic acid, 12.0 mg ml^{-1}; and syringaldehyde, 7.0 mg ml^{-1}. Three chemical treatments were evaluated to remove the microbial inhibitors from the model hydrolyzate solution: Amberlite IRA-400 (Cl form), sodium hydroxide, and calcium hydroxide. During preliminary studies, 12 chemical treatments were evaluated for the removal of syringaldehyde from a model system, and the best chemical treatments were the three described above. Some treatments not only removed syringaldehyde but also removed a portion of the xylose.

Treatment of model hydrolyzate with Amberlite IRA-400 was conducted at pH 3, 6, 9 and 12. The Amberlite resin was placed in distilled water and adjusted to the appropriate pH with concentrated HCl or 50% solution of NaOH. The resin was stirred 15 minutes and recovered from the water by filtration on Whatman No. 1 paper. The pH of the resin was adjusted in this manner for each of the four pH treatments. The moist resin was added at the rate of 0.32 g ml^{-1} of model hydrolyzate solution adjusted to the appropriate pH prior to adding the resin. The resin was stirred in the solution for 15 minutes and removed by filtration on Whatman No. 1 paper. The solution was centrifuged at 41,000 x g to remove fine particles. The model hydrolyzate solution was analyzed before and after treatment with Amberlite IRA-400 for concentrations of xylose, syringaldehyde, acetic acid, and furfural to determine the effect of the treatments. Calcium and sodium hydroxide treatments were conducted in a similar manner, except that the hydroxides were added directly to the model hydrolyzate to achieve the desired pH treatment values. Calcium hydroxide treatment was evaluated at pH 8, 9, 10 and sodium hydroxide treatment was evaluated at pH 6, 7, 8, 9, 10. The treated solution was stirred 15 min. and centrifuged at 41,000 x g to remove particulates. Following the treatments, the solutions were analyzed for concentrations of xylose and model inhibitors.

D-xylose was quantitated by the reducing sugar method of Shaffer-Somogyi.[19] A standard curve of 10 points with a range of 0.25 to 2.50 mg of xylose and a titration range of approximatley 2.5 to 22 ml of 0.005 N sodium thiosulfate was developed. The average correlation for six standard curves was 0.9977.

Syringaldehyde concentrations in the model hydrolyzate were determined by the spectrophotometric method of Berk and Schroeder.[20] Syringaldehyde concentrations from 10 to 100 μg were used to construct a standard curve. The average correlation of seven standard curves was 0.9938.

The acetic acid concentration was determined with a Varian Model 1440 gas chromatograph equipped with a glass column [2 mm (id) x 0.64 cm (od) x 2 m (L)] packed with Porapak Q 50-80 mesh and a flame ionization detector. All samples were acidified to pH 2-3 with concentrated HCl and centrifuged at 41,000 x g prior to analysis. Isopropanol was used as the internal standard. A correlation of 0.9989 was obtained for a standard curve of acetic acid over the concentration range of 2.3 to 9.24 mg ml^{-1}. The concentration of ethanol in fermented wood hydrolyzate also was determined by gas chromatography with the conditions described above, except the samples were not acidified. The correlation for the ethanol standard curve was 0.9963 over the concentration range of 1.5 to 15 mg ml^{-1}.

Furfural concentration was determined by a modification of the distillation/ spectophotometric method of Mathers and Beck.[21] A 100 to 200 μl aliquot of sample in 40 ml distilled water was distilled with a micro-Kjeldahl distillation apparatus. A volume of 50 to 100 ml of distillate was collected for furfural analysis with a Bausch and Lomb Spectronic 1001 spectophotometer. A standard curve for furfural concentrations had a correlation of 0.9999 over a range of 1 to 10 μg.

Hardwood hydrolyzate prepared by dilute sulfuric acid digestion of oakwood chips, representing first-stage hardwood hydrolyzate of wood, was supplied by the Tennessee Valley Authority, Muscle Shoals, Alabama. The hydrolyzate had 53.6 mg ml^{-1} reducing sugar as D-xylose, 6.5 mg ml^{-1} total phenolics as syringaldehyde, 0.9 mg ml^{-1} of furfural and 13.8 mg ml^{-1} of acetic acid. The pH was 1.8. The hydrolyzate was subjected to chemical treatments as described previously to evaluate their efficacy to remove compounds inhibitory to P. tannophilus. Only one pH for each chemical treatment was selected for treatment of wood hydrolyzate employed in the fermentation studies. After treatment, the wood hydrolyzate was fermented to ethanol with P. tannophilus NRRL Y-2460. The yeast inoculum was culti- vated 72 hours in minimal medium[22] with 41 mg ml^{-1} D-xylose. The cells were harvested by centrifugation and resuspended in 100-ml of treated wood hy- drolyzate to give an inoculation rate of 1.5 mg of dry yeast mass ml^{-1} of hydrolyzate (about 10^8 viable cells ml^{-1}). Prior to yeast inoculation, the hydrolyzate was supplemented with yeast growth factors which included all the nutrients used in preparation of the minimal medium except the carbo- hydrate. The hydrolyzate was filter sterilized by passing it through a sterile 0.3 μm porosity ceramic candle (Selas Flotronics). Fermentation was conducted at pH 4.0 and 5.0 with 100-ml aliquots in 250-ml Erlenmeyer flasks. The flasks were incubated in a rotary shaker water bath at 100 rpm and 32°C. The pH of the hydrolyzate was adjusted daily to pH 4.0 or 5.0 with concentrated HCl or 50% solution of NaOH. Aliquots of hydrolyzate were collected daily for analysis of ethanol and depletion of reducing sugars.

RESULTS AND DISCUSSION

Evaluation of Treatments with Model Hydrolyzate

Amberlite IRA-400 is an anion exchange resin with a quanternary ammo- nium functional group aminated with triethylamine. Thus, as the pH is raised, the functional group will have an ionic attraction for anions. Maximal removal of syringaldehyde with the resin occurred at pH 9 with 85.3% removal (Table 1). Acetic acid was removed best at pH 6 and 12. At pH 6, 27.5% of the acetic acid was removed from the model wood hydrolyzate solu- tion. About 53% of the furfural was removed at pH 3 and 6, and 97% was removed at pH 12. Although the removal of inhibitors from the model hydrol- yzate was near maximal with Amberlite IRA-400 resin, the increased loss of xylose at pH 12 would preclude treatment at pH 12.

Treatment with calcium hydroxide removed 88% of the syringaldehyde at pH 8, 9 and 10 (Table 2). Acetic acid was not removed because the acetate salt of calcium is highly soluble. About one-third of the furfural was removed. Xylose loss with calcium hydroxide treatment was minimal compared to Amberlite IRA-400 and sodium hydroxide treatments. Sodium hydroxide removed about 20% of the furfural and contributed to about an eight percent

Table 1. Amberlite IRA-400 resin treatment of model wood hydrolyzate for removal of syringaldehyde, acetic acid and furfural.[a]

	Percent Removal after Treatment			
pH	Xylose	Syringaldehyde	Acetic Acid	Furfural
3	22.2	46.4	8.3	53.0
6	20.2	62.3	27.5	54.8
9	20.2	85.3	18.3	60.5
12	35.5	76.4	26.7	97.1

[a]The model hydrolyzate contained prior to treatment 41.0 mg xylose ml^{-1}, 7.0 mg syringaldehyde ml^{-1}, 12.0 mg acetic acid ml^{-1} and 2.1 mg furfural ml^{-1}.

loss of xylose over the pH range 6 to 10 (Table 3). Syringaldehyde and acetic acid were not removed by sodium hydroxide treatment. Since syringaldehyde and furfural are not ionic species, the mechanism by which Amberlite resin and calcium hydroxide effected removal of these microbial inhibitors from the model wood hydrolyzate solution is not clear.

Treatment of Wood Hydrolyzate

Further research was conducted to evaluate the removal of inhibitors from acid hydrolyzate by chemical treatments. The hydrolyzate was treated in the same manner as the model hydrolyzate to attain the desired pH. The resin compound was added at a rate of 0.32 g ml^{-1} of hydrolyzate.

Table 2. Removal of syringaldehyde, acetic acid and furfural from a model wood hydrolyzate with caldium hydroxide.[a]

	Percent Removal after Treatment			
pH	Xylose	Syringaldehyde	Acetic Acid	Furfural
8	0.5	88.3	0	33.1
9	1.3	87.9	0	30.8
10	2.3	87.8	0	44.1

[a]See footnote, Table 1.

Table 3. Removal of microbial inhibitors from a model wood hydrolyzate with sodium hydroxide.[a]

	Percent Removal after Treatment			
pH	Xylose	Syringaldehyde	Acetic Acid	Furfural
6	5.2	0	0	19.8
7	10.8	0	0	24.8
8	8.3	0	0	29.6
9	8.2	0	0	18.8
10	8.0	0	0	22.2

[a]See footnote, Table 1.

Both the resin and hydrolyzate were adjusted to the the desired pH before they were combined. The percent removal of inhibitors and the loss of xylose are shown in Table 4. Sodium and calcium hydroxide treatments at pH 8, 9, and 10 resulted in less than 1% loss of xylose, whereas treatment with Amberlite IRA-400 resin in the pH range 6 to 9 resulted in 9 to 15% loss of xylose. At pH 8 and higher, the level of reducing substances expressed as xylose in wood hydrolyzate was increased by sodium and calcium hydroxide treatments. The resin treatment was considerably more effective in removing phenolic compounds (expressed as syringaldehyde), acetic acid and furfural than the hydroxide treatments. Resin treatment at pH 6 to 8 removed about 70% of the phenolic compounds, and the highest percent removal of acetic acid was 39% achieved with the resin in the pH range 6 to 8. The hydroxide treatments were not effective for removal of acetic acid; however, they exacerbated the level of acetic acid. The resin treatment removed 61% of the furfural at pH 9 and the hydroxides removed 33% at pH 10.

Fermentation of Wood Hydrolyzate

The effect of the chemical treatments on the production of ethanol from wood hydrolyzate by P. tannophilus was determined. Three aliquots of wood hydrolyzate with 53.6 mg ml^{-1} of reducing sugar, 6.5 mg ml^{-1} phenolic compounds, 0.9 mg ml^{-1} of furfural and 13.8 mg ml^{-1} of acetic acid were treated with either calcium hydroxide at pH 9, sodium hydroxide at pH 8 or Amberlite IRA-400 at pH 6 and centrifuged to remove insolubles. Each hydrolyzate was adjusted to pH 5 and yeast growth factors were added as described previously. Hydrolyzate from each treatment was further divided into two groups,

Table 4. Effect of treatments and pH on removal of fermentation inhibitors from wood hydrolyzate.[a]

| | | Percent Removal after Treatment | | | |
Treatment	pH	Xylose	Phenolics	Acetic Acid	Furfural
NaOH	6	6.1	15.2	+0.6	13.1
	7	15.4	15.4	+0.4	14.8
	8	0.8	12.7	+1.1	16.6
	9	+2.0	18.1	+1.9	18.0
	10	+0.4	31.2	+6.6	33.4
Ca(OH)$_2$	7	1.5	17.8	1.5	14.4
	8	+1.2	23.5	+6.6	9.6
	9	+3.5	23.1	+4.6	20.8
	10	+3.1	40.5	+3.7	34.3
IRA-400	5	14.0	66.2	33.8	48.0
	6	9.0	69.4	39.0	50.7
	7	12.4	74.4	37.8	51.0
	8	14.8	71.1	38.4	51.6
	9	9.5	82.7	33.9	61.0

[a]The hydrolyzate contained prior to treatment 55.6 mg xylose ml^{-1}, 5.1 mg phenolics ml^{-1}, 5.3 mg acetic acid ml^{-1} and 0.9 mg furfural ml^{-1}. All values representing percent removal are averages of three trials. Values preceded with a plus sign indicate percent increase.

one group adjusted to pH 4 and the other to pH 5. The hydrolyzates were filter sterilized and inoculated with P. tannophilus at the rate of 1.5 mg dry yeast mass ml^{-1} of hydrolyzate (about 10^8 viable cells ml^{-1}). Minimal medium with 41 mg ml^{-1} xylose at pH 4 and 5 was used as the control fermentation substrate. A lower level of sugar was used in the control than was quantitated in the wood hydrolyzate because hydrolyzate contains reducing substances that are not fermentable carbohydrates.

Ethanol yields after 48 hours fermentation are shown in Table 5. A higher ethanol yield was achieved in minimal medium fermented at pH 4 because the optimum pH range for ethanol production by P. tannophilus is 2.4 to 4.0.[23] Fermentation of wood hydrolyzate with P. tannophilus can be accomplished at pH 5, but attempts to ferment it at pH 4 have been unsuccessful due to the toxicity of wood hydrolyzate at pH 4.[24] The toxicity is associated with compounds such as acetic acid, phenolics and furfural in the wood hydrolyzate. The toxicity of acetic acid to microorganisms is known to be potentiated at low pH's.[9] Treatments of wood hydrolyzate which permit fermentation at pH 4, near the optimum for P. tannophilus, is viewed by the authors as one method that might improve the bioconversion of wood sugars to ethanol. Therefore, fermentations of treated wood hydrolyzate were conducted at both pH 4 and 5.

Treatment of wood hydrolyzate with sodium hydroxide to pH 8, calcium hydroxide to pH 9, and Amberlite IRA-400 ionic exchange resin to pH 6 did not alleviate the toxicity of wood hydrolyzate for P. tannophilus grown in the hydrolyzate at pH 4. With an inoculation rate of 1.5 mg yeast cells ml^{-1} of hydrolyzate (dry weight basis), the yeast died within 24 to 48 hours in the sodium hydroxide treated hydrolyzate. Survival was better in the calcium hydroxide treated hydrolyzate. The Amberlite treated hydrolyzate was least toxic to the yeast at pH 4, but ethanol production was poor, achieving less than 1.4 mg ml^{-1}. Due to poor yeast growth and ethanol production in the hydrolyzate at pH 4, these data were omitted.

Table 5. Production of ethanol from treated hardwood hydrolyzate by Pachysolen tannophilus.[a]

Treatment	pH	Carbohydrate		Ethanol	
		Initial	used	Conc.	Yield
		mg ml^{-1}		mg ml^{-1}	mg mg^{-1} sugar
Minimal	4	41.0	37.4	11.3	0.30
Medium	5	41.0	37.1	9.5	0.26
NaOH (pH 8)	5	51.0	18.6	8.0	0.43
Ca(OH)$_2$ (pH 9)	5	51.0	21.2	9.5	0.45
Amberlite IRA-400 (pH 6)	5	41.0	18.1	8.4	0.46

[a]The substrates were fermented at 32°C for 48 hours in a rotary shaker water bath (100 rpm).

Treated hydrolyzate fermented at pH 5 was not toxic to P. tannophilus (Table 5). The yeast produced 8 mg ml^{-1} of ethanol in sodium hydroxide treated hydrolyzate and a sugar-to-ethanol yield of 43%. Treatment of hydrolyzate with calcium hydroxide improved the potential for ethanol production even more. The yield of ethanol was improved to 45%, and the concentration of ethanol was identical to the concentration achieved in minimal medium at pH 5. A comparable yield of ethanol was produced in hydrolyzate treated with Amberlite IRA-400 resin. All the treatments improved the yield of ethanol from wood hydrolyzate. The yield ranged from 43% to 46% for the treated wood hydrolyzate, which was nearly double the yield for minimal medium at pH 5. Sugar utilization in the treated hydrolyzate was about half the utilization in the minimal medium after 48 hours fermentation. Therefore, the higher ethanol yields in the wood hydrolyzate were due to greater efficiencies of converting the sugars to ethanol.

The concentration of sugar in the wood hydrolyzate was affected by the treatments employed to remove the inhibitors. About 1% or less of the xylose was lost after the wood hydrolyzate was treated with sodium or calcium hydroxide at pH 8 or higher. However, the loss was 20% for xylose from the model hydrolyzate, and a 9% loss occurred for reducing sugar from wood hydrolyzate treated with the resin at pH 6. The resin treatment was the most effective for removing inhibitors from wood hydrolyzate. However, the greater loss of xylose due to the resin treatment might have offset the superior ability of the resin to remove inhibitors. Based on fermentation performance of P. tannophilus in the treated hydrolyzate, ethanol yields were similar for all hydrolyzate treatments. Therefore, treatment with calcium hydroxide performed as well as the other treatments and is more economical than the resin treatment. Following treatment the calcium could be recovered and recycled.

This study indicated that Amberlite IRA-400 ionic resin treatment of wood hydrolyzate at pH 6 removed 69% of the phenolics, 39% of the acetic acid and 51% of the furfural. After treatment, the hydrolyzate contained 2.6 mg ml^{-1} phenolics, 8.3 mg ml^{-1} acetic acid and 0.45 mg ml^{-1} furfural. The hydrolyzate was toxic to P. tannophilus at pH 4 but not at pH 5. After a 48-hour fermentation at pH 5, P. tannophilus produced 0.46 mg ethanol mg^{-1} sugar consumed. The major problem associated with the treatment was that it removed a portion of the sugar in addition to removing the inhibitors. Treatment of wood hydrolyzate with calcium hydroxide at pH 9 removed 23% of the phenolics, none of the acetic acid and 13% of the furfural. Following treatment with calcium hydroxide, the hydrolyzate had 5 mg ml^{-1} phenolics,

13.8 mg ml^{-1} acetic acid and 0.8 mg ml^{-1} furfural. Although 6 mg ml^{-1} of phenolics have been demonstrated to completely inhibit the growth of P. tannophilus in minimal medium at pH 4.0, 5.5 and 6.5, and 12 mg of acetic acid ml^{-1} suppressed ethanol production at pH 5.5, P. tannophilus achieved higher ethanol yields in calcium hydroxide treated hydrolyzate than in minimal medium without inhibitors. The mechanism by which these treatments improve the bioconversion of wood sugars is not clear. The more efficient removal of inhibitors from wood hydrolyzate, especially acetic acid, might account for the improved ethanol yield from Amberlite IRA-400 treated hydrolyzate. However, since calcium hydroxide treatment was less effective than Amberlite treatment, and it did not remove acetic acid, perhaps calcium forms soluble complexes with some of the inhibitors which reduce their toxicity.

ACKNOWLEDGEMENT

This study was supported in part by the Alcohol from Wood Project under contract TV-63512A with the Tennessee Valley Authority, Muscle Shoals, Alabama.

REFERENCES

1. T. A. McCaskey and C. P. Martin. Components of hardwood hydrolyzate inhibitory to Pachysolen tannophilus. p. 293-299. In: Proceedings of the Technical Association of the Pulp and Paper Industry. Research and Development Conference. Appleton, WS. Sept. 30-Oct. 3. 1984.
2. R. H. Leonard and G. J. Hajny. Fermentation of wood sugars to ethyl alcohol. Ind. Eng. Chem. 37:390-395. 1945.
3. G. A. Soboleva, V. I. Golubkov and A. M. Vitrinskaya. Effect of furfural on the cytochrome system of yeast. Mikrobiolgiya. 43:441-444. 1973.
4. J. Banerjee, R. Bhatnagar and L. Viswanathan. Inhibition of glycolysis by furfural in Saccharomyces cerevisiae. Eur. J. Appl. Microbiol. Biotechnol. 11:226-228. 1981.
5. I. S. Goldstein. Composition of biomass. p. 9-18. In: Organic Chemicals from Biomass. I. S. Goldstein (ed). CRC Press, Boca Raton, FL. 1981.
6. Y. Henis, H. Tagari, and R. Volcani. Effect of water extracts of carob pods, tannic acid and their derivatives on the morphology and growth of microorganisms. Appl. Microbiol. 12:204-209. 1964.
7. A. Arieta-Escobar and J. M. Belin. Effects of polyphenolic compounds on the growth and celluloytic activity of a strain of Trichoderma viride. Biotechnol. Bioeng. 24:983-989. 1982.
8. J. A. Lewis and R. L. Starkey. Vegetable tannins, their decomposition and effects on decomposition of some organic compounds. Soil Sci. 106:241-247. 1968.

9. Y. Shimazu and M. Watanabe. Effects of yeast strains and environmental conditions on formation of organic acids in must during fermentation. J. Ferment. Technol. 59:27-32. 1981.

10. P. M. Davidson. Phenolic compounds. p. 37-74. In: Antimicrobials in Foods. A. L. Branen and P. M. Davidson (eds). Marcel Dekker, Inc., New York. 1983.

11. S. Doores. Organic acids. p. 75-108. In: Antimicrobials in foods. A. L. Branen and P. M. Davidson (eds). Marcel Dekker, Inc. New York. 1983.

12. E. Leuck. <u>Antimicrobial</u> <u>Food</u> <u>Additives</u>. Springer-Verlag, New York. 1980.

13. R. Kumar and M. Singh. Tannins: Their adverse role on ruminant nutrition. J. Agr. Food Chem. 32:447-453. 1984.

14. A. Chesson. Effects of sodium hydroxide on cereal straws in relation to the enhanced degradation of structural polysaccharides by rumen microorganisms. J. Sci. Food Agr. 32:745. 1981.

15. R. D. Reichart, S. E. Fleming, and D. J. Schwab. Tannin deactivation and nutritional improvement of sorghum by anaerobic storage of H_2O-, HCl-, or NaOH-treated grain. J. Agr. Food Chem. 28:824. 1980.

16. H. G. Jung, G. C. Fahey, Jr. and J. E. Garst. Simple phenolic monomers of forages and effects of in vitro fermentation on cell wall phenolics. J. Anim. Sci. 57:1294-1305. 1983.

17. J. Zemek, B. Kosikova, J. Augustin and D. Joniak. Antibiotic properties of lignin components. Folia Microbiol. 24:483-486. 1979.

18. A. V. Tran and R. P. Chambers. Production of 2,3-butanediol and methyl ethyl ketone from mannose. p. 199-209. In: Proceedings of the Technical Association of the Pulp and Paper Industry. Research and Development Conference, Appleton, WS. Sept. 30-Oct. 3. 1984.

19. P. A. Shaffer and M. Somogyi. Copper-iodometric reagents for sugar determination. J. Biol. Chem. 98:695-713. 1932.

20. A. A. Berk and W. C. Schroeder. Determination of tannin substances in boiler waters. Ind. Eng. Chem. 14:456-459. 1942.

21. A. P. Mathers and J. R. Beck. Determination of furfural, pentoses and pentosans in distilled spirits. Assoc. Offic. Agr. Chem. 37:861-869. 1954.

22. E. J. Del Rosio, K. J. Lee and P. L. Rogers. Kinetics of alcohol fermentation at high yeast levels. Biotechnol. Bioeng. 21:1477-1482. 1979.

23. P. J. Slininger, R. J. Bothast, J. E. Cauwenberge, and C. P. Kurtzman. Conversion of D-xylose to ethanol by the yeast <u>Pachysolen</u> <u>tannophilus</u>. Biotechnol. Bioeng. 24:371-384. 1982.

24. C. P. Martin. Optimization of ethanol production from pentose rich fraction of acid hydrolyzed hardwood by <u>Pachysolen</u> <u>tannophilus</u>. Masters Thesis, Auburn University, Auburn, Alabama. 1983.

EXPERIMENTAL PRODUCTION OF ETHANOL FROM AGRICULTURAL CELLULOSIC

MATERIALS USING LOW-TEMPERATURE ACID HYDROLYSIS

J. W. Barrier, M. R. Moore, G. E. Farina, J. D. Broder,
and M. L. Forsythe

Tennessee Valley Authority
F-4 NFDC
Muscle Shoals, Alabama 35660

George R. Lightsey

Department of Chemical Engineering
Mississippi State University
State Mississippi, Mississippi 29762

ABSTRACT

A low-temperature, low-pressure acid hydrolysis process that uses
separate unit operations to convert hemicellulose and cellulose in agricultural residues and crops to fermentable sugars was evaluated in bench-
scale studies at Mississippi State University and TVA. Corn stover was used
as a feedstock. The conversion and recovery of sugars from the hemicellulose and cellulose fractions was more than 90% efficient; sugar product
concentrations of more than 10% glucose and 10% xylose were achieved.
Furfural and hydroxymethyl furfural (HMF) concentrations in the sugar product
stream was less than 0.005%.

Based on results of the bench-scale studies, TVA built a small-scale
experimental plant for producing fuel-grade ethanol 37.8 L (10 gal) of
190-proof per hour from cellulosic agricultural residues and crops. Sugars
produced by hydrolysis of the cellulosic materials are being fermented and
distilled with equipment previously used for ethanol production from starch
and sugar feedstocks. Preliminary evaluations using corn stover indicate
that conversion efficiencies in the experimental plant are as high as those
achieved in the laboratory.

Process evaluation results, including equipment performance, materials of construction performance, process efficiency and yields, and product rates for the first six months of the experimental plant operation will be discussed.

Keywords. Ethanol, low-temperature hydrolysis, acid hydrolysis, renewable energy, cellulosic crop conversion.

INTRODUCTION

Technology for producing fuel-grade ethanol from agricultural crops and residues have been developed and demonstrated by TVA in Muscle Shoals, Alabama. An experimental facility has been built to produce ethanol (190+ proof) at the rate of 37.8 lph (10 gph) using a two-stage, low-temperature, low-pressure sulfuric acid hydrolysis process. Objectives of this project are to 1) conduct experimental plant evaluations and laboratory studies required to develop and test low-temperature acid hydrolysis procedures for converting agricultural cellulosic materials to ethanol, 2) define and assess technology, production systems, and feedstocks required for developing an integrated system for producing food, feed, fuel, and chemicals from agricultural residues, and 3) coordinate activities with other organizations which have complementary objectives. This paper summarizes the design, construction, and initial testing of the experimental facility built to meet these objectives.

PROCESS BACKGROUND

Description

The process involves two-stage hydrolysis, relatively low temperatures, and a cellulose prehydrolysis treatment with concentrated sulfuric acid. Corn stover is ground and mixed with dilute sulfuric acid (about 9.5% by weight). The hemicellulose fraction of the stover is converted to pentose sugars by heating the mixture to 100°C for 1 to 6 hours in the first hydrolysis reactor. Raw corn stover has, on a dry basis, a composition of about 38% cellulose, 25% hemicellulose, 17% lignin, 5% ash, and 15% other organic materials. After initial startup, sulfuric acid for the first hydrolysis reaction is available in the product stream from the second hydrolysis step where glucose is formed by the conversion of cellulose. Since the first hydrolysis reaction is a low temperature and low pressure, degradation of glucose is negligible. Xylose is leached from the first hydrolysis reactor with large quantities of water. The leachate contains sulfuric acid, water, xylose, other reaction products, and glucose (including glucose from the second hydrolysis). The product from the leaching step is expected to have a concentration of about 5-10% glucose and 5-10% xylose.

Residue stover from the first hydrolysis reaction (hemicellulose conversion) is dewatered and prepared for the second hydrolysis step (cellulose conversion) by soaking (prehydrolysis treatment step) in sulfuric acid (about 20 to 30% concentration) for 1 to 2 hours. The residue is then screened, dewatered, and dried. Upon entering the cellulose reactor, the residue contains sulfuric acid (80% concentration) in a ratio of about 0.2 kg of acid per kg of residue. The second hydrolysis reactor (batch-type) operates at 100°C and requires a reaction time of 3 to 6 hours. The reactor product is filtered to remove solids (primarily lignin and unreacted cellulose). Composition of this stream is calculated to be about 8% sulfuric acid and 10% glucose. As described earlier, this stream is used in the first hydrolysis step of the process to supply dilute sulfuric acid for hemicellulose conversion to xylose. Residue from the reactor is washed to recover remaining sulfuric acid and sugar not removed in the filtration step. The residue is then used as boiler fuel.

The sugar product stream is neutralized with lime, filtered to remove precipitated material, and fermented to produce ethanol from the sugar components. Fermentation agents are being developed to convert the xylose component of the sugar solution to ethanol. The glucose component can be converted with conventional yeasts used for grain sugar fermentation. The total ethanol product rate is estimated to be about $0.25 - 0.33$ L kg^{-1} (60-80 gal ton^{-1}) of corn stover feed. Sulfuric acid consumption for the process is about 0.15 kg of acid per kg of stover (300 lb ton^{-1}).

Energy consumption rate for the acid hydrolysis process is calculated to be about 5.56×10^3 kJ L^{-1} (20,000 Btu's per gal) of ethanol produced. Potential for reducing energy consumption for the process is good. Heat recovery from several process streams is used to reduce overall energy requirements. Heat recovered by burring the lignocellulose residue was not considered in the energy balance.

Development

TVA's process is based on the sulfuric acid hydrolysis process developed by Dunning and Lathrop[1,2,3] at the USDA Peoria Laboratory. A flow diagram of the process is shown in Figure 1. Several variations of the Peoria process were evaluated at Purdue University's Laboratory of Renewable Resources Engineering[4,5], and other universities (Arkansas and Missouri). These studies were not sufficiently detailed to allow scale-up to the 38 L hr^{-1} (10 gph) production level for TVA's experimental plant; therefore, experiments were conducted in TVA, vendor, and Mississippi State University laboratories to verify operating conditions, equipment requirements, and design parameters. The results of these studies are summarized below.[6]

o   Corn stover ground to one inch handled well and allowed adequate hydrolysis of hemicellulose.

o   The time required for optimum hemicellulose reaction was about two hours (Table 1).

o   Formation of degradation by-products in the hemicellulose reaction was minimal. Concentrations of furfural and HMF were well below levels that could present problems during fermentation (Table 2).

o   Actual xylose yields in bench-scale studies were equal to or greater than yields estimated from literature data. Overall xylose yields of 86 and 93% were obtained in bench-scale studies at one- and three-hour reaction times, respectively. The estimated yield for this step is 92% (Tables 3 and 4).

o   Recycle leachate, dilute acid, and prehydrolysis acid solutions were stable during storage for several days.

o   Prehydrolysis was accomplished without the formation of undesirable by-products. Hemicellulose conversion should be completed during prehydrolysis if unreacted material from the hemicellulose reactor is contained in the residue to the prehydrolysis vessel (Table 5).

o   Cellulose hydrolysis was successfully accomplished by cooking stover containing 66 to 78% acid for six hours at 99 C (210°F). Yields of 75 to 99% cellulose conversion to glucose were obtained in laboratory studies (Table 6).

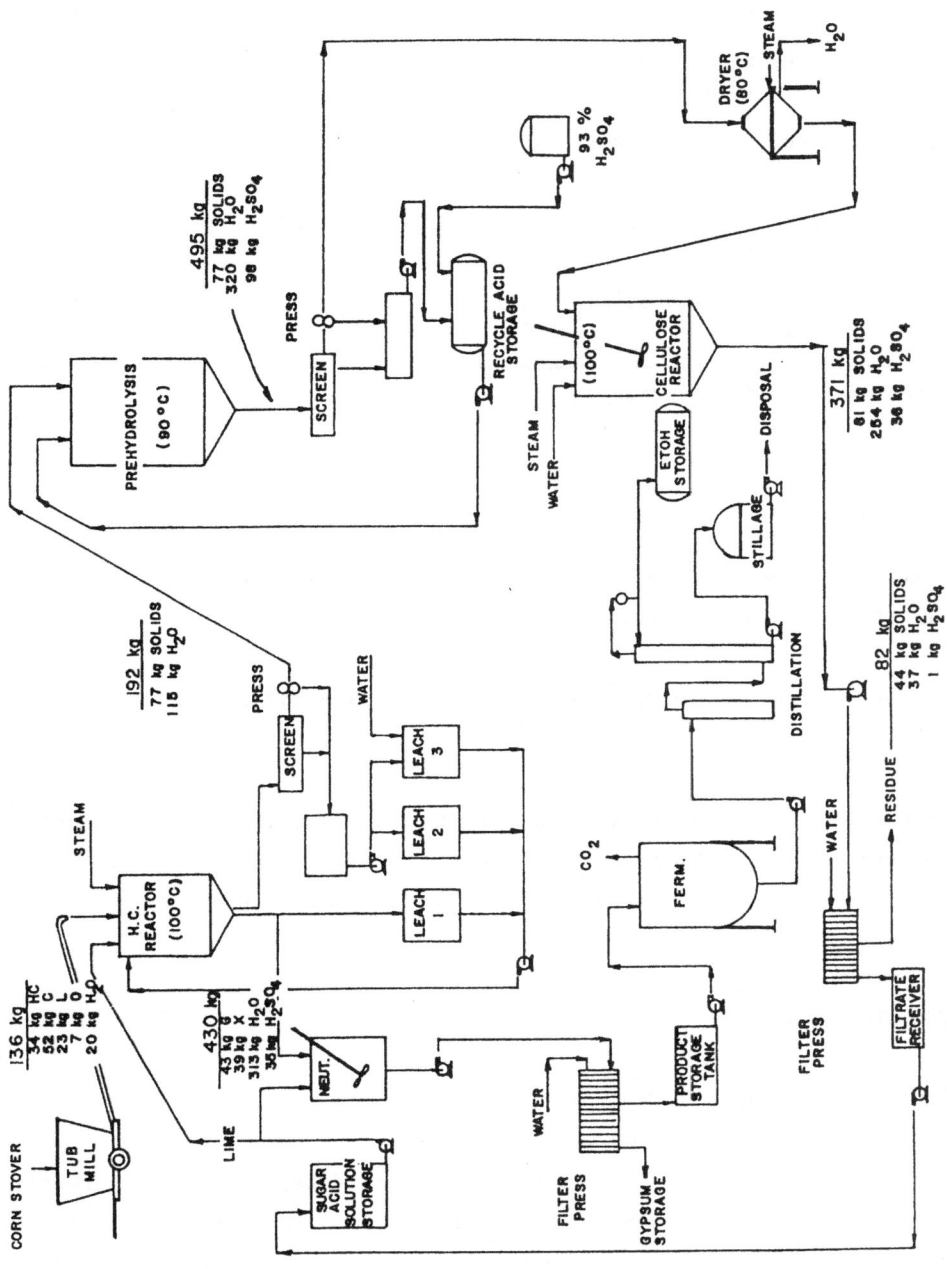

Figure 1. Flow diagram and material balance—TVA ethanol from corn stover experimental plant

591

Table 1. Weak acid hydrolysis – solid samples[1]

Sample no.	Description	Arabinose	Xylose	Mannose	Glucose	Lignin
26	Raw Stover	3.3	20.6	0.6	31.9	32.0
9	Leached Stover, after 2 hrs, Trail 1	3.2	16.3	0.5	47.2	25.1
1	Leached Stover, after 1 hr, Trail 2	1.1	9.6	2.5	39.0	21.4
2	Leached Stover, after 2 hrs, Trail 2	2.6	21.4	2.0	48.8	26.2
3	Leached Stover, after 3 hrs, Trail 2	1.5	7.1	2.1	61.5	25.6

[1] Units are in weight percent.

o   Atmospheric pressure drying equipment was not adequate for removing enough moisture from prehydrolysis residue to increase the acid content to 60–80% concentration.

o   Vacuum drying was adequate for the acid concentration drying step (Table 7).

o   Stainless steel for vessels and pipe in contact with acid-containing solutions was severely damaged by corrosion in bench-scale tests and was not recommended for use in the TVA experimental plant.

PROCESS EVALUATIONS

Facility Description

Based on the results of TVA's bench-scale studies, the existing facilities for fermentation and distillation were modified for integration with the new acid hydrolysis unit. Construction of the new facility began in September 1983 and was completed in April 1984. Acid hydrolysis equipment is housed in a 18 x 24 x 9 m (60 feet x 80 feet x 30 feet) high process building. The existing process building 9 x 12 m (30 feet x 40 feet) is used for fermentation and distillation equipment.

Table 2. Weak acid hydrolysis of hemicellulose: liquids analysis

Sample no.	Description	Sulfuric acid	Total H. C.	Xylose	Glucose	Organic acids	Furfurals
2	Excess weak acid, 4/27	9.1	1.09	0.85	0.14	0.01	0.00
3	Leachate 1, run 1, 4/27	4.7	1.16	0.96	0.15	0.10	0.08
4	Leachate 2, run 1, 4/27	1.6	0.83	0.69	0.10	0.03	0.02
5	Leachate 3, run 1, 4/27	1.0	0.42	0.35	0.08	0.14	0.02
23	Initial weak acid, Run 2, 5/2	9.1	1.01	0.81	0.13	0.02	0.00
24	Final weak acid, Run 2, 5/2	8.6	1.15	0.87	0.18	-	0.00
16	Leachate 1, Initial, Run 2 5/2	2.4	0.92	0.76	0.14	0.24	0.04
20	Leachate 1, Final, Run 2, 5/2	4.6	2.99	2.31	0.56	0.00	0.07
17	Leachate 2, Initial, Run 2 5/2	0.8	0.40	0.32	0.11	0.26	0.00
19	Leachate 2, Final, Run 2 5/2	2.7	2.21	1.81	0.33	0.09	0.03
18	Leachate 3, Run 2, 5/2	1.3	1.00	0.85	0.13	0.00	0.02

[1]Complete analysis attached in Appendix C

[2]Units are grams per 100 ml of liquid

Table 3. Weak acid hydrolysis of hemicellulose: solids analysis

Sample no.	Description	Results (Wt. percent, dry basis)					
		Arabinose	Xylose	Mannose	Glucose	Lignin	Ash
0	Original Stover	3.90	19.6	3.40	36.8	-	-
0	Original Stover	2.80	20.8	0.48	38.5	-	-
26	Original Stover	3.30	20.6	0.58	31.9	32.0	4.96
Ave	Original Stover	3.34	20.4	1.49	35.7	-	-
6	40 min. to 99°C	2.63	23.3	2.26	46.6	23.7	5.47
7	1 hr at 81°C[1/]	1.29	15.2	4.02	43.4	34.5	4.38
8	2 hrs at 65°C	1.97	17.2	0.33	41.4	27.0	4.51

[1/] Average temperature over period, e.g., 100 + 63 = 81°C average.

The new facility was designed to allow the use of relatively inexpensive materials and conventional process equipment. This minimizes capital costs and simplifies scale-up to a commercial capacity. Selection of materials of construction for the new facility was subject to certain safety constraints, because of the corrosive properties of sulfuric acid. Based on testing data, all process vessels, reactors, and tanks are constructed of FRP with Hetron 700 and 980 resins. The dryer is of carbon steel and lined with Kynar. Conveyor belts are made of acid resistant materials (PVC). Mild steel agitator shafts are coated with Kynar or Teflon. Heat exchangers are made with CPVC pipe shells and Carpenter 20 stainless steel coils. Pumps are made with non-metallic compounds, Teflon linings, or Carpenter 20 stainless steel. The two filter press units are made of polypropylene. All dewatering equipment except the hydrasieve screen is either of non-metallic materials (CPVC and FRP) or metals with an acid-resistant coating.

Metal structural supports were coated with acid-resistant paint (epoxy) to prevent corrosion of the members due to acid leaks and vapors. After six months of plant operation, no major corrosion problems have been encountered with any process components.

Process Evaluation

The process has been evaluated based on the results of plant runs made following the completion of construction in April 1984. At first, shakedown

Table 4. Weak acid hydrolysis – liquid samples[1]

Sample no.	Description	Before lab hydrolysis				After lab hydrolysis			
		Arab.	Xylose	Mann.	Glucose	Arab.	Xylose	Mann.	Glucose
1	Recycled 27% H_2SO_4 Trail 1	.03	.31	–	.06	.04	.27	.01	.07
2	Recycled 9% H_2SO_4 Trail 1	.16	.71	–	.06	.19	.85	.04	.14
3	Leachate 1 Trail 1	.15	.91	–	.09	.18	.96	.02	.15
4	Leachate 2 Trail 1	.12	.65	–	.06	.12	.69	.01	.10
102	Leachate 1, 1 hr Trail 2	.53	.31	.03	.31	–	.24	.03	.28
103	Leachate 1, 2 hr Trail 2	.65	.35	.04	.45	.55	.33	.04	.45

[1] Units are grams per 100 ml of liquid.

595

Table 5. Prehydrolysis step

Sample no.	Description	Acid[1] Conc.	H.C.[4]	Glucose	Lignin	Organic acids	Furfural[1]
1	Run 1 acid Solution 4/27	33.3	0.32[1]	0.07[1]	–	0.01	0.04
21	Run 1 acid Solution 5/2	28.5	0.27[1]	0.07[1]	–	0.00	0.05
22	Run 2 acid Solution 5/2	37.6	1.08[1]	0.24[1]	–	0.09	0.02
10	Run 1 solids After Pre-Hy 4/27	26.1[3]	12.8[2]	48.6[2]	24.2[2]	–	–
15	Run 2 solids After Pre-Hy 5/2	27.7	6.1[2]	70.1[2]	27.7[2]	–	–

[1] Grams per 100 ml.

[2] Dry weight percent.

[3] Average of samples 10, 1, 21.

[4] Includes xylose, mannose, and arabinose.

runs were performed to verify equipment performance and materials handling capabilities. Equipment performance, as stated earlier, was satisfactory though some modifications were made to facilitate materials handling.

The first complete experimental run to evaluate all process steps was conducted in November. Raw stover was added to the hemicellulose reactor followed by the adding of dilute acid (9.5%). The excess acid was drained off, and the stover was heated to 99 C (210°F) by live steam injection and allowed to cook at that temperature for two hours. Xylose product was leached from the stover by adding warm water to the stover and draining off the product through the reactor leach ports. The stover was dewatered (using the hydrasieve and roller press) and conveyed to the dryer. The cellulose prehydrolysis step was not used during this run. Instead, the dryer was used to remove some of the moisture in the stover and concentrated (85%) acid was added directly to the stover in the cellulose reactors. To minimize the rise in temperature due to the heat of dilution of the acid by the water in the moist stover, the acid was cooled to about 15.5 C (60°F). Water for hydrolysis was added to the acid/stover mixture to give a final acid concentration of 30% in the cellulose reactor. The reactor was heated to 99 C (210°F) and the contents were cooked at that temperature for four

596

Table 6. Cellulose hydrolysis reaction: concentrated acid additional samples[1]

Sample no.	Description	Adjusted "drying" acid conc. (%)	H.C. [2] wt. %	Glucose [2] wt. %	Lignin [2] wt. %	Glucose yield after cell. hyd. (%)
10	Solids after pre-hy reaction	–	12.8	48.6	24.2	–
90	Sample 10 after acid addition	58.1	30.5	35.6	61.0	–
86	Sample 90 after cell. hyd.	–	0.9	26.7	62.5	75
87	Sample 10 after acid addition	66.3	4.2	37.4	54.2	–
85	Sample 87 after cell. hyd.	–	1.3	20.7	79.3	85
89	Sample 10 after acid addition	78.0	21.3	12.2	58.9	–
88	Sample 89 after cell. hyd.	–	0.6	1.6	89.8	99

[1] Acid concentration obtained by addition of concentrated acid.
[2] Assuming constant lignin amount.

597

Table 7. Vacuum drying of acid-corn stover

Sample no.	Description of test[1]	Solids[2] (wt. %, total)	Acid[2] (wt. %, liquid)	Final sample condition
10	Initial solids	14.9	26.1	Light brown
6/9A	0.5 hrs, 90°C, 1 cm	19	41	Dark brown
6/9B	1.0 hrs, 90°C, 1cm	31[3]	79[3]	Black, burned
6/9C	2.0 hrs, 90°C, 1 cm	48[3]	162[3]	Charred
6/10A	0.5 hrs, 80°C, 1 cm	18	39	Brown
6/10B	1.0 hrs, 80°C, 1 cm	26	59	Dark brown
6/10C	1.5 hrs, 80°C, 1 cm	35[3]	92[3]	Black, burned
6/10D	0.5 hrs, 80°C, 2.5 cm	19	41	Brown
6/10E	1.0 hrs, 80°C, 2.5 cm	22	50	Dark brown
6/10F	1.5 hrs, 80°C, 2.5 cm	27[3]	63[3]	Black surface
6/14A	0.5 hrs @ 90°C, 0.5 hrs @ 80°C, 1.0 cm	27	63	Dark brown
6/16A	0.3 hrs @ 99°C, 0.3 hrs @ 80°C, 2.5 cm	30	74	Dark brown
6/16B	0.3 hrs @ 99°C, 0.3 hrs @ 80°C, 2.5 cm	26	59	Dark brown

[1] All vacuum testing at 690 mm of mercury, time, temperature, sample thickness given.
[2] Solid and acid 5 by mass balance.
[3] Mass balance suspected to be inaccurate.

Table 8. Conversion efficiencies

Plant Trial	Hemicellulose Conversion Efficiency (%)	Cellulose Conversion Efficiency (%)
7/84	74	36[1]
9/84	80	--
11/84	97	90

[1]Cellulose conversion not evaluated during this run.

hours. The remaining lignin was filtered off, and the glucose/acid mixture was added to the xylose product from the hemicellulose conversion step. Acid in the mixture was neutralized with lime and the resulting calcium sulfate was removed by filtering.

A second cellulose hydrolysis was run at about the same conditions as the first hydrolysis. Overall yields obtained in these runs were extremely encouraging. Xylose yields were more than 95% based on the final average xylose content of the solids. Average glucose yields, based on both liquid and residual solids compositions of more than 85% were obtained in these trials. Yields obtained in these inititial process runs either met or exceeded those obtained in laboratory tests (See Table 8).

Based on these initial runs, several process areas will be developed and improved in plant and laboratory studies. Technology is being investigated that would allow partial recovery and recycle of sulfuric acid from the product stream. The glucose produced will be used to produce fuel ethanol by fermentation. The xylose produced in the process may also be fermented or used to produce other materials which can be sold as co-products. One possible use of the unconverted xylose is to produce fodder yeast (single- cell protein) for animal feed.

PLANS

Plans for the TVA program include evaluation of the process as initially designed and also the investigation of alternate equipment and operating options for several process steps. Studies will identify optimum equipment requirements, by-product options, and operating conditions for the facility. Alternate feedstock evaluations will define processing differences for non-woody cellulosic feeds available in the Tennessee Valley and surrounding areas. This will involve integrating the process into fuel, feed, food, and chemical systems for crops grown on marginal lands in the Tennessee Valley.

REFERENCES

1. J. W. Dunning and E. C. Lathrop. Process for Saccharifying Plant Materials. U.S. Patent 2, 450, 586. 1948.
2. J. W. Dunning and E. C. Lathrop. The saccharification of agricultural residues. Industrial and Engineering Chemistry, Vol. 37, No. 1, p. 24-29. 1945.
3. U.S. Department of Interior. Report of Investigations 4772, Part III - Liquid Fuels from Agricultural Residues. U.S. Department of Interior. Washington, D.C. 1951.
4. M. R. Ladisch. Unpublished data. Laboratory of Renewable Resource Engineering. Purdue University. 1983.
5. G. T. Tsao, M. R. Ladisch, M. Voloch and P. Bienkowski. Production of ethanol and chemicals from cellulosic materials. Process Biochemistry. September/October, p. 34-38. 1982.
6. Tennessee Valley Authority Quarterly Technical Report. April-June. Integrated Fuel Alcohol Production Systems. Phase III. Office of Agricultural and Chemical Development. Muscle Shoals, Alabama. 1983.

A CASE STUDY OF A COMMERCIAL PLANTING AND PROCESSING

OF SWEET SORGHUM FOR ALCOHOL PRODUCTION

Mike J. Giamalva

Sugar Station
Louisiana State University Agricultural Center
Baton Rouge, Louisiana 70803

Stephen J. Clarke

Audubon Sugar Institute
Louisiana State University
Baton Rouge, Louisiana 70803

ABSTRACT

In the Louisiana sugar belt, sweet sorghum matures in late August through September. Research has shown that sweet sorghum could be processed into a raw material for alcohol production at the sugar mill. Several alcohol plants were nearing completion in Louisiana with a production capacity of over 167 million liters (44 million gallons) per year. Their primary source of raw material was sugar cane molasses, but this source was insufficient and an alternative was necessary.

In 1982, 100 hectares (250 acres) of sweet sorghum was planted at Breaux Bridge, Louisiana. Recommended fertility and cultural practices were followed. Since Breaux Bridge is a traditional sugar cane producing area, two rows, 0.36 m (14 inches) apart, were planted on each 1.8 m (6 foot) row. This row width was necessary if standard sugar cane equipment was to be used. Sweet sorghum was a new crop for these growers and processors, whose only experience was with sugar cane, and they experienced numerous problems in growing, harvesting, and processing the sorghum. Both agricultural and processing problems for this crop are discussed, especially those for which no simple solution seems forthcoming.

Keywords. Ethanol, sweet sorghum, sugar mill.

INTRODUCTION

Sugar mills in Louisiana process sugar cane into raw sugar and molasses over a period of 40-90 days per year, beginning in October and continuing into December. During the remaining 9-10 months, the mill, which represents inefficient use of a multimillion dollar investment, remains idle. Research with sweet sorghum for sugar production, which would be processed during August and September, has been conducted in Louisiana for over 100 years. Production of sugar from sweet sorghum had not been successful for several reasons, including high starch content and high aconitic acid content of the juice. A combination of these two interfered with crystallization. Also, deterioration of juice quality occurs rapidly in sweet sorghum. Thus, its leaves could not be left to dry before processing. Bagasse containing green leaves of sweet sorghum burned reluctantly; therefore, an outside source of fuel was needed for processing.

Interest in sweet sorghum for sugar has fluctuated with the price of sugar. The world market price of sugar has ranged from 6.5¢ per kg in 1945 up to $1.50 per kg in 1974. Price of sugar peaked in 1951, 1957, 1973, and 1974. During these years, there was renewed interest in sweet sorghum. In 1973, Dr. B. J. Smith[1] in Weslaco, Texas, developed a procedure for starch and calcium aconitate removal by double clarification. In 1980, Ricaud and Campbell[2] published a preliminary estimate of costs in producing sweet sorghum under farm conditions in Louisiana based on available research data. These data were partially based on similar data with sugar cane. The same equipment used in sugarcane was also used in the production of sweet sorghum, except for planting. The authors estimated a total cost of $513 per hectare plus $51 for management and $106 for land rental, or $670 total costs per hectare for sweet sorghum. This included growing, harvesting, and hauling to a sugar mill. When sugar was placed in the farm bill in 1981, assuring a fixed price, 81 hectares of sweet sorghum were grown by the Breaux Bridge Sugar Cooperative, and these were processed at the Breaux Bridge Factory in 1982. Yields of sorghum were low, around 40 Mg ha^{-1}, and sugar yield also was low, around 41 kg Mg^{-1}.

Research on the production of sweet sorghum for alcohol production has been conducted, since 1973, at the Louisiana Agricultural Experiment Station in cooperation with the Audubon Sugar Institute at L.S.U.[3,4,5] This research is still in progress.[6] Most of the research by the Audubon Sugar Institute has been with the sweet sorghum variety, Wray. Average yields were 47 Mg ha^{-1} which produced from 121 to 467 liters of 60 brix molasses per Mg of cane with this variety. In 1980, a bill was passed, by the U. S. Congress, to reduce federal taxes by $0.013 per liter of gasoline blended with 10% alcohol. This tax reduction became $0.016 on January 1, 1985. The Louisiana Legislature, in 1980, voted a tax reduction of $0.21 per liter of gasoline, alcohol blend, and in 1984, when the tax on gasoline was increased by $0.021 per liter, the alcohol blend was also exempt. Total deduction per liter of alcohol used in the blend is $.379. Louisiana has a provision that 10% of the raw product must be produced in Louisiana in order to take advantage of the tax reductions. This has led to a renewed interest in sweet sorghum as a source of locally produced raw material for alcohol production.

MATERIALS AND METHODS

In 1983, 100 hectares of sweet sorghum were planted by members of the Breaux Bridge Coop. for production of molasses, 60 brix, as feedstock for fuel alcohol. Variety Wray was planted with two lines 0.305 m apart on 1.8 m row. Planting was accomplished by using a grain drill with grain sorghum plates. At this planting rate, hills were spaced 0.305 m apart and 3 seed per hill, resulting in a plant population of 106,000 plants per hectare. Fertilizer was applied at planting time at the rate of 305 kg ha^{-1} of 13-13-13 fertilizer. Additional nitrogen at a rate of 98 kg ha^{-1} was applied as a side dressing when the plants were about 254 mm tall. Propazine was used as a preemergence herbicide at a rate of 3.3 kg ha^{-1}, and Azinphos methyl was used to control the sugar cane borer at a rate of 0.95 kg ha^{-1}. Carbofuran, 10% granules, was applied at a rate of 1.12 kg ha^{-1} at planting time to control the lesser corn stalk borer. The entire planting was cultivated twice. Harvest was accomplished with three commercial soldier type harvesters, two single-row and one two-row harvesters and a combine harvester. The crop was cut and managed similar to a typical sugar cane crop in Louisiana. Attempts were made to burn off the leaves at different periods after harvesting and before loading in wagons for transport to the mill. Cane was left in the field on 3 or 4 row heaps 6 to 52 hours before burning was attempted. Total direct costs were about $336 per hectare. After adding rent, interest on investment, depreciation, taxes and overhead costs, total costs equaled $448 per hectare.[3] The mill was not equipped to handle chopped cane.

RESULTS AND DISCUSSION

Planting sweet sorghum with a grain drill, equipped with grain sorghum plates, resulted in almost twice the desired plant population. Yield obtained averaged 41 Mg ha^{-1}. Average production of 60 brix molasses per Mg of cane was 102 liters, with a range of from 83 to 125 liters per Mg. Difficulties in harvest were encountered when using the conventional soldier type in Louisiana sugar cane harvester. This type of harvester removed most of the tops and none of the leaves. The newer, two row Louisiana harvester was the most efficient machine, especially where lodging occurred. When cut at the maximum height (2.5 m) with either type machine, difficulty was encountered in loading. The tractor, drawing the wagon being loaded, rolled over the tops or bottoms of the stalks, resulting in too many whole stalks being left in the field. Some sorghum was cut with a combine type harvester which cut the stalks into 280 mm pieces. Special wagons, fabricated from expanded steel, are required with this type machine, and losses were too high in the mill yard because existing equipment failed to pick up many pieces. Harvesting with the combine harvester was much slower than with either soldier type machine.

Because of leaves remaining on the canes, washing at the sugar mill did not remove the sand and silt. This resulted in excessive wear on the mill rolls. Leaves also decreased bulk density of the material being ground, resulting in choking due to flooding at the rolls with shredded material. To rectify this problem, preparation of the cane prior to entering the mill train was reduced, grinding speed was reduced, and pressure on the mill rolls, reduced. Most cane was harvested and processed immediately after harvest. Some cane was left on the ground for as long as 52 hours before burning. At this time a mill breakdown occurred, resulting in another 24 hour delay. From this cane, only 83 liters of molasses were produced from each Mg of cane.

Total production of molasses was 428,000 liters from 100 hectares of sweet sorghum. Total fermentable sugars per liter was about .9 kg. The price agreed upon by the cooperative and the alcohol producer was $0.21 per kg of sugar, or $0.23 per liter of molasses. With the mill retaining 40% (of gross income) and the farmer receiving 60%, income received from these molasses cannot pay the costs of production and processing.[8] The alcohol producer was able to produce 214,000 liters of alcohol. When blended with 1,926,000 liters of lead free gasoline, 2,140,000 liters of alcohol-gasoline blend was produced. With a tax reduction of 4.23¢ per liter from the state and 1.59¢ from the U. S. Government, the result was a tax value of $.58 to the alcohol producer or blender. Dr. Ricaud had demonstrated that yields of 56 Mg ha^{-1} can be successfully produced[3,4,5] from variety Wray, and the Audubon Sugar Institute reported yields of from 120 to 166 liters per Mg of cane from this variety.[6] Using these minimum figures, 42 Mg ha^{-1} and 120 per Mg, 5040 liters of molasses (60 brix) can be produced per hectare. At a sale price of $0.23 per liter, gross income per hectare would be $463 for the processor and $696 for the farmer. The grower could earn $27 per hectare. Under conditions of this study, average yield actually produced in this planting was 42 Mg ha^{-1} and 102 l Mg^{-1}. This total yield of 4284 l per hectare resulted in a net income of $985 per hectare. With 40% retained by the mill for processing ($394), the farmers share was $591. The farmer lost $79 per hectare and the mill lost a lesser amount. In order for sweet sorghum to be a profitable crop for the farmer and processor, yields would have to be higher or the sale price of fermentable sugars would have to be increased. Molasses from sweet sorghum grown solely for the production of molasses will cost more to produce and process than sugar cane molasses, which is a by-product of the sugar industry.

At the present level of tax reduction, which is essentially a subsidy, the production and blending of alcohol and gasoline is profitable at the price paid for molasses. What will happen if these subsidies are removed, either at the state or federal level, remains to be seen. An additional factor that may enhance the demand for alcohol is the announced desire of the U. S. Environmental Protection Agency to eliminate the use of lead entirely from gasoline. Either alcohol would have to be used to replace lead as an octane enhancer or low compression engines would have to be developed to run on lower octane fuel or diesel. Diesel engines and fuel would have to be used. More research with sweet sorghum production and processing of sweet sorhgum is necessary, especially with leaf removal during harvest.

REFERENCES

1. B. J. Smith, et al. Production of Raw Sugars from Sweet Sorghum Juices. Sugar Journal, Vol. 35, No. 12, p. 22-27. 1973.
2. R. Ricaud and J. R. Campbell. Preliminary Estimate of Costs in Producing Sweet Sorghum Under Farm Conditions in Louisiana Based on Available Research Data. Mimeo. 5 p. 1980.
3. R. Ricaud. Sweet Sorghum for Sugar Production in Louisiana. La. Agriculture, Winter Vol. 14, 1970-71, No. 2, p. 4-7. 1971.

4. R. Ricaud and A. Arceneaux. Sweet sorghum for sugar and biomass production in Louisiana. Project Reports of the Dept. of Agron., Center for Agriculture, L.S.U. 1982.

5. R. Ricaud and A. Arceneaux. Sweet sorghum for sugar and biomass production in Louisiana. Project Reports of the Dept. of Agronomy, Center for Agriculture, L.S.U. 1983.

6. Audubon Sugar Institute. 1983 Sweet Sorghum Studies. Mimeo. Material, Feb. 1984, 17 p. 1984.

7. S. Guilbeaux. Personal communication. Stephen Guilbeaux, Cooperative Farm Manager. Breaux Bridge Coop. Breaux Bridge, LA.

8. P. Angelle. Personal communication. Mr. Pedro Angelle, Manager, Breaux Bridge Sugar Coop. Breaux Bridge, LA.

MATERIAL AND ENERGY BALANCES FOR PROCESSING

HIGH FIBER SUGARCANE

Stephen J. Clarke

Audubon Sugar Institute
Louisiana State University
Baton Rouge, Louisiana 70803

Mike J. Giamalva

Louisiana Agricultural Experiment Station
Louisiana State University
Baton Rouge, Louisiana 70803

ABSTRACT

Sugarcane is a highly efficient converter of solar energy to biomass
and its agricultural and processing technology is well established. Recently developed high fiber sugarcane hybrids are an excellent biomass resource
with gross yields in excess of 180 Mg ha^{-1} yr^{-1} in Louisiana, more than
double the average with standard cane varieties. Optimum value from this
material requires utilization of both the soluble (fermentable) carbohydrates and the lignocellulosic fiber. Pilot and commercial scale data on
the processing characteristics and product yields of these hybrids are
presented. The high agricultural yields outweigh the relatively low content
of fermentable carbohydrates, resulting in higher potential fuel ethanol
production per hectare than with standard varieties. The very high residual
lignocellulosic (bagasse) content provides both fuel for steam for extraction and processing of the soluble carbohydrates and a significant surplus
for power generation, board or paper production, or conversion to fermentable sugars once the technology for this is established. The efficiency of
sugar and/or ethanol production is maximum when the high pressure steam
required to power extraction equipment matches the low pressure steam requirements for evaporation and distillation, etc. Detailed material, steam
and energy balances for several process alternatives for sugar and/or etha-

nol production are given, with emphasis being placed upon maximum savings of bagasse for other uses.

Keywords. Bagasse, energy, sugar, ethanol, high-fiber sugarcane.

INTRODUCTION

Current world sugar production is of the order of 100 million Mg, over 60% being from sugarcane. Over one-billion Mg of sugar containing plants are processed each year. The technology that has been developed over many years is applicable in many ways to the large scale processing of other plant materials for food, chemicals and/or energy. The purpose of this paper is to discuss the processing characteristics of sugarcane hybrids selected for their high biomass yields since similar considerations apply to the processing of any biomass crop. Sugarcane (Saccharum officinarum L.) and its relatives are considered to be prime candidates for biomass production. Major advantages of sugarcane for this purpose are its high yield and its ability to produce crops over a period of years without replanting.

Sugarcane has traditionally been grown for sugar. The bagasse (de-sugared resisual fiber) is used for fuel for the process and molasses (residual mother liquor from which no further sugar can be crystallized) is minimized. Large scale fuel alcohol production utilizes the soluble carbohydrates for fermentation and the fiber as process fuel. In both cases, the optimum level of fiber in the cane should be that required to provide sufficient fuel for process, having little or no surplus for by-product production or for generation of electric power for use external to the factory. Sugarcane varieties are usually selected on the basis of fairly low fiber levels and high sucrose content and juice purity. Varieties are now being selected for total sugars as feedstock for alcohol plants.[1] "Energy-Canes" are being studied for the production of high yields rather than for just sugar production.[2] However, it has been long recognized that high tonnage of sugarcane is usually associated with poor juice quality.[3] High fiber cane varieties have this characteristic and also have a major imbalance between the quantities of fiber to be handled at extraction and the soluble solids to be processed. For conventional cane varieties, the soluble solids to fiber ratio is about 1.3:1, but for high fiber varieties it may be as low as 0.4:1. The advantages and disadvantages of various processing schemes for this type of cane are discussed. The work described was supported by the USDA as part of the program for the study and development of renewable energy sources.

The process characteristics of the varieties have been determined on both a pilot scale facility and at the Breaux Bridge Sugar Coopertive. The data so obtained has been used to predict the overall processing performance using procedures established in the sugar industry. The extraction of the soluble carbohydrates under differing process conditions are predicted. Yields of products and energy requirements are given for the following alternatives; (i) maximum sugar production and final molasses, (ii) maximum sugar production and ethanol form molasses fermentation, (iii) partial sugar production (single crystallization from the high quality juice stream) and ethanol production from the remaining fermentable carbohydrates and (iv) complete fermentation to ethanol. The energy available for steam production from combustion of bagasse is calculated. A balance between the quantity of high pressure steam for the turbines used for extraction, boiler-feed pumps and power generation, and the quantity of low pressure steam for evaporation, juice heating, and crystallization is obtained by varying the high and low pressure steam conditions. Having obtained a balance, the high pressure steam and fuel requirement are determined, and the bagasse surplus established. The electric power that may be produced from this bagasse is calculated.

VARIETY SELECTION, YIELDS AND CHARACTERISTICS

Two high fiber cane varieties have been studied in detail and show the range of alternatives for processing and utilization. Other varieties have been studied and fit into the classes exemplified by these two varieties. CP65-357 is the standard commercial variety in Louisiana and is used for comparison with the high fiber hybrids.

L79-1002, the highest yielding variety with the highest fiber content, was selected from a cross between CP52-68 and Tianan 96, a wild saccharum species collected in Argentina. It is a very tall, thin and erect cane and shows the best potential for biomass production. L79-1003, with a moderately high fiber content, was selected from a cross between NCo-310 and US-61-21, a back cross from the USDA breeding program at Houma, Louisiana. It is tall and has a barrel diameter similar to conventional cane, but has a tendency to lodge. Yield and composition data averaged for five years for these varieties and CP65-357 are given in Table 1. The very high yield of L79-1002 is noteworthy; the yields have increased each year over five ratoons, and the variety shows resistance to RSD.

HARVESTING

Harvesting of the high fiber hybrids, especially the very tall L791002, remains a major problem. All cane in Louisiana is planted on high 1.8 m rows and is cut with a soldier harvester. The cut cane is laid across two rows so that the cane can be at most 2.5 m long. Up to four rows are piled together and the cane burned before being mechanically loaded for transport to the factory. This operation does not work well with the high fiber canes due to (i) the higher than normal levels of trash around the cane that causes some choking of the harvester, (ii) the loss of too much millable stalk (if longer than 2.5 m is cut, it gets crushed by the next passage of the tractor), (iii) the excessive quantities of tops, from both low topping and high yield, cause choking of the harvester and (iv) the large number of very tough stalks (more than 70 min^{-1}) that slows operation. A cut-chop harvester handled the cane easily (especially after it had been burned standing, but too much of the thin light stalk was lost.

PROCESSING

Pilot and Full Scale Data

Although complete sugar and ethanol processing has not been conducted with these hybrids, each critical step of the process has been carried out without much difficulty. Extraction of the juice has been studied using a sample mill, a screw press, the Audubon Sugar Institute (ASI) mill (20"x24" two-roll crusher plus three 18.5"x24" three-roll mills), and mills at the Breaux Bridge Sugar Cooperative (three 48" mills followed by three 60" mills). Preparation (preliminary disintegration of the cane before crushing) of the high fiber varieties using a single pass through the pilot scale shredder at ASI showed them to be at least as easily prepared as CP65-357.

Passing shredded cane through the screw press gave very low moisture (30%) and the bagasse was very well prepared for by-product use, but the power consumption was extremely high. Two passes of shredded cane through the ASI Farrel sample mill without imbibition (the addition of water in the mill to improve extraction) gave bagasse moistures of 48-50%. Dry milling of L79-1002 on the ASI tandem gave bagasse moisture below 50% after only two mills. Above 50% moisture, bagasse cannot be burned efficiently. These tests were conducted since one possible processing alternative is to reduce the moisture level to below 50% with minimum juice production to maximize the bagasse surplus. Results of regular milling tests at ASI and Breaux Bridge are given in Table 2. Breaux Bridge, with a nominal grinding rate of 125 Mg hr^{-1} for conventional cane, ground at a rate of only 40 Mg hr^{-1} with L79-1002. The bagasse from the mill at Breaux Bridge burned in the boilers without any problem.

No problems were encountered in the clarification (purification) of the juice, except for the unusually high quantity of lime required to neutralize the juice from L79-1002. Evaporation and crystallization presented no problems, and high crystal content massecuites were readily produced. No low grade boilings (repeated crystallization to recover the maximum sugar) have been conducted, partly due to lack of material and also because this is probably impractical with these varieties. Juice and diluted syrup from the high fiber varieties were fermented with yeast to give the ethanol yields predicted from the sugar content.

Material and Energy Balances

Eight process cases are considered, two for CP65-357, four for L79-1002 and two for L79-1003. A grinding rate of 100 Mg hr^{-1} is used in each case. The extraction achieved is based on the computer program developed at ASI to calculate the performance of varying mill schemes.[4] Standard values are taken for mill power consumption, exhaust steam requirements, bagasse calorific values and boiler efficiencies. These and the assumptions used in the calculations are listed in the Appendix. In some cases the production of only a single strike of sugar from the clarifier overflow and combination of the A-molasses with the filtrate for fermentation is a viable alternative.

Table 3 gives the data for various milling conditions. Case 1, for CP65-357 with five moderate pressure mills, is fairly standard practice in the sugar industry. Increasing the pressure on the mills increases not only the extraction but also the power requirement and, therefore, both the operational and capital costs of the milling system. The throughput of a milling tandem is determined to a major extent by the fiber content of the cane and is best expressed in terms of fiber throughput rather than cane throughput. Not included in the table and not readily predicted, but indicated from Table 2, are the relative grinding rates for the conventional and high fiber varieties. At the Breaux Bridge mill, the grinding rate for L79-1002 was less than one third that for conventional cane. Although the product yields for high fiber cane (Tables 4 and 6) are roughly half that for conventional cane, the yields per hectare (Tables 5 and 7) are better

Table 1. Yield and composition data for a standard and high fiber cane variety

Yield/composition		Commercial CP65-357	High fiber	
			L79-1002	L79-1003
Yield (Mg ha^{-1})	Gross cane (1)	58	212	94
	Net cane (b)	50	169	83
Composition (net cane) (c)	%Fiber	13.4	28.0	17.5
	%Solubles	16.2	11.1	11.8
	%Sucrose	14.0	8.4	8.9
	%Glucose	0.6	1.2	1.5
Average gross stalk wt (kg)		1.17	0.59	1.20
Average net stalk length (m)		2.2	3.3	2.6

[a] Gross cane is the total above ground material.
[b] Net cane is the remaining cane after removal of leafy trash and tops.
[c] The analytical data was obtained by standard techniques, involving disintegration of the material, pressing to extract the juice and analysis of the juice and the direct press cake. All monosaccharides are reported as glucose.

Table 2. Milling (extraction) tests of a standard and high fiber cane varieties

Test Measures	Commercial CP65-357	High Fiber	
		L79-1002	L79-1003
ASI Mill			
Extraction (% total sugar)	92.3	88.4	90.2
Bagasse Moisture (%)	47.6	48.1	49.2
Recoverable Sugar (a)	108.0	56.0	78.0
Grinding Rate (kg sec^{-1})	4.2	2.5	3.6
Breaux Bridge			
Extraction (% total sugar)	91.2	85.3	–
Bagasse Moisture (%)	50.7	57.1	–
Recoverable Sugar (a)	95.0	75(b)	–
Grinding Rate (kg sec^{-1})	34.7	11.1	–

[a] kg Mg^{-1} net cane, based on standard calculations.
[b] High since the cane was cue to standard length and the juice purity is higher at the bottom of the stalk.

Table 3. Milling (extraction) variables for standard and high fiber cane varieties under different milling conditions

Case	Variety	Milling Conditions		Extraction[c]	Power Required[d]
		Number & type of mills[a]	Imbibition[b] %	%	MW
1	CP65-357	5 C	150	91.6	2.87
2	CP65-357	5 H	200	94.6	3.12
3	L79-1002	5 H	200	89.6	5.30
4	L79-1002	3 H	200	75.9	4.25
5	L79-1002	3 H	0	57.9	4.25
6	L79-1002	Diff[e]	300	96.0	2.69
7	L79-1003	5 H	200	93.0	3.73
8	L79-1003	3 H	0	77.0	3.08

[a]The number refers to the number of units in the milling tandem and the letter to either Conventional or High power mills - see Appendix.
[b]The quantity of imbibition water added to the mill expressed as a percentage of the fiber in the cane being processed.
[c]Extraction is expressed as the percentage of the soluble solids extracted in the milling tandem.
[d]The power required by all the prime movers in the system, primarily the mills but also generators and boiler feed pumps, etc. - see Appendix.
[e]Diffusers are an alternative extraction technique, not in use in the mainland U.S.A., requiring less power for operation but more imbibition water.

for the high fiber cane. The productivity of a sugar mill, expressed in terms of quantity of product per hour, would be very much lower for high fiber cane in a mill of the same capacity. Other cases involve less power by using a smaller number of mills or replacing the mills with a diffuser. Decreasing the number of mills lowers the extraction, especially if no imbibition water is used (case 4). No imbibition and only three mills could be used if the purpose was to maximize the bagasse surplus for power generation. The bagasse surpluses and the equivalent electric power for the alternative cases are given in Table 8 on the basis of cane processed and in Table 9 on the basis of area harvested. However, the loss of sugar and/or ethanol product due to poor extraction is more significant than the addi-

Table 4. Sugar and molasses production (Mg 100Mg^{-1} cane) from standard and high fiber cane varieties

Case	Variety	Complete Sugar Production	Molasses Production	Partial Sugar Production
1	CP65-357	11.98	3.52	7.75
2	CP65-357	12.41	3.64	7.81
3	L79-1002	6.60	4.20	5.08
4	L79-1002	5.82	3.70	4.48
5	L79-1002	4.25	2.70	3.27
6	L79-1002	7.07	4.50	5.44
7	L79-1003	7.12	4.76	5.57
8	L79-1003	5.91	3.95	4.63

Table 5. Sugar and molasses production (Mg ha^{-1}) from standard high fiber cane varieties

Case	Variety	Complete Sugar Production	Molasses Production	Partial Sugar Production
1	CP65-357	5.99	1.76	3.78
2	CP65-357	6.21	1.82	3.91
3	L79-1002	11.15	7.10	8.59
4	L79-1002	9.84	6.25	7.57
5	L79-1002	7.18	4.56	5.53
6	L79-1002	11.95	7.61	9.19
7	L79-1003	5.91	3.95	4.62
8	L79-1003	4.91	3.28	3.84

Table 6. Ethanol production (Mg 100Mg^{-1} cane) from commercial and high fiber cane varieties

Case	Variety	Complete Ethanol Production	Partial Ethanol Production	Ethanol Production From Molasses
1	CP65–357	6.78	2.89	0.66
2	CP65–357	7.02	2.99	0.68
3	L79–1002	4.37	1.76	1.00
4	L79–1002	3.86	1.55	0.88
5	L79–1002	2.81	1.13	0.64
6	L79–1002	4.68	1.88	1.07
7	L79–1003	4.87	2.00	1.23
8	L79–1003	4.04	1.66	1.02

Table 7. Ethanol production (Mg ha^{-1}) from commercial and high fiber cane varieties

Case	Variety	Complete Ethanol Production	Partial Ethanol Production	Ethanol Production from Molasses
1	CP65–357	3.39	1.45	0.33
2	Cp65–357	3.51	1.50	0.34
3	L79–1002	7.39	2.97	1.69
4	L79–1002	6.52	2.62	1.49
5	L79–1002	4.75	1.91	1.08
6	L79–1002	7.91	3.18	1.81
7	L79–1003	4.04	1.66	1.02
8	L79–1003	3.35	1.38	0.85

tional electric power produced (Table 11). The need for steam efficiency to maximize bagasse surplus requires a balance between high and low pressure steam (Table 10). High pressure steam requirements can be minimized by increasing the pressure; low pressure requirements can be minimized by increasing the pressure of exhaust steam used for evaporation and by using efficient evaporation schemes (bled quadruple effects or quintuple effects) rather than relatively inefficient evaporators (simple triple effects). However, the large quantities of exhaust produced from a fairly low pressure turbine system may justify the use of triple effect evaporators. The use of higher than conventional pressures for a sugar mill, along with different milling schemes would probably require a mill of significantly different design.

Table 8. Bagasse surplus and power generation ($100Mg^{-1}$ cane) from commercial and high fiber cane varieties

Case	Variety	Bagasse Surplus		MWh Equivalent to Surplus Bagasse
		Mg	% on Bagasse	
1A	CP65-357	3.2	10.9	1.2
1B		13.2	44.4	5.1
2A	CP65-357	10.6	38.0	4.0
2B		10.6	38.0	4.0
3A	L79-1002	33.3	57.1	12.7
3B		33.3	57.1	12.7
4A	L79-1002	29.0	49.8	11.1
4B		39.3	67.5	15.0
5A	L79-1002	42.8	73.4	16.3
5B		45.9	78.7	17.5
6A	L79-1002	30.2	53.1	11.5
6B		33.7	59.3	12.9
7	L79-1003	16.4	45.0	6.3
8A	L79-1003	16.3	44.5	6.2
8B		25.3	69.3	9.7

Table 9. Bagasse surplus and power generation (ha^{-1}) from commercial and high fiber cane varieties

Case	Variety	Bagasse Surplus	MWh Equivalent to Surplus Bagasse
1A	CP65-357	1.6	0.6
1B		6.6	2.6
2A	CP65-357	5.3	2.0
2B		5.3	2.0
3A	L79-1002	56.3	21.5
3B		56.3	21.5
4A	L79-1002	49.0	18.8
4B		66.4	25.4
5A	L79-1002	72.3	27.5
5B		77.6	29.6
6A	L79-1002	51.0	19.4
6B		60.0	21.8
7	L79-1003	13.6	5.2
8A	L79-1003	13.5	5.1
8B		21.0	8.1

Electric power generation is based on production of steam at 5500 kPa using an efficient condensing turbo-alternator.[5] This electric power would be for external use since all power required for factory operation for sugar and/or ethanol production is included in the processing energy requirements. This is technically quite feasible, but the value of the electric power (Table 11) in the best alternative (case 5) is very low compared with that for sugar production, even with poor extraction. These factors suggest that sugar and/or ethanol are the optimum products from this type of cane and, thus, maximum extraction is indicated (cases 3 or 6). Neither of these alternatives, with high pressure mills or a diffuser, are in current operation and both would require a custom designed mill. It would be difficult to justify this for a crop as seasonal as sugarcane, at least in the United States. Storage of surplus bagasse for power generation would be necessary for this type of cane but is not a realistic proposition on a large scale.

Table 10. Steam conditions to extract commercial and high fiber cane
varieties

Case	Variety	Processing Conditions[a]	Absolute Steam Pressure (kPa)[b]		High/Low Ratio[c]
			High	Low	
1A	CP65–357	T P D	686	152	1.00
1B		Q C	1961	234	1.10
2A	CP65–357	Q P D	1961	234	1.00
2B		Q C	1961	234	1.12
3A	L79–1002	Q C	1961	152	1.07
3B		Q P D	1961	152	1.05
4A	L79–1002	T C	981	152	1.08
4B		T C	2500	234	1.02
5A	L79–1002	T C	2500	152	1.05
5B		T P D	5500	179	1.01
6A	L79–1002	T P D	686	179	1.02
6B		Q C	981	207	1.21
7	L79–1003	T P D	1961	207	1.00
8A	L79–1003	T C	686	152	1.03
8B		Q P D	3792	234	1.01

[a]Triple effect evaporation with no vapor bleeding;
Quadruple effect evaporator, doubly bled;
Partial sugar production with Distillery;
Complete sugar production.

[b]The low pressure steam was saturated while the high pressure steam is
superheated.

[c]The ratio of the quantity of high pressure steam required for the prime
movers to the quantity of low pressure steam required for evaporation – a
slight excess of low pressure steam is desirable.

CONCLUSIONS

Sugarcane has been shown to have an overall positive energy balance for
the production of ethanol.[6,7] High fiber sugarcanes are even more energy
productive. The principal advantages of these high fiber hybrids are their
extremely high yields. Significant problems arise when determining how such
materials can be processed to maximize the value of the various products of
processing. The high fiber and low total sugar contents drastically reduce
the productivity of the operation compared with processing regular sugarcane. The same milling operation would produce only about a quarter of

Table 11. Values of alternative products from commercial and high fiber cane varieties ($ per Mg cane/$ per hectare)

		Variety and Case		
Produce	CP65-357	----------------L79-1002----------------		
	1	3	5	6
Complete Sugar	53.9/2690	29.7/5430	19.1/3490	31.8/5820
Molasses	1.7/90	2.1/380	1.3/240	2.2/400
Total	55.6/2780	31.8/5820	20.4/3730	34.0/6220
Partial Sugar	34.9/1740	22.9/4190	14.7/2690	24.5/4480
Partial Ethanol	9.2/460	5.6/1020	3.6/660	6.0/1100
Total	45.1/2250	28.5/5220	18.3/3350	30.5/5580
Ethanol Complete	21.7/1080	14.0/2560	9.0/1650	15.0/2740
Electric Power	1.5/70	3.8/690	5.2/950	3.9/710

No value given to bagasse alone or to carbon dioxide and residual stillage from fermentation.

Based on – Sugar $450 Mg^{-1}; Molasses $50 Mg^{-1}; Ethanol $320 Mg^{-1}; Electric Power $0.03 kWh^{-1}.

the sugar and/or alcohol with the high fiber hybrid, L79-1002, as compared with conventional cane. These hybrids could become valuable only if an alternative and more valuable product than electric power could be produced from the bagasse, such as paper or board, or if an efficient lignocellulose hydrolysis scheme is developed. Also, it is doubtful whether this type of cane could be produced in the field for significantly less than the sugar and/or ethanol product value, leaving very little margin for the capital and operating costs for a processing plant designed for this type of cane. Similar problems to those encountered in trying to establish a viable processing scheme for these sugarcane hybrids will be found for any biomass crop for which sophisticated processing of this type is required.

ACKNOWLEDGMENTS

Assumptions used in the Calculations

1. Final fiber.% bagasse of 44% with conventional mill and 48% with high pressure mill.

2. 162 MJ Mg^{-1} fiber for cane preparation.

3. 63 MJ Mg^{-1} fiber for each conventional mill; 77 MJ Mg^{-1} fiber for each high pressure mill.

4. 85 MJ Mg^{-1} fiber for the diffuser.

5. Accessories (e.g. power generation for factory use) 44 MJ Mg^{-1} cane.

6. Boiler fans and feed pumps equivalent to one mill.

7. Turbine efficiencies from 60% at low pressure to 75% at high pressure.

8. Spreader-Stoker boiler efficiency based on GCV of 67.1%.

9. GCV calculations from bagasse composition based on Hersey's equation.

10. Efficiency of fermentation and distillation - 95%.

11. Losses in filter cake - 0.5% on juice solids.

12. Massecuite Brix for partial sugar - 93.0. (At a 0.60 fractional crystal content on solids, gives a molasses Brix of 84.2).

13. 85% juice overflow to clarifiers, 15% to filters.

14. Evaporation to 60 Brix syrup.

14. Distillery run on exhaust (6).

REFERENCES

1. R. P. Humbert. Growing of sugarcane for energy. Sugar Journal. 43:19-22 June. 1980.

2. G. Samuels. Production of energy cane in Puerto Rico. 3:14-17. J. Amer. Soc. Sugar Cane Tech. 1982.

3. U. K. Dass. A new approach to the problem of juice quality. p. 43-99. Proc. 15 Ann. Meeting Assoc. Hawaiian Sugar Tech. 1936.

4. H. S. Birkett et al. Factors affecting mill extraction. Vol. 5. J. Amer. Soc. Sugar Cane Tech. 1985.

5. J. M. Paturau. By-Products of the Cane Sugar Industry. Elsevier, New York. p. 58-62. 1982.

6. J. A. Polack et al. Sugarcane: positive energy source for alcohol. 77(6):62-65. Chem. Eng. Prog. 1981.

7. B. Mouris. Economics and energy balances of ethanol from sugarcane and sugarbeet. Int. Sugar Journal. 86:282-286. 1984.

PRACTICAL ASPECTS OF PRODUCTION OF FUEL

ETHANOL IN FLORIDA

J. E. Murtagh

Murtagh and Associates
2304 Wilson Boulevard
Winchester, Virginia 22601

R. A. Buening and M. J. Gaylor

Bio-Chemical Energy, Inc.
2101 U. S. 19 North, Suite 301
Palm Harbor, Florida 33563

ABSTRACT

The use of ethanol as a gasoline extender and octane enhancer has
increased very rapidly in the past decade, due mainly to the introduction of
exemptions of part of the federal and state gasoline taxes for gasoline-
ethanol blends.

Florida's first fuel ethanol plant was established near Brooksville by
Bio-Chemical Energy in 1982. It started using corn as the raw material, but
currently uses cane molasses, and its production capacity has increased from
1.8 to 7.6 million liters (500,000 to 2-million gallons) per year.

The state's second ethanol plant was commissioned by Bio-Chemical
Energy in 1984, at Bartow, with a capacity of 15.1 million liters (4-million
gallons) per year, using cane molasses and citrus molasses as raw materials.
The advantages of different feedstocks and methods of stillage disposal are
presented, together with a review of the processes involved.

The company developed a system of very rapid constructon of fuelethanol
plants, and completed the first plant in 4 months and the second plant in 5
months. The rapid, planned construction has enabled the company to take
advantage of appropriate situations as they arise.

The plants are very energy-efficient, using wood chips as the fuel source and liquid-phase molecular sieves for ethanol dehydration.

Keywords. Ethanol, fuel-ethanol, molasses, ferementation, distillation, Florida.

INTRODUCTION

During the past decade, America has realized the need to reduce dependency on imported oil fuels, which currently cost approximately $10 million every hour. In view of limited US resources of oil and other fossil fuels, alternate renewable sources of energy were sought to meet the Nation's growing requirements. One of the most practical results of this search has been the development of the fuel ethanol industry.

Ethanol has been produced intermittently for automotive fuel purposes in this country since the 1920's, but has only seen a significant increase in production capacity and public acceptance since 1978. Currently, there are approximately 75 fuel ethanol distilleries in this country, with new plant construction being planned in many states. In 1982, approximately 946 million liters (250 million gallons) of fuel ethanol were sold. That figure rose to more than 1.9 billion liters (500 million gallons) in 1984.

Two significant factors which influence the growth in sales are the use of ethanol as an octane-enhancer (to improve the combustion of gasoline in automobile engines), and the Environmental Protection Agency's phase-out of the use of lead-containing additives as octane enhancers. Sales of ethanol-enriched, unleaded gasoline in Florida have risen from about 3.8 million liters (1 million gallons) per month in 1980 to over 19.2 million liters (40 million gallons) per month by late 1984 (but still only amount to approximately 10% of total gasoline sales). The current mandate of the EPA to phase-out the use of lead in gasoline was expected to increase the demand for ethanol (or other octane-enhancers) by the equivalent of about 6 billion liters (1.6 billion gallons) in this country in 1984, while recent EPA proposals to accelerate the phase-down, to virtually eliminate lead from gasoline by 1986, could affect the situation even more substantially. The need for fuel ethanol is, therefore, very apparent.

COMPOSITION OF FUEL-GRADE ETHANOL

Fuel-grade ethanol is required to be sufficiently water-free that there
will be no phase-separation when it is blended with gasoline. The phenome-
non of phase-separation, in which an aqueous layer separates out below the
gasoline (and which can cause ignition problems and lead to frozen fuel
lines), is largely dependent on the water content of the ethanol and the
temperature of the mixture. Most states do not specify any particular
degree of concentration for fuel ethanol, but in Florida, it must contain
less than 1% by volume of water (i.e. 99% by volume ethanol, or "198 degrees
proof).

Ethanol is not only a gasoline extender, but also acts as an octane
enhancer, to increase the value of gasoline in powering normal automobile
engines without the requirement of any engine modification. Although the
percentage of ethanol used in gasoline may vary, the accepted volume proportion for use in the US is 10% ethanol to 90% gasoline. Ethanol has an
effective octane rating of about 120, so that when added to unleaded gasoline at 80-87 octane, at the 10% rate, it would boost the octane rating of
the blend by about 3.5.

GOVERNMENT INCENTIVES FOR FUEL-ETHANOL PRODUCTION AND USAGE

In recognition of the fact that ethanol is a renewable fuel (when
produced from renewable natural resources), and that it is desirable to
encourage its production and use as a gasoline extender and enhancer, the
federal government grants an exemption of 1.6 cents per liter (6 cents per
gallon) of the (9 cents per gallon) excise tax, if the gasoline contains a
minimum of 10% ethanol. The governments of over 30 states also grant similar additional exemptions from their gasoline sales taxes for gasoline
containing 10% ethanol. The exemptions vary from state to state, but in
Florida, where the exemption is near the nationwide average, it currently
amounts to 1.1 cents per liter (4 cents per gallon) of fuel. Thus, these
exemptions add up to 2.64 cents per liter (10 cents per gallon) of fuel
containing 10% ethanol, or 26.4 cents per liter ($1.00 per gallon) of etha-
nol added to the gasoline. These tax exemptions enable ethanol to be used
profitably in blends with gasoline despite the fact that ethanol sells at a
price of about 42 cents per liter ($1.60 per gallon), while the wholesale
price of gasoline is about 21 cents per liter (80 cents per gallon), because

the dealer adding the ethanol to the gasoline is, in effect, receiving a subsidy of 26.4 cents per liter ($1.00 per gallon) of ethanol used.

At present, the federal excise tax exemption on alcohol-blended fuels, which was first passed in 1979, is scheduled to remain in force at its current level of 1.6 cents per liter (6 cents per gallon), until 1992. The Florida state tax exemption, which came into being in 1981, is currently 1.1 cents per liter (4 cents per gallon), but it is scheduled to be reduced after July 1, 1985, to 0.53 cents per liter (2 cents per gallon), and to be eliminated completely after July 1, 1989.

BIO-CHEMICAL ENERGY'S EXPERIENCE IN FUEL ETHANOL PRODUCTION

Bio-Chemical Energy, Inc., was established in Florida in 1980, with the objective of constructing and operating a fuel ethanol production plant. At that time, there was considerable interest nationwide in ethanol production and a wide variety of equipment manufacturers and suppliers were attracted to the potential business. Unfortunately, many of these companies represented their equipment as performing specific functions when, in fact, in many instances they were total failures or very inadequate. At that same time, some of the larger, more reputable, established beverage ethanol, equipment manufacturing companies evaluated the fuel ethanol business and decided not to enter it at that time. This was due, in large part, to the low level of sophistication of many of the groups involved in some of the first plants and their lack of knowledge of the processes and business in which they were becoming involved.

Under those circumstances, the selection of satisfactory equipment was very difficult, but after considerable research into available technology, Bio-Chemical Energy eventually contracted to purchase, from Europe, the process equipment which would be used for a 1.8 million liters (500,000 gallons) per year production facility. Constructon of the plant at a site near Brooksville was commenced in September, 1981, and it was completed by the end of December of the same year.

Early in 1982, the plant was at full production, and investigation commenced soon into the possibility of expanding the plant's capacity. The limiting factor in the plant was found to be the ethanol dehydration stage. It should be explained that when ethanol is concentrated from aqueous solutions by distillation, a constant-boiling mixture, or "azeotrope," is obtained at about 97% by volume, making it impossible to concentrate any further by the normal distillation process. (In actual commercial practice, however, it requires an uneconomically large number of distillation stages and an uneconomically high usage of steam to distill ethanol beyond about 95% by volume, so that this lower concentration must be considered the practical economic limit). As these concentrations of ethanol would give a water phase-separation when mixed with gasoline, in a 10% ethanol and 90% gasoline blend, it is necessary to resort to special techniques to break the "azeotrope" barrier, to dehydrate the ethanol to a satisfactory level.

The drying of ethanol from 95% to the 99% concentration required by Florida law, may be achieved by various means, including azeotropic distillation using added substances such as benzene, cyclohexane, diethyl ether, or gasoline to remove the water. Alternatively, extractive distillation processes may be used with glycerine, ethylene glycol or salts as the water extractants. Each of these systems has its advantages and, more noticably, its disadvantages. The cost of construction of the facilities required is very substantial and the energy input is generally rather high, in the range of 2.2 - 2.5 MJ per liter (8,000 - 9,000 BTU's per gallon) of ethanol output.

The Brooksville plant originally used a cyclohexane system for ethanol dehydration, but after some further considerable research, a liquid-phase molecular sieve was purchased for this purpose. The design of the sieve had been modified by Bio-Chemical Energy and, when installed in December, 1982, it allowed the plant capacity to increase to 2.6 million liters (1 million gallons) per year.

The molecular sieve can be likened to a water softener in its relative simplicity. It consists, in part, of a large cylinder packed with small beads of potassium alumino-silicate. The molecular structure of the material is such that it contains "pores" of a 3-Angstrom Unit size, which is just large enough to permit the entry of water molecules, but excludes the larger ethanol molecules. Ethanol is simply passed through the sieve material to remove the water. When the sieve bed is fully loaded with water, the material can be regenerated by displacing the water with an inert gas such as nitrogen or carbon dioxide.

It is necessary to heat the gas used for sieve material regeneration, but the energy requirement amounts to a mere 0.5 MJ per liter (1,800 BTU's per gallon) of final product. The sieve system also has the advantage of requiring much less-complex control instrumentation than is necessary for azeotropic or extractive distillation.

The capacity of the Brooksville plant was further expanded to 7.6 million liters (2-million gallons) per year in 1983, by the installation of an evaporator-stripper which serves the double function of stripping ethanol from the fermented beer, and then concentrating the alcohol-free residue.

Corn is used as feedstock in most of the fuel ethanol plants in this country, as was the original feedstock for the Brooksville plant. Basically, the corn mashing and fermentation process consisted of:

(a) Weighing the corn into the plant.
(b) Feeding the corn to a hammermill, to pulverize it to pass through a fine mesh screen of selected size for optimum fermentation efficiency.
(c) Passing the ground cornmeal into a premixing tank where hot water and alpha-amylase enzymes were added, to liquefy the free starch granules.
(d) Passing the premix into a continuous cooker, where it was boiled under pressure for about 15 minutes, to liberate and gelatinize as much starch as possible.
(e) Passing the cooked mash through a series of flash tanks to cool the mash by vacuum. (The heat was recovered for heating the incoming uncooked mash.)
(f) Adding alpha-amylase and amyloglucosidase enzymes to the cooked mash, to liquefy and saccharify the starch.
(g) Pumping the mash to fermenters, where yeast was added and the temperature controlled in a suitable range for fermentation.
(h) After about 60-70 hours, the fermented mash, containing approximately 9% ethanol, was pumped to a beer stripping column for removal of the ethanol.

In the Fall of 1982, it was decided to switch from corn to molasses, due to the fluctuating prices of corn and the ready availability of molasses in Florida at that time. In making this change, the cooking section was no longer necessary, thereby saving a considerable energy input of 0.67 - 1.94 MJ per liter (5,000 - 7,000 BTU's per gallon) of ethanol produced.

In using molasses, the process consists of:

(a) Receiving the molasses in tank trucks and transferring it into storage tanks.

626

(b) Diluting the molasses with water in fermentation vessels to a suitably low concentration to allow a sufficiently rapid start to fermentation, to minimize the effects of bacterial contamination, and yet to a high enough concentration to give a reasonably high final ethanol content.

(c) Adding nutrients and a propagated yeast culture to the vessel and allowing the fermentation to proceed for 36-48 hours.

(d) On completion of fermentation, the "beer," with an ethanol content of 8-10% v/v, is pumped to the beer stripper for recovery of the ethanol.

(e) The molasses stillage by-product is evaporated and may either be sold as a syrup for blending into animal feeds, or it may be dried on suitable carriers to produce a feed ingredient.

Bio-Chemical Energy commenced the construction of their second fuel ethanol plant at Bartow in August, 1983. It was commissioned in February, 1984, with a capacity of 7.8 million liters (3 million gallons) per year, using cane molasses and citrus molasses as the feedstock. The company has developed a system for the very rapid construction of fuel ethanol plants, which involves carrying out their own project management, working very closely with a reliable contractor, and maintaining a good relationship with equipment suppliers.

The company is currently planning the construction of its third and fourth fuel ethanol plants which will be in the vicinity of Lake Okeechobee, in southern Florida. These will use cane molasses from mills to the south of Lake Okeechobee. Each plant will have a capacity of 15.2 million liters (4 million gallons) per year of ethanol, and they are scheduled for completion before the end of 1985.

The plants use wood chips for firing the boilers for steam generation. At the first plant near Brooksville, chips were used from the thinnings of pine grown in adjacent plantations. The Bartow plant uses a combination of whole-tree chips and industrial wastewood chips, mainly coming from scrapped pallets at a nearby timber yard. The two south Florida plants will use whole-tree chips which will be partly eucalyptus and partly melaleuca. The latter is an exotic species which has become a nuisance weed spreading around Lake Okeechobee, and the state is currently engaging contractors to try to control it by felling.

With the two additional production plants, the capacity will require almost all of the previously uncommitted cane molasses produced in Florida, more than 40% of the total annual production. This, coupled with the fact that the Florida legislature passed a law in 1984 requiring that fuel ethanol be made from US produced feedstocks, means that the company would not be able to build any more cane molasses ethanol plants in Florida, but will have to consider other feedstocks and probably other states. The company is advocating research into cellulose hydrolysis as a source of feedstock, and expects to be a leader in the implementation of cellulose-ethanol processes in the future.

ABSTRACTS

GEORGIA'S PROGRAM IN BIOLOGICAL RESOURCES AND BIOTECHNOLOGY

Nathan Dean
Office of the Vice-President for Research
University of Georgia
Athens, Georgia 30602

In May 1984, The University of Georgia formally established a Program
in Biological Resources and Biotechnology. The goal of the Program is to
bring together researchers from different units across the University and to
facilitate the development of new collaborations. The focus of the Program
is in the development of biological resources, including biomass, as alternative sources of both feedstock chemicals and energy; and the development
of new technologies based in the biological sciences, especially in areas
which may have an impact on biological resources. The structure, operation,
and plans of the Program will be described.

Keywords. Biomass, energy chemicals.

ALABAMA BIOMASS INVENTORY AND RESOURCE ASSESSMENT

William J. Herz
Energy Division
Alabama Department of Economic & Community Affairs
Montgomery, Alabama 36105

Preliminary data indicate that Alabama's 8.8 million ha (21.7 million
acres) of commercial forest land contain an estimated 1.407 billion Mg
(1.551 billion green tons) of above-ground biomass, an average of 160.4 Mg
ha^{-1} (71.6 tons per acre), according to USDA Forest Service data for 1982.
40% was softwood species trees and 60% hardwoods. 51% of the total was
merchantable wood, with potential use in the production of pulp, paper,
lumber, and other forest products. The remaining 49% consists of branches
and tops of merchantable trees, saplings, rough, rotten, and dead trees, and
non-commercial trees. Of these components, it is estimated that 23.7
million Mg (26.1 million green tons) per year are available for energy use.
Mill residues, produced as a by-product of the forest products industry,

totaled 8.3 million Mg (9.2 million tons), of which 4.1 million Mg (4.5 million tons) were available for energy use, giving a total of 27.8 million Mg (30.6 million tons) of forest-derived biomass available for energy production.

54,000 farms occupied 4.7 million ha (11.6 million acres) in Alabama in 1983. Five-year averages of major crops indicate an estimated annual production of 2.90 million Mg (3.2 million tons) of crop residue, 0.09 million Mg (0.1 million tons) of crop wastes, and 2.45 million Mg (2.7 million tons) of animal wastes, for a total of 5.44 million Mg (6.0 million tons) per year of agricultural biomass. Because of soil erosion potentials, only 8% or 0.18 million Mg (0.2 million tons) of crop residues are available for energy use, and only 0.72 million Mg (0.8 million tons) of animal wastes are available from confined animals. Total available agricultural biomass for energy use is therefore 1.09 million Mg (1.2 million tons) per year.

Municipal solid waste (MSW) collected daily from households, industries, offices, and institutions in Alabama totaled an estimated 11,195 Mg (12,340 tons) per day in 1984. Based on a five-day week, a total of 2.9 million Mg (3.2 million tons) of MSW is put into sanitary landfills each year, all available for energy use.

The sum of these three biomass-producing sectors is 31.8 million Mg (35.0 million tons) per year. Forest biomass contributes 88%, MSW 9%, and agriculture 3%.

Actual utilization of forest biomass is estimated at 4.4 million Mg (4.9 million tons), 16% of the total available anually. Agricultural biomass use is estimated at 0.045 million Mg (0.05 million tons), all crop wastes, for a utilization rate of 4%. MSW usage for energy in the state totals 318 Mg (350 tons) per day for a 3% utilization rate. Together, the three sectors are estimated to provide 6 million Mg (6.6 million tons) of biomass for energy production, almost all converted via direct combustion. At accepted energy rates for each biomass sector, the total contribution from biomass sources in Alabama is estimated at 56,935 x 10^9 MJ (0.054 Quads), about 3.8% of the reported energy consumption in 1982. Estimates for future use in this decade range up to 6% of the 1982 consumption rate.

Keywords. Biomass, resource assessment, inventory Alabama.

BIOMASS ENERGY FROM RESIDENTIAL WASTE

Thomas J. Laughlin and Daniel R. Coleman
Southern Research Institute
P. O. Box 54305
Birmingham, Alabama 35255

The energy potential from the biomass in residential waste is tremendous—enough to supply about 2% of the United States' energy needs.

However, technologies have not been developed adequately to harvest this energy. Instead, large amounts of land, energy, and money are consumed to bury it in landfills.

The only residential waste-to-energy technology being used commercially is mass burning, and it is not cost-effective. It has been used more for its ability to reduce landfill requirements that for its energy efficiency. Lab and pilot-scale processes have been developed to demonstrate the feasibility of producing methane, ethanol, and other useful chemicals from residential wastes, but large barriers remain to be overcome before these processes are ready for commercialization. The need for an effective residential waste-to-energy process is great, but one or more technological breakthroughs will be required to meet this need.

In this paper, we will review the lab-, pilot-, and commercial-scale processes available for the conversion of residential waste to energy. We will compare the extent of their development, and discuss the major problems yet to be solved.

Keywords. Waste-to-energy, conversion options, landfills, biomass, residential wastes.

A BIOMASS RESOURCE ASSESSMENT FOR FLORIDA

Harold M. Draper
Governor's Energy Office
301 Bryant Building
Tallahassee, Florida 32301

Chris NeSmith
Department of Botany
University of Florida/IFAS
Gainesville, Florida 32611

With the support of the Tennessee Valley Authority, Southeastern Regional Biomass Energy Program, the Florida Governor's Energy Office and the University of Florida, Institute of Food and Agricultural Sciences, are conducting a comprehensive biomass energy resource assessment for Florida. Data on the potential for solid, liquid and gaseous fuel production are being gathered for each of the 67 counties of the state. The rationale behind the biomass resource assessment is to estimate the maximum amount that biomass can contribute to Florida's energy picture under various land use and conversion process scenarios. In addition to inventorying forest energy wood and agricultural residues, the project is projecting the potential for energy crops to supplement the wastes and residues. Steps in this biomass resource assessment include: (1) grouping counties into major land resource areas in order to assist in projecting agricultural potential in counties where data are not as reliable; (2) developing county inventories

of wastes and residues and grouping the residues according to their adaptability to conversion technologies; (3) identifying promising potential
energy crops emerging from research for each county; and (4) estimating
biomass energy yields for each county. It is expected that crop residues,
unconventional aquatic plants, and new herbaceous and woody crops may hold
some potential in certain major land resource areas.

Keywords. Energy crops, crop residues, wood waste, land use, conversion options.

ETHANOL FROM SUGARCANE: THE EFFECT OF U.S. SUGAR POLICY

Clyde Kiker and Jonathan P. Gressel
Department of Food and Resource Economics
University of Florida/IFAS
Gainesville, Florida 32611

U.S. sugar program provides powerful disincentives to ethanol production from sugarcane through the setting of a cost of production based sugar
price objective, the market stabilization price (MSP). The MSP is maintained through the restriction of sugar imports and, in effect, raises the
opportunity cost of sugarcane in ethanol production beyond the price a
distillery would be willing to pay. At the lower world raw sugar price, the
opportunity cost of cane may be low enough that ethanol production from
sugarcane is financially feasible..

A partial equilibrium analysis, linking the sugar and ethanol markets
was used to devleope alternative sugar policies which may bring about
ethanol production from sugarcane in Florida.

A model of the sugar mill firm's investment decision criterion was
developed for the current and alternative sugar policies. The firm's decision criterion is the stochastic efficiency, measured in terms of the cumulative probability distribution of net present value, of the various options
open to the firm. The options open to the firm were to continue operating
the sugar mill, to invest in a stand-alone distillery, or to invest in an
annexed distillery.

A stochastic simulation model was used to evaluate the options under
the various sugar policy regimes and over a number of scenarios. The ranking of the options was performed using stochastic dominance analysis.

The current sugar policy was shown to be a disincentive to ethanol
production from sugarcane. A free market sugar policy encourages investment
in annexed distilleries as ethanol production acts as a hedge against low
world sugar prices. High import tariffs provide disincentives to investment

in distillery capacity. A policy which permits sugar mills that produce
ethanol to import sugar above the import quota, makes investment in ethanol
capacity financially feasible.

Keywords. Ethanol, sugar, policy, economics.

Woody Feedstocks Development

TISSUE CULTURE PROPAGATION OF BIOMASS SPECIES

R. L. Mott
Department of Botany
North Carolina State University
Raleigh, North Carolina 27695-7612

Given adequate water and nutrients, many species are efficient biomass
producers. Conifer forsts enjoy a traditional biomass role for marginal
sites because they persist for years and eventually capture the entire
sunlight and nutrient resource, concentrating the biomass in harvested
trees. The seasonal and maturation cycles which allow trees to persist also
present formidable obstacles to rapid breeding and vegetative propagation as
a means to improve biomass quality and quantity. Tissue culture methods are
overcoming these obstacles to permit manipulation of the valuable persistent
and concentrating features of trees.

Over 3000 tissue culture derived loblolly pines are now 1 to 3 years in
the field planted as clones at 16 southeastern sites. Methods for clone
propagation from juvenile materials are being extended to mature trees. In
vitro micropropagation from axillary buds gives an alternative to unreliable
rooted cutting technology for this and other species. The ploidy stability
of loblolly pine cell cultures and the preliminary successes with protoplast
culture and Agrobacterium infection suggest the whole of biotechnology will
be available for forest improvement.

Clones of pines in the field show that the correlative, balanced growth
of roots, trunk, limbs and needles relative to one another, is not irrevo-
cably locked in as it appears in seedlings. Cultural and genetic opportunities exist to modify these relationships to a better product.

In vitro methods to evaluate fusiform rust resistance using clones of
pine organs and cells and clones of the axenic rust in culture have brought
new precision to disease resistance improvement. Cell cultures which dif-
ferentiate to cell types, i.e. woody tracheids, can bring similar precession
to the molecular basis of wood quality.

Keywords. Forest biomass, clone propagation, fusiform rust,
juvenility/maturity, Pinus taeda.

A COMPARISON OF COPPICE AND SEEDLING BIOMASS PRODUCTION WITH TIME AND SPACING ON TWO SITES IN NORTH CAROLINA

S. J. Torreano and D. J. Frederick
School of Forest Resources
North Carolina State University
Raleigh, North Carolina 27695

Biomass production and survival for 2-year coppice and 5.5-year seedlings on two sites (Coastal Plain bottomland and Piedmont upland) are presented. The bottomland planting consisted of seven species: green ash (Fraxinus pennsylvanica Marsh.), American sycamore (Platanus occidentalis L.), water-willow oak (Quercus sp.), sweetgum (Liquidambar styraciflua L.), loblolly pine (P. taeda L.), cottonwood (Populus deltoides L.), and European black alder (Alnus glutinosa L.). The upland site was planted with sweetgum, European black alder and loblolly pine. Three spacings (.75 x 1.5m, 1.5 x 1.5m, 1.5 x 2.5m) were used on both sites. After 5.5 growing seasons, the bottomland had greatest biomass production compared to the upland for all species and spacings with the exception of loblolly pine. Total tree (seedling) yields on the bottomland, after 5.5 years, were highest for sycamore at the .75 x 1.5m and 1.5 x 1.5m spacings (29.9 Mg ha^{-1}, 29.2 Mg ha^{-1}), respectively, followed by European black alder at the .75 x 1.5m spacing (29.0 Mg ha^{-1}). Greatest biomass yields at 5.5 years on the upland were for loblolly pine at the .75 x 1.5m spacing (32.1 Mg ha^{-1}). Close spacing for all species after 5.5 years gave higher biomass yields. Wider spacing will likely produce the best yields for the fastest growing species by about age 10. At the end of the second coppice growing season, coppice survival was less than seedlings on the bottomland while being greater than that of seedlings on the upland.

Keywords. North Carolina, biomass, coppice, survival, Coastal Plain, bottomland, cutting cycles.

EVALUATION OF ROLL SPLITTING AS AN ALTERNATIVE TO CHIPPING WOODY BIOMASS

Paul E. Barnett
Forest Resources Development Program
Tennessee Valley Authority
Norris, Tennessee 37828

After testing and evaluating energy requirements of wood reducing equipment such as chippers, hammermills, and crushers, the Forest Engineering Research Institute of Canada (FERIC) selected roll splitting as an approach to be pursued toward the goal of processing woody biomass using a minimum of energy. As a result, prototype roll splitters (test benches) were constructed and the concept tested by FERIC. During the summer of 1984, the latest prototype roll splitter was tested in east Tennessee as a cooperative project between FERIC and the Tennessee Valley Authority (TVA), Division of Land and Economic Resources. Dewatering accomplished by roll splitting and the effects of roll splitting on air drying of three wood species were discussed relative to various roll speeds and pressures.

Keywords. Wood, roll splitting, chipping, wood drying.

WOODY BIOMASS CULTURE AND HARVESTING INTEGRATIONS: IMPORTANT CONSIDERATIONS FOR PLANTATION YIELD

H. G. Gibson
Department of Agricultural Engineering
Purdue University
West Lafayette, Indiana 47907

P. E. Pope
Department of Forestry and Natural Resources
Purdue University
West Lafayette, Indiana 47907

Research on the interaction of cultural and harvesting practices in short rotation tree plantations indicate a need to coordinate these practices to maximize sustained biomass yields from coppicing species. Closely spaced plantings achieve maximum yields in 3 to 5 years and sustained yields can be maintained with properly timed harvests combined with fertilization where appropriate. Mechanical harvesting in closely spaced plantings increases the probability for stump damage. Damaged stumps often die or have lower coppice productivity than undamaged stumps. Frequent harvests remove large quantities of nutrients from the site, contribute to soil compaction, i.e. increase soil bulk density, and influence the sprouting-circle of the coppicing stump. These latter two factors are important to harvester design. The degree of stump damage, resulting in death or reduced coppice yield, is often dependent on tree species but can be significantly influenced by using metal track or rubber-tire harvesters. Damage to stumps could be reduced by designing harvesters with smaller tires but soil compaction is increased by the smaller tire contact area. Information from field experiments and published research is summarized with respect to cultural and harvesting interactions that is helpful to equipment designers and biomass plantation managers.

Keywords. Harvesting, tree plantations, cultural practices.

EFFECTS OF EXTRACTION ON CHARACTERISTICS OF BARK EXTRACTS

Chia M. Chen and J. Ken Pan
School of Forest Resources
University of Georgia
Athens, Georgia 30602

As the first step of research to develop adhesives or adhesive extenders from barks, six different southeastern tree barks, freshly de-barked
from sawn logs, namely: southern pine; sweetgum; white oak; red oak; yellow
poplar; and maleluca, were obtained from sawmill in Georgia, except maleluca
bark was peeled off from living tree growing in south Florida. The chemical
analysis of bark was first carried out to examine their compositions (such
as lignin, tanin, etc.) and contents of 1% sodium hydroxide and other solvent extracts.

Air dried barks were hammer mill ground then treated and/or extracted
with various solution at various levels of concentration and temperature.
Approximately eighty different kinds of bark products were produced from the
combination of these treatments. The yields and solubility as well as
contents of polyphenol and organic substance in extracts were determined for
each treatment combination.

The highest yield of bark extracts was obtained from sodium hydroxide
extraction. The extracts of sodium carbonate extraction yielded the lowest
amount and jelled in a couple of days at room temperature.

Keywords. Bark extracts, southern pine, oaks, sweetgum, poplar,
melaleuca, polyphenol.

Agricultural Feedstocks Development

HARVESTING SYSTEMS FOR AGRICULTURAL BIOMASS

Gordon E. Monroe and Harold R. Sumner
USDA-ARS
Southern Agricultural Energy Center
Tifton, Georgia 31793

Most biomass has low density, high moisture content, and consequently a
high handling cost. These factors can reduce the fuel value by 50% or more.
However, the production of biomass has promise as a substantial source of
energy. Only about 20% of the estimated potential for biomass is currently
being produced and used. Wood biomass contributes the highest percentage of
energy, and probably will continue to do so for several years.

Concentrated sources of residues, such as from sugar mills (bagasse),
cotton gins, nut shelling plants, and lumber plants, have the advantage that
they require little or no additional recovery cost. Seasonality is important for utilization of residues, especially those from most field crops.

Packaging of the materials is also an important consideration. Large 1.2m x 1.2m x 2.4m rectangular bales (4' x 4' x 8') that weigh about 454 Kg (1000 lb), for example, make an acceptable payload on standard trucks.

Large-stem grasses are an important source of biomass and include sugarcane, sweet sorghum, and Napiergrass. Sugarcane production systems are well established and specialized. Sweet sorghum systems have received considerable research effort, but total working systems have not been established. Napiergrass systems must be developed, especially for harvesting, handling, and reducing field moisture content.

Root crops and non-conventional or 'exotic' crops are the last main categories of biomass deserving mention and attention. These are both widely diversified categories. Conventional equipment exists for harvesting sugar beets and potatoes, and may have application for other root crops that do not grow deep in the soil or have widely varying sizes and shapes. Most of the exotic crops need considerable research, with the possible exception of jojoba, for which researchers in Israel have developed a harvesting system.

Considerable additional research is needed before attaining a major portion of the potential for biomass production and utilization.

Keywords. Energy, systems, wood, residue, grasses, root crops, non-conventional crops.

BIOMASS YIELD OF TISSUE CULTURE AND VEGETATIVELY PROPAGATED NAPIERGRASS HYBRIDS

K. Rajasekaran, C. Burns, and I. K. Vasil
Department of Botany
University of Florida/IFAS
Gainesville, Florida 32611

S. C. Schank
Department of Agronomy
University of Florida/IFAS
Gainesville, Florida 32611

Several hundred plants of a sexually sterile triploid Pennisetum hybrid (Pennisetum americanum x P. purpureum Selection #3) were regenerated from embryogenic callus cultures obtained from segments of young inflorescences. The plants were successfully transferred to soil (survival rate 100%) and were later planted out in the field in replicated trials along with vegetatively propagated plants of the hybrid, and plants of N-75, the male parent.

The first harvest was made 80 days after planting. The plants of tissue culture origin consistently gave higher biomass yield (5426 kg dry weight ha^{-1}) compared to the vegetatively propagated clone (3426 kg) and N-75 (3156 kg). Results from additional field plantings indicated a similar

637

trend. However, the initial difference in biomass yield became insignificant at second harvest. Tissue culture derived plants had significantly more tillers (56.6 plant^{-1}) than vegetatively propagated plants (31.0 plant^{-1}). The method of propagation, either tissue culture or vegetative, did not affect the percent dry matter content.

Cytological analysis on root tip cells indicated stability of the triploid status (3x=21) in all the regenerants except one plant which was hexaploid (6x=42). Morphological analysis, on the other hand, showed 22 variants, most of them being dwarf and non-flowering. These variants could be grouped into 8 distinct morphological groups, indicating a very low level of variability. Regeneration of Napiergrass from embryogenic callus cultures is a very reliable and attractive procedure for rapid clonal propagation.

Keywords. Tissue culture, Pennisetum, Napiergrass, biomass yield.

INTERSPECIFIC HYBRIDIZATION AND CLONAL SELECTION OF PENNISETUM TO INCREASE BIOMASS PRODUCTION

S. C. Schank
Department of Agronomy
University of Florida/IFAS
Gainesville, Florida 32611

Pearlmillet (Pennisetum americanum (L.) K. Schum.) is an annual forage grass that has been used in the South for decades. Pearlmillet hybridizes readily with Napiergrass (Pennisetum purpureum Schum.). The F_1 hybrids of this cross are triploids and therefore sterile. However, with clonal selection, many desirable F_1 genotypes have been identified and evaluated. Interspecific hybrids studied in 1983 at the University of Florida were between a dwarf pearlmillet, 23DA (female parent) and dwarf Napiergrass N75 (male parent). Over 1500 seedlings from that cross were evaluated, and selections of the most vigorous dwarf types were chosen for further testing. In a genetic study from some of the seedlings above, the dwarf x dwarf cross produced 670 tall plants compared to 54 dwarf plants (close to a 15:1 ratio).

Interspecific hybrids planted in 1984 utilized 23A and Gahi-3 pearlmillet as the female parents and various tall Napiergrasses, including PI 300086, an extremely robust biomass producer as the male parent. A wide spectrum of phenotypes were evaluated in the F_1 population of approximately 12,000 hybrid plants. The most vigorous plants were selected and increased for biomass production in 1985.

Keywords. Pearlmillet, hybridization.

ECONOMIC ANALYSIS OF NUTRIENT AVAILABILITY FOR WATERHYACINTH PRODUCTION

Carolyn Fonyo, William G. Boggess and Clyde F. Kiker
Department of Food and Resource Economics
University of Florida/IFAS
Gainesville, Florida 32611

The purpose of this study was to assess the effects of nutrient application strategies on the economic feasibility of waterhyacinth production in Lake Apopka. The lake has been in a state of eutrophication for the past thirty years due to a steady loading of nutrients from land runoff and other sources. Nutrient uptake by aquatic macrophytes and subsequent export through harvesting will decrease available nutrients in the lake water column and sediments over time. A point will be reached at which an external nutrient source will be required to maintain maximum production and biomass yields. Therefore, there may exist an economic trade-off between the amount of biomass produced and the degree of water quality improvement. The potential economic gains of enhanced water quality may outweigh the gain of sustained maximum yield. The BIOMET (Biomass to Methane) simulation model was used to determine the effects of different production strategies on biomass yields and water quality. An equilibrium point may be maintained by selecting the optimal strategy for fertilizer application and biomass harvesting.

Keywords. Economic analysis, nutrients, waterhyacinth.

THE BROWN SEAWEED SARGASSUM AS A MARINE SOURCE OF BIOMASS FROM THE WEST COAST OF FLORIDA

Clinton J. Dawes
Department of Biology
University of South Florida
Tampa, Florida 33620

The respiratory and photosynthetic rates of a number of subtropical and tropical seaweeds common to the coasts of Florida have been studied with relation to effects of light, temperature and salinity as well as nutrients. The physiological studies on seaweed ecology include species of the brown alga Sargassum, the red algae Hypnea, Eucheuma, Gracilaria, Bostrychia, and the green algae Caulerpa and Ulva. Although a number of species of these genera are now grown in field or tank culture as sources of food or phycolloids, no mariculture scheme has yet been found to be successful for biomass production. Field and laboratory studies on distinct populations of Sargassum filipendula and S. pteropleuron from the west coast of Florida suggest the genus could be used in field mariculture studies for biomass production. Both populations have high photosynthetic rates under low (100 μE m^{-2}s^{-1})

light and do not show strong photoinhibition to high ($2400\,\mu E\ m^{-2}s^{-1}$) light. Thus they could be grown in dense floating cultures in the field. Caloric values do not change (kcalories g dry wt^{-1}) after the plant is mature in July or August and so the plants could be harvested in September, before reproduction begins. The information is obtained from comparisons of population growth cycles, proximate constituents and caloric values, and physiological processed including photosynthetic-light intensity curves and tolerances to temperature and salinity regimes.

Keywords. Sargassum, Hypnea, Eucheuma, Gracilaria, Bostrychia, Caulerpa, Ulva.

Conversion Processes -- Thermochemical

TWO-STAGE BIOMASS COMBUSTION FOR DIRECT DRYING APPLICATIONS

P. K. Chandra and F. A. Payne
Department of Agricultural Engineering
Clemson University
Clemson, South Carolina 29631

A validated computer-aided simulation of a two-stage biomass combustion system was used to analyze the effects of biomass moisture and degree of insulation around the gasifier-combustor on its turndown ratio. The size of the gasifier-combustor was also varied to examine its influence combined with other factors described earlier on the turndown ratio and heat loss from the system. The turndown ratio was found to decrease with the increase of biomass moisture. The effect of insulation was more prominent for smaller units than the larger ones. Turndown ratio was found to increase with the increase in thermal resistance. The maximum fraction of heat loss varied between 16 and 39% for biomass moistures of 0.5 and 0.1, respectively, for any size with any level of insulation.

Keywords. Biomass, gasifier, combustor, heat loss, insulation, biomass moisture, turndown ratio.

AGRICULTURAL BIOMASS FOR RICE DRYING

Lalit R. Verma
Department of Agricultural Engineering
Louisiana State University Agricultural Center
Baton Rouge, Louisiana 70803

Rice accounts for almost 22% of the total dollar value of all agricultural crops in Louisiana. With the rice straw and other agricultural biomass being available, but not presently used as an energy source, a research study has been initiated to determine the feasibility of using these sources for rice drying in Louisiana. An indirect-fired biomass furnace will be used to evaluate the efficiency of converting residues into heat energy to dry rice. The furnace is installed at the St. Gabriel Research Station of the Louisiana Agricultural Experiment Station. This furnace uses a heat exchanger which heats the air blown through the grain. The heated air is delivered to the floor plenum of the batch-type bin dryer. The furnace is being operated with different biomass fuels to determine the temperature rise at the desired airflow setting for grain drying. Temperature rise and airflow through the furnace is being recorded to calculate heat energy and efficiency of conversion to heat. A calorimeter is used to determine the theoretical energy content of the biomass fuels. Large round bales of the biomass fuels are used to evaluate the effect on combustion duration and time to reload the furnace.

Preliminary testing of large round bales has given mixed results. Two tests of rice straw have failed to reach the desired drying temperature for rice (30°C) and no drying energy was observed. Calorimeter tests of core samples indicated lower heat energy than reported in published data. One test with Dallas-bermudagrass produced desired temperatures as well as showed potential for greater temperature rise in drying other grains. The drying energy observed was about half of the theoretical energy determined by calorimeter tests. Wet rice will subsequently be dried in the bin using this biomass furnace heat. Quality of the resulting rice will be evaluated. Energy thus used in rice drying will be compared to the energy obtained from low temperature on-farm rice drying for the last five years.

Keywords. Biomass, furnace, rice, straw, drying, energy.

ON-FARM PROCESSING OF VEGETABLE OIL AS FUEL

John Goodrum
Department of Agricultural Engineering
University of Georgia
Athens, Georgia 30602

Peanuts and other oilseeds have potential to supply farm diesel fuel substitutes for fuel extenders. Oilseed producers may purchase small scale screw expellers 189 l day^{-1} (50GPD and up) to produce crude vegetable oil and by-product oilseed meal. Crude oil may be filtered and used as diesel fuel. The meal is a valuable feed for livestock. For farm use the expeller possesses great advantages of cost, simplicity, and fire safety, compared to commercial solvent extraction systems.

A farm scale oil expelling system has been constructed and tested. It was designed for simple, automatic operation to minimise skill and labor demands. Key system components were the preheater/conveyer, the expeller, and a microprocessor control system for pressure and temperature control.

Under optimal conditions of temperature, pressure, and moisture; the system has demonstrated 92% oil recovery from peanuts. The product oil has about double the storage life of commercial solvent extracted oil due to retention of natural antioxidants.

Expeller operation may be modified to give a higher value by-product meal. This includes (1) adjustment of preheat temperature for roasting effects and (2) reducing the oil recovery, leaving higher nutritional energy in the meal.

Keywords. Oilseeds, fuel substitutes, fuel extenders, on-farm processing.

Conversion Processes -- Biological Gasification

GRI BIOMASS PROGRAMS IN THE SOUTHERN USA

Peter H. Benson
Gas Research Institute
8600 West Bryn Mawr Avenue
Chicago, Illinois 60631

The southern USA provides a setting conducive to developing biomass-to-energy systems. The environment enhances biomass growth, the area supports a large agribusiness and has a favorable industrial, political and technical infrastructure. As a result, the Gas Research Institute (GRI) supports 3 major biomass-to-energy R&D efforts in the southern USA through its Methane from Biomass and Waste Program.[1,2] The objective of this program is to provide an alternative supply of low-cost, pipeline quality methane from wastes in the near-term and from biomass in the longer term to supplement declining natural gas resources past the year 2000.

Within the near-term Methane from Waste area, studies are conducted at an integrated water hyacinth, wastewater treatment/energy production facility located at Walt Disney World, Orlando, Florida.[3] The technical and economic feasibility of converting biomass/sewage sludge mixtures to methane is being evaluated by GRI at this facility which consists of channels, where water hyacinths are grown on primary effluent to achieve secondary treatment; harvesting equipment and an anaerobic digestion experimental test unit to produce methane.

A preliminary A&E analysis indicated that $2.37-4.55 \text{ GJ}^{-1}$ ($2.50-4.80 mm Btu^{-1}) methane could be produced from $38-190 \times 10^3 \text{ m}^3 \text{ day}^{-1}$ (10-50 MGD) hyacinth treatment plants if research performance goals are achieved. To date, secondary treatment standards have been maintained at high sewage loadings, hyacinths yields in excess of 67.2 tons $\text{ha}^{-1} \text{ yr}^{-1}$ (30 dry tons $\text{acre}^{-1} \text{ yr}^{-1}$) have been achieved and methane yields of 0.30-0.31 SCM kg^{-1}

(4.8-5 SCF lb^{-1}) volatile solids added have been obtained with 2:1 water hyacinth/primary sludge blends which represents good progress towards our goals.

As part of its long-range supply strategy GRI has implemented two Regional Methane from Biomass Programs in the southern USA. Since biomass resources, energy demands and fuel prices and availability vary by region and location, a regional approach permits the flexibility to respond to local energy demand changes with local resources at a local scale. These multidisciplinary programs, which utilize teams of researchers and an integrated systems approach, are equally cofunded and comanaged by GRI and its cosponsors.

The initial regional program with the University of Florida Institute of Food and Agricultural Sciences has focused on two feedstocks, Napiergrass and water hyacinth, and a high-solids bioconversion process. An important part of this program is the development and application of advanced biotechnologies to further improve system efficiencies and reduce gas cost. Numerous projects in this program will be discussed and visited during this conference. The second regional effort with the University of Texas A&M Agricultural Experiment Station is focused upon developing a single energy crop: high-energy sorghum. On-going systems modeling and analysis is applied to both programs to ensure optimization of the many interrelated parameters and a determination of major areas of cost sensitivity. Independent A&E analyses indicate that cost-competitive gas can be produced from biomass without major technological breakthroughs and that these costs can be further reduced through applied biotechnology.

Good progress has been demonstrated in both programs toward achieving R&D performance goals. Biomass and methane yields in small test plots and laboratory scale reactors have equaled or surpassed R&D goals. Biomass yields up to 61 tons ha^{-1} yr^{-1} (27 dry tons $acre^{-1}$ yr^{-1}) have been achieved with Napiergrass and methane yields of 0.38 SCM kg^{-1} (6.1 SCF lb^{-1} VS) added have been obtained with a variety of sorghum, which corresponds to a reduction in volatile organic matter of 94%, the highest achieved to date for a biomass feedstock.

Keywords. Biomass, waste energy system, anaerobic digestion, methane, wastewater treatment.

REFERENCES

1. P. H. Benson, J. R. Frank and H. R. Isaacson. Gas Research Institute Programs on Methane from Biomass and Wastes. p. 1505-1519. In: Symposium Papers, Energy from Biomass and Wastes, VIII. Institute of Gas Technology. Chicago, IL. 1984.
2. P. H. Benson, H. R. Isaacson and J. R. Frank. A Regional Approach to Producing Methane from Biomass and Wastes. p. 683-690. In: Pro-

ceedings of the 19th Annual Intersociety Energy Conversion Engineering Conference: Advanced Energy Systems, Vol 2. San Francisco, CA.
1984.

3. GRI. Methane from Biomass and Wastes Research: Renewable Resources for
Localized Energy Production. GRI Brochure. Chicago, IL. p.11.
1984.

DECOMPOSITION OF FRESH AND ANAEROBICALLY-DIGESTED WATER HYACINTH ADDED TO SOIL

K. K. Moorhead, D. A. Graetz and K. R. Reddy
Department of Soil Science
University of Florida/IFAS
Gainesville, Florida 32611

The anaerobic digestion of biomass generates residue which must be
disposed of in an environmentally-safe manner. The potential use of the
residue as a source of plant nutrients and as a soil amendment needs to be
investigated. During decomposition of the residue nitrogen and other nutrients become available for plant growth. Decomposition rates must be evaluated to determine proper soil application rates.

A laboratory incubation study was conducted to evaluate the decomposability of fresh and anaerobically-digested water hyacinth. Nitrogen-15
labeled water hyacinth plant tissue, with either 1 or 4% total Kjeldahl
nitrogen (TKN), was digested in batch digesters for 4 months. The digester
residues and the fresh water hyacinth plant material were freeze-dried and
characterized for lignin, cellulose, hemicelluloses, total solids, volatile
solids, TKN, P, Ca, K, Mg, Na, Fe and Zn.

The above materials were added to a sandy soil at rates of 0 and 10 Mg
ha^{-1} (dry weight) and incubated for 90 days at 26°C. Decomposability was
evaluated by CO_2 evolution. Soil samples were analyzed at 0, 30, 60 and 90
days for KCl-extractable NH_4 and NO_3, TKN, double acid-extractable P, Ca,
Mg, K, Na, Fe and Zn, organic matter and pH.

After 90 days, less than 17% of the added carbon of the digested residues had evolved as CO_2 compared to 31% and 41% of the 1% and 4% TKN fresh
water hyacinth, respectively. Addition of digested and fresh water hyacinth
residues resulted in increases in soil organic matter, TKN and double acid-
extractable P, Ca, Mg, K, Na, Fe and Zn. Net NO_3-N mineralization was
observed for both digested residues and the 4% TKN fresh water hyacinth.

Keywords. CO_2 evolution, nitrogen mineralization, energy crop, soil
amendment.

RECYCLING ANAEROBIC DIGESTER EFFLUENT IN THE PRODUCTION OF TERRESTRIAL
AND AQUATIC BIOMASS CROPS

D. A. Graetz and P. A. Krottje
Department of Soil Science
University of Florida/IFAS
Gainesville, Florida 32611

Digestion of organic materials for the production of methane produces
an effluent which may contain relatively large quantities of plant nutrients
in a readily-available form. Disposal of this effluent thus becomes a
significant consideration in an integrated "methane from biomass" system.
Our research was concerned with the recycling of this effluent in terrestrial and aquatic systems to produce additional biomass for methane production.

The effect of anaerobic digester effluent on the growth of two promising biomass crops, a Napiergrass x pearl millet hybrid (NPH) and water
hyacinth, was investigated in 90 liter microcosms. The feedstock for the
digesters was effluent from a lagoon receiving swine manure. Effluent
equivalent to 0, 150, 3000, and 600 kg inorganic N ha^{-1} was applied initially and after each 50 day harvest period for the terrestrial system. For
the aquatic system, effluent equivalent to 300 and 600 kg N ha^{-1} was applied
initially and after each 21 day harvest period.

Application of effluent increased NPH yield at all application rates
although some plant mortality was noted at the highest application rate.
Dry matter yields ranged from 1 to 15 Mg ha^{-1} per 50 day harvest period for
the 0 and 600 kg N ha^{-1} application rates respectively. Water hyacinth
yield averaged 7 Mg ha^{-1} per 21 day harvest period for both the 300 and 600
kg N ha^{-1} application rates. Results indicate that both terrestrial and
aquatic systems, if managed properly, can be used to recycle nutrients in an
integrated "Energy from Biomass System".

Keywords. Energy crop, Napiergrass, pearl millet, water hyacinth,
methane digester.

FEASIBILITY OF STORING INTERMEDIATE ACIDS IN TWO STAGE ANAEROBIC DIGESTION
OF STARCH

Mohammed K. Habbal
Department of Chemical Engineering
Louisiana State University
Baton Rouge, Louisiana 70803

The study was conducted on a process involving two stage anaerobic
digestion of starchy feeds to methane gas. The need to have a constant flow
into the second (methane) reaction necessitates the storage of fatty acid
products from the first (acid) reactor. The main goal of this study was to

monitor the potential of methane production from stored slurry under different conditions of storage. Storage variables investigated were: temperature, starch concentration, and pH.

The acid concentration remained constant at all temperatures studied from 4°C to 55°C. A slight increase in acid concentrations accompanied by a drop in pH was noted upon the addition of 0.2 to 1.0% starch to the stored slurry. Raising the pH of the slurry to 8 or lowering it to 4 from the initial 4.7 had little effect. In both cases the pH dropped slightly and then stabilized.

In all of these studies the main acids were lactic, acetic, and butyric. Little off gas was produced. Component carbon and hydrogen balances indicated that the potential for methane production remained very stable for these acids slurries, and this may be a feasible method for insuring a constant reliable feed for a methanogenic reactor.

Keywords. Anaerobic digestion, acid storage, methane production, storage temperature, substrate concentration, pH level.

SORGHUM LEACHING FOR METHANE PRODUCTION

Richard P. Egg, Charlie G. Coble and Martin G. Tibbets
Department of Agricultural Engineering
Texas A & M University
College Station, Texas 77843

A two stage anaerobic digestion scheme is being evaluated for producing methane from sorghum. The first stage is used for both storage and hydrolysis of sorghum, with the leachate of the first stage being fed to a high rate methane reactor. A laboratory study was conducted to evaluate the rate of volatile fatty acid (VFA) and chemical oxygen demand (COD) production in the first stage by leaching chopped sorghum with fresh water. Leachate was removed from the sorghum at three rates: weekly, at pH 5, and pH 6. Total COD production in the leachate averaged 537 mg g^{-1} of initial sorghum VS with no difference among the treatments. The sorghum was leached for a period of 241 days. In all treatments, lactic acid was the predominant acid in the leachate, followed by acetic and small amounts of propionic, butyric, and isobutyric. Leachate volumes produced were 11.2, 28.6, and 12.8 volumes leachate per reactor volume for leaching once per week, at pH 6, and pH 5, respectively. VS reductions in the sorghum ranged from 20.3 to 30.1%.

Keywords. COD, volatile fatty acids, sorghum, anaerobic digestion.

CHARACTERIZATION OF CARBOHYDRATES IN SORGHUM FOR POTENTIAL METHANE PRODUCTION

G. G. McBee, R. L. Monk and F. R. Miller
Department of Soil and Crop Sciences
Texas A & M University
College Station, Texas   77843

Sorghum is a plant that produces biomass containing high levels of carbohydrates. Our studies have shown that stems of various sorghum culti-vars contain 70-80% carbohydrates, partitioned between structural (SC) and nonstructural (NSC) forms. Feedstock of this material offers high potential for use in production of methane.

Generally, the sorghum cultivars have been divided into three cate-gories for study: grain, high energy and sweet or forage types. An example from the grain type, ATx623 x RTx430, has produced biomass composed of a maximum of 27% NSC and 54% SC. The "high energy" types normally produce more biomass than the grain types, plus acceptable grain yields. Carbo-hydrate values for an example from this group, ATx623 x BMR-12, were 40.7% NSC and 44.9% SC. Comparative values for a selection from the sweet types, such as Brandes were 43.1% NSC and 43.6% SC. Ranges among the sorghum categories for cellulose, hemicellulose, and lignin have been 21.2 - 34.8, 13.6 - 25.4, and 4 - 10.5% respectively. Volatile solids normally exceed 95%.

Additional studies have indicated that entries containing higher levels of the mono- and disaccharides normally contain higher levels of starch whereas lower starch levels have been associated with low levels of NSC. A highly significant negative correlation (-0.95) exists for partitioning between NSC and SC. Closer spacing of plants has resulted in an increase of percent NSC and decrease of SC.

Many of these cultivars are currently being evaluated for potential methane production using bench scale techniques.

Keywords. High energy sorghum, structural carbohydrates, non-structural carbohydrates, sorghum biomass, methane potential.

EXPERIENCE WITH WOOD MEDIA FOR ANAEROBIC FIXED BED REACTORS

R. A. Nordstedt
Department of Agricultural Engineering
University of Florida/IFAS
Gainesville, Florida 32611

A major cost in the construction of an anaerobic fixed bed reactor is the media on which the biofilm is attached. Commercially manufactured media of plastic, metal and ceramic are available. Although they are expensive, they may also be durable, have few clogging problems, and possess high surface to volume ratios. Wood has been suggested for use as a media. Wood has a rough surface which should promote rapid biofilm growth and result in less sloughing of biofilm. Wood chips are approximately 3 percent of the cost of some plastic media on a volume basis. Although wood media may degrade, methane would be produced.

Oak, cypress and pine wood blocks were compared with three types of plastic media in 5 liter liquid volume reactors. The wood block media were superior to the plastic media in L gas L^{-1} liquid volume. However, because of the lower porosity of the wood media, the L gas L^{-1} total reactor volume was less. After two years of continuous operation, the wood block media showed no visual signs of deterioration. Additional tests compared startup characteristics of pine wood chip media with plastic ring media using a whey and cellulose feedstock. With pH control, the wood chip and plastic media achieved nearly the same gas production rates with in 30 days. A 20 cubic meter anaerobic fixed bed reactor filled with cypress wood chips has also been monitored for approximately 8 months.

Keywords. Anaerobic fixed bed reactors, wood media, biofilm.

ANAEROBIC DIGESTION OF CELLULOSIC WASTES

D. D. Lee and T. L. Donaldson
Chemical Technology Division
Oak Ridge National Laboratory
Oak Ridge, Tennessee 37831

Anaerobic digestion is a potentially attractive technology for volume reduction of radioactive cellulosic wastes. A substantial fraction of the waste is converted to off-gas, and a relatively small volume of biologically stabilized sludge is produced.

Three runs lasting 36, 90, and 423 days have been made with a mixed cellulosics feed using a 75-L digester under batch and batch-fed conditions. Solids solubilization and gas production rates and total solids destruction have met or exceeded the target values of 0.6 g cellulose per L of reactor per day, 0.5 L off-gas per L of reactor per day, and 80% destruction of solids, respectively. Successful start-up procedures have been developed, and a simple dynamic process model has been constructed. The three-culture model successfully simulates both short-term and long-term dynamic phenomena under stable and stressed process conditions.

Keywords. Simulation, anaerobic digestion, low-level radioactive waste, cellulose, dynamic model, volume reduction.

648

MULTIPHASE ANAEROBIC DIGESTION OF BIOMASS

Paul H. Smith
Department of Microbiology and Cell Science
University of Florida/IFAS
Gainesville, Florida 32611

Anaerobic digestion of complex organic matter to methane gas requires a
cascade of biochemical conversions catalyzed by different physiological
groups of interacting microorganisms. A conventional continuously mixed
digestion system contains a more or less homogeneous mixture of organisms
which catalyze each phase of the cascade. The reaction rates obtainable in
different phases are markedly different, with the net result of the overall
process becoming rate restricted by the slowest phase of the cascade. This
observation has led to the development of two-phase anaerobic digestion
processes designed to obtain a higher overall rate by optimizing two major
phases of the cascade. This is done by separating the acid forming phase
from the methane forming phase. However, the cascade can be conceptualized
as consisting of multiple phases, each characterized by the predominance of
unique microbial communities. This suggests that the overall process may be
optimally regulated using a multi-phase digestion system. Each phase of the
digestion system would be regulated to optimize the catalytic activity of
it's unique microbial communities. The multiphases would be integrated into
a total system. Such a system is described.

Keywords. Anaerobic digestion, methane.

BIOSOLAR CONVERSION OF HEXOSE TO HYDROGEN BY CYANOBACTERIA

Hart Spiller and K. T. Shanmugam
Department of Microbiology and Cell Science
University of Florida/IFAS
Gainesville, Florida 32611

Hydrogen gas is a clean fuel since its oxidation product with O_2 is
water. The photosynthetic bacterium Anabaena variabilis is capable of
converting the reducing power available in sugars such as fructose to H_2,
using solar energy. We have used a mutant strain derepressed for nitro-
genase to study the evolution of hydrogen. Hydrogen evolution as a function
of fructose concentration revealed a biphasic pattern: from 0 to 10 mM
fructose, conversion efficiency was 6.4 mol per mol fructose, from 10 - 20
mM the conversion efficiency was 8.2 mol H_2 per mol fructose or 68%. Rates
of greater than 6 mol H_2 evolved per mol fructose supplied indicated that
all fructose was fermented to H_2 and CO_2. On the other hand, the parent
strain converted fructose to hydrogen only at a maximum efficiency of 5.0.

We have further studied the distribution pattern of U-^{14}C-fructose in
liquid and gas phases of wild type A. variabilis cultures in relation to

acetylene reduction to understand the mechanism of hydrogen production.
After several days 25% of the fructose carbon assimilated had been respired
to CO_2, whereas 45% was fixed in cell material. For every 7 mol CO_2 generated 20.6 mol acetylene was reduced. Conversion efficiency was 4.1 mol C_2H_2
reduced per mol fructose or 34%. Our data suggest that nitrogenase- derepressed mutants of cyanobacteria can be used to convert hexoses to hydrogen
at efficiency factors approaching 100%.

Keywords. Solar energy, cyanobacteria, photohydrogen production,
hexose fermentation.

Conversion Processes -- Biological Liquification

BIOLOGY AND BIOCHEMISTRY OF ALCOHOL FERMENTATIONS

Lars G. Ljungdahl
Department of Biochemistry
University of Georgia
Athens, Georgia 30602

Ethanol, considered a substitute for gasoline, can be produced by
fermentations of carbohydrates obtained from biomass, a renewable source.[1]
Of the carbohydrates present in biomass roughly 60% is cellulose, which
contains only glucose, the rest is mainly hemicellulose, which consists of
xylose, mannose, and smaller amounts of galactose, uronic acid and arabinose.[2] To efficiently produce ethanol from biomass, all of these sugars
should be fermented.

Conventional fermentations with yeast (Saccharomyces species) and
Zymomonas species have limitations. These microorganisms do not directly
utilize the polysaccharides cellulose, hemicellulose and starch, and also
after hydrolysis of the polysaccharides to monosaccharides by chemical or
enzymic means they are limited to the fermentation of glucose. Anaerobic
bacteria including Clostridium thermocellum and Clostridium thermohydrosulfuricum as well as some thermophilic non-spore forming species such as
Thermoanaerobacter ethanolicus ferment all sugars mentioned above and also
directly utilize the polysaccharides.[3,4,5,6] With these bacteria ethanol
yields of 90% of theoretical are obtained with free sugars and starch as
substrates and with cellulose the yield may be about 75%.[1] However, these
bacteria are sensitive to substrate concentrations over 1% and generally do
not produce a "beer" with more than 1% of ethanol. Enzyme studies indicate
that the fermentations are tightly controlled[7] and work to develop mutants
strains has led to the isolation of a T. ethanolicus strain that yields a
beer with 4% ethanol. The cellulolytic enzyme system in C. thermocellum
differs from the cellulase systems of aerobic fungi in that it is a very

high molecular weight polypeptide complex which binds tightly to cellulose.[9]

Keywords. Ethanol fermentations, bacteria, anaerobes, thermophiles, cellulase.

REFERENCES

1. L. H. Carreira and L. G. Ljungdahl. Production of ethanol from biomass using anaerobic thermophilic bacteria. In: Liquid Fuel Developments. D. L. Wise, Ed. p. 1-29. CRC Press, Boca Raton, Florida. 1983.
2. N. S. Thompson. Hemicellulose as a biomass resource. In: Wood and Agricultural Residues. p. 101-119. E. J. Soltes, Ed. Academic Press, New York. 1983.
3. J. Wiegel, L. G. Ljungdahl and J. R. Rawson. Isolation from soil and properties of the extreme thermophile Clsotridium thermohydrosulfuricum J. Bacteriol. 139:800-810. 1979.
4. J. G. Zeikus. Chemical and fuel production by anaerobic bacteria. Ann. Rev. Microbiol. 34:423-464. 1980.
5. J. Wiegel. Formation of ethanol by bacteria. A pledge for the use of extreme thermophilic anaerobic bacteria in industrial ethanol fermentation processes. Experientia 36:1434-1446. 1980.
6. J. Wiegel and L. G. Ljungdahl. Thermoanaerobacter ethanolicus gen. nov., spec. nov., a new, extreme thermophilic anaerobic bacterium, Arch. Mirbobiol. 128:343-348. 1981.
7. L. H. Carreira, L. G. Ljungdahl, F. Bryant, M. Szulczynski and J. Wiegel. Controls of products formation with Thermoanaerobacter ethanolicus, enzymology and physiology. In: Proceedings IVth. International Symp. Genetics of Industrial Microorganisms. Y. Ikeda and T. Beppu, Eds. Kodansha Ltd. Tokyo. p. 351-356. 1982.
8. L. H. Carreira, J. Wiegel and L. G. Ljungdahl. Production of ethanol from biopolymers by anaerobic, thermophilic, and extreme thermophilic bacteria. I. Regulation of carbohydrates utilization in mutants of Thermoanaerobacter ethanolicus Biotech. Bioeng. Symp. 13:183-191. 1983.
9. L. G. Ljungdahl and K. E. Eriksson. Ecology of Microbial cellulose degradation. Adv. Microbial Ecology. 8:237-299. 1985.

IMMOBILIZED YEAST IN BIOREACTOR FOR ALCOHOL FERMENTATION

M. K. Hamdy and K. Kim
Department of Food Science
University of Georgia
Athens, Georgia 30602

Mutant of Saccharomyces cerevisiae (ASTY-81) was developed using Co-60 source. Cells (stationary phase) were immobilized onto sterile, channeled alumina beads were immobilized onto sterile, channeled alumina beads and packed into bioreactor column under controlled temperature. Feedstock (FS) containing substrate and nutrients was fed into bioreactor at specific rates. Beads with greatest porosity and surface area produced the most ethanol. Factors affecting ethanol productivity included: temperature, pH, flow rate, nutrients and substrate in FS. Sterile FS containing 4% glucose such as malt-YE broth, synthetic medium, aqueous solution of essential mineral and deionized water showed ethanol yields ranging from 62.6 to 66.5%

after 4 h operation. Productivity of 27.0 g ethanol $L^{-1} h^{-1}$ was achieved using malt-YE broth containing 10% glucose at dilution rate of 1.3 h^{-1}. Stability of bioreactor (25°C) indicated that after reactivation following intermittent storages for 1, 2, and 2 months, productivities were 18.50, 19.08, and 19.08 g ethanol $L^{-1} h^{-1}$, respectively, using malt-YE broth (10% glucose and flow rate of 160 ml h^{-1}). Scanning electron microscopic examination showed immobilized-yeast onto internal and external surfaces of beads.

Keywords. Ethanol, feedstock, fermentation, Saccharomyces.

KINETICS OF HEMICELLULOSE CONVERSION TO ETHANOL BY PACHYSOLEN TANNOPHILUS

M. J. Beck
Division of Chemical Developement
Tennessee Valley Authority
Muscle Shoals, Alabama 35660

The utilization of hemicellulose for ethanol production can enhance the value of many lignocellulosic substrates as sources of liquid fuels. Hemicellulose-derived monosaccharides are recoverable from biomass sources at a higher efficiency than are those derived from cellulose. For example, hemicellulose-derived monosaccharides have been reported to compose approximately half of all sugars derived from a particular hardwood.

Microorganisms have been identified which can convert xylose, the predominate sugar of hardwood hemicellulose, to ethanol. One such yeast is Pachysolen tannophilus. It has a number of characteristics which make ethanol productivity and yields from xylose less than those obtainable from glucose and sucrose by conventional yeasts.

This study shows productivities and yields obtainable with P. tannophilus in dilute-acid hydrolyzates of hemicellulose from hardwoods. A conversion efficiency of 0.35 gram ethanol per gram sugar consumed has been achieved. A retention time of three days was optimal under a certain set of fermentation conditions. Improvements in productivities are being studied which will make processing more attractive.

Keywords. Hemicellulose, ethanol production, cellulose, Pachysolen.

FUEL ETHANOL FROM POPLAR WOOD FOLLOWING HYDROFLUORIC ACID SOLVOLYSIS

C. M. Ostrovski and J. C. Aitken
Allose Manitoba Ltd.
1100-213 Notre Dame Avenue
Winnipeg, Canada R3B IN3

Gaseous hydrofluoric acid (HF) was used to break down poplar chips into

sugars for fermentation to fuel ethanol. Conversion of wood to sugars has
surpassed that of current technologies, consistently producing yields of
hexose sugars in excess of 90% of the theoretical (tests have obtained 96%),
and 75% of the theoretical pentose yields (87% was the highest). Since the
HF is used as a gas, it is removed as a gas and available for recycling in
the process. In excess of 99% of the HF absorbed is desorbed.

Ongoing fermentation studies have concentrated on the pentose sugars,
and have repeatedly shown that with the organisms isolated, 5% ethanol is
produced from a 10% xylose solution in less than 24 hours. It is expected
that 300-400 litres of ethanol will be produced from a metric ton of oven
dry poplar. The test facility work is continuing and is giving empirical
support to the planning of the pilot plant.

Keywords. Hydrofluoric acid, ethanol, hexose, pentose, poplar,
cellulose.

ETHANOL RECOVERY FROM LOW-GRADE BEERS BY SOLVENT EXTRACTION AND EXTRACTIVE
DISTILLATION

D. W. Tedder
School of Chemical Engineering
Georgia Institue of Technology
Atlanta, Georgia 30332-0100

This research focuses on the use of solvent extraction, coupled with
extractive distillation, to recover ethanol economically from low-grade
beers. Studies at Georgia Tech indicate that water coextracts with ethanol
from fermentates, but the resulting extracts can be dehydrated by extractive
distillation. Subsequently, the dehydrated extracts can be regenerated to
produce fuel grade ethanol even from low-grade (i.e. less than 1 wt%) beers.
The use of solvent extraction in this case permits a more economical and
energy efficient recovery of low molecular weight metabolites. It provides
a way of utlizing thermophyllic bacteria to convert cellulose, for example,
without requiring high ethanol concentrations in the fermentate. Moreover,
the cost and energy savings, compared to optimized distillation, increase as
the beer quality decreases. The fact that the solvents used are not toxic
to a variety of microorganisms, including Zymomonas Mobilis, suggests that
this technology can also be coupled with continuous fermentation to enhance
production rates.

Keywords. Ethanol, cellulose, distillation, beers, Zymomonus.

EVALUATION OF SEVERAL FERMENTATION SCHEMES FOR ETHANOL PRODUCTION FROM A
HEMICELLULOSIC HARDWOOD HYDROLYZATE

R. C. Strickland and M. J. Beck
Development Branch
Tennessee Valley Authority
Muscle Shoals, Alabama 35660

The Tennessee Valley Authority (TVA) is investigating ethanol production from acid-catalyzed hardwood hydrolyzates. A two-stage process with
dilute sulfuric acid is used to sequentially hydrolyze hemicellulose and
cellulose to produce two sugar streams enriched in xylose and glucose,
respectively. TVA has emphasized the conversion of hemicellulosic hydrolyzates using Pachysolen tannophilus. Hemicellulosic hydrolyzate conversions
are slower and less efficient than desired despite improvements in ethanol
production through hydrolyzate pretreatments, nutrient additions, and the
use of large yeast populations. The sugar composition of hydrolyzates could
influence conversion because P. tannophilus utilizes fermentable hexoses
before xylose. Therefore, ethanol production from a hemicellulosic hydrolyzate was compared using three fermentation schemes which could indicate a
preferred method of conversion. Hydrolyzate was converted using: (1) P.
tannophilus alone, (2) a coculture of Saccharomyces uvarum and P. tannophilus, and (3) S. uvarum alone followed by P. tannophilus (with and without
an intermediate ethanol removal step). Tests were performed using 8 and 4
percent reducing sugar and several levels of nutrients (yeast extract, urea,
and phosphate). Ethanol production was influenced by sugar concentration
and nutrients. Based on ethanol production and nutrient requirements,
conversion using P. tannophilus alone was the superior fermentation scheme.
Reduced ethanol production using the sequential or coculture conversion
schemes was not the result of nutrient limitations or inhibitors in the
hydrolyzate.

Keywords. Hemicellulosic hydrolyzate, Pachysolen tannophilus,
Saccharomyces uvarum, xylose, glucose.

PROCESSING SWEET SORGHUM SOLIDS TO PRODUCE ETHANOL

William L. Bryan
USDA-ARS
Southern Agricultural Energy Center
Tifton, Georgia 31793-0748

From agronomic and harvesting considerations, sweet sorghum [Sorghum
bicolor (L.) Moench] has favorable characteristics to become a future cash
crop, particularly in the Southern states, as a source of readily fermentable carbohydrates and lignocellulosic biomass. Carbohydrate yields per ha
are more than twice those of corn. The sugar and starch fractions, which

constitute about half the stalk dry matter, may be readily fermented into ethanol, acetic acid or other chemicals. The lignocellulosic fraction could also be used as a feedstock for production of chemicals or animal feed and/or used as fuel. One limitation to sweet sorghum utilization as a fermentation feedstock is the lack of low-cost processing technology that could be used instead of the capital-intensive compound imbibition or counter-current leaching or extraction processes used in the sugar industry to separate sugar in high yields from bagasse.

An alternate approach has been used to ferment the sugar in chopped sweet sorghum stalks, or bagasse following juice expression with a 3-roll mill, into ethanol during a solid-phase fermentation process. Slurries of 1 to 10% dried activated yeast in water were mixed with chopped sweet sorghum stalks or bagasse (0.1 to 0.2% w/w yeast) and stored for 3-4 days under anaerobic conditions. Ethanol fermentation yields were 93-97% of theoretical. Most of the fermented juice was recovered by pressing in a 15 cm diameter screw press, and the total potential yields of ethanol in fermented juice were 75-78% of theoretical.

Keywords. Ethanol, fermentation, juice expression, solid-phase fermentation, sweet sorghum.

CHEMICALS FROM LIGNINS BY HYDROGENATIONS REACTIONS

Irvin A. Jefcoat and Fawaz H. Jabali
Department of Chemical Engineering
The University of Alabama
University, Alabama 35486

A series of screening studies on the batch hydrogenation of Kraft and organosolv lignins was performed at temperatures from 200°C to 325°C and hydrogen pressures of 350 psig to 2000 psig. The Kraft lignin results of Benigni and Goldstein were duplicated with the additional observation of a large number of EPA designated priority pollutants. An analysis of the product gas from the reactor showed as much as 50% of the organosolv lignin charged was converted to methane gas. The increase of reaction temperature between 200°C and 325°C resulted in the conversion of the phenolic compounds into cyclohexyl derivatives. The bulk of the organosolv hydrogenation products were formed after three hours and the increase of treatment time to seven hours had little effect on the major products.

The high pressure reactions greatly increased the formation of cyclopentanol, cyclohexanol, and substituted cyclohexanols. In the absence of catalyst, little hydrogenation took place and no methane gas was present in

the product gas from the reactor. However, the addition of two grams of
Raney nickel catalyst increased the consumption of hydrogen five fold and
the formation of methane was very significant.

Keywords. Lignin, organosolv, wood, methane.

INDEX

Acetate, 423, 424
Acetic acid, 343, 344, 565, 570,
    573-581, 646
Acetivibrio cellulolyticus, 417
Acetoanaerobium noterae, 418
Acetobacterium woodii, 418
Acetogenic bacteria, 417
Acetogenium Kivui, 418
Acetoin, 565
Acetylene reduction, 650
Acid hydrolysis of wood, 551-560
    and sugar yield, 553-555
Acidogenic bacteria, 416-417
Aconitic acid, 602
Acutrac tractor, 370-372
Adhesive from tree bark, 634
Aerobacter aerogenes,
            see Klebsiella pneumoniae
Aerospace Research Corporation
            (ARC) 21
    electricity from wood fuel, 21
Africa food supply, critical, 6
Agency for International Devel-
            opment (AID), 18
Agriculture
    area, 4
    global, 5
    product yield, 4
    surplus crops, 1
AID, *see* Agency for International
            Development
Alcaligenes eutrophus, 13
Alcohol, *see* Ethanol
Alfalfa, 169
Algae as biomass, *see* seperate
            genera of algae
Alginate, 304, 305, 308
Alginate lyase
    of bacteria, 311
        of marine bacteria, 303-320
    digestion of *Sargassum*
            filipendula tissue, 313,
            316

Alginate lyase (continued)
    isolation from bacteria, 303-320
    reaction, 306
    substrate, 311-313, 315
Alginic acid, 306
Alnus glutinosa, 87, 634
Alteromanas sp., 303-320
    and alginate lyase, 303-320
    isolation, 308
    properties, 308, 310
Amaranthus australis, 207, 210, 214
    A. hybridus, 207, 209, 211-213, 215
Ambrosia artemisiifolia, 207, 210,
            214, 215
American sycamore, *see Platanus*
            occidentalis
α-Amylase of *Bacillus subtilis,*
            542
Anabaena variabilis, 649
α-Angelica lactone, 334, 337
    and levulinic acid, 334, 337
Angiosperm, 164
Animal waste
    and biogas, 504
    and electricity on the farm,
            504, 512-513
    manure
        collection, 505-507
        handling, 507
        preparation, 507
    and methane production, 504
Arabinose, 567-570, 592, 594-596
Arrowleaf sida, *see Sida rhombifolia*
Aspergillus niger hydrolase, 542
Attachment, bacterial, and
            degradation, 421
Agur, 123-124
Azinphos, 603

Bacardi Corporation
    anaerobic treatment process,
            518-525

Bacardi Corporation (continued)
  treatment plant, diagram of,
    520
Bacillus subtilis α-amylase,
    542
Bacteria, *see* separate genera
  interaction, 423–425
Bacteroides sp., 416–419
  B. *amylophilus*, 420
  B. *cellulosolvens*, 417
  B. *ruminicola*, 420
  B. *succinogenes*, 420
Bagasse, 160, 602, 610–611, 616,
    617, 655
Bantu beer, 548
Beech, *see Fagus* sp.
Belts, opposing, 125, 126
Bermudagrass, *see Cynodon dactylon*
Beta vulgaris, 166
Bio-Chemical Energy Inc.
  corn as feedstock, 626
  dehydration of ethanol, 625
  molasses as feedstock, 626, 627
  plants 624, 627
Bioenergy in
  Canada, 9
  Latin America, 8
  Scandinavia, 9
  U.S.A., 9
Bioenergy production
  consumption, 2
  perspective, futuristic, 1/15
  thermochemical conversion, 321–331
  see Carbonization, Combustion,
        Gasification, Tarification
Biofuel, 9–10
Biogas, 162 *see* Hydrogen, Methane
  digester
    cross-section diagram of,
      508, 509
    liquid-solid separation,
      513–514
    operation, 508–512
  electricity on the farm, 504
  from manure, 504, 507
Biogasification of water hyacinth,
    469–586
  digester, experimental unit, 482, 483
Biologue, newsletter, 22
Biomass
  from algae of the ocean, 241–257
  aquatic, 259–286
    harvesting systems, 259–274
  from bagasse, 636
  from bark extract, 636
  from *Colosia esculenta*, 185–196

Biomass (continued)
  from conversion, thermochemical,
      161
    units, 156
  coppice production, 634
  from cotton gin, 636
  from crop residue, 630
  from cruciferous crops, 173–184
  definition of, 5
  electricity from, 349–361
  energy from, 217–227, 349–361
  equation of, 147
  from *Eucalyptus grandis*, 103–110,
      143–156
  in Florida, 636
  from forest, 629
  as fuel, 324, 349–375
  from grass, 637 *see* Napiergrass
      below
  harvesting, 636
  from herbaceous plants, 163
  hydrolysis of, 337–341
  inventory of
    forest area, 57
    procedure for, 59
    tree population, 59
  from jojoba, 637
  and levulinic acid, 333–348
  liquefaction, direct of, 10
  from lumber plants, 636
  from marine algae, 241–257
  from municipal waste, 630
  from Napiergrass, 643, 645
  from non-woody plants, 160–161
  in North Carolina, 634
  from nut-shelling, 636
  perspective, futuristic, 1–15
  production
    in gravel bed hydroponics,
        287–302
    systems, 5–7
  programs, 119–131, 157–171, 629,
        631, 634, 642
    regional, 20–27
  regression, 57–69
    cluster sampling, 57–69
  from residential waste, 630–631
  from rice, 640
  from root crops, 173–184, 637
  from slash/brush, 363–375
  from sorghum, 643
  from swine manure, 645
  technology, 17–27
  from timber, *see* wood
  from tree plantation, 635
  from water hyacinth, 275–286
      643–645
  from weeds, 207–216